Progress in Mathematics
Volume 94

Series Editors

J. Oesterlé
A. Weinstein

Effective Methods in Algebraic Geometry

Edited by
Teo Mora
Carlo Traverso

Springer Science+Business Media, LLC

Teo Mora
Dipartimento de Matematica
Università di Genova
Via L. B. Alberti 2
I–16100 Genova
Italy

Carlo Traverso
Dipartimento de Matematica
Università di Pisa
Via Buonarroti 2
56127 Pisa
Italy

Library of Congress Cataloging-in-Publication Data
Effective methods in algebraic geometry / edited by Teo Mora, Carlo
  Traverso.
    p. cm. – (Progress in mathematics ; 94)
  „[Papers from] the symposium ‚MEGA-90 – Effective Methods in
Algebraic Geometry' . . . held in Castiglioncello (Livorno, Italy), in
April 17–21, 1990" – Foreword.
  Includes bibliographical references.
  ISBN 978-1-4612-6761-4   ISBN 978-1-4612-0441-1 (eBook)
  DOI 10.1007/978-1-4612-0441-1
  1. Geometry, Algebraic–Congresses.   I. Mora, Teo.   II. Traverso,
Carlo, 1945–   III. Series: Progress in mathematics (Boston,
Mass.) ; vol. 94.
QA564.E33   1991
516.3'5 – dc20

© Springer Science+Business Media New York 1991
Originally published by Birkhäuser Boston in 1991
Softcover reprint of the hardcover 1st edition 1991

ISBN 978-1-4612-6761-4

Camera-ready copy prepared by the editors.

9 8 7 6 5 4 3 2 1

FOREWORD

The symposium "MEGA-90 – Effective Methods in Algebraic Geometry" was held in Castiglioncello (Livorno, Italy) in April 17–21 1990.

The themes — we quote from the "Call for papers" — were the following:

- Effective methods and complexity issues in commutative algebra, projective geometry, real geometry, algebraic number theory
- Algebraic geometric methods in algebraic computing

Contributions in related fields (computational aspects of group theory, differential algebra and geometry, algebraic and differential topology, etc.) were also welcome.

The origin and the motivation of such a meeting, that is supposed to be the first of a series, deserves to be explained.

The subject — the theory and the practice of computation in algebraic geometry and related domains from the mathematical viewpoint — has been one of the themes of the symposia organized by SIGSAM (the Special Interest Group for Symbolic and Algebraic Manipulation of the Association for Computing Machinery), SAME (Symbolic and Algebraic Manipulation in Europe), and AAECC (the semantics of the name is varying; an average meaning is "Applied Algebra and Error Correcting Codes"). Papers on the subject are scattered in (more or less) non-specialized journals in mathematics and computer science. Only recently the appearance of the Journal of Symbolic Computation (in particular the special issue on computational commutative algebra) has provided a sufficiently specialized forum on the subject. (And even more recently the AAECC Journal — Applicable Algebra in Engineering, Communications and Computing — constitutes another possibility.)

Several conferences, seminars, etc had shown that the subject raises increasing interest, and has an audience of its own. The success of the series of conferences "Computers and mathematics" in the U. S. A. (on a wider theme) has suggested that a similar interest could be created by an European conference with themes of the same kind, even if more specialized.

The decision to start was eased by the fact that in 1990 both ISSAC (International Symposium on Symbolic and Algebraic Computation - the current name of ACM symposia on the subject) and AAECC are in Japan. The same reason has led to the organization of DISCO-90 (Design and Implementation of Symbolic Computer Systems), that corresponds to another specialized — and companion — subject inside ISSAC. The start to both was given in Roma, in the occasion of ISSAC-AAECC 1988.

A promoting and programme committee was gathered, composed by A. Conte (Torino), J. H. Davenport (Bath), A. Galligo (Nice), D. Yu. Grigoriev (Leningrad), J. Heintz (Buenos Aires), W. Laßner (Leipzig), D. Lazard (Paris), H. M. Möller (Hagen), T. Mora (Genova), M. Pohst (Düsseldorf), T. Recio (Santander), J. J. Risler (Paris), M. F. Roy (Rennes), R. Schoof (Utrecht), C. Traverso (Pisa), and a call for papers was prepared, with a deadline for submission of November 1 1989. We received 54 submissions, and 31 papers were accepted for the conference after a refereeing procedure organized as follows: at least two referee reports were sought for each paper, including referees from outside the P. C.; in a meeting in January 6–8 1990 the P. C. decided for the inclusion in the conference programme and in the proceedings. For some papers, which were either incomplete, or incompletely refereed, or needed revisions, but for which a positive first report existed, the decision on the inclusion in the proceedings was delayed to the end of the conference. All the authors were allowed to submit to the proceedings revised versions.

Six invited speakers were planned; unfortunately Bruno Buchberger had a serious – but temporary — illness, and was unable to attend, so we had five invited talks, given by R. Benedetti, G. van der Geer, P. Milman, G. Pfister and N. Vorobev; the papers derived from their talks appear in the the proceedings.

These papers too were refereed, with the only exception of the paper of Bierstone and Milman, that contains results that were partly proved after the conference, and in some sense as a consequence of the conference. It was impossible to referee this paper that was submitted only when the proceedings were almost ready.

During the conference, an informal session was organized, in which the participants were allowed to give short talks on current research. A list of such talks, together with the titles of official talks that do not correspond to items of the proceedings (because the final version was not submitted, or was not accepted) appears at the end of this introduction.

The conference was a success — at least, no complaints reached us up to now — and towards the end of the conference a meeting of the programme committee together with the invited speakers decided to organize a second conference on the same themes, to be held in Nice in April 1992.

Here is a list of communications held at the conference and not included in the proceedings, either because the paper in final form was not submitted, or because it was not accepted, or because the communication was held in an informal session.

A. G. Alexandrov, A. V. Bocharov, V. L. Kistlerov: *On using FLAC as an implementation language for differential algebraic algorithms*

M. Alonso, T. Mora, M. Raimondo: *Complexity of algebraic power series*

M. Alonso, T. Mora, M. Raimondo: *Local decomposition algorithms*

J. Backelin: *A graph-component algorithm for eliminating S-polynomials: is it new?*

G. Carrà Ferro, W. Sit: *On term-orderings and rankings*

I. Del Corso: *Factorization of prime ideal extensions in number fields*

T. Dubé: *Homogenizing prime sequences.*

M. Fiorentini: *Curve spaziali k-Buchsbaum che non sono quasi-completa intersezione*

G. Gentili: *Some remarks on rational geometry in space*

M. C. Gontard: *Amélioration de la méthode de Wu — Implementation en Scratchpad*

P. Milman: *A combinatorial analogue of desingularization*

G. Moreno: *Climbing generic stairs*

F. Ollivier: *Standard bases of differential ideals*

R. Ramanakoraisina: *Complexity of Nash functions*

B. Ruitenburg: *Lie groups and power series*

C. Traverso: *Intersection algorithms*

Wang D.: *A method for determining the finite basis of an ideal from its characteristic set, with an application to irreducible decomposition of algebraic varieties*

We want to conclude thanking the Consiglio Nazionale delle Ricerche, the Ministero della Ricerca Scientifica e Tecnologica and the Università di Pisa, whose contribution has made possible the conference; the Comune di Rosignano Marittimo and the Unione Albergatori di Castiglioncello for their hospitality.

A special thanks to Giuseppina Donato, whose efficient cooperation made everything run smoothly during the conference; to the members of the programme committee and the referees, whose work has made the conference and the proceedings possible; and finally to Donald Knuth, the inventor of TEX, to whom we owe the typographical excellence of the proceedings.

Teo Mora
Carlo Traverso

The language issue.

A long discussion on the official languages of the (present and future) MEGA conferences was held inside the programme committee (you can find a trace of this discussion in one of the papers). The committee was unanimous to stress that the aim to pursue was to ensure the better communication of the mathematical and cultural issues, but the ideas were different about the "practical rule" that would follow. It was too late to state a rule for the current conference, and the practical outcome was that no submission was rejected because of the language chosen or because of a poor conformity to literary standards. A policy was stated for the next issue: "mathematical english", (in the sense of the today de-facto mathematical esperanto — or pidgin, as you like) is the official language, with the aim to maximize the international communication. Exceptions will be authorized by the Programme Committee if they comply to the stated aim.

The MEGA Programme Committee

TABLE OF CONTENTS

x

Addresses of the contributors

Riccardo Benedetti
Dipartimento di Matematica
Via F. Buonarroti 2
I-56100 Pisa
ITALIA
diparmat@icnucevm.cnuce.cnr.it

Edward Bierstone
University of Toronto
Toronto
CANADA M5S 1A1
bierston@math.toronto.edu

Leandro Caniglia
Instituto Argentino de Matemática
(IAM-CONICET)
Viamonte 1636
(1055) Buenos Aires
ARGENTINA
atina!mate!caniglia@uunet.uu.net

Marc Chardin
Centre de Mathématiques
et Laboratoire d'Informatique
Ecole Polytechnique
F-91128 Palaiseau Cedex
FRANCE
cfmcha@frpoly11.BITNET

Felipe Cucker
Dept. L.S.I., Facultat d'Informàtica
U. P. C.
Pan Gargallo 5
Barcelona 08028,
ESPAÑA
cucker@lsi.upc.es

Alicia Dickenstein
Departamento de Matemática - FCEyN
Universidad de Buenos Aires
Ciudad Universitaria - Pabellón I
(1428) Buenos Aires
ARGENTINA
alidick@mate.dcfcen.edu.ar
atina!mate!alidick@uunet.uu.net

André Galligo
Département de Mathématiques
Université de Nice
Parc Valrose
F-06034 NICE cedex
FRANCE galligo@mirsa.inria.fr

Giovanni Gallo
Courant Institute
of Mathematical Sciences
New York University
Mercer Street 251.
New York, NY 10012 USA
or
Università di Catania
Dipartimento di Matematica
I-95125 Catania
ITALIA
gallo@esd4.cs.nyu.edu

Tatiana Gateva-Ivanova
Institute of Mathematics
Bulgarian Academy of Sciences
Sofia 1090
BULGARIA

Gerard van der Geer
Faculteit Wiskunde en Informatica
Universiteit van Amsterdam
Plantage Muidergracht 24
1018 TV Amsterdam
NEDERLAND
geer@fwi.uva.nl

Marc Giusti
Laboratoire d'Informatique
Ecole Polytechnique
91128 Palaiseau Cedex
FRANCE
giusti@cmep.polytechnique.fr
cfmagi@frpoly11.BITNET

Laureano González Vega
Dept. Matem. Estad. y C.C.
Universidad de Cantabria
Santander 39071,
ESPAÑA
g_vega@ccuvx.unican.es

Dimitri Yu. Grigor'ev
Leningrad Department
of Mathematical V.A.Steklov Institute
Academy of Sciences of the USSR
Fontanka 27, Leningrad
191011, USSR

Jorge A. Guccione
Instituto Argentino de Matemática
(IAM-CONICET)
Viamonte 1636
(1055) Buenos Aires
ARGENTINA

Juan J. Guccione
Instituto Argentino de Matemática
(IAM-CONICET)
Viamonte 1636
(1055) Buenos Aires
ARGENTINA

Akli Haddak
Centre d'Enseignement et de Recherche
en   Mathématiques   Appliquées
(CERMA),
Ecole Nationale des Ponts et Chaussées
(ENPC),
La   Courtine,   93167   Noisy-le-Grand
Cedex
FRANCE

Joos Heintz
Instituto Argentino de Matemática
(IAM-CONICET)
Viamonte 1636
(1055) Buenos Aires
ARGENTINA
atina!mate!nfitchas@uunet.uu.net
nfitchas@mate.edu.ar
atina!dcfcen!mate!nfitchas

Teresa Krick
Instituto Argentino de Matemática
(IAM-CONICET)
Viamonte 1636
(1055) Buenos Aires
ARGENTINA

Lakshman Y. N.
Deptartment of Computer and Informa-
tion Sciences
University of Delaware
103, Smith Hall
Newakk, DE 19716
U. S. A.
lakshman@math.udel.edu

Daniel Lazard
Informatique, Université Paris VI
4, place Jussieu,
75252 Paris Cedex 05
FRANCE
dl@frunip11.BITNET
lazard@litp.ibp.fr

Lars Langemyr
Wilhelm-Schickard-Institut
für Informatik
Universität Tübingen
Auf dem Sand 13
D-7400 Tübingen
DEUTSCHLAND
larsl@nada.kth.se

Alessandro Logar
Università degli Studi di Trieste
Dipartimento di Scienze Matematiche
Piazzale Europa, 1
34121 Trieste
ITALIA
ti2tsg23@icineca2.BITNET

Henri Lombardi
Mathématiques
UFR des Sciences et Techniques
Université de Franche-Comté
25 030 Besançon cédex
FRANCE

Gennadi I. Malashonok
Institute for Applied Problems
of Mechanics and Mathematics
Academy of Sciences
Vul. Naukova 3-b
290053 Lvov
USSR

Roger Marlin
Université de Nice
Parc Valrose
F-06034 NICE cedex
FRANCE
rm@cerisi.cerisi.fr

Jean-François Mestre
Université de Paris VII
2 place Jussieu
75005 Paris
FRANCE
et
Département de Mathématiques
et d'Informatique
École normale supérieure
45 rue d'Ulm,
75230 Paris Cedex 05
FRANCE
mestre@frmap711.BITNET
mestre@frulm11.BITNET

Pierre D. Milman
University of Toronto
Toronto
CANADA M5S 1A1
milman@math.toronto.edu

Bhubaneswar Mishra
Courant Institute
of Mathematical Sciences
New York University
Mercer Street 251.
New York, NY 10012
USA
mishra@nyu.edu

H. Michael Möller
FB Mathematik und Informatik
der FernUniversität
D-5800 Hagen 1
DEUTSCHLAND
ma105@dhafeu11.BITNET

José L. Montaña
Departamento de Matemáticas,
Estadística y Computación
Facultad de Ciencias
Universidad de Cantabria
Avda. de los Castros, s/n
39071- Santander
ESPAÑA

Teo Mora
Dipartimento di Matematica
Via L. B. Alberti 2
I-16100 Genova
theomora@IGECUNIV.BITNET

Bernard Mourrain
Centre de Mathématiques
Ecole Polytechnique
F-91128 Palaiseau Cedex
FRANCE
mourrain@cmep.polytechnique.fr

François Ollivier
Laboratoire d'Informatique de l'X
École Polytechnique
F-91128 Palaiseau Cedex
FRANCE
cffoll@frpoly11.BITNET
ollivier@cmep.polytechnique.fr

Luis M. Pardo
Departamento de Matemáticas,
Estadística y Computación
Facultad de Ciencias
Universidad de Cantabria
Avda. de los Castros, s/n
39071- Santander
ESPAÑA
pardo@ccucvx.unican.es

Gerhard Pfister
Humboldt-Universität zu Berlin
Sektion Mathematik
108 Berlin, Unter den Linden 6
DEUTSCHLAND

Jean-François Pommaret
Centre d'Enseignement et de Recherche
en Mathématiques Appliquées
(CERMA),
Ecole Nationale des Ponts et Chaussées
(ENPC),
La Courtine
93167 Noisy-le-Grand Cedex
FRANCE

Tomás Recio
Departamento de Matemáticas,
Estadística y Computación
Facultad de Ciencias
Universidad de Cantabria
Avda. de los Castros, s/n
39071- Santander
ESPAÑA

Daniel Richardson
Department of Mathematics
University of Bath
Bath BA27AY,
U.K.
dsr@maths.bath.ac.uk

Francesc Rosselló
Dept. Àlgebra i Geometria
Universitat de Barcelona
Barcelona 08007,
ESPAÑA
d3agfr10@eb0ub011.BITNET

Fabio Rossi
Università degli Studi di Trieste
Dipartimento di Scienze Matematiche
Piazzale Europa, 1
34121 Trieste
ITALIA
ti27sg23@icineca2.BITNET

Marie Françoise Roy
IRMAR Université de Rennes
Campus de Beaulieu
35 042 Rennes CEDEX
FRANCE
costeroy@frcicb81.BITNET

Walter Spangher
Università degli Studi di Trieste
Dipartimento di Scienze Matematiche
Piazzale Europa, 1
34121 Trieste
ITALIA

Carmen Sessa
Departamento de Matemática - FCEyN
Universidad de Buenos Aires
Ciudad Universitaria - Pabellón I
(1428) Buenos Aires
ARGENTINA

Kiyoshi Shirayanagi
NTT Software Laboratories
3-9-11 Midori-cho, Musashino-shi
Tokyo 180
JAPAN
shirayan%ntt-20.ntt.jp@relay.cs.net

Elisabetta Strickland
II Università di Roma "Tor Vergata"
Roma
ITALIA

Aviva Szpirglas
CNRS U.R.A 742
Université Paris-Nord
Avenue Jean-Baptiste Clément
93 430 Villetaneuse
FRANCE

Carlo Traverso
Dipartimento di Matematica
Via Buonarroti 2
56127 Pisa
ITALIA
traverso@dm.unipi.it

Nikolaj N. Vorobjov (Jr.)
Leningrad State University
Universitetskaya emb., 7/9
Leningrad, 199164
USSR

On Lack of Effectiveness
in Semi-algebraic Geometry

RICCARDO BENEDETTI

Introduction

It is a fact and a motivation of this meeting, that many interesting constructions in semi-algebraic geometry can be *effectively* done, that is, roughly speaking, by means of algorithms with computably bounded complexity. It is also a fact that semi-algebraic objects have some *effective* finiteness properties. Hence the interest in finding reasonably "fast" algorithms or "sharp" bounds. Neverthless, on the other side of the main-stream, one can be interested in *essential lack of effectiveness or finiteness*. We are going to see some examples and to discuss shortly some possible *basic sources* of such a lack.

To be a little more precise, let us fix some notations.

We write $X \in s(n, r, d)$ to mean that X is a semialgebraic set in $\mathbb{R}^n$ with a *given* representation of the form:

$$X = \bigcup_{i=1}^{k} \bigcap_{j=1}^{s_i} \{f_{ij} \star_{ij} 0\}$$

where, for every i and j, $f_{ij} \in \mathbb{R}[x_1, \ldots, x_n]$, $\star_{ij} \in \{=, >\}$ and $\sum_i s_i \leq r$, $\max_{ij}\{degree f_{ij}\} \leq d$.

Of course the triple (n, r, d) is a rough measure of the "complication" of X (better of the given representation of X) but enough to our aim.

As basic example of *effective construction* I should take the *triangulation of* (compact) *semialgebraic sets*. By this we mean that there exists an algorithm starting from any couple $((n, r, d), X) \in \mathbb{N}^3 \times S(n, r, d)$, providing that X is a compact set, and producing:

a) A *triangulation* of X, that is a semialgebraic homeomorphism

$$f : |K| \to X$$

K being a finite simplicial complex in $\mathbf{R}^n$

b) A function $\phi : \mathbf{N}^3 \to \mathbf{N}^3$, $\Phi(n, r, d) = (R, D, h)$ such that the graph of f, $\Gamma_f \in S(n^2, R, D)$ and the number of simplexes of K, $\#K \leq h$.

As basic example of *effective finiteness property*, I should take the following: for every $(n, r, d) \in \mathbf{N}^3$ the quotient set of $S(n, r, d)$ up to semi-algebraic homeomorphism is a finite set and the number of its elements can be computably bounded in terms of (n, r, d). Note that it could be actually obtained as a corollary of the triangulation theorem.

Let us come to some examples of essential lack of effectiveness.

1. Nabutovski's examples

I will limit to very roughly recall the simplest example. A nice idea is the use of P.S.Novikov's result (see later) in the contest of semialgebraic geometry.

Let us call $A(n, d)$ the set of compact nonsingular hypersurfaces S in $\mathbf{R}^n$ defined as $S = \{p = 0\}$ for some degree $\leq d$ polynomial $p \in \mathbf{R}[x_1, \ldots, x_n]$, such that grad $p|_x \neq 0$ $\forall x \in S$. Let us denote by $\hat{A}(n, d)$ the subset of $A(n, d)$ of those S as before but smoothly isotopic to the standard $(n-1)$-sphere S^{n-1}. Given $S \in \hat{A}(n, d)$ a D-isotopy between S and S^{n-1} is a smooth isotopy S_t, $t \in [0, 1]$, such that $S_0 = S$, $S_1 = S^{n-1}$, for each t, $S_t \in \hat{A}(n, D)$.

1.1. Theorem. *([Na])) a) There exists a function $D = D(n, d)$ such that for every $S \in \hat{A}(n, d)$ there exists a D-isotopy between S and S^{n-1}.*

b) Any such a function D cannot be upper bounded by any computable function of (n, d), providing $n \geq 6$.

Rough idea of proof: For fixed D the fact that $S \in S(n, D)$ could be isotopic to S^{n-1} by means of a D-isotopy, can be reduced to the problem of path-connecting points in a suitable semialgebraic "parameter space" belonging to some $S(N, p, q)$ with (N, p, q) computable in terms of (n, D). It is well known that it can be solved effectively.

Assume that the function $D = D(n, d)$ should be upper bounded by a computable function of (n, d); then the problem of deciding if any given $S \in A(n, d)$ is isotopic to S^{n-1} could be realized by means of an algorithm. By the h-cobordism theorem, S is isotopic to S^{n-1} if and only if S is diffeomorphic to S^{n-1}, providing $n \geq 6$. Thus the computability assumption about D, contradicts the following P.S.Novikov's result (which we formulate in two ways convenable to our aim):

Theorem (P. S. Novikov). *For $n \geq 6$, there is a sequence of manifolds $\{M_i\}$ of dimension n, with boundary ∂M_i, embedded in $\mathbf{R}^n$ such that:*

(a) Every M_i can be effectively realized as an element of some $A(n, d)$ or as $M_i = |K_i|$ K_i being a simplicial complex in $\mathbf{R}^n$.

(b) There does not exist any algorithm deciding if any M_i is diffeomorphic (or P. L. homeomorphic) to the n-ball, or equivalently (use h-cobordism) if ∂M_i is isomorphic to S^{n-1}.

1.2 *Remarks*: 1) For Novikov's theorem see [V.K.F]. Actually the proof is simply an integration of known results about the decision problem for finitely presented groups and the Smale's h-cobordism theorem.

2) The "parameter space" evoked in the sketch of proof carries a natural stratification. Points in the same stratum are isotopic by a D-isotopy which can be effectively constructed. The non computability of $D = D(n, d)$ comes from the fact that points of different strata could be actually isotopic. One way to construct a path, should be to consider the closure of the parameter space, consider path passing though singular S_t and then try, if possible, to resolve the singularity. The result says that there is no hope to do it effectively.

2 About the effectiveness of the Semialgebraic Hauptvermutung

Most of the material of this paragraph comes from [A-B-B]. Let us recall the semialgebraic Hauptvermutung due to Shiota-Yokoi ([S-Y]).

"If two compact polyhedra P, Q are semialgebraically homeomorphic, then they are P. L. (= piecewise linear) homeomorphic".

This important result, actually holds under more general hypotheses, for example in the sub-analytic category. However, in my opinion, it is a *critical* example to be studied under the point of view of the opposition "effectiveness non-effectiveness", in the semialgebraic case.

2.1 **(E.H.)$_n$ (Effective Hauptvermutung problem in dimension n).**
Does there exists any algorithm starting from any semialgebraic homeomorphism $f : |K_1| \to |K_2|$ such that K_1, K_2 are finite simplicial complexes in some $\mathbf{R}^m$, $\dim |K_i| = n$, $\Gamma_f \in S(m^2, r, d)$, $\#K_i \leq h$, $i = 1, 2$ and producing:
(a) a simplicial isomorphism

$$g : K_1' \to K_2'$$

K_i' being a subdivision of K_i, $i = 1, 2$
(b) a function $\Psi : \mathbf{N}^5 \to \mathbf{N}$ such that

$$\#K_i' \leq \Psi(n, m, r, d, h).$$

If we forget the point (b) we shall say that it is the *weak* effective Hauptvermutung problem (shortly (W. E. H)$_n$). Actually (E. H)$_n$ in its generality is still open. We shall see later some partial or related results.

Before it, I will spend few time considering the Shiota-Yokoi proof, showing how it is not (at least not immediately) effective. As a by-product I will state some problems, any answer should be very interesting.

(i) First, one can *effectively* reduce to the following situation: (it is a consequence of the effective triangulation theorem recalled at the beginning): solve the problem for $0 \leq m \leq n$, assuming that $P_1 = |K_1|$ and $|K_2| = P_2$ are polyhedra of dimensions m in $\mathbf{R}^m$ (*same m*), and that P_2 is actually the *standard m-simplex* Δ_m with its natural triangulation $K_2 = K_\Delta$. (We shall write P instead of P_1)

(ii) Then one can *effectively* (Whitney)-stratify the semialgebraic homeomorphism $f : P \to \Delta_m$.

(iii) The proof is then achieved by two basic "moves": the "reduction to the smooth case" and the "Alexander trick". Let us explain it in the simplest case:

Assume that we have only two strata X_1, X_2, $X_2 \leq \overline{X}_1$ in P corresponding to the strata $Y_2 \subset \overline{Y}_1$ in Δ_m. Consider the corresponding "distance" functions d, d' along X_2 and Y_2. Using it one constructs two small "tubes" T, T' along X_2 and Y_2 : $T(T')$ is a local product with "conical" fibre; $\overline{X_1 \setminus T}$ $(\overline{Y_1 \setminus T'})$ is a smooth manifold with boundary. Any triangulation of $\overline{X_1 \setminus T}$ $(\overline{Y_1 \setminus T'})$ can be extended by "coning" (Alexander trick) to $T(T')$. The problem is that $T'' = f^{-1}(T)$ does not coincide with T.

Note that $T'' = \{h \leq \epsilon\}$ for a suitable $\epsilon > 0$, where $h = d' \circ f$. Thus we have, locally along X_2, two positive semialgebraic functions (d, h) having X_2 as zero-set and we want to isotopy the respective "tubes" T, T'' one onto the other. This can be done (using a trick essentially due to Milnor) by means the vector field $|\mathrm{grad}h| \, \mathrm{grad}d + |\mathrm{grad}d| \mathrm{grad}\, h$; the crucial fact is that, along X_1, $\mathrm{grad}d$ and $\mathrm{grad}h$ never point in apposite directions (curve selection lemma). Actually we can construct a diffeomorphism $g : \overline{X_1 \setminus T} \to$ $(\overline{Y_1 \setminus T''})$ and then apply to g the Hauptvermutung for C^∞-triangulations (Whitehead [M]) and conclude. In presence of several strata we have certain not trivial technical complications, but the essential moves remain the same.

It should be clear now what are the possible sources of non effectiveness in Shiota-Yokoi proof.

A) *To control the "complexity" of diffeomorphism obtained by integrating semialgebraic (even polynomial) vector fields.*

Probably the simplest problem in this order of idea is the following.

2.2. Let $P \in \mathbf{R}[x_1, \dots, x_n]$, $P \geq 0$, $\{P = 0\} = \{0\}$ degree $P \leq d$; let $\epsilon_0 > 0$ small enough such that $\mathrm{grad}P \neq 0$ on $U = \{P \leq \epsilon_0\} \setminus \{0\}$ and $\forall 0 < \epsilon \leq \epsilon_0$ $\{P = \epsilon\}$ is a compact non singular hypersurface; consider the vector field on U $\frac{-\mathrm{grad}P}{||\mathrm{grad}P||^2}$ and normalize the solution such that any integral curve starts at time 0 from the lever surface $X_0 = \{P = \epsilon_0\}$ so that it arrives at time t to the level surface $X_t = \{P = \epsilon_0 - t\}(t \in [0, \epsilon_0[)$.

Fix two algebraic set $Y_0 \subseteq X_0$ and $Y_t \subseteq X_t$ defined by polynomials of degree $\leq d$. Let Y_{0t} the image of Y_0 in X_t by the flow. Set $b(Y_t \cap Y_{0t})$ to be the number of the connected components of $Y_t \cap Y_{0t}$.

Questions:
(a) Is $b(Y_y \cap Y_{0t})$ finite and does there exist a function $b = b(n,d)$ such that

$$b(Y_y \cap Y_{0t}) \leq b.$$

(b) Is such a function $b = b(n,d)$ (if any) computably bounded in terms of (n,d).

B) *to control the "complexity" of the Hauptvermutung for C^∞-triangulations*. Remark that a main tool to do it (see [M]) is the use of the *secant approximation* to approximate diffeomorphisms by means of P.L. homeomorphisms. This suggests the following question for general semialgebraic homeomorphisms.

2.3: Let $f : P \to \Delta_m$ be a semialgebraic homeomorphism (not necessarily smooth). Assume that $\Gamma_f \in S(m^2, r, d)$, $P = |K|$, $\#K \leq h$.
 (i) Does there exist a triangulation H of $P, P = |H|$ such that the *secant approximation of f associated to H, sec (f,H)* (recall that it coincides with f or the vertices of H and then it is extended by linearity), is a simplicial isomorphism onto Δ_m?
 (ii) Does there exists a function $P = P(m,r,d,h)$ such that

$$\#H \leq P(m,r,d,h)?$$

 (iii) (effectiveness) Does there exist an algorithm producing
    (a) H
    (b) A function $q = q(m,r,d,h)$ such that $\#H \leq q$?

2.4 *Remarks*: 1) The problem 2.2 and any similar one involving integration of vector fields (even polynomial) seem to be hard. Note that the isotopy of the tube T'' onto T, evoked above, can be seen as a sort of "unicity of tubular neighbourhood" extending the classical one of differential topology. However the analogy should be treated carefully. In fact we have a hard distorsion of the conical structure of T dong X_2 (with the previous notations).
    The following very simple example (Shiota) could clarify what I mean.

2.5 *Example*: Consider $P = (X^2 + Y)^2 + \frac{Y^2}{10}$, $P \geq 0$, $\{P = 0\} = \{0\}$. It is easy to see that for every $\epsilon > 0$ there exists a line $\{y = ax\}$ intersecting $\{P_\epsilon\}$ transversely at 4 points, so that there exists a line $\{y = a'x\}$ tangent to $\{P = \epsilon\}$. This, in particular, implies that the map which associates to each $x \in \{P = \epsilon_0\}$ (ϵ_0 fixed) the "limit tangent" at 0 of the integral line of $-\mathrm{grad}P/||\mathrm{grad}P||^2$ starting from x (recall that a famous, still open,

Thom's conjecture states that such a limit tangent always exists, for every analytic gradient vector fields; the conjecture holds for fields on the plane), is not continuous. In fact a direct analysis shows that all the integral lines but $\{x = 0\}$ (up to parametrization) have the line $\{y = 0\}$ as limit tangent line.

A similar example can be obtained by the pull-back of the standard function $\alpha^2 + \beta^2$, via the semialgebraic homeomorphism $\alpha = x^3 + y$, $\beta = \frac{y}{3}$.

2) The problem 2.3 (i) has positive answer if f is also smooth (this is a main point of [M]). In this case we fix any triangulation H of P; then we give a suitable way to subdivide H, obtaining τH. Then one shows that there exists s such that $\tau^s H$ satisfies the required property. τ is suitable in the sense that it preserve the "thicknees" of the simplexes of H. s depends on *local* geometric invariants of f, that is of its graph, such as "curvatures".....

The questions 2.3 (ii) and (iii) are not clear also in the smooth case. Note that if $\Gamma_f \in S(m^2, r, d)$ we have, naturally, bounds for the *integral geometry* of Γ_f (volume, total curvature....) but not any control on the local geometric invariants. Of course a positive answeer to (iii) should imply an effective direct solution of $(E.H.)_m$.

3) The question (iii) becomes more plausible working on **Z** and introducing also the size of the coefficients as measure of the complexity of Γ_f.

My feeling is that Shiota-Yokoi proof is "natural" and that any attempt to prove the Hauptvermutung starting from a given semialgebraic homeomorphism $f : P \to \Delta_m$, necessarily should involve some of the problems sketched above and the related difficulties with respect to the effectiveness.

In [**ABB**] we have tried to get $(E.H.)_n$ by a different approach, forgetting f, and using only that we know P is a PL-ball, that is it is $P.L.$ homeomorphic to Δ_m. This suggests the following problem:

$(E.T.)_m$ *(Effective trivialization of P. L.-balls in dimension m)*: Does there exist an algorithm starting from any P. L.-m-ball $P = |K|$, K being a simplicial complex in $\mathbf{R}^m$, $\#K \le h$ and producing:

(a) Simplicial subdivisions $K' \lhd H$, $H \lhd K_\Delta$ and a simplicial isomorphism

$$g : K' \to K_\Delta$$

(b) a function $\theta : \mathbf{N} \to \mathbf{N}$ such that

$$\#K' \le \theta(h).$$

Forgetting the point (b) we have the *weak* trivialization problem (shorthly $(W.E.T.)_m$).

One can prove that $(E.T.)_m$ is equivalent to the following

$(E.S.)_m$ *(Effective shelling of PL-balls in dimension m)*: Does there exist an algorithm starting from any $P = |K|, h$ as above and producing

(a) A subdivision $K' \triangleleft K$ and a *shelling* of K' (i.e. an ordering $\delta_1, \ldots, \delta_k$ on the m-simplexes of K', such that, setting

$$B_0 = P \quad B_1 = \overline{P \backslash \delta_1}, \ldots, B_n = \overline{B_{n-1} - \delta_n} \ldots$$

each B_i $i = 0, \ldots, k$ is a *P.L.-m*-ball).

(b) $\theta : \mathbf{N} \to \mathbf{N}$ such that $\#K' \leq \theta(h)$?

Analogously we have the weak problem $(W. E. S.)_m$ forgetting (b), and $(W. E. T.)_m \Longleftrightarrow (W. E. S)_m$.

It is clear from the above discussion that

$$(E.T)_m \quad \text{for} \quad m \leq n \implies (E.H)_n$$

$$(W.E.T)_m \quad \text{for} \quad m \leq n \implies (W.E.H)_n$$

I summarize some results of ([**ABB**]).

2.6 Theorem.

1) $(E. T.)_m$ (hence $(E. H)_m$) holds for $0 \leq m \leq 3$.

2) $(W. E. T.)_m$ (hence $(W. E. H.)_m$ holds for every m.

3) For $m \geq 6$ $(E. T.)_m$ does not hold. More precisely there exists a function $\rho = \rho(h)$ such that any P. L. m-ball $P = |K| \subseteq \mathbf{R}^m$, with $\#K \leq h$, admits a subdivision $K' \triangleleft K$, with $\#K' \leq \rho(h)$, and a simplicial isomorphism $g : K' \to K_\triangle$. On the other hand, any such a function ρ cannot be bounded by any computable function of h.

2.7 *Remarks*: 3) holds also for $m = 5$ allowing that the P. L. 5-balls are embedded in $\mathbf{R}^6$.

4) The proof of 2) is based on the effective *listing* of all the subdivisions of a given K having no more that a fixed number of simplexes (up to simplicial isomorphism). The proof of 3) is based on such a listing and on Novikov's result recalled at the beginning.

5) the case $m = 4$ is open (for (E. T.)). (E. H)$_m$ is open for $m \geq 3$.

Theorem 2.6 (especially the point (iii)) has some interesting corollaries. We limit ourselves to recall the following:

2.8. **Corollary.** *In general, collars along the boundary of PL-balls cannot be effectively trivialized.*

(by a trivialized collar of ∂P in P we mean a P. L. homeomorphism h between a polyhedral neighborhood U of ∂P in P, and $\partial P \times [0, 1]$, such that $h|_{\partial P}$ is the identy map onto $\partial P \times \{0\}$).

This corollary should contrast with the fact that, on the contrary, *regular neighbourhoods* of ∂P in P with their *P.L.* retraction onto ∂P, can be effectively (and classically) constructed. On the other hand, any "constructive cut and pasting" actually needs trivialized collars.

2.9. Corollary. *(a) There exists a map* $p = (p_1, p_2) : \mathbf{N}^2 \to \mathbf{N}^2$ *such that for every* $P = |K|, h$ *as in the statement of* $(E.T)_m$ *there exists a semialgebraic homeomorphism* $f : P \to \Delta_m$ *with* $\Gamma_f \in S(m^2, d, r)$ $d \leq p_1(m, h)$, $r \leq p_2(m, h)$.

(b) At least one (providing $m \geq 6$ *):*

(i) $(E.H)_m$ *fails.*

(ii) Any such functions p_1, p_2 *cannot be bounded by any computable functions of* (m, h)

Consider $S(n, r, d)$; It is not hard to see, using the local triviality theorem of semialgebraic maps, that there exists a semialgebraic map $\Phi : \mathcal{B} \to \mathcal{P}$, effectively associated to $S(n, r, d)$ such that:

1) $S(n, r, d)$ is in $1 - 1$ correspondente with the fibres of Φ.

2) There exists an effective partition of $\mathcal{P}$ by a finite number of semialgebraic subsets $\{A_i\}$, such that, $\forall i$ $\phi| : \phi^{-1}(A_i) \to A_i$ is semialgebraically trivial; in particular there exists an *effective* semialgebraic homeomorphism between any two given fibres $\phi^{-1}(x_0), \phi^{-1}(x_1) \in \phi^{-1}(A_i)$.

3) Actually over each A_i $\phi|\phi^{-1}(A_i)$ is a very tame map without "blow-up" (in the sense of Thom; roughly this means that no fibre has strictly higher dimension than the fibres in a neighbourhood) which can be "triangulated" so that the trivialization and the construction of the homeomorphism between fibres is translated into a problem of $P.L$ geometry. The presence of blow-up makes necessary such a partition of $\mathcal{P}$.

4) The lack of effectiveness evocked by 2.9 (or 2.6 (iii)) depends on the fact that one could have isomorphic fibres belonging to different bundles $\phi^{-1}(A_i)$ $\phi^{-1}(A_j)$ $i \neq j$.

3. On lack of finiteness

Topological types of polynomial maps offers a very classical example of *lack of finiteness* in real algebraic geometry. Let $f_1, f_2 : X \to Y$ semialgebraic maps. We say that they have the same *topological type* if there exist homeomorphisms $\phi : X \to X$, $\sigma : Y \to Y$ such that $f_2 \circ \phi = \sigma \circ f_1$.

If we take *semialgebraic* homeomorphisms, then we have the same *semialgebraic type*. It is known since Thom's work ([T]), see [N] for more general results) that topological types of polynomial maps $f : \mathbf{R}^n \to \mathbf{R}^m$ ($m \geq 2$) of bounded degree are not finite and actually depend on *continuous moduli*. The basic source of this fact is the existence of "blow-up" in polynomial maps. On the other hand, Fukuda ([F]) proved that for polynomial functions $f : \mathbf{R}^n \to \mathbf{R}$, the number of *topological types* is finite, providing thath the degree is bounded. The proof in [F] is not elementary and is based on the theory of Whitney Stratifications and Thom's isotopy lemmas. In [B-S] we have improved Fukuda's result as follows:

3.1 **Proposition.** *The number of semialgebraic types of polynomial functions* $f : \mathbf{R}^n \to \mathbf{R}$ *of degree* $\leq d$ *if finite and effectively bounded in terms of* (n, d)

Our proof, which is elementary, is, in fact, a corollary of an "*effective triangulation theorem of semialgebraic functions*" extending the one of *sets* recalled at the beginning. Recall that a triangulation of a semialgebraic continuous function $f : X \to \mathbf{R}$, providing X is compact, is a semialgebraic triangulation $\tau : K \to X$ of X, such that $f \circ \tau$ is linear on each simplex of K. A famous Thom's conjecture states (the question is open) that "maps without blow-up can be triangulated".

Moreover it is known that generic (smooth) maps are without blow-up. Thus, in the semialgebraic setting, it is natural to strenghten Thom's conjecture as follows:

"Semialgebraic maps without blow-up can be *effectively* triangulated".

Apply it, for instance, to the polynomial maps (of bounded degree) we should have the following conjectural picture: there is a "nice" effective semialgebraic stratification of the space of polynomial maps of bounded degree; on the generic strata we have effectively triangulable maps and hence effective finiteness of the semialgebraic types. On the other strata "blow-up" does occur; this is the source of the topological instability and, hence, of the lack of finiteness.

This picture is quite similar the one of remark 2.10 (see also 1.2). All these facts suggest a link between (lack of) effectiveness, finiteness and "triangulability" indicating the "blow-up" in maps as a basic source of such a lack.

REFERENCES

[ABB] F. Acquistapace, R. Benedetti, F. Broglia, *Effectiveness — non Effectiveness in semialgebraic and P. L. geometry*, Inventiones Math. (to appear).

[BS] R. Benedetti, M. Shiota, *Finiteness of semialgebraic types of polynomial functions*, to appear.

[F] T. Fukuda, *type topologique des polynômes*, Publ. Math. I.H.E.S. **46** (1976), 87–106.

[M] J. R. Munkres, "Elementary Differential Topology," Am. math. Studies, n.61, Princeton Univ. Press, 1968.

[N] I. Nakai, *On topological types of polynomial mappings*, Topology **23** (1984), 45–66.

[Na] A. Nabutovski, *Isotopies and non-recursive functions in real algebraic geometry*, in "Real analytic and algebraic geometry," Springer Lect. Notes in Math. n. 1420, 1990.

[SY] M. Shiota, M. Yokoi, *Triangulations of subanalytic sets and locally subanalytic manifolds*, Trans. AMS **286** (1984), 727–750.

[T] R. Thom, *La stabilité topologique des applications polynomiales*, . Einseign. Math. **8** (1962), 24–33.

[VKF] I. A. Volodin, V. E. Kuztenov, A. T. Fomenko, *The problem of discriminating algorithmically the standard 3-sphere*, Russian Math. Surveys **29 (5)** (1974), 71–172.

Riccardo Benedetti
Dipartimento di Matematica
Via F. Buonarroti 2
I-56100 PISA

A simple constructive proof of Canonical Resolution of Singularities

EDWARD BIERSTONE   PIERRE D. MILMAN

0. Foreword

In these notes, we describe some of the main features of an explicit proof of canonical desingularization (of algebraic varieties or analytic spaces X) in characteristic zero. Full details will appear in [7]. The proof is a variation on our proof of local desingularization ("uniformization") [4], [5], and justifies the philosophy that "a sufficiently good local choice [of centre of blowing-up] should globalize automatically" [5, p. 901]. The final version is surprisingly elementary; these notes, for example, include an essentially self-contained presentation of the hypersurface case. The general case involves a "reduction to the hypersurface case" result from [5].

The computation in §3 below is a reworking of the hypersurface case from [5] to produce a local measure of singularity "$\operatorname{inv}_X(a)$" which completely determines a smooth centre of blowing-up, locally. $\operatorname{inv}_X(a)$ is a finite sequence of discrete numerical characters which, in a strikingly simple way, can be seen to be an *invariant*, (§4 below) and thus to determine a global centre.

The possibility of resolution of singularities in characteristic zero was established by Hironaka, first for algebraic varieties [8] and later for complex analytic spaces [3]. Hironaka's work does not, however, give an explicit resolution algorithm, despite the fact that desingularization of X is achieved as a composite of simple quadratic operations (blowings-up with smooth centres). A renewed interest in resolution of singularities in the last decade, particularly from the points of view of constructiveness or explicit computation, can be traced in works of Abhyankar [1], Hironaka and Spivakovsky [11], Villamayor [12], Youssin [13], [14] and Moh [10], as well as in [4], [5], [6].

Research partially supported by NSERC operating grant OGP 0009070 and A8849.

We are indebted to Orlando Villamayor for pointing out a useful technique of Hironaka [9, pp. 68-70] which Villamayor himself uses in [12]. (A variant of Hironaka's idea appears earlier in Abhyankar (1966); see [2].)

1. Introduction

In this section we introduce our resolution strategy in the central context of "embedded desingularization". Throughout these notes, M denotes an analytic (respectively, algebraic) manifold and X denotes a closed analytic subspace (respectively, algebraic subvariety) of M. Our spaces are defined over a field $\mathbf{K}$ of characteristic zero (with a complete valuation, in the analytic case). For simplicity of presentation, we concentrate on the analytic situation, though all statements make sense for algebraic varieties (with regular submanifolds as the centers of blowing up).

Desingularization is achieved by a sequence of blowings-up of the ambient space. The centres of the blowings-up will be determined by a local invariant of singularity of X, defined over a sequence of blowings-up: Consider a sequence of blowings-up

$$\cdots \xrightarrow{\sigma_j} M_j \longrightarrow \cdots \xrightarrow{\sigma_1} M_1 = M$$

with smooth centres $C_j \subset M_j$. Put $X_1 = X$ and $E_1 = \emptyset$. For each $j \geq 1$, let X_{j+1} denote the strict transform of X_j by σ_j, and let E_{j+1} denote the exceptional divisor $E_{j+1} = \sigma_j^{-1}(C_j \cup E_j)$. Each $C_j \subset X_j$. (Definitions of the basic notions involved in resolution of singularities – analytic space, blowing-up, strict transform, normal crossings, etc. – will be recalled in §2 below.) Our invariant $\mathrm{inv}_X(a)$, $a \in M_j$, will be defined inductively with respect to j, assuming that all centres C_i, $i < j$, are *admissible* in the sense that

(1) C_i and E_i simultaneously have only normal crossings.

(2) $\mathrm{inv}_X(\cdot)$ is locally constant on C_i.

(1) guarantees that E_{i+1} has only normal crossings. $\mathrm{inv}_X(a)$ depends, in fact, only on a and on the tower of local blowings-up near the successive images of a.

$\mathrm{inv}_X(a)$ is a "word" of the form

$$\left(H_{X_j,a}, s_1(a); \ \nu_2(a), s_2(a); \ \ldots, \ s_t(a); \ \nu_{t+1}(a)\right),$$

where $t \leq \dim_a M$, $H_{X_j,a}$ is the *Hilbert-Samuel function* of X_j at a, i.e.,

$$H_{X_j,a}(k) = \dim_{\mathbf{K}} \frac{\mathcal{O}_{X_j,a}}{\mathfrak{m}_{X_j,a}^{k+1}}, \qquad k \in \mathbf{N}$$

(where $\mathfrak{m}_{X_j,a}$ denotes the maximal ideal of the local ring $\mathcal{O}_{X_j,a}$ of X_j at a), each $s_r(a)$ is a positive integer reflecting the history of the successive exceptional divisors, $\nu_{t+1}(a) = 0$ or ∞, and $\nu_2(a), \ldots, \nu_t(a)$ are each positive rational "multiplicities". Moreover, $e_{r-1}!\nu_r(a) \in \mathbf{N}$, $r = 2,\ldots,t$, where e_1 is the minimal integer such that $H_{X_j,a}(k)$ coincides with a polynomial if $k \geq e_1$, and $e_r = \max\{e_{r-1}!, e_{r-1}!\nu_r(a)\}$, $r > 1$.

$\mathrm{inv}_X(a)$ can be determined by an explicit local construction following our proof of uniformization in the hypersurface case [5, §3]. This local construction will be presented in §3 below, in the case of a hypersurface. The general case uses [5, Thm 7.3] (or an algebraic variant) in a simple way; see [7] for details.

"Words" like $\mathrm{inv}_X(a)$ can be ordered lexicographically. The local construction reveals that, with respect to this ordering, $\mathrm{inv}_X(a)$ is analytic (or algebraic) Zariski upper-semicontinuous as a function of $a \in M_j$, and that $\mathrm{inv}_X(a) \leq \mathrm{inv}_X(\sigma_{j-1}(a))$ for all a. (Of course, $\mathrm{inv}_X(a) = \mathrm{inv}_X(\sigma_{j-1}(a))$ if $\sigma_{j-1}(a) \notin C_{j-1}$.) These are the key properties used in proving invariance of inv_X, as described in §4 below.

Our local construction shows immediately that the germ at a of

$$S_X(a) = \{x \in M_j : \mathrm{inv}_X(x) = \mathrm{inv}_X(a)\}$$

has only normal crossings and, moreover, that each component Z of $S_X(a)$ can be identified as the intersection of $S_X(a)$ with those hyperplanes in E_j which pass through a and contain Z. ($S_X(a)$ is smooth in the case that $\nu_{t+1}(a) = \infty$.)

Assume, for simplicity, that M is compact (or that we are in the algebraic case). Let Max_j denote the (finite) set of maximal values of $\mathrm{inv}_X(a)$, $a \in M_j$. Then $S_j = \{a \in M_j : \mathrm{inv}_X(a) \in \mathrm{Max}_j\}$ is a union of global smooth components which has only normal crossings. Our local construction determines an additional invariant $\mu_X(a) \geq 1$ ($e_t!\mu_X(a)$ is a positive integer) such that, taking any component of S_j as the centre C_j of σ_j, we have

$$\big(\mathrm{inv}_X(a'), \mu_X(a')\big) < \big(\mathrm{inv}_X(a), \mu_X(a)\big)$$

for all $a \in C_j$ and $a' \in \sigma_j^{-1}(a)$.

Our embedded desingularization theorem follows. Suppose that our sequence of blowings-up $\sigma_{j+1}\colon M_{j+1} \to M_j$ is *locally finite* (over M); i.e., the images of at most finitely many centres C_j intersect a given compact subset of M. Then the composition $\pi\colon M' \to M$ of the sequence, the (final) strict transform $X' \subset M'$ of X by π, and the exceptional divisor E' of M' make sense (as inverse limits). Clearly, E' coincides with the set of critical points of the mapping π.

Let $|X|$ denote the underlying point-set of X. Define

$\mathrm{Sing}\,|X| = \{a \in |X| : |X|$ is not smooth at $a\}$,

$\mathrm{Sing}_H X = \{a \in |X| : H_{X,x}$ is not locally constant on $|X|$ near $a\}$,

$\mathrm{Sing}_{\dim} X = \{a \in |X| : \dim_a |X| < \dim_a X\}$

(where $\dim_a X$ is the Krull dimension of $\mathcal{O}_{X,a}$). If $\mathbf{K}$ is algebraically closed, then $\mathrm{Sing}_{\dim} X = \emptyset$ and $\mathrm{Sing}\,|X|$, $\mathrm{Sing}_H X$ are closed analytic (or algebraic) subsets of $|X|$, but these properties are not true in general. Put

$$\Sigma = \mathrm{Sing}_H X \cup \mathrm{Sing}_{\dim} X .$$

It is not difficult to prove that Σ is a closed analytic (or algebraic) subset of $|X|$, $\mathrm{Sing}\,|X| \subseteq \Sigma$, and, if C is a closed connected submanifold of X (connected in the sense of Zariski topology, in the algebraic case) on which $H_{X,x}$ is constant, then either $C \subset \Sigma$ or $C \cap \Sigma = \emptyset$. (See [7], [5, Rmks 8.1]).

It follows that any component C of S_j (defined above) satisfies one of the following three conditions:

(1) $C \subset \mathrm{Sing}_H X_j \cup \mathrm{Sing}_{\dim} X_j$;

(2) $C \subset X_j \cap E_j$;

(3) C coincides with $|X|$ in germs at every point of C.

(By modifying the definition of $\mathrm{inv}_X(a)$, we can replace (2) above by the following more precise condition: (2′) $C \cap \mathrm{Sing}|X_j| = \emptyset$, $C \subset X_j \cap E_j$, but $|X_j|$ and E_j do not simultaneously have only normal crossings at any point of C.) In either of cases (1) and (2), $\dim_x C < \dim_x X$, for all $x \in C$. Components of S_j satisfying (3) can be ignored. We then obtain the following desingularization theorem (essentially the theorem of Hironaka).

Theorem. *There is a locally finite sequence of blowings-up $\sigma_{j+1}\colon M_{j+1} \to M_j$ with smooth admissible centres C_j such that:*

(1) Each component of C_j lies either in $\mathrm{Sing}_H X_j \cup \mathrm{Sing}_{\dim} X_j$ or in $X_j \cap E_j$.

(2) Let X' and E' denote the (final) strict transform of X and exceptional divisor, respectively. Then:

(i) $H_{X',x}$ is locally constant on $|X'|$.

(ii) $|X'|$ is smooth.

(iii) $|X'|$ and E' simultaneously have only normal crossings.

It follows that, if $\pi\colon M' \to M$ denotes the composite of the sequence of blowings-up σ_j, then

$$\mathrm{Sing}_H X \cup \mathrm{Sing}\,|X| \subseteq \pi(E') \subseteq \mathrm{Sing}_H X \cup \mathrm{Sing}_{\dim} X .$$

In particular, π restricts to an isomorphism $X' \to X$ over the smooth part of X. (If $\mathbf{K}$ is algebraically closed, then both inclusions above are equalities.)

Finally, the components of the loci $S_X(a) = \{x : \mathrm{inv}_X(x) = \mathrm{inv}_X(a)\}$ can be ordered (because of the way they have been described above) to give a *universal* choice of the centre of blowing up. A generalization of the invariant presented here can be used to deduce *canonical* resolution of singularities for (not necessarily compact) analytic spaces X: Any isomorphism between open subspaces of X lifts to isomorphisms throughout the entire history of blowings-up involved in desingularization.

2. Blowing up and the strict transform

Let M be an analytic manifold (or a smooth algebraic variety) defined over $\mathbf{K}$, and let C be a closed submanifold (of the same class).

The *blowing-up* of M with (smooth) *centre C* is a mapping $\sigma: M' \to M$, where M' is a manifold (of the same class) which can be defined locally in the following way: Let U be a coordinate chart of M such that

$$U \cap C = \{x \in U : x_\ell = 0,\ \ell \in J\},$$

where $x = (x_1, \ldots, x_n)$ are the coordinates of U and $J \subseteq \{1, \ldots, n\}$ with $\#J$ equal to the codimension p of C in M. Let $P^{p-1}(\mathbf{K})$ denote the $(p-1)$-dimensional projective space with homogeneous coordinate system $\xi = [\xi_\ell]_{\ell \in J}$. Then

$$\sigma^{-1}(U) = \{(x, \xi) \in U \times P^{p-1}(\mathbf{K}) : x_\ell \xi_m = x_m \xi_\ell,\ \ell, m \in J\}$$

and $\sigma(x, \xi) = x$. It follows that $U' = \sigma^{-1}(U)$ can be covered by coordinate charts

$$U'_\ell = \{(x, \xi) \in U' : \xi_\ell \neq 0\}, \quad \ell \in J,$$

where each U'_ℓ admits a coordinate system $y = (y_1, \ldots, y_n)$ given by

$$
\begin{aligned}
y_\ell &= x_\ell \\
y_m &= \xi_m / \xi_\ell, & m \in J - \{\ell\} \\
y_m &= x_m, & m \notin J.
\end{aligned}
$$

In these coordinates, then, σ is given by

$$
\begin{aligned}
x_\ell &= y_\ell \\
x_m &= y_\ell y_m, & m \in J - \{\ell\} \\
x_m &= y_m, & m \notin J.
\end{aligned}
$$

Therefore

(1) $$U'_\ell \cap \sigma^{-1}(C) = \{y \in U'_\ell : y_\ell = 0\}$$

and, for any $m \neq \ell$,

(2) $$U'_\ell \cap \mathrm{cl}\, \sigma^{-1}(\{x \in U : x_m = 0\} - C) = \{y \in U'_\ell : y_m = 0\}$$

(where cl denotes closure). The preceding local definition determines M' and the blowing-up $\sigma': M' \to M$ uniquely. Clearly, σ is proper and surjective, σ restricts to an isomorphism $M' - \sigma^{-1}(C) \to M - C$, and $\sigma^{-1}(C)$ is a hyperplane (smooth hypersurface); $\sigma^{-1}(C)$ is called an *exceptional divisor*.

We will also consider *local blowings-up over* open subspaces of M: Let V be an open subspace of M and C a smooth closed submanifold of V. The *local blowing-up* $\pi\colon N' \to M$ over V with *centre* C is the composite of the blowing up $\sigma\colon N' \to V$ with centre C, and the inclusion $V \hookrightarrow M$.

Suppose that S and E are subsets of M, where E is a union of hyperplanes. We say that S and E *simultaneously have only normal crossings* if, locally, M admits a coordinate system $(x_1, \ldots, x_n)$ in which E is a union of coordinate hyperplanes $x_\ell = 0$, and S is a coordinate subspace (i.e., given by $x_m = 0$, for certain m). If $S = \emptyset$, we say that E *has only normal crossings*.

Suppose that $E \subset M$ is a union of hyperplanes H (we write $H \in E$) and that E has only normal crossings. Let $\pi\colon N' \to M$ be a local blowing up, as above, with centre C. By (1) and (2) above, if C and E simultaneously have only normal crossings, then $E' = \pi^{-1}(C \cup E)$ is the union of the hyperplanes $\pi^{-1}(C)$ and $H' = \mathrm{cl}\,\pi^{-1}(H - C)$, and E' has only normal crossings.

Let X be a closed analytic subspace of M (closed algebraic subvariety, in the algebraic case); X consists of an underlying point-set $|X|$ and a sheaf $\mathcal{O}_X$ of germs of "regular functions" on $|X|$. Both $|X|$ and $\mathcal{O}_X$ are determined by a coherent sheaf of ideals $\mathcal{I}_X$ in $\mathcal{O}_M$ ($\mathcal{O}_M$ is the sheaf of germs of regular, i.e., analytic or regular rational, functions on M): $|X|$ is the support of $\mathcal{O}_M/\mathcal{I}_X$ and $\mathcal{O}_X$ the restriction of $\mathcal{O}_M/\mathcal{I}_X$ to $|X|$.

The *strict transform* X' of X by a blowing-up (or local blowing-up) $\sigma\colon M' \to M$ with smooth centre C is the smallest closed analytic subspace of $\sigma^{-1}(X)$ which coincides with $\sigma^{-1}(X)$ outside $\sigma^{-1}(C)$. (By definition, $\sigma^{-1}(X)$ is defined by the sheaf of ideals $\sigma^{-1}(\mathcal{I}_X)$ in $\mathcal{O}_{M'}$ generated by the pull-back $\sigma^*(\mathcal{I}_X)$.)

For example, suppose that X is a *hypersurface*; i.e., $\mathcal{I}_X$ is a principal ideal. Thus $\mathcal{I}_X$ is generated near every point a by a regular function f. Suppose that $a \in C$. Put $f' = y_{\mathrm{exc}}^{-d}\, f \circ \sigma$ at $a' \in \sigma^{-1}(a)$, where y_{exc} is a local generator of the ideal of $\sigma^{-1}(C)$ and d is the largest integer such that f' is regular. (It follows that d is the *multiplicity* $\mu_{C,a}(f)$ of f *along* C at a: $\mu_{C,a}(f) = \max\{p : f \in \mathcal{I}_{C,a}^p\}$.) In this case it is easy to see that $\mathcal{I}_{X',a'}$ is generated by f'.

For a general closed analytic subspace X of M, it is also not difficult to prove (cf. [7]):

Lemma 1. $\mathcal{I}_{X',a'}$ *is generated by* f', *for all* $f \in \mathcal{I}_{X,a}$.

Suppose that X is a hypersurface. Let $a \in M$. The *multiplicity* $\mu_a(X)$ of X at a is the largest integer d such that $\mathcal{I}_{X,a} \subseteq \mathfrak{m}_a^d$, where $\mathfrak{m}_a$ denotes the maximal ideal of $\mathcal{O}_{M,a}$. If f generates $\mathcal{I}_{X,a}$, then $\mu_a(X) = \mu_a(f)$, where $\mu_a(f)$ is the order of vanishing of f at a. It follows that $a \mapsto \mu_a(X)$ is an (analytic) Zariski upper-semicontinuous function.

Lemma 2. *Let $\sigma: M' \to M$ be a blowing-up with smooth centre C. Let $a \in C$. If f is a regular function such that $\mu_x(f)$ is constant on C near a, then $\mu_{a'}(f') \leq \mu_a(f)$ for all $a' \in \sigma^{-1}(a)$.*

Proof: Put $d = \mu_a(f)$. We can choose coordinates $(x_1, \ldots, x_n)$ at $a = 0$ such that $(\partial^d f / \partial x_n^d)(a) \neq 0$ and (using the implicit function theorem), such that $\partial^{d-1} f / \partial x_n^{d-1} \sim x_n$ at a (where $\sim$ means equal up to an invertible factor). Thus there is a neighbourhood U of a in which $f(x)$ can be expressed

$$f(x) \ \sim \ c_d(x)x_n^d + \sum_{q=0}^{d-1} c_q(\tilde{x})x_n^q \ ,$$

where $\tilde{x} = (x_1, \ldots, x_{n-1})$, $c_d(x) \neq 0$ in U, and $c_{d-1}(\tilde{x}) \equiv 0$. Since $c_{d-1}(\tilde{x}) \equiv 0$, it follows that $\mu_x(f) = d$ if and only if $x_n = 0$ and $\mu_x(c_q) \geq d - q$, $q = 0, \ldots, d-2$. In particular, after a change in the coordinates $\tilde{x}$, we can assume that

$$U \cap C \ = \ \{x \in U : x_n = 0, \ x_\ell = 0, \ \ell \in I\} \ ,$$

where $I \subseteq \{1, \ldots, n-1\}$.

Therefore (in the notation for blowing-up above) $\sigma^{-1}(U) = \bigcup_{\ell \in J} U_\ell'$, where $J = I \cup \{n\}$. First consider f' in U_ℓ', where $\ell \in I$: Then

$$f'(y) \ = \ y_\ell^{-d}(f \circ \sigma)(y)$$

$$= \ c_d'(y)y_n^d + \sum_{q=0}^{d-1} c_q'(\tilde{y})y_n^q \ ,$$

where $\tilde{y} = (y_1, \ldots, y_{n-1})$ and $c_q' = y_\ell^{-d+q} c_q \circ \sigma$, $q = 0, \ldots, d$. Since $c_d'(y) \neq 0$ on U_ℓ', $\mu_y(f') \leq d$, for all $y \in U_\ell'$.

On the other hand, if $y \in U_n' - \bigcup_{\ell \in I} U_\ell'$, then $\mu_y(f') = 0$, as follows:
$U_n' - \bigcup_{\ell \in J} U_\ell' = \{(y_1, \ldots, y_n) \in U_n' : y_\ell = 0, \ell \in I\}$. In U_n',

$$f'(y) \ = \ y_n^{-d}(f \circ \sigma)(y)$$

$$= \ c_d'(y) + \sum_{q=0}^{d-1} c_q'(y) \ ,$$

where $c_q' = y_n^{-d+q} c_q \circ \sigma$, $q = 0, \ldots, d$. Now $c_d'(y)$ vanishes nowhere. For each $q = 0, \ldots, d-1$, $\mu_x(c_q) \geq d - q$, $x \in C$, so that c_q belongs to the $(d-q)$'th power of the ideal generated by x_ℓ, $\ell \in I$. It follows that c_q' belongs to the $(d-q)$'th power of the ideal generated by y_ℓ, $\ell \in I$, so that $c_q'(y) = 0$, $y \in U_n' - \bigcup_{\ell \in I} U_\ell'$, $q = 0, \ldots, d-1$. Q. E. D.

Let X be a closed subspace of M. The *Hilbert-Samuel function* of X at a,

$$H_{X,a}(k) = \dim_{\kappa} \frac{\mathcal{O}_{M,a}}{\mathcal{I}_{X,a} + \mathfrak{m}_a^{k+1}}$$

is a discrete local invariant which, in the case that X is a hypersurface, measures singularity in the same way as the multiplicity $\mu_a(X)$. This is because, when X is a hypersurface,

$$H_{X,a}(k) = \binom{k+n}{n}, \qquad k < \mu_a(X)$$

$$H_{X,a}(k) = \binom{k+n}{n} - \binom{k-\mu_a(X)+n}{n} \qquad k \geq \mu_a(X).$$

The set of functions $H: \mathsf{N} \to \mathsf{N}$ can be partially ordered as follows: $H_1 \leq H_2$ if $H_1(k) \leq H_2(k)$ for all $k \in \mathsf{N}$. With this partial ordering, one can show:

Lemma 3. *The function $a \mapsto H_{X,a}$ is (analytic) Zariski upper-semicontinuous on M.*

Lemma 4. *If $H_{X,x}$ is constant on the centre C of a blowing-up $\sigma: M' \to M$, then $H_{X',y} \leq H_{X,\sigma'(y)}$, for all $y \in M'$.*

(See [5] for simple proofs.)

3. Local computation of inv_X and the effect of blowing-up with admissible centre

Let

$$\xrightarrow{\pi_j} N_j \longrightarrow \cdots \xrightarrow{\pi_1} N_1 = M$$

be a sequence of local blowings-up $\pi_j: N_{j+1} \to N_j$ (each over an open subspace V_j of N_j). Each π_j is the composite of a blowing-up $\sigma_j: N_{j+1} \to V_j$ with smooth centre C_j, and the inclusion $V_j \hookrightarrow N_j$. Put $E_1 = \emptyset$ and let $X_1 = X$ be a closed analytic subspace (or algebraic subvariety) of M. For each $j \geq 1$ let X_{j+1} denote the strict transform of X_j by π_j, and let $E_{j+1} = \pi_j^{-1}(C_j \cup E_j)$. Let $a \in N_j$. In this section, we introduce

$$\mathrm{inv}_X(a) = \big(H_{X_j,a}, s_1(a); \ldots, s_t(a); \nu_{t+1}(a) \big),$$

where $\nu_{t+1}(a) = 0$ or ∞, by a non-invariant calculation, inductively over the sequence of local blowings-up (with admissible centres, as defined in §1).

The entries of $\mathrm{inv}_X(a)$ will themselves be defined successively, by induction. Let us write inv_r for inv_X truncated after the entry s_r (with the convention that $\text{inv}_r(a) = \text{inv}_X(a)$ if $r > t$). Once $(\text{inv}_r; \nu_{r+1})$ is defined, s_{r+1} can be introduced immediately, *in an invariant way*:

Write $\pi_{ij} = \pi_{j-1} \circ \cdots \circ \pi_i$, $i = 1, \ldots, j-1$, and $\pi_{jj} = \text{id}$. Recall that each E_i, $i \leq j$, is a union of smooth hyperplanes which has only normal crossings, by the inductive assumption that each C_i is admissible, $i < j$. Put

$$E_a = \{H \in E_j : a \in H\} .$$

Set $a_i = \pi_{ij}(a)$, $i = 1, \ldots, j$. Suppose that i is the smallest index k such that $H_{X_j,a} = H_{X_k,a_k}$, and set

$$E_a(1) = \{H \in E_a : H \text{ is the strict transform}$$
$$\text{of some hyperplane in } E_{a_i}\} .$$

We define

$$s_1(a) = \#E_a(1) .$$

In general, suppose that $(\text{inv}_r(a); \nu_{r+1}(a))$ has been defined. Suppose that i is the smallest index k such that $(\text{inv}_r(a); \nu_{r+1}(a)) = (\text{inv}_r(a_k); \nu_{r+1}(a_k))$. Set

$$E_a(r+1) = \{H \in E_a - \bigcup_{q \leq r} E_a(q) : H \text{ is the strict}$$
$$\text{transform of some hyperplane in } E_{a_i}\} .$$

We define

$$s_{r+1}(a) = \#E_a(r+1) .$$

The crucial point, then, is to introduce $\nu_{r+1}(a)$ assuming that $\text{inv}_r(a)$ has already been defined. It is good enough, in fact, to define ν_{r+1} inductively over the tower of local blowings-up, assuming that inv_r has been defined and that all centres C_i are *r-admissible* in the sense that:

(1) C_i and E_i simultaneously have only normal crossings;

(2_r) inv_r is locally constant on C_r.

We will implicitly use the following observation: If $\text{inv}_r(x) = \text{inv}_r(a)$, where x is sufficiently close to a, then, for each $q = 1, \ldots, r$, $E_x(q) = \{H \in E_a(q) : x \in H\}$.

We will assume, for simplicity, that X is a hypersurface, so that each ideal $\mathcal{I}_{X_,}$ is principal. Then the Hilbert-Samuel function $H_{X_j,a}$ can be replaced by the multiplicity $\nu_1(a) = \mu_a(X_j)$ of X_j at a.

Our construction involves associating to the local defining equation of X certain collections "$\mathcal{F}$" of auxiliary functions with assigned "multiplicities". In the following, we allow these multiplicities to be rational. We could, in fact, reduce each "$\mathcal{F}$" to the case of a single integral assigned multiplicity by raising all functions in the collection to suitable powers. This would have some advantage in simplifying the notation and exposition. But our formulation here might be of interest from a computational viewpoint because it is closer to a practical calculation.

Let f be a local defining function for X at $a_1 = \pi_{1j}(a)$ (i.e., a local generator of the ideal of X at a_1). Then

$$(f \circ \pi_{1j})(x) \sim D(x) \cdot g(x) ,$$

where $g(x)$ is a local defining function for X_j at a, and $D(x) = D_{11}(x) \cdot D_{12}(x)$, with $D_{11}(x) = 0$ and $D_{12}(x) = 0$ local equations of $E_a - E_a(1)$ and $E_a(1)$, respectively. We first claim there is a chart U with coordinates $(x_1, \ldots, x_n) = (\tilde{x}, x_n)$ at $a = 0$, such that:

(a)
$$D_{11}(x) = \tilde{x}^\eta ;$$
$$D_{12}(x) = \ell_1(x)^{n_1} \cdots \ell_s(x)^{n_\bullet} ,$$

where $s = s_1(a)$ and

$$\ell_p(x) = a_{p1}(x)x_n + a_{p0}(\tilde{x}) , \qquad p = 1, \ldots, s ;$$

and

$$g(x) = c_d(x)x_n^d + \sum_{q=0}^{d-1} c_q(\tilde{x})x_n^q ,$$

where $d = \nu_1(a)$.

(b) Either
$$a_{p1}(x) \neq 0 \quad \text{but} \quad a_{p0}(\tilde{x}) \equiv 0 ,$$

for some p, or

$$c_d(x) \neq 0 \quad \text{but} \quad c_{d-1}(\tilde{x}) \equiv 0 .$$

(c) The coordinate functions x_ℓ such that $\eta_\ell \neq 0$ are local defining functions for the $H \in E_a - E_a(1)$, and the $\ell_p(x)$ are local defining functions for the $H \in E_a(1)$.

To get the local representation above, let i be the smallest index k such that $\nu_1(a) = \nu_1(a_k)$. Put $b = a_i$. We first write $f \circ \pi_{1i}$ in a way similar to the above in suitable local coordinates at $b \in V_i$, but with "D_{11}" $= 1$ and $s' \geq s$ exceptional locus factors "ℓ", s of which correspond to those exceptional hyperplanes in $E_b = E_b(1)$ whose strict transforms at a are the $\ell_1, \ldots, \ell_s$ above. The coordinates are chosen so that x_n is a regular direction either for one of the s distinguished "ℓ"'s or for g (cf. the proof of Lemma 2, §2). The formulas above follow from the transformation rules in the proof of Lemma 2.

From the local representation above, it is clear that the "1-admissible set",

$$\{x \in U : \mathrm{inv}_1(x) = \mathrm{inv}_1(a)\}$$
$$= \{x : \mu_x(g) = d, \ \mu_x(\ell_p) = 1, \ p = 1, \ldots, s ,$$
$$\mu_x(f \circ \pi_{1j}) \geq \mu\}$$
$$= \{x : x_n = 0, \ \mu_{\tilde{x}}(h) \geq \mu_h , \quad \text{for all}$$
$$(h, \mu_h) \in \mathcal{F}\} .$$

where $\mu = 0$ and $\mathcal{F}$ denotes the following collection of pairs (h, μ_h), each consisting of a function $h = h(\tilde{x})$ with an assigned "multiplicity" μ_h:

$$\mathcal{F} = \{(a_{p0}, 1), \quad p = 1, \ldots, s \, ;$$
$$(c_q, d - q), \quad q = 0, \ldots, d - 1 \, ;$$
$$(b, \mu_b)\} \, ,$$

with $b(\tilde{x}) \equiv 0$ and $\mu_b = 1$ if $\mu \leq d + \Sigma n_p$, and $b(\tilde{x}) = \tilde{x}^n$ and $\mu_b = \mu - d - \Sigma n_p$ if $\mu > d + \Sigma n_p$. Set $e_1 = \max\{\mu, d\}$. Then $e_1!/\mu_h \in \mathbb{N}$ and $h^{e_1!/\mu_h}$ is a regular function for every $(h, \mu_h) \in \mathcal{F}$. (Of course, since $\mu = 0$, the condition "$\mu_x(f \circ \pi_{1j}) \geq \mu$" is vacuous, $e_1 = d$, and $\mu < d + \Sigma n_p$, but we include these conditions here to establish a pattern that runs through the induction.)

(The adjective "regular" above is used loosely in the algebraic case: Each $h^{e_1!/\mu_h}$ is, more precisely, the restriction of a regular function in a Zariski open neighbourhood of a in N to a subvariety that is locally the graph of a function $x_n = x_n(x_1, \ldots, x_{n-1})$ which is algebraic in the sense of Artin; i.e., "Nash".)

Let us consider the effect of a blowing-up $\sigma: U' \to U$ with 1-admissible smooth centre C containing a. Clearly, $C \subset \{x_n = 0\}$. Moreover, since C and E_j simultaneously have only normal crossings, we can assume that

$$C = \{x : x_n = 0, \ x_\ell = 0, \ \ell \in I\} \, ,$$

where $I \subseteq \{1, \ldots, n - 1\}$. We write $U' = \bigcup_{\ell \in I \cup \{n\}} U'_\ell$, as in §2 above. Then, for all $a' \in U'_n - \bigcup_{\ell \in I} U'_\ell$, $\text{inv}_1(a') < \text{inv}_1(a)$ (cf. proof of Lemma 2). On the other hand, in U'_ℓ, $\ell \in I$, we have:

$$(f \circ \pi_{1j} \circ \sigma)(y) \sim \tilde{y}^{n'} \ell'_1(y)^{n_1} \cdots \ell'_s(y)^{n_s} \cdot g'(y) \, ,$$

where g' and ℓ'_p denote the strict transforms of g and ℓ_p, respectively; thus,

$$\ell'_p(y) = a'_{p1}(y) y_n + a'_{p0}(\tilde{y}) \, , \qquad p = 1, \ldots, s \, ,$$
$$g'(y) = c'_d(y) y_n^d + \sum_{q=0}^{d-1} c'_q(\tilde{y}) y_n^q \, ,$$

with

$$a'_{pq} = y_\ell^{1-q} a_{pq} \circ \sigma \, , \qquad q = 0, 1, \quad p = 1, \ldots, s \, ,$$
$$c'_q = y_\ell^{d-q} c_q \circ \sigma \, , \qquad q = 0, \ldots, d \, .$$

Moreover,

$$\eta'_m = \eta_m \, , \qquad m \neq \ell$$
$$\eta'_\ell = \sum_{m \in I} \eta_m + d + \Sigma n_p \, .$$

In particular, if $\mathrm{inv}_1(a') = \mathrm{inv}_1(a)$, where $a' \in \sigma^{-1}(a) \cap U'_\ell$, and if we denote by $\mathcal{F}'$ the collection of pairs analogous to $\mathcal{F}$, then

$$\mathcal{F}' = \{(a'_{p0}, 1), (c'_q, d-q), \ q \leq d-1, \ (b', \mu_{b'})\}$$

and $\mathcal{F}'$ is obtained from $\mathcal{F}$ by the following transformation law:

$$\mathcal{F}' = \{(h'(\tilde{y}), \mu_{h'}) = \left(y_\ell^{-\mu_h} \cdot h(\sigma(y)), \mu_h\right) : (h, \mu_h) \in \mathcal{F}\} .$$

We define $\nu_2(a) = \infty$ if $h = 0$ for all $(h, \mu_h) \in \mathcal{F}$. Otherwise: Let $D_2(\tilde{x})^{e_1!}$ be the greatest common divisor of the $h^{e_1!/\mu_h}$, where $(h, \mu_h) \in \mathcal{F}$, which is a monomial in the x_H, $H \in E_a - E_a(1)$. Define

$$\nu_2(a) = \min\{\frac{1}{\mu_h}\mu_a(h) - \mu_a(D_2) : (h, \mu_h) \in \mathcal{F}\}$$

(where $\mu_a(D_2)$ has an obvious meaning as a rational number).

If $\nu_2(a) = 0$ or ∞, we put

$$\mathrm{inv}_X(a) = \left(\nu_1(a), s_1(a); \nu_2(a)\right) .$$

Otherwise, $e_1!\nu_2(a) \in \mathbf{N}$ and, because of the formulas for transformation by a blowing-up with 1-admissible centre, we can repeat the construction above with $\mathcal{F}$ in place of $(f \circ \pi_{1j}, \mu)$, where $\mu = 0$:

For each $(h, \mu_h) \in \mathcal{F}$, let D_{2h} denote the greatest divisor of $D_2^{-\mu_h} \cdot h$ which is a monomial in the x_H, where $H \in E_a(2)$. Let $\mathcal{G} = \{(g, \mu_g)\}$ be the collection of functions $g = D_{2h}^{-1} D_2^{-\mu_h} h$ together with assigned "multiplicities" $\mu_g = \mu_h \cdot \nu_2(a) - \mu_a(D_{2h})$, for all $(h, \mu_h) \in \mathcal{F}$. Clearly, each $g = g(\tilde{x})$ is a regular function times a monomial with rational exponents in the x_H, $H \in E_a - (E_a(1) \cup E_a(2))$.

If $\mu \in \mathbf{Q}$, let (μ) denote the smallest integer $\geq \mu$. Put $s_2 = s_2(a)$. Write $D_2(\tilde{x}) = D_{21}(\tilde{x}) \cdot D_{22}(\tilde{x})$, where $D_{21}(\tilde{x})$ is the greatest divisor of $D_2(\tilde{x})$ which is a monomial in the x_H, $H \in E_a - (E_a(1) \cup E_a(2))$. After a change in the coordinates $\tilde{x} = (\tilde{\tilde{x}}, x_{n-1})$, we can assume:

(a)

$$D_{21}(\tilde{x}) = \tilde{\tilde{x}}^{\eta_2};$$
$$D_{22}(\tilde{x}) = \ell_{2p}(\tilde{x})^{n_{21}} \cdots \ell_{2s_2}(\tilde{x})^{n_{2s_2}},$$

where

$$\ell_{2p}(\tilde{x}) = a_{2p1}(\tilde{x})x_{n-1} + a_{2p0}(\tilde{\tilde{x}}) , \qquad p = 1, \ldots, s_2 ;$$

and, for all $(g, \mu_g) \in \mathcal{G}$,

$$g(\tilde{x}) = c_{g,(\mu_g)}(\tilde{x})x_{n-1}^{(\mu_g)} + \sum_{q=0}^{(\mu_g)-1} c_{g,q}(\tilde{\tilde{x}})x_{n-1}^q .$$

(b) Either

$$a_{2p1}(\tilde{x}) \neq 0 \qquad \text{but} \qquad a_{2p0}(\tilde{\tilde{x}}) \equiv 0 \, ,$$

for some p, or

$$c_{g,(\mu_g)}(\tilde{x}) \neq 0 \qquad \text{but} \qquad c_{g,(\mu_g)-1}(\tilde{\tilde{x}}) \equiv 0 \, ,$$

for some $(g, \mu_g) \in \mathcal{G}$ such that $\mu_g = (\mu_g)$.

(c) The coordinate functions x_ℓ such that $\eta_{2\ell} \neq 0$ are local generators $x_\ell = x_H$ of the ideals of the $H \in E_a - \big(E_a(1) \cup E_a(2)\big)$, and the $\ell_{2p}(\tilde{x})$ are local defining functions for the $H \in E_a(2)$.

The proof is similar to the first step. Then the 2-admissible set,

$$\{x : \mathrm{inv}_2(x) = \mathrm{inv}_2(a)\}$$
$$= \{x : x_n = x_{n-1} = 0, \ \mu_{\tilde{x}}(h) \geq \mu_h \, , \quad \text{for}$$
$$\text{all } (h, \mu_h) \in \mathcal{F}_3\}$$

where $\mathcal{F}_3 = \{h(\tilde{\tilde{x}}), \mu_h)\}$ is given by

$$\mathcal{F}_3 = \{(a_{2p0}, 1), \ p = 1, \dots, s_2;$$
$$(c_{g,q}, \mu_g - q), \ q = 0, \dots, (\mu_g) - 1, \ (g, \mu_g) \in \mathcal{G};$$
$$(b_2, \mu_{b_2})\} \, ,$$

with $b_2 = 0$ and $\mu_{b_2} = 1$ if $1 \leq \nu_2(a) + \mu_a(D_{22})$, and $b_2(\tilde{x}) = \tilde{\tilde{x}}^{\eta_2}$ and $\mu_{b_2} = 1 - \nu_2(a) - \mu_a(D_{22})$ if $1 > \nu_2(a) + \mu_a(D_{22})$. ("$\mathcal{F}_2$" is $\mathcal{F}$ above.)

Set $e_2 = \max\{e_1!, e_1!\nu_2(a)\}$. Then $e_2!/\mu_h \in \mathbf{N}$ and $h^{e_2!/\mu_h}$ is a regular function, for all $(h, \mu_h) \in \mathcal{F}_3$: First consider $h = b_2 = D_{21}$ and $\mu_h = 1 - \nu_2(a) - \mu_a(D_{22}) > 0$. Both $D_{21}^{e_1!}$ and $D_{22}^{e_1!}$ are regular; therefore, $e_1!\mu_h \in \mathbf{N}$. Since $e_1!\mu_h \leq e_1! \leq e_2$, it follows that $h^{e_2!/\mu_h}$ is regular and $e_2!/\mu_h \in \mathbf{N}$. It remains to consider $h = c_{g,q}(\tilde{x})$ and $\mu_h = \mu_g - q$, $q = 0, \dots, (\mu_g) - 1$, $(g, \mu_g) \in \mathcal{G}$. Now $g = D_{2\theta}^{-1} D_2^{-\mu_\theta} \theta$ and $\mu_g = \mu_\theta \cdot \nu_2(a) - \mu_a(D_{2\theta})$, for some $(\theta, \mu_\theta) \in \mathcal{F}$. Since $\theta^{e_1!/\mu_\theta}$ is regular, it follows from the definitions of D_2 and $D_{2\theta}$ that $D_{2\theta}^{e_1!/\mu_\theta}$ and $g^{e_1!/\mu_\theta}$ are regular. Therefore, $h^{e_1!/\mu_\theta} = c_{g,q}^{e_1!/\mu_\theta}$ is regular (because g is a regular function times a monomial in $\tilde{x}$ with rational coefficients), and also $(e_1!/\mu_\theta)\mu_g = e_1!\nu_2(a) - (e_1!/\mu_\theta)\mu_a(D_{2\theta}) \in \mathbf{N}$. Since $e_1!/\mu_\theta \in \mathbf{N}$, $(e_1!/\mu_\theta)\mu_h = (e_1!/\mu_\theta)(\mu_g - q) \in \mathbf{N}$. But $e_2 \geq e_1!\nu_2(a) \geq (e_1!/\mu_\theta)\mu_h$, so that $h^{e_2!/\mu_h}$ is regular and $e_2!/\mu_h \in \mathbf{N}$.

Now consider the effect of blowing up with 2-admissible smooth centre C containing a. Clearly, we can assume that

$$C = \{x : x_n = x_{n-1} = 0, \ x_\ell = 0, \ \ell \in I\} \, ,$$

where $I \subseteq \{1, \ldots, n-2\}$. We write $U' = \bigcup_{\ell \in I \cup \{n-1,n\}} U'_\ell$, as before. Then, for all $a' \in U' - \bigcup_{\ell \in I} U'_\ell$, $\mathrm{inv}_2(a') < \mathrm{inv}_2(a)$. In U'_ℓ, $\ell \in I$,

$$\{y : \mathrm{inv}_2(y) = \mathrm{inv}_2(a)\}$$
$$= \{y : y_n = y_{n-1} = 0, \ \mu_y(h') \geq \mu_{h'} \, ,$$
$$\text{for all } (h', \mu_{h'}) \in \mathcal{F}'_3\} \, ,$$

where $\mathcal{F}'_3$ is obtained from $\mathcal{F}_3$ by the following transformation law:

$$\mathcal{F}'_3 = \{(h'(\tilde{y}), \mu_{h'}) = \left(y_\ell^{-\mu_h} h(\sigma(y)), \mu_h\right) : (h, \mu_h) \in \mathcal{F}_3\} \, .$$

We continue in the same way, to define $\nu_j(a), \ldots$. Finally, we reach $t \leq n$ such that $0 < \nu_r(a) < \infty$ if $r = 1, \ldots, t$ and $\nu_{t+1}(a) = 0$ or ∞. Define

$$\mathrm{inv}_X(a) = \left(\mathrm{inv}_t(a); \nu_{t+1}(a)\right) \, .$$

(Then the t-admissible set is already admissible.)

In the case that $\nu_{t+1}(a) = \infty$, the t-admissible set $\{x : \mathrm{inv}_t(x) = \mathrm{inv}_t(a)\}$ is

$$\{x : x_n = x_{n-1} = \cdots = x_{n-t+1} = 0\} \, .$$

We have seen that if $\sigma: U' \to U$ is the blowing-up with centre C equal to this set, then $\mathrm{inv}_X(y) < \mathrm{inv}_X(a)$, for all $y \in U'$.

If $\nu_{t+1}(a) = 0$, then $h = D_{t+1}^{\mu_h}$, for some $(h, \mu_h) \in \mathcal{F}_{t+1}$, and the t-admissible set is

$$\{x : x_n = \cdots = x_{n-t+1} = 0, \ \mu_x(D_{t+1}) \geq 1\} \, .$$

Now, $D_{t+1}(x_1, \ldots, x_{n-t})$ is a monomial $x_1^{\Omega_1} \cdots x_{n-t}^{\Omega_{n-t}}$ with rational exponents. (Each x_ℓ such that $\Omega_\ell \neq 0$ is a local defining function $x_\ell = x_H$ for some $H \in E_a - \bigcup_{r < t} E_a(r)$.) Therefore, the t-admissible set is a union of smooth components $\bigcup_I Z_I$, where

$$Z_I = \{x : x_n = \cdots = x_{n-t+1} = 0, \ x_\ell = 0, \ \ell \in I\}$$

and the union is over the minimal subsets I of $\{1, \ldots, n-t\}$ such that $\sum_{\ell \in I} \Omega_\ell \geq 1$; equivalently, over the subsets I of $\{1, \ldots, n-t\}$ such that

$$0 \leq \sum_{m \in I} \Omega_m - 1 < \Omega_\ell , \qquad \text{for all } \ell \in I \, .$$

Consider the blowing-up $\sigma: U' \to U$ with centre one of the Z_I. Let $a' \in \sigma^{-1}(a)$. Suppose that $\mathrm{inv}_X(a') = \mathrm{inv}_X(a)$. Then $a' \in \bigcup_{\ell \in I} U'_\ell$. If $a' \in U'_\ell$, then $\mathcal{F}'_{t+1}$ includes the pair (h', μ_h), where

$$h' = y_\ell^{-\mu_h} \cdot D_{t+1}^{\mu_h} \circ (\sigma | U'_\ell) = (y_1^{\Omega'_1} \cdots y_{n-t}^{\Omega'_{n-t}})^{\mu_h}$$

and

$$\Omega'_m = \Omega_m , \qquad m \neq \ell ,$$

$$\Omega'_\ell = \sum_{m \in I} \Omega_m - 1 < \Omega_\ell .$$

Therefore, $1 \leq \sum \Omega'_m < \sum \Omega_m$. In particular, $\mu_X(a') < \mu_X(a)$, where $\mu_X(a) = \mu_{t+1}(a)$ is defined as

$$\mu_{t+1}(a) = \min\{\frac{1}{\mu_h}\mu_a(h) : (h, \mu_h) \in \mathcal{F}_{t+1}\} .$$

Since $e_t!/\mu_h \in \mathbf{N}$ and $h^{e_t!/\mu_h}$ is regular, for every $(h, \mu_h) \in \mathcal{F}_{t+1}$, it follows that $e_t!\mu_{t+1}(a) \in \mathbf{N}$ and that $\mathrm{inv}_X(a)$ must decrease after at most $e_t!\mu_{t+1}(a)$ blowings-up with admissible centres as described.

We will see in §4 that μ_{t+1} is an invariant, as is each exponent Ω_{rH} of the monomial $D_r = \prod_{H \in E_a - \cup_{q < r} E_a(q)} x_H^{\Omega_{rH}}$, $r = 1, \ldots, t$ (and $r = t + 1$ if $\nu_{t+1} = 0$). It is possible to choose coordinates $(x_1, \ldots, x_n)$ in our construction above such that each $H \in E_a$ is a coordinate hyperplane and each $H \in E_a(r)$ is $\{x_\ell = 0\}$ for some $\ell \leq n - r + 1$, $r = 1, \ldots, t$.

This section already provides an alternative proof of the uniformization theorem of [5]. But it includes a feature which is crucial for globalization: Once we show that inv_X is an invariant, a closed admissible set $S_X(a) = \{x : \mathrm{inv}_X(x) = \mathrm{inv}_X(a)\}$ is necessarily a union of *global* smooth subspaces. This is because each nontrivial factor $x_H^{\Omega_H}$ of the monomial D_{t+1} above defines a (global) exceptional hyperplane H.

4. Invariance

We follow the notation of §3. In this section, we give a key part of our proof that $\mathrm{inv}_X(a)$ and $\mu_X(a)$, $a \in N_j$, are invariants. Of course, the Hilbert-Samuel function $H_{X_j,a}$ is an invariant. In §3, the $s_r(a)$ were introduced in an invariant way. Therefore, to prove that $\mathrm{inv}_X(a)$ is invariant, it suffices to prove that $\nu_{r+1}(a)$ is an invariant, assuming the invariance of $\mathrm{inv}_r(a)$, $r = 1, \ldots, n$. As in §3, we can work with a tower of local blowings-up

$$\xrightarrow{\pi_j} N_j \longrightarrow \cdots \xrightarrow{\pi_1} N_1 = M$$

with smooth centres which are r-admissible.

It follows from the local construction used to define inv_r in §3, together with Lemmas 2, 3 and 4 of §2, that, for each i:

(1) inv_r is (analytic) Zariski upper-semicontinuous on N_i.

(2) $\mathrm{inv}_r(x) \leq \mathrm{inv}_r(\pi_i(x))$, for all $x \in N_{i+1}$.

We will again suppose that X is a hypersurface. We can assume that $0 < \nu_q(a) < \infty$, $q = 1, \ldots, r$. (Otherwise, $\mathrm{inv}_X(a)$ is already determined completely.) Let us begin by recalling certain features of the local construction of §3 which are crucial to the definition of $\nu_{r+1}(a)$: In suitable local coordinates $(x_1, \ldots, x_n)$ at $a = 0$, we have $\mathcal{F}_{r+1} = \{(f, \mu_f)\}$, a finite collection of functions $f = f(x_1, \ldots, x_{n-r})$ with assigned rational multiplicities $\mu_f > 0$. (Each $\mu_f \leq \mu_a(f)$.) We put $\nu_{r+1}(a) = \infty$ if $f = 0$ for all $(f, \mu_f) \in \mathcal{F}_{r+1}$. Otherwise: Each f is a regular function times a monomial in the x_H, $H \in E_a - \bigcup_{q \leq r} E_a(q)$, with rational exponents. Moreover, for all $(f, \mu_f) \in \mathcal{F}_{r+1}$, $e_r!/\mu_f \in \mathbf{N}$ and $f^{e_r!/\mu_f}$ is regular (i.e., the monomial factor has integral exponents). We let $D_{r+1}(x_1, \ldots, x_{n-r})$ denote the greatest common divisor of the f^{1/μ_f} which is a monomial in the x_H, $H \in E_a - \bigcup_{q \leq r} E_a(q)$, with rational exponents. (Each such $x_H = x_\ell$, for some $\ell = 1, \ldots, n - r$.) Then we put

$$\nu_{r+1}(a) = \min\left\{ \frac{1}{\mu_f}\mu_a(f) - \mu_a(D_{r+1}) : (f, \mu_f) \in \mathcal{F}_{r+1} \right\}.$$

If $\nu_{r+1}(a) = \infty$, put $\mu_{r+1}(a) = \infty$. Otherwise: Let

$$\mu_{r+1}(a) = \min\left\{ \frac{1}{\mu_f}\mu_a(f) : (f, \mu_f) \in \mathcal{F}_{r+1} \right\}$$

(so that $\mu_{r+1}(a) \geq 1$). We can write

$$D_{r+1} = \prod_{H \in E_a - \bigcup_{q \leq r} E_a(q)} x_H^{\Omega_{r+1,H}(a)},$$

where each $\Omega_{r+1,H}(a)$ is a nonnegative rational number. Thus,

$$\nu_{r+1}(a) = \mu_{r+1}(a) - \sum_{H \in E_a - \bigcup_{q \leq r} E_a(q)} \Omega_{r+1,H}(a).$$

It suffices to show that $\mu_{r+1}(a)$ is an invariant and, if $\mu_{r+1}(a) < \infty$, then each $\Omega_{r+1,H}(a)$ is an invariant. In fact, each of these numerical characters can be defined abstractly in terms of blowings-up of the product $N_j \times \mathbf{K}$.

We recall that (the germ at a of) the r-admissible set, $\{x : \mathrm{inv}_r(x) = \mathrm{inv}_r(a)\}$ is

(*)
$$\{x : x_n = \cdots = x_{n-r+1} = 0,$$
$$\mu_x(f) \geq \mu_f, \text{ for all } (f, \mu_f) \in \mathcal{F}_{r+1}\}.$$

It follows that $\mu_{r+1}(a) = \infty$ if and only if the r-admissible set at a is a submanifold of codimension r in N_j. Hence it remains to show that $\mu_{r+1}(a)$ is an invariant in the case that $\mu_{r+1}(a) < \infty$.

We recall also that if $a' \in \pi_j^{-1}(a)$ and $\mathrm{inv}_r(a') = \mathrm{inv}_r(a)$, then the set $\mathcal{F}'_{r+1}$ analogous to $\mathcal{F}_{r+1}$ at the point a' is given by the transformation law $\mathcal{F}'_{r+1} = \{(f', \mu_f) : (f, \mu_f) \in \mathcal{F}_{r+1}\}$, where each $f' = y_{\mathrm{exc}}^{-\mu_f}(f \circ \pi_j)$ and y_{exc} denotes a local generator of the ideal of $\sigma_j^{-1}(C_j)$ at a'.

We will prove here that $\mu_{r+1}(a)$ is an invariant, using properties of the local construction of §3 recalled above and the observation that the data obtained by taking the product of our original data with $\mathbf{K}$ admit the same local construction. A similar argument can be used to show that each $\Omega_{r+1,H}(a)$, $H \in E_a - \bigcup_{q \le r} E_a(q)$, is an invariant provided that $\mu_{r+1}(a) > 1$. (See [7].) In the case that $\mu_{r+1}(a) = 1$, invariance of $\Omega_{r+1,H}(a)$ can be obtained by applying the same considerations to the *total transform* (inverse image) of $X_j \times \mathbf{K}$ by the blowing-up σ with centre $H \times \{0\}$. The point is that, even though $H \times \{0\}$ is not r-admissible, we still have a local description as in §3 for the transformed data (at $c = \gamma'(0)$, where γ' denotes the lifting by σ of the arc $\gamma(t) = (a,t)$) relative to the sets "$E_c(q)$" given by the strict transforms of the $H' \times \mathbf{K}$, $H' \in E_a(q)$. (See [7] for details.)

To prove that $\mu_{r+1}(a)$ is an invariant, assuming $\mu_{r+1}(a) < \infty$: Consider the sequence of local blowings-up induced by $\times \mathbf{K}$,

$$N_j \times \mathbf{K} \xrightarrow{\pi_{j-1} \times \mathrm{id}} \cdots \longrightarrow N_1 \times \mathbf{K} = M \times \mathbf{K},$$

whose centres are clearly r-admissible with respect to the closed subspace (subvariety) $X \times \mathbf{K}$ of $M \times \mathbf{K}$. Put $P_1 = N_j \times \mathbf{K}$ and $c_1 = (a,0) \in P_1$. Let $\gamma_1(t)$ denote the arc $\gamma_1(t) = (a,t)$ with image $\{a\} \times \mathbf{K}$, and consider the sequence of blowings-up,

$$\xrightarrow{\sigma_\beta} P_\beta \longrightarrow \cdots \longrightarrow P_2 \xrightarrow{\sigma_1} P_1,$$

with successive centres $c_\beta = \gamma_\beta(0)$, where the γ_β are the successive liftings of γ_1 to the P_β. Then $\mathrm{inv}_r(c_\beta) = \mathrm{inv}_r(a,0) = \mathrm{inv}_r(a)$ for all β, because inv_r is upper-semicontinuous and $\mathrm{inv}_r(y) \le \mathrm{inv}_r(\sigma_\beta(y))$, $y \in P_{\beta+1}$, for all β.

We introduce a subset S_r of $\mathbf{N} \times \mathbf{N}$ as follows: First, let us say that $(\beta, 0) \in S_r$, $\beta \ge 1$, if, after β blowings-up as above, there exists (a germ at $c_{\beta+1}$ of) an r-admissible smooth subspace (subvariety) W_1 of codimension r in the last exceptional hyperplane $\sigma_\beta^{-1}(c_\beta)$. In this case, we can blow up $P_{\beta+1}$ locally with r-admissible centre W_1. Put $Q_1 = P_{\beta+1}$, $d_1 = c_{\beta+1}$, and $\delta_1 = \gamma_{\beta+1}$. Continuing inductively, we say that $(\beta, \alpha) \in S_r$, $\alpha \ge 1$, if $(\beta, \alpha - 1) \in S_r$ and there exists again at $d_{\alpha+1} = \delta_{\alpha+1}(0)$ (where $\delta_{\alpha+1}$ denotes the lifting of δ_α by the local blowing-up $\tau_\alpha \colon Q_{\alpha+1} \to Q_\alpha$ with centre W_α), an r-admissible smooth subspace $W_{\alpha+1}$ of codimension r in the last exceptional hyperplane $\tau_\alpha^{-1}(W_\alpha)$.

Lemma 5. $S_r = \emptyset$ *if and only if* $\mu_{r+1}(a) = 1$. *If* $S_r \neq \emptyset$, *then*

$$S_r = \{(\beta,\alpha) \in \mathbb{N} \times \mathbb{N}: \ \beta(\mu_{r+1}(a)-1) - \alpha \geq 1\}.$$

Lemma 5 specifies $\mu_{r+1}(a)$ uniquely; in the case that $1 < \mu_{r+1}(a) < \infty$, as

$$\mu_{r+1}(a) = 1 + \sup_{(\beta,\alpha)\in S_r} \frac{\alpha+1}{\beta}.$$

Proof of Lemma 5: We use the local coordinate construction of §3, as recalled above. If $(x_1,\ldots,x_n,x_0)$ are coordinates for $P_1 = N_j \times \mathbf{K}$ centered at $c_1 = (a,0)$, then σ_1 is given in the chart "U_0'" of P_2 by

$$x_0 = y_0$$
$$x_\ell = y_0 y_\ell, \qquad \ell = 1,\ldots,n;$$

c_2 is the origin of U_0' and $\gamma_2(t) = (0,\ldots,0,t)$ in U_0'. Since $\mathrm{inv}_r(c_2) = \mathrm{inv}_r(c_1)$, $\mathcal{F}_{r+1} = \{(f,\mu_f)\}$ at $c_1 = (a,0)$ transforms to $\mathcal{F}_{r+1}' = \{(f',\mu_f)\}$ at c_2, where

$$f' = y_0^{-\mu_f} \cdot (f \circ \sigma_1),$$

for each f. Therefore, if $\mu_{r+1}(a) < \infty$, the $(f')^{e_r!/\mu_f}$ (which are regular) admit $y_0^{(\mu_{r+1}(a)-1)e_r!}$ as their greatest common divisor which is a power of the local defining function $y_0 = y_{\mathrm{exc}}$ for $\sigma_1^{-1}(c_1)$. Write

$$f' = y_0^{(\mu_{r+1}(a)-1)\mu_f} \tilde{f}',$$

for all $(f',\mu_f) \in \mathcal{F}_{r+1}'$, and put

$$\tilde{\mu}_{r+1}(c_2) = \min\left\{\frac{1}{\mu_f}\mu_{c_2}(\tilde{f}')\right\}.$$

Then $\tilde{\mu}_{r+1}(c_2) \leq \mu_{r+1}(c_1) = \mu_{r+1}(a)$, by Lemma 2. But $\tilde{\mu}_{r+1}$ is upper-semicontinuous on the lifting γ_2 of γ_1, and $\tilde{\mu}_{r+1}(\gamma_2(t)) = \mu_{r+1}(c_1)$ if $t \neq 0$. Therefore, $\tilde{\mu}_{r+1}(c_2) = \mu_{r+1}(a)$.

It follows that, after β blowings-up $\sigma_1,\ldots,\sigma_\beta$ as above, the transform $\mathcal{F}_{r+1}' = \{(f',\mu_f)\}$ of $\mathcal{F}_{r+1}$ at $c_{\beta+1}$ satisfies the following condition: Each

$$f' = y_0^{\beta(\mu_{r+1}(a)-1)\mu_f} \tilde{f}',$$

where the $\tilde{f}'^{e_r!/\mu_f}$ do not admit $y_0 = y_{\mathrm{exc}}$ as a common factor. (This formula is in "U_0'"-coordinates centered at $c_{\beta+1}$, where the y_0-axis is the lifting $\gamma_{\beta+1}$ of γ_1.) According to (*) above, the only possible r-admissible smooth submanifold of codimension r at $c_{\beta+1}$ in $\{y_0 = 0\}$ is

$$W_1 = \{y \in U_0': \ y_n = \cdots = y_{n-r+1} = 0, \ y_0 = 0\},$$

and, moreover, W_1 is r-admissible if and only if $\mu_y(f') \geq \mu_f$, for all $y \in W_1$ and $(f', \mu_f) \in \mathcal{F}'_{r+1}$. Since each f' depends only on $(y_0, y_1, \ldots, y_{n-r})$ and some $\tilde{f}'$ does not vanish identically on W_1 near $c_{\beta+1} = 0$, it follows that W_1 is r-admissible if and only if

$$\beta(\mu_{r+1}(a) - 1) \geq 1 .$$

In particular, $\mu_{r+1}(a) = 1$ if and only if such W_1 does not exist after any number β of blowings-up as described.

Now suppose that W_1 above is r-admissible, and consider the local blowing-up $\tau_1 \colon Q_2 \to Q_1 = P_{\beta+1}$ with centre W_1. In the chart "U_0'" of Q_2, τ_1 is defined by

$$\begin{aligned} x_0 &= y_0 \\ x_\ell &= y_\ell , & \ell = 1, \ldots, n - r , \\ x_\ell &= y_0 y_\ell , & \ell = n - r + 1, \ldots, n . \end{aligned}$$

The transformation from $\mathcal{F}_{r+1} = \{(f, \mu_f)\}$ at $d_1 = c_{\beta+1}$ to $\mathcal{F}'_{r+1}$ at d_2 is given by $\mathcal{F}'_{r+1} = \{(f', \mu_f) = (y_0^{-\mu_f}(f \circ \tau_1), \mu_f)\}$, and each $\tilde{f}' = \tilde{f} \circ \tau_1$. (In fact, $\tau_1 = $ identity on the relevant variables.) Thus

$$f' = y_0^{(\beta(\mu_{r+1}(a)-1)-1)\mu_f} \tilde{f}' ,$$

for all $(f', \mu_f) \in \mathcal{F}'_{r+1}$, where the $\tilde{f}'^{e_r!/\mu_f}$ do not admit $y_0 = y_{\text{exc}}$ as a common factor.

After α such blowings-up $\tau_1, \ldots, \tau_\alpha$, $\mathcal{F}'_{r+1}$ consists of pairs (f', μ_f), where each

$$f' = y_0^{(\beta(\mu_{r+1}(a)-1)-\alpha)\mu_f} \tilde{f}' ,$$

and the $\tilde{f}'$ do not admit $y_0 = y_{\text{exc}}$ as a common factor. As above, the only possible (germ at $d_{\alpha+1}$ of an) r-admissible smooth subspace $W_{\alpha+1}$ of codimension r in the last exceptional hyperplane $\tau_\alpha^{-1}(W_\alpha)$ is

$$W_{\alpha+1} = \{y \in U_0' : y_n = \cdots = y_{n-r+1} = 0, \ y_0 = 0\} ,$$

and $W_{\alpha+1}$ is r-admissible if and only if

$$\beta(\mu_{r+1}(a) - 1) - \alpha \geq 1 .$$

Q. E. D.

REFERENCES

[1] S.S. Abhyankar, "Weighted expansions for canonical desingularization," Lecture Notes in Math. No. 910, Springer, Berlin-Heidelberg-New York, 1982.

[2] S.S. Abhyankar, *Good points of a hypersurface*, Adv. in Math. **68** (1988), 87–256.

[3] J.M. Aroca, H. Hironaka and J.L. Vicente, *Desingularization theorems*, Mem. Math. Inst. Jorge Juan No. 30, Consejo Superior de Investigaciones Científicas, Madrid, 1977.

[4] E. Bierstone and P.D. Milman, *Semianalytic and subanalytic sets*, Publ. Math. I.H.E.S. **67** (1988), 5–42.

[5] E. Bierstone and P.D. Milman, *Uniformization of analytic spaces*, J. Amer. Math. Soc. **2** (1989), 801–836.

[6] E. Bierstone and P.D. Milman, *Arc-analytic functions*, Invent. Math. **101** (1990), 411–424.

[7] E. Bierstone and P.D. Milman, *Canonical desingularization in characteristic zero: a simple constructive proof*, (to appear).

[8] H. Hironaka, *Resolution of singularities of an algebraic variety over a field of characteristic zero: I, II*, Ann. of Math. (2) **79** (1964), 109–326.

[9] H. Hironaka, *Idealistic exponents of singularity*, in "Algebraic Geometry," J.J. Sylvester Sympos., John Hopkins Univ., Baltimore, Md.,1976, John Hopkins Univ. Press, Baltimore, Md., 1977, pp. 52–125.

[10] T.T. Moh, *Canonical resolution of hypersurface singularities of characteristic zero*, preprint, Purdue University, 1990.

[11] M. Spivakovsky, *A solution to Hironaka's polyhedra game*, "Arithmetic and geometry, Vol. II," Prog. Math. No. 36, Birkhäuser, Boston, Mass., 1983, pp. 419–432.

[12] O. Villamayor, *Constructiveness of Hironaka's resolution*, Ann. Scient. Ecole Norm. Sup. (4e série) **22** (1989), 1–32.

[13] B. Youssin, *Newton polyhedra without coordinates*, Mem. Amer. Math. Soc. **433** (1990), 1–74.

[14] B. Youssin, *Newton polyhedra of ideals,*, Mem. Amer. Math. Soc. **433** (1990), 75–99.

Edward Bierstone
Pierre D. Milman
University of Toronto
Toronto, Canada M5S 1A1
bierston@math.toronto.edu
milman@math.toronto.edu

Local Membership Problems
for Polynomial Ideals

LEANDRO CANIGLIA   JORGE A. GUCCIONE
JUAN J. GUCCIONE

Introduction

Let K be a field, $R := K[X]$ the ring of polynomials in the indeterminates $X_1, \ldots, X_n$ over K and J an ideal of R. In this work we consider the following *Localization Problem (LP)*: Given $f, f_1, \ldots, f_t \in R$,

i) decide *efficiently* whether a non negative integer N exists verifying

$$f \cdot \mathrm{rad}(J)^N \subseteq (f_1, \ldots, f_t)$$

ii) If the condition in i) holds, then for each $g \in \mathrm{rad}(J)^N$, construct *efficiently* polynomials $a_1, \ldots, a_t$ such that:

$$f \cdot g = \sum_{1 \leq j \leq t} a_j f_j$$

Note that (LP) generalizes the well known Membership Problem (MP), of deciding whether a polynomial belongs to an ideal given by its generators. In fact (LP) also includes the Representation Problem (RP), of representing *effectively* a polynomial as a linear combination of given generators (provided the necessary membership relation).

Well known lower degree bounds for (RP) imply that, in order to obtain single exponential bounds for the complexity of (LP), some additional hypothesis must be imposed to the involved polynomials. In this paper we give a *single exponential* bound for the complexity of an algorithm which solves (LP) when the following condition is satisfied,

(*) $f_1, \ldots, f_t$ form a complete intersection "outside $V(J)$"

where $V(J)$ is the zero set defined by J in the affine space $\mathbf{A}^n$ over an algebraic closure of K.

We also study certain variations of (LP) and relate them with to-day known results about Effective Nullstellensatz Problems. We discuss applications to Algebraic Geometry as computation of Chow Froms and equations for the Projective Closure of affine algebraic sets.

The main ideas (and results) of this paper come from (and point to) Brownawell's "prime power product version" of the radical Nullstellensatz (see [Br1]), as well as the "localization problem" introduced in [Ca1]. (See (1.4) and the beginning of section 2 for more extended explanations about this fact).

1. Local Membership problems

Let K be a field, $R := K[X]$ the ring of polynomials in the indeterminate $X_1, \ldots, X_n$ over K and J an ideal of R. In this work we consider the following

(1.0) *Localization Problem (LP)*: Given $f, f_1, \ldots, f_t \in R$,
  i) decide efficiently whether a non negative integer D exists verifying

$$f \cdot \mathrm{rad}(J)^D \subseteq (f_1, \ldots, f_t).$$

  ii) If the condition in i) holds, then for each $g \in \mathrm{rad}(J)^D$, construct *efficiently* polynomials $a_1, \ldots, a_t$ such that:

$$f \cdot g \sum_{1 \leq j \leq t} a_j f_j$$

Note that (LP) generalizes: (a) all known results about Effective Nullstellensatz Problems; (b) the so called Membership Problem (MP), of deciding whether a polynomial belongs to an ideal given by its generators; (c) the Representation Problem (RP), of representing *effectively* a polynomial as a linear combination of given generators (provided the necessary membership relation).

When we say "decide" or "construct efficiently" we are implicitly speaking about *algorithmic procedures* with sequential and parallel complexity bounds as low as possible. The parameters with which we measure these complexities are the number of indeterminates and the degrees of the input polynomials.

The bounds for (RP) have been shown to be the intrinsically *double* exponential in the number of indeterminates (see [MM]), when the "worst" case is considered. From the point of view of the classification of algorithmic problems in terms of (their intrinsic) complexity, it is interesting (at leat for us) to find additional hypothesis the input polynomials must verify in order to obtain single exponential bounds.

In this paper we give a solution of (LP) when the polynomials $f_1, \ldots, f_m$ form a "complete intersection outside the zero set $V(J)$".

(1.1) *Notations*: From now on J will denote an ideal of R.
  Let $\mathbf{A}^n$ be the affine space of dimension n over an algebraic closure of K. We introduce the zero set $V(J) := \{\alpha \in \mathbf{A}^n : g(\alpha) = 0 \text{ for all } g \in J\}$.

(1.2) *Definition*: We say that the polynomials $f_1, \ldots, f_m \in R$ define a *complete intersection outside* $V(J)$ when each irreducible component of the zero set $\{\alpha : f_1(\alpha) = 0, \ldots, f_m(\alpha) = 0\}$, not included in $V(J)$, has dimension $n - m$.

Note that the above complete intersection condition holds trivially when $\{\alpha : f_1(\alpha) = 0, \ldots, f_m(\alpha) = 0\} \subseteq V(J)$. This is what we could call the Nullstellensatz case , i.e. the case in which $\mathrm{rad}(J) \subseteq \mathrm{rad}(f_1, \ldots, f_m)$.

The reader should observe that an equivalent definition may be obtained by requiring each isolated prime component of $f_1, \ldots, f_m$ not containing J to have height m.

Now we can state the main result of this paper which solves the first part of (LP) under the additional hypothesis we have discussed above. The second (i.e. the representation) part of (LP) will be deduced from this Theorem in the third section.

(1.3) **Theorem.** *Let the notations be the same as before and assume the* $\deg(f_2) \geq \cdots \geq \deg(f_m) \geq \deg(f_1)$, *and* $\deg(f_j) \neq 2$ *for* $j > 1$. *Then, if* $f_1, \ldots, f_m$ *form a complete intersection outside* $V(J)$, *given a polynomial* $f \in R$, *the following conditions, are equivalent:*

  i) $f \cdot \mathrm{rad}(J)^D \subseteq (f_1, \ldots, f_m)$ *for some* $D \geq 0$
  ii) $f \cdot P_1^{m_1} \cdots P_r^{m_1} \subseteq (f_1, \ldots, f_m)$ *for non-negative integers* $e_1, \ldots, e_r$ *such that* $\sum e_j \leq D := \deg(f_1) \cdots \deg(f_u)$

where $u := \min\{n, m\}$ *and* $P_1, \ldots, P_r$ *are the isolated prime components of* J.

(1.4) The reader should note that (1.3) generalizes the Brownawell's "prime power product version" of the radical Nullstellensatz, as well as the "localization problem" given in [**Ca1**]. Our proof is based on the refinements introduced in [**Br**] to the ideas of Kollar [**Ko**], but the techniques we adopt are those of N. Fitchas and A. Galligo [**FG**] who use Ext functors. Of course, appropriate changes make it easy to formulate our arguments in terms of Local Cohomology or Homology of Kozul complexes (cf. [**Phi**]).

2. Proof of the theorem (1.3)

In what follows we shall freely use the following three properties of elemental Commutative Algebra:

  1) If a prime ideal P includes the intersection of a finite family of ideals, then P includes some of the members of the family.
  2) If an ideal is contained in the set-theoretic union of a finite family of prime ideals, then it is included in some of the members of the family.
  3) (Krull's Hauptidealsatz) If h is neither a zero divisor nor a unit of a noetherian ring A, then every minimal prime ideal P containing h has height 1.

For technical reasons, we first introduce the notion of "regular sequence with respect to J" which is stronger than that of "complete intersection outside $V(J)$" of (1.2). Our definition generalizes the notion of "regular sequence outside a hypersurface" used in [**Ca1**] and that of "rather regular sequence" given in [**Br**]. Note moreover that this idea was present in other papers on the subject (e.g. [**Bri**], [**CGH**], etc.).

(2.0) *Definition:* Let $h_1, \ldots, h_t$ be a sequence of polynomials in R. We say that $h_1, \ldots, h_t$ is *regular with respect to J* (w.r.t. J for short) if for $k = 0, \ldots, t-1$, the two following conditions are verified:
  a) h_{k+1} does not lie in any associated prime component P of $H_k :=$ $(h_1, \ldots, h_k)$ unless $J \subseteq P$ (where $H_0 := (0)$)
  b) there exists some associated prime component P of H_k such that $J \not\subseteq P$.

Note that we obtain for $J = R$, the classic definition of regular sequence; for $J = \mathrm{rad}(h_1, \ldots, h_k)$, the notion of rather regular sequence, and for J the principal ideal generated by g, the notion of regular sequence outside the hypersurface $\{g = 0\}$.

It is easy to see that a regular sequence w.r.t. J defines a complete intersection outside $V(J)$. Although the converse is not true, Lemma (2.4) below shows that these two notions are closed related, at least when K is infinite. We need before some useful characterizations and basic properties.

(2.1) *Notation and Remarks:* Let $h_1, \ldots, h_t$ be a sequence of polynomials of R. For each k, $1 \le k < t$, let $\mathbf{P}(k)$ be the set of all associated prime components P of $H_k := (h_1, \ldots, h_k)$ such that $J \not\subseteq P$. Let S be the subset of $\mathrm{rad}(J)$ consisiting of those elements s lying outside each $P \in \mathbf{P}(k)$, $1 \le k < t$; i.e.

$$S := \mathrm{rad}(J) \backslash U\{P : P \in \mathbf{P}(k) \,,\; 1 \le k < t\}$$

Then S is not empty (cf. Lemma (1.7) below) and the following conditions are equivalent:
  i) The sequence $h_1, \ldots, h_t$ is regular w.r.t. J.
  ii) For all $s \in S$, the image of the sequence $h_1, \ldots, h_t$ in the localization $R[s^{-1}]$ is regular.
  iii) There exists $s \in S$ such that the sequence $h_1, \ldots, h_t$ is regular in the localization $R[s^{-1}]$.

(2.2) **Proposition.** *Let the notations be as before. Then* $\mathrm{rad}(J)$ *is generated by S.*

Proof: We prove a somewhat stroger property: Given an ideal J and a finite family P of prime ideals P not containing J, the set S of elements of J lying outside all $P \in \mathbf{P}$ is a system of generators of J.

Let J' be the ideal generated by S; then $J' \subseteq J$. Suppose that $J' \ne J$ and let $x \in J \backslash J'$. Let P' be the subfamily of P consisting of those prime ideals P such that $x \in P$ and $\mathbf{P}''$ the subfamily of $\mathbf{P}$ of those prime ideals Q such that $x \notin Q$ and $Q \subseteq \bigcup\{P : P \in \mathbf{P}'\}$. Since $J' \cap (\cap \{Q : Q \in \mathbf{P}''\})$ is not included in $\bigcup\{P : P \in \mathbf{P}'\}$, there exists $y \in J' \cap (\cap\{Q : Q \in \mathbf{P}''\})$ lying outside all prime ideals $P \in \mathbf{P}'$. It is easy to see that the element $x + y$ belongs to S. But since $x + y \notin J'$, this fact contradicts the definition of J'.

(2.3) *Remark*: If $h_1, \dots, h_t$ is rather regular w.r.t. J, then $h_1, \dots, h_s$, where s is the height of J, is a regular sequence in R.

   The following Lemma will allow us to replace our original sequence $f_1, \dots, f_m$ with an other $h_1, \dots, h_t$ which is regular w.r.t. J. This result combines ideas and techniques which appear in the literature, e.g. [**Bri**, Théorème 1], [**CGH2**, Proposition 3], [**Br5**, Lemma 0].

(2.4) **Lemma.** *Assume that K is infinite and let $f_1, \dots, f_m$ define a complete intersection outside $V(J)$. Then there exist polynomials $h_1, \dots, h_t$, $t \leq u := \min\{m, n + 1\}$, with the following properties:*
   1) $h_1, \dots, h_t$ is regular w.r.t. J
   2) $h_1 = f_1$ and for all j, $2 \leq j \leq t$, $h_j \in (f_2, \dots, f_0)$
   3) for each k, $1 \leq k \leq t$, there exists j_k, $1 \leq j_k \leq m$, with $\deg(h_k) = \deg(h_{j_k})$. Moreover $j_k \neq j_{k'}$ if $k \neq k'$.
   4) If $J \not\subseteq \mathrm{rad}(f_1, \dots, f_0)$, then $t = m$ and $(h_1, \dots, h_m) = (f_1, \dots, f_m)$.
   5) If $J \subseteq \mathrm{rad}(f_1, \dots, f_m)$, then $J \subseteq \mathrm{rad}(h_1, \dots, h_t)$.

Proof: Recall that we are assuming $\deg(f_2) \geq \cdots \geq \deg(f_m)$. The sequence $h_1, \dots, h_t$ is constructed recurrently. Let $h_1 := f_1$ and suppose that for some k we have defined a sequence $h_1, \dots, h_k$ verifying the conditions 1), 2) and 3) for k.

   Let $\mathbf{P} = \mathbf{P}(k)$ be, as in (2.1), the set of all associated prime components P of $H_k := (h_1, \dots, h_k)$ such that $J \not\subseteq P$.

First case: $J \subseteq \mathrm{rad}(f_1, \dots, f_m)$.

   It is easy to see that in this case $\mathbf{P}$ is not empty. Let $P \in \mathbf{P}$. From (2.1) we deduce that P has height k. If $k < m$, then the complete intersection hypothesis implies that $(f_1, \dots, f_m) \not\subseteq P$. Since $f_1 \in H_k \subseteq P$, we see that there exists j, $1 < j \leq m$, verifying $(f_j, \dots, f_m) \not\subseteq P$. Let j_P be the largest of such j's. We choose j_{k+1} to be $\min\{j_P : P \in \mathbf{P}\}$.

   For each $P \in \mathbf{P}$ let us consider the K-linear variety:

$$E(P) := \{\alpha \in K^{m-j_{k+1}} : h(\alpha) \in P\}$$

where $h(\alpha) := f_{j_{k+1}} + \sum_{i > j_{k+1}} \alpha_i f_i$.

Claim. $E(P) \neq K^{m-j_{k+1}}$

Proof of the Claim: Suppose that $E(P) = K^{m-j_{k+1}}$. Then $f_{j_{k+1}}$ and $f_{j_{k+1}} + f_1$ belong to P for each $i > j_{k+1}$. Thus we get $f_{j_{k+1}}, \dots, f_m \in P$. This contradicts the choice of j_{k+1}.

The Claim implies that there exists $\alpha \in K^{m-j_{k+1}}$ such that $h(\alpha) \notin \bigcup\{P : P \in \mathbf{P}(k)\}$; let $h_{k+1} := h(\alpha)$.

Now we prove that $j_{k+1} \neq j_i$ for all $i \leq k$. Assume that $j := j_{k+1} = j_i$ for some $i \leq k$. Since $h_i \in H_{k+1}$, we see that $h_{k+1} \in (H_k, h_{k+1} - h_i) \subseteq (H_k, f_{j+1}, \ldots, f_m)$. Let $P \in \mathbf{P}$. Since h_{k+1} does not belong to P, we obtain that not all of $f_{j+1}, \ldots, f_m$ belong to P. This contradicts the definition of j_{k+1}.

Thus, we obtain a sequence $h_1, \ldots, h_m$ of polynomials verifying 1), 2) and 3).

In order to show that $(h_1, \ldots, h_m) = (f_1, \ldots, f_m)$ it suffices to remark that, after a suitable reordering of the indices, the system of equations expressing the polynomials h's as K-linear combinations of the polynomials f's is upper triangular with 1's in the diagonal.

Second Case: $J \subseteq \mathrm{rad}(f_1, \ldots, f_m)$.

While $\mathbf{P}(k) \neq \emptyset$ we may repeat the construction of the First Case and add one more polynomial h_{k+1} to the sequence $h_1, \ldots, h_k$ in such a way that the new sequence $h_1, \ldots, h_{k+1}$ verifies properties 1), 2) and 3). Since all prime ideals in $\mathbf{P}(k)$ have height k, we see that $t := \min\{k : \mathbf{P}(k) = \emptyset\} \leq n + 1$. Thus $J \subseteq \mathrm{rad}(h_1, \ldots, h_t)$. Obviously, property 3) implies $t \leq m$.

In the following Lemma we condense the cohomological tools which have been shown to be powerful for the kind of problems we are concerned with. The related references we know are: [**Ko**, Lemma 3.4] (this author first introduced the basic ideas), [**FG**] (the present Ext version in inspired on this paper), [**Phi**] (where Homology of Koszul complexes are used), [**Br5**, Lemma 1 and 2] (where the structure of our presentation was taken from).

(2.5) **Lemma** (see also [**Br5**, Lemma 1 and 2]). *Let $B \subseteq R$ be an ideal and $h \in R$ a non-zero divisor of R/B. Let $(B, h) = Q_1 \cap \cdots \cap Q_s \cap Q_{s+1} \cap \cdots \cap Q_r \cap E$, where the Q's are isolated primary components of (B, h) and E is an intersection of embedded primary ideals. Let $B' := Q_1 \cap \cdots \cap Q_s$ and $\Gamma : Q_{s+1} \cap \cdots \cap Q_r$. Assume that for some ideal $N \subseteq R$ and some family $\mathbf{U}$ of ideals U with $ht(B) < ht(U)$,*

$$ N \cdot Ext^k(R/U, R/B) = 0 $$

for all $U \in \mathbf{U}$ and all k, $0 \leq k < ht(U) - ht(B)$. Then,

1) *If $E \neq R$ and $E \in \mathbf{U}$, $\Gamma \cdot N^3 \cdot Ext^k(R/U, R/B') = 0$ for all $u \in \mathbf{U}$ and all k, $k < ht(u) - ht(B, h)$.*

2) *If $E = R$, $\Gamma \cdot N^2 \cdot Ext^k(R/U, R/B') = 0$ for all $u \in \mathbf{U}$ and all k such that $k < ht(u) - ht(B, h)$.*

3) *$N \cdot ((B' \cap \Gamma)/(B, h)) = 0$ if $E \neq R$ and $E \in \mathbf{U}$ or $E = R$.*

Proof: Let $I := B' \cap \Gamma$. The short exact sequences:

$$0 \longrightarrow R/B \longrightarrow R/B \longrightarrow R/(B, h) \longrightarrow 0$$

and

$$0 \longrightarrow I/(B, h) \longrightarrow R/(B, h) \longrightarrow R/I \longrightarrow 0$$

give rise to the long exact sequences:

$$\dots Ext^k(R/U, R/B) \longrightarrow Ext^k(R/U, R/(B, h)) \longrightarrow Ext^{k+1}(R/U, R/B) \dots$$

and

$$\dots Ext^k(R/U, R/(B, h)) \to Ext^k(R/U, R/I) \to Ext^k(R/U, I/(B, h)) \dots$$

The hypothesis and the first long sequence imply that:

$$(*) \qquad N^2 \cdot Ext^k(R/U, R/(B, h)) = 0$$

if $k+1 < ht(U) - ht(B)$; i.e. if $k < ht(U) - ht(B, h)$. In particular, if $E = R$, $N^2 \cdot Ext^k(R/U, R/I) = 0$ for all $u \in U$ and k such that $k < ht(U) - ht(B, h)$.

Claim. $Ext^0(R/E, R/B) = 0$

Proof of the Claim: Let $\Phi \in Hom(R/E, R/B)$. Let $cl(1)$ be the class of 1 in R/E and let $f := \Phi(cl(1))$. Then $f \cdot E = 0$ in R/B. In particular, $fh = 0$ in R/B. Since h is not a zero-divisor in R/B, this shows that $\Phi = 0$.

From our claim and the long exact sequences we get:

$$Ext^0(R/E, I/(B, h)) \subseteq Ext^0(R/E, R/(B, h)) \subseteq Ext^1(R/E, R/B)$$

Suppose that $E \neq R$ and that $E \in U$. Since $ht(E) - ht(B) > 1$, we obtain that N annihilates $Ext^0(R/E, I/(B, h))$. Since there is a natural inclusion $I/(B, h) \subseteq Ext^0(R/E, I/(B, h))$, we deduce that $N \cdot (I/(B, h)) = 0$ proving 3).

Consequently N annihilates all the modules $Ext^*(R/U, I/(B, h))$. Thus, from $(*)$ and the second long exact sequence we obtain

$$N^3 \cdot Ext^k(R/U, R/I) = 0$$

for all $U \in U$ and all k such that $k < ht(U) - ht(B, h)$.

Now consider the short exact sequence:

$$0 \longrightarrow B'/I \longrightarrow R/I \longrightarrow R/B' \longrightarrow 0$$

Given U and k such that $B' \subseteq U$ and $0 \leq k < ht(U) - ht(B')$, the above short exact sequence induces a long one:

$$\dots Ext^k(R/U, R/I) \longrightarrow Ext^k(R/U, R/B') \longrightarrow Ext^{k+1}(R/U, B'/I) \dots$$

Since multiplication by Γ annihilates B'/I and consequently all the modules $Ext^*(R/U, R/B')$, we see that:

$$\Gamma \cdot Ext^k(R/U, R/B') \subseteq \quad \text{image of} \ \ Ext^k(R/U, R/I)$$

Now for $d = 2$ or 3, depending on the case $E = R$ or $E \neq R$ and $E \in U$, we obtain:

$$\Gamma \cdot N^d \cdot Ext^k(R/U, R/B') = 0$$

(2.6) *Notations and Remarks*: Up to now we have said nothing about homogenization; but perhaps it is time to point out that, soon or later, we shall need a Bezout Theorem bounding certain exponents which happen to appear. We have two possibilities: a) to homogenize our input data so that we can use a "projective" Bezout Equality, or b) to use directly an "affine" Bezout Inequality. In the first case some caution must be regarded: the ideal J must be replaced by $J' := X_0 \cdot J^*$, where J^* is the ideal generated by homogenizations of all the polynomials of J. This carefulness is needed to ensure the sequence obtained from $h_1, \ldots h_t$ via homogenization to be regular w.r.t. J'. Although the procedure commonly used in the literature is a), we have presented all statements in such a way that the "affine" approach is also viable. The reader should choose the option he considers better taking into account that, for the representation part of (LP), homogenization will be indispensable.

Let $h_1, \ldots, h_t$ be the sequence of Lemma (2.4). Let $B_0 := (0)$ and for each j, $1 \leq j \leq t$, define B_j, Γ_j, E_j inductively by grouping the components of a primary decomposition of (B_{j-1}, h_j) to obtain:

$$(2.6.1) \qquad (B_{j-1}, h_j) = B_j \cap \Gamma_j \cap E_j$$

where: B_j is the intersection of those isolated primary components corresponding to prime ideals which do not contain J; Γ_j is the intersection of the isolated pimary components whose corresponding prime ideals contain J, and E_j is the intersection of the remaining (embedded) components. We define Γ_j or $E_j = R$ in the absence of the corresponding primary components. Note that we may have $B_t = R$. In any case, B_t is the ideal we are interested in, as shown in item 7) of Lemma (2.7) below.

(2.7) **Lemma.** *Let S be the set introduced in (2.1),. Then for all j, $1 \leq j \leq t$, and all $s \in S$,*
  1) $B_j R[s^{-1}] = (h_1, \ldots, h_j) R[s^{-1}]$
  2) $B_j = B_j R[s^{-1}] \cap R$
  3) h_{j+1} is not a zero divisor of R/B_j.
  4) B_j is unmixed of height j
  5) $\Gamma_j = R$ or Γ_j is unmixed of height j
  6) Each prime component of E_j includes J.
  7) $B_t = \bigcup_{D \geq 0} \{ f \in R : f \cdot \mathrm{rad}(J)^D \subseteq (h_1, \ldots, h_j) \}$

Proof 1) and 6): The proof works by induction on j. Let $s \in S$. Since $\Gamma_j R[s^{-1}] = R[s^{-1}]$ by the definition of Γ_j, and $(B_{j-1}, h_j) R[s^{-1}] = (h_1, \ldots, h_j) R[s^{-1}]$ by the inductive hypothesis, we have:

$$(h_1, \ldots, h_j) R[s^{-1}] = B_j R[s^{-1}] \cap E_j R[s^{-1}]$$

As we have remarked in item ii) of (2.1), the image of the sequence $h_1, \ldots, h_j$ in $R[s^{-1}]$ is regular. Since R is Cohen-Macaulay, the left part of the previous equality is an unmixed ideal. Therefore $E_j R[s^{-1}] = R[s^{-1}]$ showing property 1). Moreover, the last equality implies that s belongs to each of the associated primes of E_j. Since $s \in S$ was arbitrary chosen and by Proposition (2.2) S generates $rad(J)$, we get that each prime component of E_j includes J.

2) Let P be a prime component of B_j. By definition P does not include J. Thus, by Proposition (2.2), we can choose $s \in S \backslash P$. Now 1) implies that $PR[s^{-1}]$ is a prime component of $(h_1, \ldots, h_j)R[s^{-1}]$. So P is a component of $(h_1, \ldots, h_j)$. Since $J \nsubseteq P$, by definition of S, we get $S \cap P = \emptyset$. This shows that $B_j = B_j R[s^{-1}] \cap R$.

3) and 4) These properties follows easily from 1), 2) and the fact that the image of the sequence $h_1, \ldots, h_j$ in $R[s^{-1}]$ is regular.

5) By 3), 4) and the Principal Ideal Theorem, we know that each isolated prime component of (B_{j-1}, h_j) has height j. In particualr, the components of Γ_j have the same property.

7) It follows immediately from 1), 2) and Proposition (2.2).

(2.8) **Lemma** (see [**Br5**] Lemma 2). *Let the notations be the same as in (2.6). Then for all j, $1 \leq j < t$, the ideal $N_j := \Gamma_1^{3^{d-1}} \Gamma_2^{3^{d-2}} \cdot \Gamma_j^{3^{d-j}}$ verifies*

$$N_j (B_{j+1} \cap \Gamma_{j+1}) \subseteq (B_j, h_{j+1})$$

Proof: By item 3) of Lemma (2.7) and item 3) of Lemma (2.5), it sufficies to show that, for $j = 0, \ldots, t - 1$, $N_j (N_0 := R)$ annihilates $\mathbf{Ext}^k(R/U, R/B_j)$ for all ideal U such that $B_j \subseteq U$ and all k, $0 \leq k < ht(U) - ht(B_j)$. We prove this property by induction on j. In this case we must show that
$$\mathbf{Ext}^k(R/U, R) = (0)$$
for all ideal U and all $k < ht(U)$. Since R is Cohen-Macaulay of depth $dim(R)$ and $ht(U) + dim(R/U) \leq dim(R)$, this case is consequence of [**Ma**] (15.E) Lemma 2 via localization.

The inductive step follows directly from items 1) and 2) of Lemma (2.5), because the prime components of E_{j+1} have height greater that $ht(B_{j+1})$ (recall that they were embedded components of (B_j, h_{j+1})).

(2.9) **Proposition.** *With the notations of (2.6), the ideal $N := \Pi_{1 \leq j \leq t} \Gamma_j^{(3^{t-j}+1)/2}$ satisfies $B_t N \subseteq (h_1, \ldots, h_t)$.*

Proof: It follows easily from Lemma (2.8).

(2.10) **Proposition** (See e.g. [**Gr**, 143.7]). *For all j, $1 \leq j \leq t$, we have*

$$\sum_{1 \leq k \leq j} \deg(\Gamma_k) \cdot \deg(h_{k+1}) \cdots \deg(h_j) + \deg(B_j) < \deg(h_1) \cdots \deg(h_j)$$

(2.11) End of the proof of Theorem (1.3): Recall that $P_1, \ldots, P_r$ are the isolated prime components of J. For each j, $1 \leq j \leq t$, let $\Gamma_j = Q_{j1} \cap \cdots \cap Q_{jm(j)}$ be a primary decomposition of Γ_j with $P_{jk} := \text{rad}(Q_{jk})$. Let $u_{jk} := \min\{u : (P_{jk})^u \subseteq A_{jk}\}$.

It is easy to see that $\sum_{1 \leq k \leq s(j)} \deg(P_{jk}) \cdot u_{jk} \leq \deg(\Gamma_j)$. Since each P_{jk} contains some P_i, there exist non-negative integers $e_1, \ldots, e_r$ such that:

$$P_1^{e_1} \cdots P_r^{e_r} \subseteq \prod_{j,k} P_{jk}^{(3^{t-j}+1)/2} \subseteq \prod_{1 \leq j \leq t} \Gamma_j^{(3^{t-j}+1)2} = N$$

and

$$\Sigma e_1 = \sum_{j,k} u_{jk}(3^{t-j}+1)/2 \leq \sum_j \deg(\Gamma_j) \cdot (3^{t-j}+1)/2$$

hence by Proposition (2.10),

$$\sum e_1 \leq \deg(h_1) \cdots \deg(h_t) -$$
$$\sum_{1 \leq k < t} \deg(\Gamma_k) \cdot (\deg(h_{k+1}) \cdots \deg(h_t) - (3^{t-k}+1)/2) - \deg(B_t)$$

If $t \leq n$, the last inequality and item 7) of (2.7) complete the proof. If $t = n+1$, then $B_t = B_{n+1} = R$ and so J is included in $\text{rad}(f_1, \ldots, f_m)$ and the statement of our Theorem is the same as that of [**Br**].

3. Some Consequences

In order to show the role this Theorem plays in the present context, let us see some of its consequences. In the Corollaries below we shall suppose *implicitly*, for the sake of simplicity of the statements, that n and m are greater that 1 and that the polynomials $f_2, \ldots, f_m$ have degree greater than 2. The bound D is the same as in Theorem (1.3). In what follows we shall use the following

(3.0) *Notations*: $S := K[X_0, X_1, \ldots, X_n]$; given $h \in R = K[X_1, \ldots, X_n]$, $H \in S$ will denote the homogenization of h; for an ideal $J \subseteq R$, J^* will be the ideal of S generated by $\{H \in S : h \in J\}$.

The first Corollary gives us a solution for the representation part of (LP). In fact this result can be used to find a representation like in item ii) of (1.0), by any "comparison of coefficients" K-linear algebra method.

(3.1) Corollary (cf. [**Ca1**], [**Br5**]). *Let the sequence $f_1, \ldots, f_m$ define a complete intersection outside $V(J)$. The for all $f \in R$ the following conditions are equivalent:*

    i) $f \cdot \text{rad}(J)^N \subseteq (f_1, \ldots, f_m)$ *for some $N \geq 0$*

ii) There exist non negative integers $e_1, \ldots, e_r$ with $\Sigma e_k \leq D$, such that given $g \in P_1^{m_1} \cdots P_r^{m_r}$ there are polynomials $a_1, \ldots, a_m \in R$ verifying

$$fg = \sum_{1 \leq j \leq m} a_j f_j$$

where $\max\{\deg(a_j f_j) : 1 \leq j \leq m\} \leq \deg(f) + \deg(g) + D.$

Proof: We only need to show that i) implies ii). One easily checks that $F_1, \ldots, F_m$ define a complete intersection outside $J' := X_0 \cdot J^*$ and that the membership condition i) implies the new one $F \cdot \mathrm{rad}(J')^N \in (F_1, \ldots, F_m)$ for some $N \geq 0$. By the Theorem,

$$F \cdot (X_0)^{e_0} \cdot (P_1^*)^{e_1} \cdots (P_r^*)^{e_r} \subseteq (F_1, \ldots, F_m)$$

for some non negative integers $e_0, \ldots, e_r$ with $\Sigma e_k \leq D$. Thus there is a representation

$$F \cdot (X_0)^{e_0} \cdot G = A_1 F_1 + \cdots + A_m F_m$$

with homogeneous A_j's satisfying: $\deg(A_j F_j) = \deg(f) + e_0 + \deg(g)$. Now we can finish the proof dehomogenizing the last representation.

(3.2) Corollary (cf. [Ca1]). *Let $g \in R \backslash K$ with irreducible factors $g_1, \ldots, g_r$. If the sequence $f_1, \ldots, f_m$ define a complete intersection outside $V(g)$, then for all $f \in R$ the following conditions are equivalent:*

i) $f \cdot g^N \subseteq (f_1, \ldots, f_m)$ for some $N \geq 0$

ii) There exist a non negative integers $e_1, \ldots, e_r$ with $\Sigma e_k \leq D$, and polynomials $a_1, \ldots, a_m \in R$ verifying

$$fg_1^{e_1} \cdots f_r^{e_r} = \sum_{1 \leq j \leq m} a_j f_j$$

and

$$\max\{\deg(a_j f_j) : 1 \leq j \leq m\} \leq \deg(f) + D \cdot \max\{\deg(g_j) : 1 \leq j \leq r\}$$

(3.3) Corollary (cf. [DFGS] and [Ca1]). *Let $f, f_1, \ldots, f_m \in R$ polynomials such that $f_1, \ldots, f_m$ define a complete intersection in A^n. Then the following conditions are equivalent:*

i) $f \in (f_1, \ldots f_m)$

ii) There exist polynomials $a_1, \ldots, a_m \in R$ such that

$$f = \sum_{1 \leq j \leq m} a_j f_j$$

with $\max\{\deg(a_j f_j) : 1 \leq j \leq m\} \leq \deg(f) + D$

In particular, in the case of complete intersections, the Representation Problem (RP) can be solved in sequential time single exponential and parallel time polynomial with respect to the number of variables, using any method of comparison of coefficients.

(3.4) Corollary (cf.   [Br1,2] [CGH1,2], [Ko], [Shi], [Phi].   [FG], etc.). *Given $g, f_1, \dots, f_m \in R$, $g \notin K$, the following conditions are equivalent:*

i) *$g \in \mathrm{rad}(f_1, \dots, f_m)$*

ii) *There exist non negative integrs $e_1, \dots, e_r$ with $\Sigma e_k \leq D$, and polynomials $a_1, \dots, a_n \in R$ verifying*

$$g_1^{e_1} \cdots g_r^{e_r} = \sum_{1 \leq j \leq m} a_j f_j$$

where $g_1, \dots, g_r$ are the irreducible factors of g and

$$\max \deg(a_j f_j) : 1 \leq j \leq m\} \leq D \cdot \max\{\deg(g_j) : 1 \leq j \leq r\}$$

(3.5) Corollary (cf. Idem (3.4)). *Let $f_1, \dots, f_m \in R$ be polynomials defining in A^n the zero set $V := \{\alpha : f_1(\alpha) = 0, \dots, f_m(\alpha) = 0\}$. Then the following conditions are equivalent;*

i) *$V = \emptyset$*

ii) *There exist $a_1, \dots, a_m \in R$ such that*

$$1 = a_1 f_1 + \cdots + a_m f_m$$

with $\deg(a_j f_j) \leq D$ for all j, $1 \leq j \leq m$.

4. Applications

The results about Effective Nullstellensatz Problems have given single exponential sequential complexity bounds to an important number of (parallelizable) algorithms in Commutative Algebra and Algebraic Geometry as: Computation of Gröbner basis in the zero dimensional case, membership tests for unmixed ideals, computation of dimension, effective versions of Noether's Normalization Lemma, equations for the projective closure of affine sets, equidimensional decomposition of algebraic sets, computation of Chow Forms, etc. ([CGH1], [DS], [CGH3], [Lo], [DFGS], [GH], [Ca1], [Ca2], etc.). The reader should note that those bounds can be refined with the help of recient stronger versions of the basic theorems (e.g. [Ca1], [Br5], Theorem (1.3) and its Corollaries in this paper), essentially by replacing $\max\{\deg(f_j)\}^n$ with the bound D of (1.3)). On this direction, some remarks may be taken into account:

Remark 1: In order to apply Theorem (1.3) and its Corollaries to an input data, say: $J, f, f_1, \dots, f_m$, it is necessary to know first whether the polynomials $f_1, \dots, f_m$ define a complete intersection outside $V(J)$. So it would be interesting to have a parallelizable algorithm, with single exponential sequential complexity bounds, deciding the validity of the complete intersection condition. The algorithm we give here, has the particularity of being based on Corollary (3.4). (Thus one of the consequences of Theorem (1.3) is that it permits to decide whether its hypothesis are fulfilled!)

In the algorithm, the d-dimensional stratum of $V(f_1,\ldots,f_m)$ is the closed algebraic subset of $V(f_1,\ldots,f_m)$ which is the union of all irreducible components of dimension d. For the computation of equations defining each of these strati in single exponential time see [GH].

Test for the complete intersection hypothesis:
```
INPUT   J := (g_1,...,g_r), f_1,...,f_m
A=TRUE
WHILE   n − m + 1 ≤ d ≤ n and A is TRUE   DO:
   FIND   h_1,...,h_s defining the d-dimensional stratum of V(f_1,...,f_m)
   WHILE   1 ≤ j ≤ r AND A is TRUE   DO:
      IF   g_j ∉ rad(h_1,...,h_m)
      THEN   A=FALSE
      ELSE   j = j + 1
   LOOP
   d = d + 1
LOOP
IF   A is TRUE
THEN PRINT
```
"$f_1,\ldots,f_m$ defines a complete intersection outside $V(J)$"
```
ELSE PRINT
```
"$f_1,\ldots,f_m$ does not define a complete intersection outside $V(J)$"

Remark 2: Sometimes the "degree" of algebraic sets (both, affine and projective) is used as an alternative bounding parameter. (Of course, the notion of degree is essential in proofs because of the Bezout-Inequalities).

Let $V := \{\alpha \in A^n : f_1(\alpha) = 0,\ldots,f_m(\alpha) = 0\}$ be the zero-set defined by the polynomials $f_1,\ldots,f_m$ in the n-dimensional affine space A^n over an algebraic closure of K. Let $V = V_1 \cup \ldots \cup V_r$, be the decomposition of V in irreducible components. Following [He] we define the *degree* of V as:

$$\deg(v) = \sum_{1\leq j\leq r} \deg(V_j)$$

Assume that $\deg(f_2) \geq \cdots \geq \deg(f_m) \geq \deg(f_1)$ and let $D := \deg(f_1)\cdots\deg(f_u)$, where $u := \min\{n,m\}$. Then

$$\deg(V) \leq D$$

This fact follows from [He, Theorem 1], since by Lemma (2.4) we may replace the sequence $f_1,\ldots,f_m$ with an other $h_1,\ldots,h_u$ defining the same algebraic set V in A^n.

Note moreover that, with the notations of Theorem (1.3), each e_j bounds the *exponent* of the corresponding P_j-primary component of J. (If Q is P-primary, the *exponent* of Q is $\min\{E : P^E \subseteq Q\}$).

REFERENCES

[BGS]   D. Bayer, A. Galligo, M. Stillman, *Primary Decomposition*, Preprint 1989.

[Bri]   J. Briancon, *Sur le degré des relations entre polynômes*, C. R. Acad. Sc. Paris **297** (1983).

[Br1]   W. D. Brownawell, *Bounds for the degree in the Nullstellensatz*, Ann. Math. **126** (1987), 577-591.

[Br2]   W. D. Brownawell, *Borne effective pour l'exposant dans le théorème des zéros*, C. R. Acad. Sci. Paris **305** (1987), 287-290.

[Br3]   W. D. Brownawell, *Local diophantine Nullstellen inequalities*, J. Am. Math. Soc. **1** (1988), 311-322.

[Br4]   W. D. Brownawell, *Note on a paper of P. Philippon*, Manuscript.

[Br5]   W. D. Brownawell, *A prime power product version of the Nullstellensatz*, Manuscript 1989.

[Ca1]   L. Caniglia, *Complejidad de Algoritmos en Geometria Algebraica Computacional*, Thesis Univ. Buenos Aires, 1989.

[Ca2]   L. Caniglia, *How to compute the Chow Form of an unmixed polynomial ideal in single exponential time*, Manuscript. 1989.

[CGH1] L. Caniglia, A. Galligo, J. Heintz, *Some new effectivity bounds in computational geometry*, in "Proc. of the 6th Intern. Conference AAECC," Rome 1988, Springer LN Comp. Sci. 357, 1989, pp. 131-151.

[CGH2] L. Caniglia, A. Galligo, J. Heintz, *Borne simple exponentielle pour les degrés dans le théorème des zéros sur un corps de caractéristique quelconque*, Comptes Rendus de l'Academie des Sciencies **307** (1988).

[CGH3] L. Caniglia, A. Galligo, J. Heintz, *Equations for the projective closure of affine algebraic varietes*, Algebraic Algorithms and Error Correcting Codes, AAECC-7, Toulouse 1989, Discrete Applied Mathem (to appear).

[DFGS] A. Dickenstein, N. Fitchas, M. Giusti, C. Sessa, *The membership problem for unmixed polynomial ideals is solvable in single exponential time*, Algebraic Algorithms and Error Correcting Codes, AAECC-7, Toulouse 1989, Discrete Applied Mathem. (to appear).

[DS]   A. Dickenstein, C. Sessa, *An effective residual criterion for the membership problem in* $C[Z_1, \dots, Z_n]$, in "Proc. IX ELAM, Santiago de Chile, 1988" (to appear).

[Fi]   N. Fitchas, *Catania Lecture Notes on Complexity in Elementary Geometry*, Manuscript, October, 1988.

[FG]   N. Fitchas, A. Galligo, *Nullstellensatz effectiv et Conjecture de Serre*, Preprint Seminaire Structures Algébriques Ordonnées, 1987–1988, Univ. Paris VII, Math. Nachrichten (to appear).

[FGM] N. Fitchas, A. Galligo, J. Morgenstern, *Algorithmes rapides en séquentiel et en parallele por l'élimination de quantificateurs en géométrie élémentaire*, Manuscript 1987.

[GT]   A. Galligo, C. Traverso, *Practical determination of the Dimension of an algebraic variety*, in "Computers and Mathematics," Kaltofen-Watt ed., Springer, 1989.

[GH]   M. Giusti, J. Heintz, *Un algorithme - disons 'rapide' - pour la décomposition équidimentionelle d'une varété algébrique*, (These proceedings).

[Gr]   W. Göbner, "Moderne Algebraische Geometrie," Springer Verlag, Wien-Innsbruk, 1949.

[He]   J. Heintz, *Definability and fast quantifier elimination in algebraically closed fields*, Theoret. Comput. Sci. **24** (1983), 239–277.

[Ko]   J. Kollar, *Sharp effective Nullstellensatz*, J. Am. Math. Soc. 1 (1988), 963–975.

[Lo]   A. Logar, *A computational proof of the Noether's Normalization Lemma*, in "Proc. AAECC-6," LN Comput. Sci., Springer.

[Ma]   H. Matsumura, "Commutative Algebra," Second Edition, Benjamin/Cummings, 1980.

[MM]   E. Mayr - A. Meyer, *The complexity of the word problem for commutative semigroups and polynomial ideals*, Advances in Math. **46** (1982), 305–329.

[Ph1] P. Philippon, *Théorème des zéros effectif d'après J. Kollar*, Publ. Math. Univ. Paris VI, N.88. Probl. Dioph. 1988-89.

[Ph2] P. Philippon, *Dénominateurs dans le théorème des zéros*, Manuscript.

[Sh]   B. Shiffman, *Degree bounds for the division problem in polynomial ideals*, Manuscript.

Leandro Caniglia
  (Working Group Noaï Fitchas)
Jorge A. Guccione
Juan J. Guccione
Instituto Argentino de Matemática
(IAM-CONICET)
Viamonte 1636
1er Piso, 1er Cuerpo
(1055) Buenos Aires
ARGENTINA

Un Algorithme pour le Calcul des Résultants

MARC CHARDIN

Abstract. We here give a method to calculate the resultant of three polynomials in terms of a square-free decomposition and resultants of two polynomials. After that, we show how the subresultant algorithm enables us to avoid many calculations.

In the last part, we study the possible extension to the general case of n homogeneous polynomials in n variables.

Le Résultant: rappel

La condition pour que n polynômes homogènes en n variables à coefficients dans un corps aient un zéro non trivial en commun dans une extension du corps, est donnée par un unique polynôme en les coefficients de ces n polynômes; ce polynôme appelé résultant est un "objet universel" en un sens précisé ci-dessous. Nous nous contenterons ici de rappeler le résultat suivant qui résume les principales propriétés du résultant et en constitue une définition, ce théorème classique est démontré dans thèse ([Th]) à laquelle nous renvoyons pour plus de détails sur ce qu'est le résultant et les différents autres moyens de le calculer.

Théorème. *Pour tout anneau A, tout entier non nul n et tout n-uplet de polynômes $(P_1, \ldots, P_n)$, homogènes, de degrés respectifs $d_1, \ldots, d_n$ de $A[X_1, \ldots, X_n]$, on définit un élément de A noté $Res(P_1, \ldots, P_n)$, vérifiant les propriétés suivantes, dont les quatre premières en constituent une définition.*

(1) Si $\phi : A \longrightarrow B$ est un homomorphisme d'anneaux, et $\tilde{\phi}$ son prolongement naturel en un homomorphisme de $A[X_1, \ldots, X_n]$ dans $B[X_1, \ldots, X_n]$, alors

$$Res(\tilde{\phi}(P_1), \ldots, \tilde{\phi}(P_n)) = \phi(Res(P_1, \ldots, P_n)).$$

(2) Le résultant universel[1] est irréductible.

---

[1]On appelle *résultant universel* le résultant des polynômes P_i =

(3) Si A est un corps et C une clôture algébrique de A, $Res(P_1, \ldots, P_n)$ est nul si et seulement si les polynômes P_i ont un zéro commun non trivial dans C^n.

(4) $Res(X_1^{d_1}, \ldots, X_n^{d_n}) = 1$.

(5)

$$Res(P_1, \ldots, P_{n-1}, X_n^{d_n}) = (Res(P_1^\star, \ldots, P_{n-1}^\star))^{d_n}$$

où $P_i^\star(\mathbf{X}) = P_i(X_1, \ldots, X_{n-1}, 0)$.

(6) Si $P_i = Q_i R_i$, alors

$$Res(P_1, \ldots, P_n) =$$
$$Res(P_1, \ldots, P_{i-1}, Q_i, P_{i+1}, \ldots, P_n) Res(P_1, \ldots, P_{i-1}, R_i, P_{i+1}, \ldots, P_n).$$

(7) Si $M \in \mathbf{M}_n(A)$, $M = (a_{i,j})_{(i,j) \in [1,n]^2}$ et

$$Q_k(X_1, \ldots, X_n) = P_k \left(\sum_{j=1}^{n} a_{1,j} X_j, \ldots, \sum_{j=1}^{n} a_{n,j} X_j \right)$$

on a:

$$Res(Q_1, \ldots, Q_n) = (\det M)^{d_1 \cdots d_n} Res(P_1, \ldots, P_n).$$

(8) Si $\sigma \in S_n$ est une permutation de signature $\epsilon(\sigma)$, alors

$$Res(P_{\sigma(1)}, \ldots, P_{\sigma(n)}) = (\epsilon(\sigma))^{d_1 \cdots d_n} Res(P_1, \ldots, P_n).$$

(9) Si $A_1, \ldots, A_{n-1}$ sont des polynômes nuls ou homogènes de degrés $d_n - d_i$,

$$Res \left(P_1, \ldots, P_{n-1}, P_n + \sum_{i=1}^{n-1} A_i P_i \right) = Res(P_1, \ldots, P_n).$$

(10) Si A est un corps, il existe une extension algébrique finie K de A, et des éléments $\alpha_j^{(i)}$ $(i = 1, \ldots, d_1 \cdots d_{n-1} \, ; \, j = 1, \ldots, n)$ de K tels que pour tout polynôme Q homogène on ait:

$$Res(P_1, \ldots, P_{n-1}, Q) = \prod_{i=1}^{d_1 \cdots d_{n-1}} Q(\alpha_1^{(i)}, \ldots, \alpha_n^{(i)}).$$

Si $P_1, \ldots, P_{n-1}$ forment une suite régulière, alors $(\alpha_1^{(i)} : \cdots : \alpha_n^{(i)})$ pour i de 1 à $d_1 \cdots d_{n-1}$ sont les coordonnées homogènes des points constituant la variété des zéros de l'idéal $(P_1, \ldots, P_{n-1})$; sinon les $\alpha_j^{(i)}$ peuvent être choisis tous nuls.

(11) Le résultant universel est un polynôme multihomogène de degré $\prod_{j \neq i} d_j$ en chaque paquet de variables $U_{(I,i)}$ (i fixé).

I Algorithme théorique

I.1 Description de l'algorithme.

$\sum_{|I|=d_i} U_{(I,i)} X^I$ où les variables $U_{(I,i)}$ sont algébriquement indépendantes, avec les notations condensées usuelles.

Entrées: P_1, P_2, P_3 trois polynômes homogènes non nuls de $A[X_1, X_2, X_3]$ de degrés respectifs d_1, d_2 et d_3, A un corps.
   Sortie: $Res(P_1, P_2, P_3)$.

Etape 1 (facultative): On élimine les facteurs monomiaux des polynômes, en calculant les facteurs du résultant qu'ils donnent grâce aux formules (5) – qui amène à calculer un résultant de deux polynômes – et (6); les problèmes de signe dus aux permutations nécessaires sont reglés par les formules (7) – avec M matrice d'une permutation σ, donc $\det M = \epsilon(\sigma)$ – et (8). Et l'on passe à l'étape 2.

Etape 2: On regarde si l'un des polynômes fait apparaître une puissance pure de X_3.

   – Si oui, après une permutation éventuelle, de telle manière que P_1 ait cette propriété, on passe à l'étape 3.

   – Sinon $Res(P_1, P_2, P_3) = 0$.

Etape 3: On fait la spécialisation $X_1 = 1$ et on calcule $\Phi = Res_{X_3}(P_1, P_2)$. Si $\Phi = 0$, $Res(P_1, P_2, P_3) = 0$, sinon on passe à l'étape 4.

Etape 4: On factorise Φ;

$$\Phi = a \prod_{i=1}^{k} \Phi_i^{e_i}$$

avec Φ_i irréductible et unitaire en X_2 et $e_1 \geq e_2 \geq \cdots \geq e_k$.
   On pose $e_0 = d_1 d_2 - \deg \Phi$ et on passe à l'étape 5.

Etape 5:
[1] Si $e_0 \geq 1$, calculer $g_0 = pgcd(P_1(0, 1, X_3), P_2(0, 1, X_3))$, que l'on prendra unitaire en X_3.
   – Si $\deg g_0 = e_0$ calculer $r_0 = Res_{X_3}(g_0, P_3(0, 1, X_3))$ et passer au [2].
   – Si $g_0 = (X_3 - a_0)^{f_0}$ calculer $r_0 = (P_3(0, 1, a_0))^{e_0}$ et passer au [2].
   – Sinon (ce qui ne peut être le cas que si $e_0 \geq 3$), on fait le changement de variables $X_1 \longmapsto X_1 + X_2$ et on retourne à l'étape 2.[2]
[2] Une fois calculé r_0, pour chaque i de 1 à k:
On calcule le pgcd (que l'on prendra unitaire en X_3) g_i de $P_1^{(i)}$ et $P_2^{(i)}$, images de P_1 et P_2 dans $(A[X_2]/\Phi_i)[X_3]$ par la surjection canonique.
   – Si $\deg_{X_3} g_i = e_i$ on choisi un relèvement G_i de g_i dans $A[X_2, X_3]$ et on calcule:
$$r_i = Res_{X_2}(\Phi_i, Res_{X_3}(G_i, P_3)).$$

---

[2]En caractéristique p positive, tous les $p - 1$ changements de variables de ce type, il faut en faire un de la forme $X_i \longmapsto X_i + \lambda_l X_{i+1}$ ($i = 1$ ou 2 suivant que l'on est en [1] ou [2]), les coefficients λ_l successifs vérifiant $\lambda_l \notin \bigoplus_{i=1}^{l-1} \lambda_i \mathbf{F}_p$; ce qui peut nécessiter d'étendre le corps de base dans le cas des corps finis.

- S'il existe f_i tel que $g_i = (X_3 - a_i(x_2))^{f_i}$; choisissant A_i relèvement de a_i dans $A[X_2]$, on calcule

$$r_i = (Res_{X_2}(\Phi_i, P_3[X_2 = A_i]))^{e_i}.$$

- Sinon (ce qui ne peut être le cas que si $e_i \geq 3$), faire le changement de variables $X_2 \longmapsto X_2 + X_3$ et retourner à l'étape 2.[3]
  Une fois tous les r_i calculés, on passe à l'étape 6.

Étape 6: Soit $d_2' = \deg_{X_3} P_2$.

On pose $b = (-1)^{\deg \Phi} a\gamma^{d_2 - d_2'}$ où γ est le coefficient de $X_3^{d_1}$ dans P_1.

On a alors:

$$Res(P_1, P_2, P_3) = b^{d_3} \prod_{i=0}^{k} r_i$$

I.2 Quelques remarques géométriques.

(1) A l'étape 2 on choisit deux polynômes n'ayant pas le point $(0:0:1)$ comme zéro commun.

(2) L'étape 3 est une projection de la variété des zéros de (P_1, P_2) de centre $(0:0:1)$; le nombre e_0 est la somme des multiplicités des points de coordonées homogènes $(0:1:\alpha)$; ce sont les facteurs correspondant à ces points qui donnent des termes constants dans la spécialisation $X_1 = 1$ du résultant en X_3 de P_1 et P_2, d'où la formule pour e_0.

(3) Les polynômes Φ_i correspondent à un, ou éventuellement plusieurs si $e_i \geq 2$, ensemble(s) de conjugués. Si l'on note $E_{i,j}$, $j = 1, \ldots, s_i$ ces ensembles de conjugués, on a $\#E_{i,j} = c_{i,j} \deg \Phi_i$ et si $m_{i,j}$ est la multiplicité de l'un quelconque des points de $E_{i,j}$, on a pour e_i la formule $e_i = \sum_{j=1}^{s_i} c_{i,j} m_{i,j}$.

(4) A l'étape 5, d'après ce que l'on vient de voir à la remarque précédente, on sera toujours dans le premier cas si tous les points sont de multiplicité 1. En revanche on ne sera pas dans ce cas si l'un des points correspondant à Φ_i est de multiplicité supérieure au minimum des degrés de P_1 et P_2, nous verrons au lemme 4 un résultat plus précis sur cette question.

On sera dans le deuxième cas lorsque $s_i = 1$, c'est-à-dire pour presque tout choix de coordonnées[4] tous sauf un nombre fini d'où la troisième alternative, à laquelle on a recours quand un point de grande multiplicité a même projection qu'un ou plusieurs autres points.

I.3 Cette procédure se termine.

Il nous suffit de remarquer que:

- Pour presque toute valeur de N, le changement de variables $X_1 \longmapsto X_1 + N X_2$ nous amène, pour chaque couple de polynômes possible, à ne plus avoir de zéro commun non trivial avec $X_1 = 0$.

---

[3]Voir la remarque précédente.

[4];

– Pour presque toute valeur de M, deux points distincts de la forme $(1 : \alpha : \beta)$ ont des projections distinctes. En effet pour chaque couple de points $((1 : \alpha : \beta), (1 : \alpha' : \beta'))$, il existe au plus une valeur de M telle que $\alpha + M\beta = \alpha' + M\beta'$.

I.4 Cet algorithme calcule effectivement le résultant.

Le détail de la démonstration est assez long, nous nous contenterons donc d'en indiquer ici le principe et renvoyons au chapitre 3 de notre thèse ([Th]) pour une démonstration détaillée.

Rappellons que:

Lemme 1. *Soit $a_1, \ldots, a_r$ des éléments d'un anneau A et $P \in A[X]$, alors*

$$Res_X(\prod_{i=1}^{r}(X - a_i), P) = \prod_{i=1}^{r} P(a_i)$$

Définition: Si P_1 et P_2 sont deux polynômes homogènes de degrés respectifs d_1 et d_2 d'un anneau de polynômes $A[X_1, \ldots, X_r, Y]$, notons $B = A[X_1, \ldots, X_r]$.

On appelle alors déterminant de Sylvester associée à P_1 et P_2 relativement à la variable Y le déterminant de l'application B-linéaire de $(\bigoplus_{i=0}^{d_2-1} BY^i) \bigoplus (\bigoplus_{i=0}^{d_1-1} BY^i)$ dans $\bigoplus_{i=0}^{d_1+d_2-1} BY^i$ définie par

$$(F, G) \longmapsto FP_1 + GP_2$$

on note ce déterminant $Sylv_Y(P_1, P_2)$.

De façon équivalente, si $d_i = \deg P_i$ et $d'_i = \deg_Y P_i$:

$$Sylv_Y(P_1, P_2) = \begin{cases} Res_Y(P_1, P_2) & \text{si} \quad d_1 = d'_1 \quad \text{et} \quad d_2 = d'_2 \\ t(P_1)^{d_2 - d'_2} Res_Y(P_1, P_2) & \text{si} \quad d_1 = d'_1 \quad \text{et} \quad d_2 > d'_2 \\ t(P_2)^{d_1 - d'_1} Res_Y(P_1, P_2) & \text{si} \quad d_1 > d'_1 \quad \text{et} \quad d_2 = d'_2 \\ 0 & \text{sinon} \end{cases}$$

où $t(P_1)$ et $t(P_2)$ sont respectivement les coefficients dominants de P_1 et P_2, vus comme polynômes en Y.

Les deux lemmes importants sont les suivants:

Lemme 2. *Soit A un anneau et P_1, P_2 deux polynômes de $A[X_1, X_2, X_3]$ $(\subset A[p, q, r][X_1, X_2, X_3])$, et soit $L = pX_1 + qX_2 + rX_3$.*
Notons $\Psi(p, q, r) = Res(P_1, P_2, L)$; alors:

$$\Psi(-X_2, X_1, 0) = Sylv_{X_3}(P_1, P_2)$$

Démonstration: Il suffit de se placer dans le cas universel, c'est-à-dire: $A = \mathbf{Z}[U_{I,1}, U_{J,2}]_{|I| = \deg P_1, |J| = \deg P_2}$, $P_1 = \sum_I U_{I,1} X^I$, $P_2 = \sum_J U_{J,2} X^J$.

De la propriété (11) du théorème, on déduit immédiatement que $\Psi(-X_2, X_1, 0)$ est un polynôme trihomogène de degrés $d_1 d_2$ en (X_1, X_2), d_2 en $U_{I,1}$ et d_1 en $U_{J,2}$; on vérifie également sans difficultés sur la matrice de Sylvester que $Sylv_{X_3}(P_1, P_2)$ possède la même propriété.

D'autre part, la nullité de l'un ou de l'autre de ces deux polynômes en un point de $\mathbf{P} = \mathbf{P}_{D_1} \times \mathbf{P}_{D_2} \times \mathbf{P}_1(\overline{\mathbf{Q}})$ (où $\overline{\mathbf{Q}}$ désigne une clôture algébrique de $\mathbf{Q}$ et $D_1 = \binom{d_1+2}{2} - 1$, $D_2 = \binom{d_2+2}{2} - 1$) est équivalente à la condition suivante:

"$P_1(\mathbf{u}_1; \mathbf{u}_2; 0, 0, 1) = P_2(\mathbf{u}_1; \mathbf{u}_2; 0, 0, 1) = 0$, ou il existe $x_3 \in \overline{\mathbf{Q}}$ tel que: $P_1(\mathbf{u}_1; \mathbf{u}_2; x_1, x_2, x_3) = P_2(\mathbf{u}_1; \mathbf{u}_2; x_1, x_2, x_3) = 0$".

On vérifie facilement que $\Psi(-X_1, X_2, 0)$ est irréductible, donc également $Sylv_{X_3}(P_1, P_2)$ pour des raisons de degré. Ils sont donc égaux à une constante près. Or on a,

$$
\begin{aligned}
Res(X_3^{d_1}, X_2^{d_2}, L) &= [Res(X_3, X_2, L)]^{d_1 d_2} \\
&= [Res(X_3, X_2, pX_1)]^{d_1 d_2} \\
&= [-Res(pX_1, X_2, X_3)]^{d_1 d_2} \\
&= (-p)^{d_1 d_2},
\end{aligned}
$$

et si U ne dépend pas de X_3, alors $Sylv_{X_3}(X_3^{d_1}, U) = U^{d_1}$, donc $Sylv_{X_3}(X_3^{d_1}, X_2^{d_2}) = X_2^{d_1 d_2}$, d'où l'égalité annoncée. \hfill Q. E. D.

Lemme 3. *Soit A un corps, P_1 et P_2 deux polynômes de $A[X,Y]$ sans facteur commun et M de coordonnées (u,v) un zéro commun de P_1 et P_2.*
  Notons $\mathcal{O}_M$ l'anneau local de M, $\mathcal{M}_M$ l'idéal $(X-u, Y-v)$ et:
  - *$I(M, P_1 \cap P_2) = m_M$ la multiplicité d'intersection de P_1 et P_2 en M, ou multiplicité du point M (i.e. la longueur de $\mathcal{O}_M$).*
  - *e_M l'exposant de l'idéal $\mathcal{M}_M$-primaire $\mathcal{Q}_M$ apparaissant dans la décomposition primaire de l'idéal (P_1, P_2).*
  - *r_M la multiplicité du facteur $(p + qu + rv)$ dans $Res(^h P_1, ^h P_2, L)$, où $^h P_1$ et $^h P_2$ désignent les homogénéisés de P_1 et P_2 et $L = pT + qX + rY$.*
  - *r'_M la multiplicité du facteur $(Y-v)$ dans $Res_X(P_1, P_2)$.*
  - *g_M la multiplicité du facteur $(Y-v)$ dans $pgcd(P_1(u,Y), P_2(u,Y))$.*
  On a:
$$ g_M \leq e_M \leq m_M = r_M \leq r'_M $$
toutes les inégalités pouvant être strictes.

Démonstration:
  - $r_M \leq r'_M$: c'est un corollaire du lemme 3.
  - $r_M = m_M$: les propriétés (6) à (10) du théorème montrent que $r_M = r_M(P_1, P_2)$ vérifie les axiomes de définition de la multiplicité d'intersection, d'où l'égalité (voir [Fu] Ch. 3, §3, Theorem 3).
  - $e_M \leq m_M$: m_M est également la longueur maximale d'une chaine d'idéaux $\mathcal{M}_M$-primaires distincts entre $\mathcal{M}_M$ et $\mathcal{Q}_M$, or on a
$$ \mathcal{M} \supset \mathcal{M}^{(2)} \supset \cdots \supset \mathcal{M}^{(e_M)} \supseteq \mathcal{Q}_M $$
d'où l'inégalité.
  - $g_M \leq e_M$: localisons-nous au point M, on a $(P_1, P_2)_{\mathcal{M}_M} = (\mathcal{Q}_M)_{\mathcal{M}_M}$ et, puisque $\mathcal{M}_M^{e_M} \subseteq \mathcal{Q}_M$, on a $(Y-v)^{e_M} \in (\mathcal{Q}_M) \subset (P_1, P_2)_{\mathcal{M}_M}$.

Ainsi, il existe trois polynômes A B et D de $A[X, Y]$, avec $D(u, v) \neq 0$, tels que $D(Y - v)^{e_M} = AP_1 + BP_2$, d'où

$$D(u, Y)(Y - v)^{e_M} = A(u, Y)P_1(u, Y) + B(u, Y)P_2(u, Y)$$

puisque $D(u, v) \neq 0$, $(Y - v)^{g_M}$ divise $(Y - v)^{e_M}$ et on a bien $g_M \leq e_M$. Q. E. D.

Expliquons maintenant la démonstration proprement dite.

Les notations sont celles utilisées en **I.1**.

Soit $L = pX_1 + qX_2 + rX_3$, d'après le (10) du théorème il existe une extension algébrique $K = A[\alpha_j^{(u)}]$ de A telle que:

$$Res(P_1, P_2, L) = \prod_{u=1}^{d_1 d_2} (p\alpha_1^{(u)} + q\alpha_2^{(u)} + r\alpha_3^{(u)})$$

et

$$Res(P_1, P_2, P_3) = \prod_{u=1}^{d_1 d_2} P_3(\alpha_1^{(u)}, \alpha_2^{(u)}, \alpha_3^{(u)})$$

grâce à la préparation faite aux étapes 1 et 2, on a toujours $\alpha_1^{(u)} \neq 0$ ou $\alpha_2^{(u)} \neq 0$.

D'après le lemme 2,

$$\gamma^{d_2 - d_2'} Res_{X_3}(P_1, P_2) = \prod_{u=1}^{d_1 d_2} (-X_2 \alpha_1^{(u)} + X_1 \alpha_2^{(u)})$$

Supposons pour fixer les idées que $\alpha_1^{(u)} = 0$ pour $u \leq e_0$ et $\alpha_1^{(u)} \neq 0$ sinon, on a alors:

$$\prod_{i=1}^{k} \Phi_i^{e_i} = \prod_{u=e_0+1}^{d_1 d_2} \left(X_2 - \frac{\alpha_2^{(u)}}{\alpha_1^{(u)}}\right) \quad \text{et} \quad b = \left(\prod_{u=1}^{e_0} \alpha_2^{(u)}\right)\left(\prod_{u=e_0+1}^{d_1 d_2} \alpha_1^{(u)}\right).$$

Les zéros des g_i ($i > 0$) correspondent à la coordonnée en X_3 des points au dessus des zéros de Φ_i exprimée comme fonction rationelle ou algébrique de X_2 (ayant posé $X_1 = 1$). En substituant donc cette valeur à X_3 dans P_3 par un résultant, puis en remplaçant X_2 par les racines de Φ_i – également par un résultant – on retrouve la formule exprimant le résultant de trois polynômes comme produit du troisième sur les racines communes aux deux premiers.

Le lemme **3** assure que les racines ont étés comptées avec la bonne multiplicité et le fait d'avoir imposé aux polynômes d'êtres unitaires entraîne que la normalisation est la bonne, au vu des expressions pour Φ_i et b.

Les zéros correspondant à g_0 se traitent de manière similaire.Q. E. D.

II Algorithme pratique

Nous supposerons ici que l'anneau de base A est factoriel, et resterons le plus souvent dans A pour les calculs. C'est le cas lorsque A est l'anneau des entiers, un corps, ou un anneau de polynômes sur un tel anneau.

Nous allons rappeler ici l'algorithme des sous-résultants qui permet le calcul de l'étape 3 et monter comment l'utiliser pour éviter les calculs de pgcd de l'étape 5.

Nous montrerons également comment éviter la factorisation de l'étape 4, suivant l'idée utilisée par le logiciel D5 (voir [D5]). Commençons par rappeler l'algorithme des sous-résultants dont une version correcte si les deux polynômes sont de degrés distincts se trouve dans [Loos] (p. 129), pour le cas restant et une étude plus détaillée nous renvoyons à [Lom]. Je tiens à remercier Henri Lombardi pour ses remarques sur une première version érronée que j'avais fourni de cet algorithme, la version ci-dessous est la conséquence de ses conseils.

[SR] Algorithme des sous-résultants.

Entrées: $F, G \in A[X]$ non nuls, A intègre.

Sorties: L liste ordonnée de j tels que $S_j \neq 0$, $S_j, R_j = t(S_j)$ pour $j \in L$ $(S_0 = Res_X(F, G))$.

(1) $m \leftarrow \deg F$, $n \leftarrow \deg G$.
   - Si $m < n$, appliquer **[SR]**(G,F), puis $S_j \leftarrow (-1)^{(m-j)(n-j)} S_j$, $R_j \leftarrow (-1)^{(m-j)(n-j)} R_j$, fin.
   - Si $m > n$, $j \leftarrow n$, $S_{j+1} \leftarrow F$, $S_j \leftarrow G$, $R_{j+1} \leftarrow 1$, $L \leftarrow \emptyset$, aller en (2).
   - Si $m = n$, $S_{n-1} \leftarrow reste(t(G)F, G)$, $w \leftarrow \deg S_{n-1}$.

   Si $w = n - 1$, $S_{n-2} \leftarrow \frac{reste(t(S_{n-1})F, S_{n-1})}{t(F)}$.   Si $w < n - 1$, $S_w \leftarrow t(S_{n-1})^{n-1-w} S_{n-1}$, $S_{w-1} \leftarrow (-1)^{n-1-w} \frac{reste(t(S_{n-1})^{n-w+1}F, S_{n-1})}{t(F)}$.
   $R_w \leftarrow t(S_w)$, $L \leftarrow w$, $j \leftarrow w - 1$, aller en (2).

(2) $r \leftarrow \deg S_j$, $s \leftarrow \deg S_{j+1}$, $L \leftarrow ajout(r, L)$.

(3) Si $0 \leq r < j$, $S_r \leftarrow \frac{t(S_j)^{j-r} S_j}{R_{j+1}^{j-r}}$.

(4) Si $r \leq 0$, fin.

(5) $S_{r-1} \leftarrow \frac{reste(t(S_j)^{s-r+1} S_{j+1}, S_j)}{(-R_{j+1})^{j-r+2}}$.

(6) $j \leftarrow r - 1$, $R_{j+1} \leftarrow t(S_{j+1})$, aller en (2).

Notations:
   - $t(P)$ désigne le coefficient dominant de P.
   - $reste(P, Q)$ désigne le reste de la division euclidienne de P par Q.
   - On fait la convention $\deg 0 = -1$.
   - $ajout(j, L)$, ajoute j au début de L.

Résumons quelques propriétés utiles de la suite $S_0, \dots, S_n$.

Proposition 1. *Soit ϕ est un homomorphisme de A dans un corps k, $\tilde{\phi}$ son extension naturelle en un homorphisme de $A[X]$ dans $k[X]$, et supposons que $\phi(t(F)) \neq 0$ ou $\phi(t(G)) \neq 0$.*

Alors si d est le degré du pgcd de $\tilde{\phi}(F)$ et $\tilde{\phi}(G)$, on a:

- $d = \min\{i \mid \tilde{\phi}(S_i) \neq 0\} = \min\{i \mid \phi(t(S_i)) \neq 0\}.$
- $\operatorname{pgcd}(\tilde{\phi}(F), \tilde{\phi}(G)) = \tilde{\phi}(S_d).$

Démonstration: Voir [Loos] p. 122.

Une fois connue la suite des sous-résultants, on connaît donc les pgcd recherchés; la proposition suivante montre que l'on connaît alors également le résultant du quotient des polynômes par leur pgcd.

Proposition 2. *Soit A un anneau et F, G, P trois polynômes de A[X] de degrés respectifs m, n et p, alors:*

$$S_{k+p}(PF, PG) = t(P)^{m+n-2k-1} S_k(F, G) P$$

et

$$R_{k+p}(PF, PG) = t(P)^{m+n-2k} R_k(F, G)$$

pour tout $k < \min\{m, n\}$.

Démonstration: Voir [Th] Chap. 4, deuxième partie, fait 1.

On opère de la manière suivante:

Etape 3': **[SR]** (P_1, P_2).

Etape 4': Faire une décomposition sans facteur carré de S_0:

$$S_0 = a' \prod_{i=1}^{r} \Psi_i^{h_i}$$

avec: $h_1 > h_2 > \cdots > h_r > 0$, Ψ_i primitif, sans facteur carré et tel que $\operatorname{pgcd}(\Psi_i, \Psi_j) = 1$ pour $i \neq j$.

Etape 5':
[1] : inchangé.
[2] : opérer comme suit,
[2a] *Factorisation minimale.*
(1) $i \leftarrow 1$, $L' \leftarrow \emptyset$, $S \leftarrow \emptyset$.
(2) $j \leftarrow premier(L)$, $U \leftarrow \Psi_i$.
(3) Si $j \notin S$, $T_i \leftarrow p(R_j).p$
(4) $U \leftarrow$ **[SRPGCD]**(U, T_i), $V_{i,j} \leftarrow p(U)$, $u \leftarrow c(U)$.
(5) $T_i \leftarrow T_i / V_{i,j}$.
(6) Si $\deg V_{i,j-1} \neq \deg V_{i,j}$:
    (6a) $\Psi_{i,j} \leftarrow V_{i,j-1}/V_{i,j}$, $d_{i,j} \leftarrow u/t(V_{i,j})^{\deg T_i + \deg \Psi_{i,j-1}}$.
    (6b) $L' \leftarrow ajout(L', (i, j))$, $S \leftarrow S \cup \{j\}$.
    (6c) Si $h_i = 1$ ou $h_i = j$, aller en (7).

(6d) Si $[\mathbf{T}](S_j(X_3), \Psi_{i,j})$=non, aller en (10).

(6e) $F_{i,j} \leftarrow F$.

(7) Si $\deg U > 0$, $j \leftarrow s(j, L)$, aller en (3).

(8) Si $i = r$ fin.

(9) $i \leftarrow i + 1$, aller en (2).

(10) Faire le changement de variables $X_2 \longmapsto X_2 + X_3$, et retourner à l'étape 2.

[T] Test.

Entrées: $S = \sum_{l=0}^{j} K_l X^l \in k[Y][X]$ avec $K_j \neq 0$, $\Psi \in k[Y]$, k un corps.

Sortie: $F \in k[Y][X]$ de degré 1 en X tel que $S \equiv K_j F(X)^j$ modulo Ψ, s'il existe un tel élément dans $k[Y]$, non sinon.

Si k est de caractéristique 0:

(1) $l \leftarrow 1$.

(2) Si $reste((l + 1)jK_{j-l-1}K_j - (j - l)K_{j-l}K_{j-1}; \Psi) \neq 0$, non, fin.

(3) Si $l < j - 1$, $l \leftarrow l + 1$, aller en (2).

(4) $F \leftarrow X + [j^{-1}K_{j-1}inverse(K_j, \Psi)]$, fin.

Si k est de caractéristique $p > 0$:

(1) Décomposer j en $j = p^e j'$ avec $(p, j') = 1$, $q \leftarrow p^e$.

(2) S'il existe i non multiple de q tel que $reste(K_i, \Psi) \neq 0$, non, fin.

(3) $K_i' \leftarrow K_{iq}$, $S' \leftarrow \sum_{i=1}^{j'} K_i' X^i$.

(4) $I \leftarrow inverse(K_{j'}', \Psi)$, $A \leftarrow IK_{j'-1}'$.

(5) $A_i \leftarrow reste(Y^i, \Psi)$, $\psi \leftarrow \deg \Psi$.

(6) Si $A \notin Vect_k(A_0, \ldots, A_{\psi-1})$, non, fin.

(7) Déterminer $a_0, \ldots a_{\psi-1}$ tels que $A = \sum_{i=0}^{\psi-1} a_i A_i$.

(8) $A \leftarrow \sum_{i=0}^{\psi-1} a_i X^i$.

(9) Si $e \neq 1$, $e \leftarrow e - 1$, aller en (6).

(10) $F \leftarrow X + j'^{-1}A$.

Notations:

- $p(P)$ désigne la partie principale du polynôme P.
- $c(P)$ désigne le contenu du polynôme P.
- $s(j, L)$ retourne le succéseur de j dans L s'il y en a un et fin sinon.
- $premier(L)$ retourne le premier élément de la liste L.
- $append(L, e)$ ajoute e à la fin de la liste L.
- $[\mathbf{SRPGCD}](P, Q)$ désigne le sous-résultant non nul de plus petit ordre, c'est donc $S_{premier(L)}(P, Q)$, l'algorithme $[\mathbf{SRPGCD}]$ se déduit donc de manière évidente de $[\mathbf{SR}]$.
- $Vect_k(P_1, \ldots, P_r)$ désigne le k espace vectoriel engendré par $P_1, \ldots, P_r$
- $inverse(P, \Psi)$ retourne un polynôme I tel que $IP \equiv 1$ modulo Ψ.

On a ainsi:
- $p(R_j) = \prod_i V_{i,j}$, pour les j donnant naissance à un pgcd, le produit étant pris sur les i pour lesquels $V_{i,j}$ est défini.
- $\Psi_i = \prod_j \Psi_{i,j}$, le produit étant pris sur les j pour lesquels $\Psi_{i,j}$ est défini.
- Les $\Psi_{i,j}$ définis (donc non constants) correspondent aux facteurs de Ψ donnant un pgcd de degré j, car $\prod_{i<j} V_{i,j}$ est inversible modulo $\Psi_{i,j}$.
- $Res(\Psi_{i,j}, R_j) = c(R_j)^{\deg \Psi_{i,j}} d_{i,j} \prod_{l<j} Res(\Psi_{i,j}, V_{l,j})$, d'après la proposition 2.

[2b] Calcul des $r_{i,j}$.
(1) Pour $j \in S$, $N_j \leftarrow Res_{X_3}(S_j, P_3)$, $n_j \leftarrow \deg N_j$.
(2) $(i,j) \leftarrow premier(L')$.
(3) $\Psi \leftarrow \Psi_{i,j}$, $\psi \leftarrow \deg \Psi$.
(4) $\omega \leftarrow [h_i/j]$, $\varpi \leftarrow h_i - \omega$.
(5) $D \leftarrow c(R_j)^\psi d_{i,j} \prod_{l<j} Res(\Psi, V_{l,j})$, $N \leftarrow Res(\Psi, N_j)$.
(6) $r_\omega \leftarrow N/t(\Psi)^{n_j} D^{d'_3}$.
(7) Si $\varpi \neq 0$, $M \leftarrow Res_{X_3}(F_{i,j}, P_3)$, $r_\varpi \leftarrow Res(\Psi, M)/t(\Psi)^{\deg M}$.
(8) $r_{i,j} \leftarrow r_\omega^\omega r_\varpi^\varpi$.
(9) $(i,j) \leftarrow s((i,j), L')$, aller en (3).

[2c] Fin.
(1) $b \leftarrow (-1)^{\deg S_0} a' \gamma^{d_2 - d'_2}$.
(2) $Res(P_1, P_2, P_3) \leftarrow b^{d_3} r_0$.
(3) Si $\deg S_0 = 0$, fin.
(4) $(i,j) \leftarrow premier(L')$.
(5) $Res(P_1, P_2, P_3) \leftarrow r_{i,j} Res(P_1, P_2, P_3)$.
(6) $(i,j) \leftarrow s((i,j), L')$, aller en (5).

III Généralisation possible au cas de n polynômes

Nous allons commencer par étendre le lemme 3 de I.3 au cas de n polynômes homogènes.

Lemme 2 bis. *Soit A un anneau et $P_1, \ldots, P_{n-1}$, $n-1$ polynômes homogènes de $A[X_1, \ldots, X_n] = A[X_1, X_2][X_3, \ldots, X_n]$.*
Notons:
- *$\hat{P}_k$ le polynôme homogénéisé de $P_k \in A[X_1, X_2][X_3, \ldots, X_n]$ dans $A[X_1, X_2][T, X_3, \ldots, X_n]$:*

$$\hat{P}_k(X, T) = P_k(TX_1, TX_2, X_3, \ldots, X_n)$$

- *$\Psi(p_1, \ldots, p_n)$ le résultant de $P_1, \ldots, P_{n-1}$ et $L = p_1 X_1 + \cdots + p_n X_n$, une forme linéaire générique; tous étant vus comme polynômes de $A[p_1, \ldots, p_n][X_1, \ldots, X_n]$:*

$$\Psi = Res_{X_1, \ldots, X_n}(P_1, \ldots, P_{n-1}, L) \in A[p_1, \ldots, p_n].$$

On a alors:

$$\Psi(X_2, -X_1, 0, \ldots, 0) = Res_{T, X_3, \ldots, X_n}(\hat{P}_1, \ldots, \hat{P}_{n-1}).$$

Esquisse de la démonstration: La preuve utilise les mêmes arguments que pour le lemme 2; à savoir que:

(1) Dans le cas universel, on trouve des polynômes ayant les mêmes multidegrés à savoir:

– $D = d_1 \cdots d_{n-1}$ en X_1, X_2;

– D/d_i en $U_{I,i}$.

(2) Ces polynômes sont nuls en un point $(\mathbf{u}_1; \ldots; \mathbf{u}_{n-1}; x_1, x_2)$ de $\mathbf{P} = \mathbf{P}_{D_1} \times \cdots \times \mathbf{P}_{D_{n-1}} \times \mathbf{P}_1(\overline{\mathbf{Q}})$ (où $\overline{\mathbf{Q}}$ désigne une clôture algébrique de $\mathbf{Q}$ et $D_i = \binom{d_i + n - 1}{n-1} - 1$, sous la condition suivante:

"Il existe $\mathbf{y} \in \mathbf{P}_n$ avec $y_1 = y_2 = 0$ ou $\{y_1 = x_1$ et $y_2 = x_2\}$ tel que: $\forall i, \ P_i(\mathbf{u}_1; \ldots; \mathbf{u}_{n-1}; \mathbf{y}) = 0$."

On vérifie ensuite que $Res_{T,X}$ est irréductible, et l'on conclut en observant que:

$$Res(L, X_2^{d_2}, \ldots, X_n^{d_n}) = Res(p_1 X_1, X_2, \ldots, X_n)^{d_2 \cdots d_n}$$
$$= p_1^{d_2 \cdots d_n}$$

et

$$Res_{T, X_3, \ldots, X_n}(X_2^{d_2} T^{d_2}, X_3^{d_3}, \ldots, X_n^{d_n} = (X_2^{d_2})^{d_3 \cdots d_n}$$

$$\text{Q. E. D.}$$

Signalons que, plus généralement, on a la formule:

Formule 1:

$$X_1^{(r-1)d_1 \cdots d_{n-r}} Res_{T, X_{r+2}, \ldots, X_n}(\hat{P}_1, \ldots, \hat{P}_{n-r}) = F_{(P_1, \ldots, P_{n-r})}(M)$$

avec

$$M = \begin{pmatrix} X_2 & -X_1 & 0 & \cdots & \cdots & \cdots & \cdots & 0 \\ X_3 & 0 & -X_1 & 0 & & & & \vdots \\ \vdots & \vdots & \ddots & \ddots & \ddots & & & \vdots \\ X_{r+1} & 0 & \cdots & 0 & -X_1 & 0 & \cdots & 0 \end{pmatrix}.$$

Les notations étant les suivantes:

– $\hat{P}_k(X, T) = P(TX_1, \ldots, TX_{r+1}, X_{r+2}, \ldots, X_n)$,

– $F_{(P_1, \ldots, P_{n-r})} = Res_X(L_1, \ldots, L_r, P_1, \ldots, P_{n-r})$ où $L_i = \sum u_j^{(i)} X_j$ est une forme linéaire générique.

– $F_{(\ldots)}(M)$ est la spécialisation de $F_{(\ldots)}$ correspondant à remplacer la matrice des $u_j^{(i)}$ par M.

Soit V_r et V_r' les sous-variétés de $\mathbf{P}_{n-1}$ définies par les idéaux $(P_1, \ldots, P_r)$ et $(P_1, \ldots, P_{r-1}, P_{r+1})$. Si elles ne rencontrent pas le $(r-1)$-plan H_r d'équation $X_1 = \cdots = X_{n-r+1} = 0$ pour $1 < r < n$

– c'est "presque toujours" le cas puisque $\mathrm{codim} V_r + \mathrm{codim} H_r = n + 1 > n - 1$ – on peut alors calculer par hypothèse de récurrence:

$$\Phi(1) = P_1$$
$$\Phi'(1) = P_2$$
$$\Phi(2) = Res(\hat{P}_1, \hat{P}_2)$$
$$\Phi'(2) = Res(\hat{P}_1, \hat{P}_3)$$
$$\vdots \qquad \vdots$$
$$\Phi(n-2) = Res(\hat{P}_1, \ldots, \hat{P}_{n-3}, \hat{P}_{n-2})$$
$$\Phi'(n-2) = Res(\hat{P}_1, \ldots, \hat{P}_{n-3}, \hat{P}_{n-1})$$
$$\Phi(n-1) = Res(\hat{P}_1, \ldots, \hat{P}_{n-1})$$

Remarque: Si il n'y a pas de composante de dimension plus grande ou égale à 1 dans la variété des zéros de l'idéal défini par les P_i, ce qui se teste par un résultant d'ordre inférieur, on peut se ramener, par changement linéaire de variables, à la situation précédente.

$\Phi(n-1) \in A[X_1, X_2]$ se décompose en:

$$\Phi(n-1) = a \prod_{i=1}^{k} \Psi_i^{e_i}$$

où $a \in A$ et les Ψ_i sont irréductibles et unitaires en X_2, le cas où $\Psi_i = X_1$ se traîtant séparément de manière similaire au cas de 3 variables, nous l'exclurons dans cette brève description.

Comme dans le cas de trois polynômes on calcule les pgcd Φ_i des images de $\Phi(n-2)$ et $\Phi'(n-2)$ par la surjection canonique $A[X_1, \ldots, X_3] \rightarrow A[1, X_2, X_3]/\Psi_i(1, X_2) = A(i)[X_3]$.

On factorise les Φ_i dans $A(i)[X_3]$: $\Phi_i = \prod_{j=1}^{r_i} \Psi_{i,j}^{e_{i,j}}$.

Puis on calcule le pgcd des images de $\Phi(n-3)$ et $\Phi'(n-3)$ par la surjection canonique de $A[X_1, \ldots, X_4]$ dans $A[1, X_2, \ldots, X_4]/(\Psi_i, \Psi_{i,j}) = A(i,j)[X_4]$, et caetera...

On obtient ainsi un arbre $\mathcal{A}$ de profondeur $n - 1$ donnant une décomposition triangulaire de l'ensemble des zéros communs de $P_1, \ldots, P_{n-1}$. Les noeuds du niveau k sont constitués des $\Psi_{i_1, \ldots, i_k}$.

Le problème des multiplicités est le même que précedemment, on peut également y remédier par changement linéaire de coordonnées. Si l'on ne se préoccupe pas des multiplicités le polynôme R suivant est nul si et seulement si les P_i ont un zéro commun et coïncide avec le résultant dans le cas où

$$\sum_{B \in \text{Branches}(\mathcal{A})} \left[\prod_{\Psi_N \in \text{Noeuds}(B)} \deg \Psi_N \right] = d_1 \cdots d_{n-1},$$

c'est-à-dire si les zéros sont simples.

Notons que la somme ci-dessus est le degré ensembliste de la variété des zéros et est donc toujours majoré par $d_1 \cdots d_{n-1}$.

En supposant les Ψ_N unitaires (ce sont des polynômes en une variable), R est donné par la formule suivante:

$$R = a^{d_n} \prod_{B \in \text{Branches}(\mathcal{A})} Res_{X_2}(\Psi_{N_1}, Res_{X_3}(\ldots(\tilde{\Psi}_{N_{n-2}},(Res_{X_n}(\tilde{\Psi}_{N_{n-1}},P_n)))\ldots))$$

où N_k correspond au noeud du k-ième niveau de B, $\Psi_{N_k} = \Psi_{i_1,\ldots,i_k}$ et $\tilde{\Psi}_{N_k}$ est un relèvement de Ψ_{N_k} dans $A(i_1,\ldots,i_{k-2})[X_k,X_{k+1}]$ pour $k > 1$.

La démonstration du fait que R est le résultant suit exactement les mêmes étapes que dans le cas de trois variables.

Un exemple

Nous travaillons ici avec des polynômes non homogènes, la variable d'homogénéisation étant considérée comme première variable, x comme deuxième variable et y comme dernière variable.

$$P_1(x,y) = -3y^3 + (-x^2 - 5x + 1)y + x^3 + x - 1$$
$$P_2(x,y) = y^3 + y^2 - xy + x^3 - 5x + 7$$
$$P_3(x,y) = y^2 - 5xy + 2x^3 - x - 2$$
$$Res_y(P_1,P_2) = -65x^9 + 613x^7 - 1321x^6 - 2301x^5$$
$$+9146x^4 - 3832x^3 - 17295x^2 + 22583x - 8877$$
$$pgcd(\overline{P_1},\overline{P_2}) = y + \frac{18210335}{633423819}x^8 + \frac{18468580}{1900271457}x^7 - \frac{950235277}{3800542914}x^6$$
$$+ \frac{308795812}{633423819}x^5 + \frac{1326508}{1264319}x^4 - \frac{693872678}{211141273}x^3$$
$$+ \frac{154755823}{542934702}x^2 + \frac{11777980426}{1900271457}x - \frac{5743291441}{1266847638}$$
$$P_3[y = A(x)] = \frac{95877535}{3800542914}x^8 + \frac{68631485}{1266847638}x^7 - \frac{964507987}{3800542914}x^6$$
$$+ \frac{61850473}{633423819}x^5 + \frac{2313311}{1083702}x^4 - \frac{11567256973}{3800542914}x^3$$
$$- \frac{8072110544}{1900271457}x^2 + \frac{70856141555}{3800542914}x - \frac{3419877955}{180978234}$$
$$Res(P_1,P_2,P_3) = (-1)^9 \times (-10364532010628741437778125)/(-65)^5$$
$$= -893270339974413$$
$$= -3^4 \times 79 \times 139595302387$$

REFERENCES

[Ar] J.-M. Arnaudiès, *Thèse à l'Université de Toulouse*, 1989.

[BA] N. Bourbaki, "Algèbre," Ch. 4 à 7, Masson, 1981.

[BAC] N. Bourbaki, "Algèbre Commutative," Ch. 1 à 9, Masson, 1983, 1985.

[Bo] Borchart, "Zur theorie der Elimination und Kettenbruch Entwicklung," Mathematische abhandlungen der Akademie der Wissenchaften zu Berlin, 1878.

[Th] M. Chardin, *Thèse de l'Université Pierre et Marie Curie*, (Paris VI) 1990; Prépublication du Centre de Mathématiques de l'Ecole Polytechnique.

[Co] Collins, *The calculation of multivariate polynomial resultants*, Journal of the ACM **18** (1971), 515–532.

[De] M. Demazure, *Une définition constructive du résultant*, Notes informelles de calcul formel 2, Prépublication de l'Ecole Polytechnique, 1984.

[D5] J. Della Dora, C. Dicrescenzo et D. Duval, *About a new Method for Computing in Algebraic Number Fields*, in "Proc. EUROCAL 85," Springer Lecture Notes in Computer Science Vol. 204, Springer Verlag, 1985, pp. 289-290.

[Fu] W. Fulton, "Algebraic Curves," Benjamin, 1969.

[Grö] W. Gröebner, "Moderne Algebraische Geometrie," Springer-Verlag, Wien und Innsbruck, 1949.

[Ha] W. Habicht, *Zur inhomogenen Eliminationstheorie & Eine Verallgemeinerung des Sturmschen Wurzelzählverfahrens*, Comm. Math. Helvetici **21** (1948), 79–116.

[Had] J. Hadamard, *Mémoire sur l'élimination*, Acta Mathematica **20** (1896).

[Jo1] J.-P. Jouanolou, *Idéaux résultants*, Adv. in Math. **37** (1980), 212-238.

[Jo2] J.-P. Jouanolou, "Le formalisme du résultant," Publ. IRMA 417/P-234, Université de Strasbourg. A paraître dans Progress in Math., Birkhäuser.

[Jo3] J.-P. Jouanolou, *Notes sur le résultant*, (communication personnelle).

[La] D. Lazard, *Résolution des systèmes d'équations algébriques*, Theorical Computer Science **15** (1981), 77–110.

[Lom] H. Lombardi, *Algèbre élémentaire en temps polynomial*, Thèse à l'Université de Nice, 1989.

[Loos] R. Loos, *Generalized Polynomial Remainder Sequences*, in "Symbolic and Algebraic computation," (Computing Supplementum 4), ed. B. Buchberger, G. E. Collins et R. Loos, Springer Verlag, Wien New-York, 1982.

[Mac1] F. S. Macaulay, "The Algebraic Theory of Modular Systems," Cam-
        bridge University Press, 1916 reprinted by Stechert-Hafner Service
        Agency, New-York and London, 1964

[Mac2] F. S. Macaulay, *Some formulæ in elimination*, Proc London Math.
        Soc. **1, 35** (1903), 3–27.

[No]   D.G. Northcott, "Lessons on Rings Modules and Multiplicities,"
        Cambridge University Press, 1968.

[Ph]   P. Philippon, *Critères pour l'indépendance algébrique*, Publ. Math.
        de l'I.H.E.S. **64** (1984).

[Se]   J.-P. Serre, "Algèbre locale et multiplicités," Lectures Notes in
        Math. **11**, 1965.

[vW]   van der Waerden, "Modern Algebra Vol. 2 (deuxième édition),"
        Frederick Ungar Publishing, 1953.

[Weyl] H. Weyl, "The Classical Groups," Second Edition with Suppl., Prin-
        ceton Univ. Press, 1953.

[ZS]   O. Zariski et P. Samuel, "Commutative Algebra," Springer-Verlag,
        1960.

[Za]   O. Zariski, *Generalized weight properties of the resultant of $n + 1$
        polynomials in n indeterminates*, in "Collected papers."

Marc Chardin
Centre de Mathématiques
et Laboratoire d'Informatique
Ecole Polytechnique
F-91128 Palaiseau Cedex
CFMCHA@FRPOLY11.BITNET
Unité de recherche associée au CNRS D.0169

On algorithms for real algebraic plane curves

FELIPE CUCKER   LAUREANO GONZALEZ VEGA
FRANCESC ROSSELLO

Introduction

During the last years several researchers have considered the problem
of finding polynomial–time sequential algorithms for the computation of the
topology of a real algebraic plane curve. Up till now, we can divide these
algorithms into two main groups: those coding real algebraic numbers by
means of isolating intervals (see for instance [3] and [7]) and those coding
them à la Thom (see [9]). The aim of this note is to survey the state of
the art as far as the second group is concerned. In particular we introduce
two new such algorithms, one of which is, to our knowledge, the algorithm
computing the topology of a real algebraic plane curve with the lowest
running time, while the other one has been successfully implemented.

Let $F = \sum a_{i,j} X^i Y^j \in \mathbf{Z}[X,Y]$, and let $\mathcal{C} \subset \mathbf{R}^2$ be the algebraic set
defined by the equation $F(X,Y) = 0$. Let p be the (total) degree of F,
$|F| = \log \sqrt{\sum a_{i,j}^2}$ and $n = \max\{p, |F|\}$.

To our knowledge, the fastest algorithm using isolating intervals, which
computes the topology of $\mathcal{C}$, is the one given in [3], which runs in time
$O(n^{30})$ (unless otherwise stated, all running times are given with respect to
classical arithmetic). On the other side, and again to our knowledge, the
fastest algorithm using isolating intervals which produces an F–invariant
cylindrical algebraic decomposition of $\mathbf{R}^2$ is the one given in [2], which
runs in time $O(n^{21})$ (in [2] it is shown to run in time $O(n^{14} \log n)$ using
fast arithmetic). We want to remark that both algorithms only work for
regular curves.

In [9], M. F. Roy gives the first algorithm for the computation of the
topology of a (possibly singular) curve which uses Thom codes, algorithm

First A. partially supported by DGICyT PB 860062 and the ESPRIT BRA
Program of the EC under contract no. 3075, project ALCOM.
Second A. partially supported by DGICyT PA 86/471 and PB 86/6.
Third A. partially supported by DGICyT PB 870137.

which we shall refer to as TOP_1. In [11] it is proved that it runs in time $O(n^{28})$, and also that an F-invariant c.a.d. of $\mathbf{R}^2$ can be obtained by a similar algorithm, which we shall call CAD_1, in time $O(n^{23})$.

Recently, it has been noted that the ground algorithm used to code real algebraic numbers à *la Thom* has a complexity smaller than the one used for the complexity estimates given in [11]. In the first part of this note we give a proof of this fact and, using it, we obtain new upper bounds for the running times of the auxiliar procedures introduced in [10] and [11], and which we use in our algorithms. Using these new bounds one can show that Roy's algorithm TOP_1 actually runs in time $O(n^{21}(\log n)^4)$ and that CAD_1 runs actually in time $O(n^{17}(\log n)^3)$. Notice that, in both cases, the running times are much lower than those of the corresponding fastest algorithms using isolating intervals.

In the second part of this note we introduce some modifications to Roy's algorithms, obtaining: i) an algorithm CAD_0 which computes an F-invariant c.a.d. of $\mathbf{R}^2$ in time $O(n^{16}(\log n)^3)$; ii) an algorithm TOP_0 which computes the topology of $\mathcal{C}$ in time $O(n^{19}(\log n)^3)$; and iii) another algorithm for the computation of the topology, TOP_2, whose worst–case running time is in $O(n^{25})$, but which has been implemented. From the running times we have been giving in this Introduction we can conclude that, right now, CAD_0 is the fastest algorithm computing an F-invariant c.a.d. of $\mathbf{R}^2$, while TOP_0 is the fastest algorithm computing the topology of $\mathcal{C}$.

We want to remark that, although they all are similar in spirit, there is a great difference between Roy's algorithms and TOP_2 on one side, and CAD_0 and TOP_0 on the other side: while the first ones freely use points of the real spectrum of $\mathbf{R}[X, Y]$, the second ones only use real algebraic numbers to get the same information. (In fact, the subscripts in the names of the different algorithms TOPs and CADs refer to nothing but the maximum dimension of the points from real spectra which they use.) Our feeling is that the use of points of real spectra makes the algorithms clearer and getting more to the point, and thus easier to implement, but also that it makes them more expensive.

1. Ground tools

1.1. Sturm–Habicht sequences and systems of equalities and inequalities.

We begin by introducing some notations and terminology that will be useful for the rest of this note. In the sequel we shall call *sign condition*, or simply *sign*, one of the following three: < 0, > 0 and $= 0$. We shall denote them by $-, +$ and 0 respectively. We shall call *generalized sign condition* any sign condition as well as ≤ 0 and ≥ 0.

ALGORITHMS FOR REAL CURVES

Given two k-tuples of sign conditions

$$\epsilon = (\epsilon_1, \ldots, \epsilon_k) \text{ and } \epsilon' = (\epsilon'_1, \ldots, \epsilon'_k),$$

we shall say that ϵ is *compatible* with ϵ' when

$$\epsilon_i \neq \epsilon'_i \implies \epsilon_i = 0, \qquad i = 1, \ldots, k.$$

Given a polynomial $F = \sum a_{ij} X^i Y^j \in \mathbf{Z}[X, Y]$ we recall that its *norm* is defined as $\sqrt{\sum a_{ij}^2}$, and that its *size* $|F|$ is defined as the logarithm of its norm.

Now, let A be an ordered domain, $K(A)$ its field of fractions and $\overline{K(A)}$ the real closure of $K(A)$. Given polynomials $P, Q_1, \ldots, Q_k \in A[X]$ and a k-tuple of sign conditions $\epsilon = (\epsilon_1, \ldots, \epsilon_k)$, let

$$c_\epsilon(P; Q_1, \ldots, Q_k) = \text{card}\{x \in \overline{K(A)} \mid P(x) = 0, \ Q_i(x)\epsilon_i \ \ 1 \leq i \leq k\}$$

We shall say that ϵ is *satisfied* (by $Q_1, \ldots, Q_k$ on the roots of P) when $c_\epsilon(P, Q_1, \ldots, Q_k) \neq 0$.

In [8] a new tool for computing the c_ϵ's is introduced: the Sturm–Habicht sequence of two polynomials. This sequence is based on the general theory of subresultants and has the following main features:

— it yields the same information as the generalized Sturm sequence
— it can be computed in polynomial time
— it has good specialization properties.

We don't pretend to introduce here this tool; the interested reader can get acquainted with it by looking at [8]. We just recall that if $P, Q \in A[Y]$ (A not necessarily ordered) are polynomials of degree p and q respectively, then the Sturm–Habicht sequence associated to P and Q is a sequence of polynomials $F_{p+q} = P, F_{p+q-1} = P'Q, F_{p+q-2}, \ldots, F_0$ in $A[Y]$. Each polynomial F_j in this sequence has degree $d_j \leq j$. One calls the j-*th* *Sturm–Habicht principal coefficient* of P and Q the coefficient of degree j of F_j, $j = 0, \ldots, p + q$. In the case $Q = 1$, we shall talk about the Sturm–Habicht sequence of P and the Sturm–Habicht principal coefficients of P.

If $A = \mathbf{Z}[X]$, then F_0 agrees with the resultant w.r.t. Y of P and $P'Q$.

If A is ordered, and given $a, b \in A \cup \{-\infty\} \cup \{\infty\}$ with $a < b$, we denote by $Sth^A_{a,b}(P, Q)$, or simply by $Sth_{a,b}(P, Q)$, the difference of sign variations taken by the Sturm–Habicht sequence associated to P and Q at a and b. We shall denote $Sth_{-\infty,\infty}(P, Q)$ just by $Sth(P, Q)$.

Theorem 1.1 (See [8] and [11]). *Let $P, Q \in A[Y]$ be arbitrary polynomials of degrees p and q respectively (A not necessarily ordered in (i) and (ii)).*

   i) *In the Sturm–Habicht sequence associated to P and Q there appear at most $O(p)$ non-zero polynomials.*

   ii) *The Sturm–Habicht sequence of P and Q can be computed in $O(p(p + q))$ arithmetic operations over A. Moreover, if $A = \mathbf{Z}[X]$ and $\bar{p}$ and $\bar{q}$ denote the degrees of P and Q with respect to the variable*

X, *then the sizes of the coefficients ($\in \mathbf{Z}[X]$) appearing in the Sturm–*
Habicht sequence of P and Q are bounded by $O((p+q)(\log(\bar{p}+\bar{q}) +$
$|P|) + p|Q|)$, *and their degrees are at most $O(\bar{p}(p+q) + \bar{q}p)$.*

iii) *Let $a, b \in A$ with $a < b$. Assume that neither a nor b are roots of*
  P. Given a sign condition ϵ, set $c_\epsilon(P;Q;a,b) = \text{card}\{x \in \overline{K(A)} \mid$
  $P(x) = 0$, $Q(x)\epsilon$, $a < x < b\}$. Then $Sth_{a,b}(P,Q) = c_+(P;Q;a,b) -$
  $c_-(P;Q;a,b)$. In particular $Sth(P,Q) = c_+(P;Q) - c_-(P;Q)$.

iv) *$Sth(P,Q)$ can be obtained by looking at the signs of the Sturm–Habicht*
  principal coefficients of P and Q.

Let A be now a computable ordered domain (for instance, $\mathbf{Z}$). The
last theorem allows us to compute the values of $c_0(P;Q)$, $c_+(P;Q)$ and
$c_-(P;Q)$ for $P, Q \in A[X]$ by just solving the following matrix equation

$$\begin{pmatrix} 1 & 1 & 1 \\ 0 & 1 & -1 \\ 0 & 1 & 1 \end{pmatrix} \begin{pmatrix} c_0(P;Q) \\ c_+(P;Q) \\ c_-(P;Q) \end{pmatrix} = \begin{pmatrix} Sth(P,1) \\ Sth(P,Q) \\ Sth(P,Q^2) \end{pmatrix}.$$

The SI procedure, which we now describe, allows us to get the c_ϵ's
when several polynomials Q's are involved. For details about it see [10].

Procedure SIadd:
  The *input* of this procedure is the following set of data
— polynomials $P, Q_1, \ldots, Q_k \in A[X]$
— the list of k–tuples of sign conditions $\epsilon_1, \ldots, \epsilon_{r(k)}$ satisfied by the polynomials $Q_1, \ldots, Q_k$ on the roots of P
— the vector c_k of non–zero values

$$c_{\epsilon_1}(P;Q_1,\ldots,Q_k), \ldots, c_{\epsilon_{r(k)}}(P;Q_1,\ldots,Q_k)$$

— polynomials $R_1, \ldots, R_{r(k)}$ of the form $Q_1^{\nu_1} \cdots Q_k^{\nu_k}$, $0 \le \nu_1, \ldots, \nu_k \le 2$
— a vector v_k whose elements are the values $Sth(P, R_j)$, $j = 1, \ldots, r(k)$
— an invertible $r(k) \times r(k)$ matrix $A(k)$ satisfying the equation $A(k) \cdot c_k = v_k$
— a new polynomial $Q_{k+1} \in A[X]$.
  Its *output* is
— the list of polynomials $P, Q_1, \ldots, Q_{k+1}$
— the list of satisfied $(k+1)$–tuples of sign conditions $\epsilon_1, \ldots, \epsilon_{r(k+1)}$
— the vector c_{k+1} of non–zero values

$$c_{\epsilon_1}(P;Q_1,\ldots,Q_{k+1}), \ldots, c_{\epsilon_{r(k+1)}}(P;Q_1,\ldots,Q_{k+1})$$

— polynomials $R_1, \ldots, R_{r(k+1)}$ of the form

$$Q_1^{\nu_1} \cdots Q_{k+1}^{\nu_{k+1}}, \ 0 \le \nu_1, \ldots, \nu_{k+1} \le 2$$

— a vector v_{k+1} whose elements are the values

$$Sth(P, R_j), \quad j = 1, \ldots, r(k+1)$$

— an invertible $r(k+1) \times r(k+1)$ matrix $A(k+1)$ satisfying the equation

$$A(k+1) \cdot c_{k+1} = v_{k+1}$$

The procedure performs the following steps:

1) Let $H_j = R_j$, $H_{r(k)+j} = Q_{k+1}R_j$ and $H_{2r(k)+j} = Q_{k+1}^2 R_j$, for $j = 1, \ldots, r(k)$. We compute the $3r(k)$–dimensional vector v whose l^{th} component is the number $Sth(P, R_l)$.

2) We now define a list of $3r(k)$ $(k+1)$-tuples of sign conditions $\{\epsilon_i'\}$ whose first $r(k)$ elements are the input k–tuples followed by 0, the second $r(k)$ elements are the same k–tuples followed by $+$, and for the last $r(k)$ elements we add a $-$.

3) Let

$$A = \begin{pmatrix} A(k) & A(k) & A(k) \\ 0 & A(k) & -A(k) \\ 0 & A(k) & A(k) \end{pmatrix}.$$

We compute the $3r(k)$–dimensional vector c satisfying the equation $A \cdot c = v$. Notice that the jth component of c is equal to $c_{\epsilon_j'}(P; Q_1, \ldots, Q_{k+1})$.

4) We get a new vector c_{k+1} by deleting the zero components of c, a new matrix $A'(k+1)$ by deleting in A the columns corresponding to these components and a new list of $r(k+1)$ $(k+1)$-tuples of sign conditions by deleting the non satisfied ones.

5) We get an invertible square submatrix $A(k+1)$ of $A'(k+1)$ by conserving the first $r(k+1)$ linearly independent rows, a new vector v_{k+1} by conserving in v the components corresponding to these rows, and a new list of polynomials $R_1, \ldots, R_{r(k+1)}$ by conserving the polynomials H's corresponding to these components.

Procedure SI.

The *input* of this procedure consists of a sequence of polynomials $P, Q_1, \ldots, Q_k$ with coefficients in a computable ordered domain.

Its *output* is formed of

— the list of satisfied k–tuples of sign conditions $\epsilon_1, \ldots, \epsilon_{r(k)}$

— the vector c_k of non–zero values

$$c_{\epsilon_1}(P; Q_1, \ldots, Q_k), \ldots, c_{\epsilon_{r(k)}}(P; Q_1, \ldots, Q_k)$$

— polynomials $R_1, \ldots, R_{r(k)}$ of the form $Q_1^{\nu_1} \cdots Q_k^{\nu_k}$, $0 \leq \nu_1, \ldots, \nu_k \leq 2$

— a vector v_k whose elements are the values $Sth(P, R_j)$, $j = 1, \ldots, r(k)$

— an invertible $r(k) \times r(k)$ matrix $A(k)$ satisfying the equation

$$A(k) \cdot c_k = v_k$$

The procedure performs k times the SIadd procedure.

Theorem 1.2. *Let r be the number of roots of P in $\overline{K(A)}$. The polynomials R_i $(i = 1, \ldots, r(k))$ in the output of $SI(P; Q_1, \ldots, Q_k)$ are products of at most $\log r$ polynomials Q_j or Q_j^2 $(j = 1, \ldots, k)$.*

Since our proof of this result is, although quite elementary, too involved, in order to avoid losing the thread of the exposition we postpone it until an Appendix at the end of the first part of this note.

Proposition 1.3. *Let $A = \mathbf{Z}$, and let r be the number of real roots of P, n the maximum of the degrees of the polynomials $P, Q_1, \ldots, Q_k$, and S the maximum of their sizes. The procedure SI runs in time $O(krn^4(\log r)^3 S^2)$.*

Proof: We perform k times the procedure SIadd. In the first step of each performance, we compute at most $2r$ Sturm–Habicht sequences associated to P and polynomials R which are products of at most $O(\log r)$ polynomials Q_i $(i = 1, \ldots, k)$. So, the degree of such a polynomial R is at most $O(n \log r)$ and its size is at most $O(S \log r)$. It follows that the sizes of the coefficients of the corresponding Sturm–Habicht sequences are bounded by $O(nS \log r)$. Thus, each one of these sequences costs $O(n^4(\log r)^3 S^2)$. Since the cost of the remaining steps of the procedure SIadd is negligible, we get the stated total time by multiplying the latter cost by kr.   Q. E. D.

For the sake of simplicity, in the sequel we shall bound the number of real roots of a polynomial $P \in \mathbf{Z}[X]$ by its degree. Notice however that, in average, the number of real roots is much smaller than the degree (see [6]).

1.2. Real algebraic numbers

In [5] Coste and Roy introduce a way of coding real algebraic numbers which relies upon the preceding algorithms, as well as on the following result:

Thom's lemma. *([5]) Let $P \in A[X]$ be a polynomial of degree n and $\epsilon = (\epsilon_0, \ldots, \epsilon_n)$ an $(n+1)$-tuple of generalized sign conditions. Let*

$$A(\epsilon) = \{x \in \overline{K(A)} \mid P^{(i)}(x)\epsilon_i, \ i = 0, \ldots, n\}.$$

Then: i) $A(\epsilon)$ is either empty or semi-algebraically connected; ii) if $A(\epsilon)$ is not empty, then its closure is $A(\underline{\epsilon})$ where $\underline{\epsilon}$ is obtained from ϵ by relaxing the sign conditions.

Setting $\epsilon_0 = 0$ we get a way of individualizing the roots of P:

Proposition 1.4. (Cf. Prop. 2.1 in [5].)
    Let $P \in A[X]$ be a polynomial of degree $n \geq 1$.
    i) Let $\epsilon_1, \ldots, \epsilon_n$ be sign conditions. There is at most one element $x \in \overline{K(A)}$ such that $P(x) = 0$ and $P^{(k)}(x)\epsilon_k$ for every $k = 1, \ldots, n$.

ii) Let ξ_1 and ξ_2 be two elements of $\overline{K(A)}$, and set $\epsilon_j = \text{sign } P^{(j)}(\xi_1)$, $\delta_j = \text{sign } P^{(j)}(\xi_2)$ $(j = 0, \ldots, n)$. If the $(n+1)$-tuples of sign conditions $(\epsilon_0, \ldots, \epsilon_n)$ and $(\delta_0, \ldots, \delta_n)$ are different, then we can deduce from them the relative position of ξ_1 and ξ_2 in the following way:

Let j be the largest index such that $\epsilon_j \neq \delta_j$. It is clear that $j \leq n-1$ and that $\epsilon_{j+1} = \delta_{j+1}$ is different from 0.

a) If $\epsilon_{j+1} = +$, then $\xi_1 > \xi_2$ iff $P^{(j)}(\xi_1)$ is greater than $P^{(j)}(\xi_2)$, and this can be checked on ϵ_j and δ_j.

b) If $\epsilon_{j+1} = -$, then $\xi_1 > \xi_2$ iff $P^{(j)}(\xi_1)$ is smaller than $P^{(j)}(\xi_2)$, and this can be checked on ϵ_j and δ_j.

Remark: This Proposition allows us to code each root of P by the n-tuple of signs taken on it by the derivatives of P. This n-tuple of signs is called the *Thom code* of the corresponding root. Notice that the rules given in point (ii) of the last Proposition allow us to deduce the relative position of all these roots from their codes.

We want also to remark that we can always sort a root ξ of P and another real number ζ by comparing the code $(\epsilon_1, \ldots, \epsilon_n)$ of ξ (as a root of P) and the $n+1$-tuple of signs taken by P and its derivatives on ζ (if $\zeta \neq \xi$, then this $(n+1)$-tuple of signs is different from $(0, \epsilon_1, \ldots, \epsilon_n)$). This observation allows one to simplify the algorithm for the comparison of two real algebraic numbers given in [10]. In particular, we can always sort a root ξ of P and a root ζ of a derivative of P, let's say $P^{(k)}$, by means of their respective codes, since ξ's code includes the signs taken by the derivatives of $P^{(k)}$ on ξ.

Procedure RAN:

Its *input* is a polynomial P with integer coefficients.

Its *output* is the list of Thom codes of its roots, together with a matricial equation like in SI. The codes are sorted in increasing order of the corresponding roots.

If n is the degree of P, then the algorithm performs the procedure $SI(P; P^{(n-1)}/(n-1)!, \ldots, P')$.

Procedure RANI:

Its *input* is the output of $RAN(P)$ and a new polynomial Q.

Its *output* is the sign Q takes on the roots of P.

The algorithm just applies SIadd to the output of $RAN(P)$ and Q. We shall denote RANI(RAN(P),Q) simply by RANI(P, Q)

Proposition 1.5. *Let d and N be the degree and the size of P, respectively.*

i) $RAN(P)$ runs in time $O\big(d^6(\log d)^3(d+N)^2\big)$.

ii) If now Q is a polynomial of degree q and size M, the computation RANI(P,Q) takes time

$$O\big(d^2(q + d\log d)(d(d+N)\log d + dM + qN)^2\big).$$

Proof:

i) Recall that the size of $P^{(i)}/i!$ is bounded by $i + N$. We then just apply Proposition 1.3 with $k = n = d$ and $S = d + N$.

ii) As in Proposition 1.3, the dominant step is the first one, in which we compute at most $2d$ Sturm–Habicht sequences of P and products RQ, where R is a product of at most $O(\log d)$ derivatives of P. The degree of each RQ is bounded by $O(q + d \log d)$ and its size by $O((d + N) \log d + S)$.

Each Sturm–Habicht sequence then takes time $O\big(d(q + d \log d)(d(d + N) \log d + dS + qN)^2\big)$, so we get the stated total computation time multiplying it by d.   Q. E. D.

1.3. Specialized computations.

As far as we are concerned in this note, three main procedures are needed for specialized computations. The first two ones count and code respectively the roots $\{\eta_j\}$ of a bivariate integer polynomial F one of whose indeterminates is specialized at a root ξ of a given integer polynomial D, whilst the third one determines the signs taken by an integer polynomial G on the points $\{(\xi, \eta_j)\}$.

Let $F, G \in \mathbf{Z}[X, Y]$ and $D \in \mathbf{Z}[X]$, and let $\xi_1, \ldots, \xi_r$ denote the real roots of D, that we shall suppose coded by RAN.

Procedure St_1:

Its *input* is the pair (T, S) where
— T is the output of RAN(D),
— S is the Sturm–Habicht sequence of F as a polynomial in Y.

Its *output* is the sequence $n_1, \ldots, n_r$, where n_j is the number of real roots of $F(\xi_j, Y)$.

The procedure just performs RANI(D, c), for every Sturm–Habicht principal coefficient c of $F(\xi_j, Y)$; it obtains every $n_j = Sth(F(\xi_j, Y), 1)$ from these signs (see Theorem 1.1). We shall denote by $St_1(D, F)$ the invocation $St_1(T, S)$.

Procedure RAN_1:

Its *input* is the pair $((D, \xi_j), F)$ where (D, ξ_j) denotes the output of RAN applied to D as well as the choice of a particular root ξ_j of D.

Its *output* is the list of Thom codes of the roots of $F(\xi_j, Y)$, as well as a data structure similar to the one produced by RAN, but with bivariated polynomials instead of univariated ones.

The procedure is based on the same principle as RAN. The only difference is that, in the present case, the Sturm–Habicht queries, which form the vector appearing as the right member of the matricial equality, must be computed in the same way as the values $Sth(F(\xi_j, Y), 1)$ in St_1, i. e. using RANI.

Procedure $RANI_1$:

Its *input* is the triple $((D, \xi_j), T, G)$ where
— (D, ξ_j) is as in RAN$_1$,
— T is the output of RAN$_1((D, \xi_j), F)$.
Its *output* is the list of signs taken by $G(\xi_j, Y)$ on the roots of $F(\xi_j, Y)$. The procedure is based on the same principle as RANI, but with the same kind of modifications as in RAN$_1$. We shall denote by RANI$_1((D, \xi_j), F; G)$ the invocation RANI$_1((D, \xi_j), T, G)$.

Remark: Notice that, while St$_1$ computes its output for all the roots of D simultaneously, this is not the case with RAN$_1$ and RANI$_1$, where a single root must be distinguished. The reason is that, while each specialized Sturm–Habicht query of the form $Sth(F, R)$ reduces to several applications of RANI for the principal coefficients of the sequence and this procedure is executed simultaneously for all the roots of D, the forking process that generates the products of derivatives to be used in these queries varies with each root.

For further details about these procedures, the reader can look up [11]. As far as the running time of them goes, we have the following result:

Proposition 1.6. *Let d and N be the degree and the size of $D \in \mathbb{Z}[X]$ respectively. Likewise, let p and M be the total degree and the size of $F \in \mathbb{Z}[X, Y]$ respectively. We have*
   i) $St_1(D, F)$ runs in time $O\big(pd^2(p^2 + d \log d)\big(d(d + N) \log d + dp(M + \log p) + p^2 N)^2\big)$.
   ii) Let ξ be a root of D. Then $RAN_1((D, \xi), F)$ runs in time

$$O\big(p^{10}(\log p)^5(p + M)^2 +$$
$$p^3 d^2(p^2 \log p + d \log d)\big(d(d + N) \log d + dp(p + M) \log p + p^2 N \log p)^2\big)$$

   iii) Let $G \in \mathbb{Z}[X, Y]$, and let q and S denote its total degree and size, respectively. Set $L = p \log p + q$. Then $RANI_1((D, \xi), F; G)$ runs in time

$$O\big(p^4 L^3\big(p(S + p \log p) + L(M + \log L)\big)^2$$
$$+ p^2 d^2(d \log d + pL)\big(d(d + N) \log d + L(pN + d(M + \log L))$$
$$+ dp(p \log p + S))^2\big).$$

Proof: i) We must call RANI(D, c), for every principal Sturm–Habicht coefficient c in the input sequence. Each such a c has degree $O(p^2)$ and size $O(p(M + \log p))$. Applying Proposition 1.3 (ii) we see that each run of such a RANI takes time

$$O\big(d^2(p^2 + d \log d)(d(d + N) \log d + dp(M + \log p) + p^2 N)^2\big)$$

and therefore we get the total time by multiplying this bound by p (since $O(p)$ is an upper bound for the number of c's).
   ii) We perform $O(p)$ iterative steps. Each one of them computes at most $O(p)$ Sturm–Habicht sequences of polynomials of degrees $O(p)$ and

$O(p \log p)$, carrying out in this way $O(p^3 \log p)$ arithmetical operations with univariated polynomials.

These univariated polynomials have their sizes bounded by $O(p(p + M) \log p)$ and their degrees bounded by $O(p^2 \log p)$. Thus, the computation of these sequences takes

$$O\left(p^9 (\log p)^5 (p + M)^2\right) \tag{1}$$

bit operations.

Once computed the Sturm–Habicht sequences, the RANI procedure must be called at most $O(p)$ times per queried Sth value. Each RANI has as input $RAN(D)$ and a polynomial of degree $O(p^2 \log p)$ and size $O(p(p + M) \log p)$. Thus, each such a RANI costs

$$O\left(d^2(p^2 \log p + d \log d)(d(d + N) \log d + dp(p + M) \log p + p^2 N \log p)^2\right).$$

It follows that, once the corresponding Sturm–Habicht sequences are known, the vector of queries in each iterative step is computed in time

$$O\left(p^2 d^2(p^2 \log p + d \log d)(d(d + N) \log d + dp(p + M) \log p + p^2 N \log p)^2\right). \tag{2}$$

Summing up (1) and (2) and multiplying the result by p we get the stated total time.

The proof of point (iii) is similar.   Q. E. D.

1.4. Computations with infinitesimals

In this paper we shall also need procedures similar to those studied in the previous paragraph, but for points of real spectra instead of real algebraic points. We devote this paragraph to introduce the facts on real spectra that we shall use in the sequel, as well as the aforementioned algorithms.

Let us recall that, given a commutative ring R, its real spectrum $\mathrm{Spec}_r R$ parametrizes all couples $(\wp, <)$, where $\wp$ is a prime ideal of R and $<$ is an ordering on the residue field $k(\wp)$. We shall say that such a $(\wp, <)$ is d–dimensional when $\dim k(\wp) = d$. Given $p \in R$ and $\alpha = (\wp, <) \in \mathrm{Spec}_r R$, the sign taken by p on α, which we shall denote by $\mathrm{sign}\, p(\alpha)$, is the sign, w.r.t. the ordering $<$, of the element in $k(\wp)$ corresponding to p.

We have a natural imbedding $i : \mathbf{R}^n \hookrightarrow \mathrm{Spec}_r \mathbf{R}[X_1, \ldots, X_n]$: for every point $(\xi_1, \ldots, \xi_n) \in \mathbf{R}^n$, $i((\xi_1, \ldots, \xi_n))$ is the 0–dimensional point $(\wp, <)$ with $\wp = (X_1 - \xi_1, \ldots, X_n - \xi_n)$ and $<$ the usual ordering on $\mathbf{R}$. We shall identify a point of $\mathbf{R}^n$ and its image by i.

We can associate to every real number $\xi \in \mathbf{R}$ the following two 1–dimensional points in $\mathrm{Spec}_r \mathbf{R}[X]$, which we shall call ξ^+ and ξ^-: in both cases we take $\wp = (0)$, and in ξ^+ we take the ordering $<$ in $k(\wp) = \mathbf{R}(X)$ given by $\xi < X < \xi + \varepsilon$ for all real positive numbers ε, whilst in ξ^- we take the ordering given by $\xi - \varepsilon < X < \xi$ for all real positive numbers ε. We shall denote by $\mathbf{R}(\xi^+)$ and $\mathbf{R}(\xi^-)$, respectively, the corresponding ordered field $(\mathbf{R}(X), <)$.

Given a polynomial $Q \in \mathbf{Z}[X]$, it is not difficult to see that $\text{sign}\,Q(\xi^+)$ (resp. $\text{sign}\,Q(\xi^-)$), agrees with the sign Q takes on $]\xi, \xi+\varepsilon[$ (resp. $]\xi-\varepsilon, \xi[$), with $\varepsilon > 0$ such that the sign of Q does not change on this interval. In particular, $\text{sign}Q(\xi)$ is compatible with $\text{sign}Q(\xi^+)$ and $\text{sign}Q(\xi^-)$.

Let now β denote a real half–branch of an algebraic curve $\mathcal{C}$ through (x_0, y_0) and let us consider the Puiseux expansion of this half–branch

$$\begin{cases} X = x_0 + at^q \\ Y = y_0 + bt^p(1 + c_1 t^{q_1} + c_2 t^{q_2} + \ldots) \\ t < 0 \quad \text{or} \quad t > 0 \end{cases}$$

If $a \neq 0$ (i.e., if β does not lie on the line $X = x_0$), we shall say that β lies over x_0^- (resp. over x_0^+) when it lies on the left (resp. on the right) of $X = x_0$.

We associate to β the 1–dimensional point $\wp$ in $\text{Spec}_r\mathbf{R}[X, Y]$ given by the prime ideal corresponding to the (unique) irreducible curve containing this half–branch, and the ordering $<$ on $k(\wp) \subset \mathbf{R}((t))$ is determined by the sign of t. If $a \neq 0$ then we shall call this point β^- or β^+, depending on β lies on x_0^- or x_0^+, respectively. We can think of β^- or β^+ as lying on β and "infinitely close" to (x_0, y_0).

Given $G \in \mathbf{Z}[X, Y]$, one easily proves that $\text{sign}G(\beta^-)$ (resp. $\text{sign}G(\beta^+)$) agrees with $\text{sign}G(x, y)$, for (x, y) belonging to the graph of β and $x \in]x_0 - \varepsilon, x_0[$ (resp. $]x_0, x_0 + \varepsilon[$), $\varepsilon > 0$ and small enough to guarantee that this sign is independent of x. In particular, $\text{sign}\,G(x_0, y_0)$ is compatible with $\text{sign}\,G(\beta^-)$ and $\text{sign}\,G(\beta^+)$.

Remark: Given a polynomial $F \in \mathbf{R}[X, Y]$, a real point ξ, and $\epsilon \in \{+, -\}$, there is a bijection between the half–branches of the curve corresponding to F lying over ξ^ϵ and the roots of $F(\xi, Y) = 0$ in the real closure of $\mathbf{R}(\xi^\epsilon)$. We want to attract the reader's attention to the fact that Theorem 1.1, Procedure SI, Proposition 1.2 and Proposition 1.4 are valid in such a $\mathbf{R}(\xi^\epsilon)$. This fact will be strongly used, with no further mention, in the procedures we explain below, as far as in the algorithm TOP_1.   Q. E. D.

To simplify the notations, we shall refer to points of the form ξ^+, ξ^-, β^+ and β^- generically as *infinitesimals*.

No more information on real spectra is needed for our purposes. We advise the reader interested in further details on the general theory to look up [4].

Now, let $D, Q \in \mathbf{Z}[X]$, $F, G \in \mathbf{Z}[X, Y]$, and let $\mathcal{C}$ be the algebraic curve defined by the equation $F(X, Y) = 0$. Let $\xi_1 < \ldots < \xi_r$ be the real roots of D, which we suppose coded by RAN.

Procedure RANI_{inf}:

Its *input* is the output of $\text{RAN}(D)$ and the (non–zero) polynomial Q. Its *output* is the list of signs Q takes at ξ_i^+ and ξ_i^-, $1 \leq i \leq r$.

The algorithm calls $\text{RANI}(P, Q^{(k)}/k!)$ in increasing order of k, beginning with $k = 0$ and until for all ξ_i we have found a k_i such that $Q^{(k_i)}(\xi_i) \neq 0$. Then it applies the following rule:

If $Q^{(k)}$ $(0 \leq k \leq \deg Q)$ is the first derivative of Q which does not vanish at ξ_i, then $\text{sign } Q(\xi_i^+) = \text{sign } Q^{(k)}(\xi_i)$ and $\text{sign } Q(\xi_i^-) = (-1)^k \text{sign } Q^{(k)}(\xi_i)$.

We shall denote the call RANI_{inf} $(\text{RAN}(D),Q)$ simply by RANI_{inf} (D,Q).

Procedure St_2:

Its *input* is the same as in St_1.

Its *output* is the list of numbers $\{m_i^\epsilon\}$, $1 \leq i \leq r$ and $\epsilon \in \{+, -\}$, where m_i^ϵ is the number of real half–branches of C over ξ^ϵ.

Since every m_i^ϵ agrees with $Sth^{\mathbb{R}(\xi_i^\epsilon)}(F(\xi_i, Y), 1)$, the procedure is similar to St_1, using RANI_{inf} instead of RANI.

Procedure RAN_2:

Its *input* is the same as in RAN_1.

Its *output* is the list of n–tuples of sign conditions satisfied by the derivatives of F on the infinitesimals corresponding to half–branches of C over ξ_j^+ and ξ_j^-, together with a data structure similar to the one produced by RAN_1.

The procedure is similar to RAN_1, using RANI_{inf} instead of RANI.

Procedure $RANI_2$:

Its *input* is the pair $((D, \xi_j), F)$ as in RAN_1, the output of RAN_2, and the polynomial G.

Its *output* is the list of signs G takes on the infinitesimals corresponding to half–branches of C over ξ_j^+ and ξ_j^-.

The algorithm is similar to $RANI_1$, using RANI_{inf} instead of RANI.

For more details on all these algorithms, the reader can look up again [11].

Proposition 1.7. *Let d and N be the degree and the size of $D \in \mathbb{Z}[X]$ respectively. Likewise, let p and M be the total degree and the size of $F \in \mathbb{Z}[X, Y]$ respectively.*

i) If q and S denote the degree and the size of Q, respectively, then $RANI_{\text{inf}}$ (D, P) runs in time

$$O\big(qd^2(q + d \log d)((d \log d + q)(d + N) + dS)^2\big).$$

ii) $St_2(D, F)$ runs in time $O\big(p^3 d^2(p^2 + d \log d)(d(d+N) \log d + dp(M + p) + p^2 N)^2\big)$.

iii) Let ξ be a root of D. Then $RAN_2((D, \xi), F)$ runs in time

$$O\big(p^{10}(\log p)^5(p + M)^2 +$$

$$p^5 d^2 \log p(p^2 \log p + d \log d)\big(d(d+N) \log d + dp(p+M) \log p + p^2 N \log p\big)^2\big)$$

iv) Let $G \in \mathbf{Z}[X,Y]$, and let q and S denote its total degree and size, respectively. Set $L = p\log p + q$. Then $RANI_2((D,\xi), F; G)$ runs in time

$$O\big(p^4 L^3 \big(p(S + p\log p) + L(M + \log L)\big)^2$$

$$+ p^3 d^2 L(d\log d + pL)\big(d(d+N)\log d + L(p(d+N) + d(M+\log L)) + dpS\big)^2\big).$$

Proof: To prove (i), just note that we apply at most $O(q)$ times $RANI(D,)$ to polynomials with their degrees bounded by $O(q)$ and their sizes bounded by $O(q + S)$.

The proofs of (ii), (iii) and (iv) are similar to those given in Proposition 1.6, replacing RANI by $RANI_{inf}$. Q. E. D.

1.5. Appendix

We devote this appendix to prove the following result:

Theorem 1.2. *Let $P \in A[X]$ and let r be the number of roots of P in $\overline{K(A)}$. The polynomials R_i ($i = 1, \ldots, r(k)$) which are the output of $SI(P; Q_1, \ldots, Q_k)$ are products of at most $\log r$ polynomials Q_j or Q_j^2 ($j = 1, \ldots, k$).*

To begin with, let us recall that for every $n \in \mathbf{N}$, $A(n)$ is the matrix returned by SI after n calls to SIadd, and $A'(n)$ the resulting after step (4) in the nth call of SIadd. Let

$$M(1) = \begin{pmatrix} 1 & 1 & 1 \\ 0 & 1 & -1 \\ 0 & 1 & 1 \end{pmatrix}$$

and

$$M(n) = \begin{pmatrix} M(n-1) & M(n-1) & M(n-1) \\ 0 & M(n-1) & -M(n-1) \\ 0 & M(n-1) & M(n-1) \end{pmatrix}$$

if $n > 1$, and let $M'(n)$ be the submatrix of $M(n)$ obtained by deleting those columns that do not appear in $A'(n)$. Notice that $A(n) \subset A'(n) \subset M'(n) \subset M(n)$.

Given $1 \leq i \leq 3^n$ we denote by r_i^n the ith row of $M'(n)$, and by H_i the product $Q_n^{\nu_n} \cdots Q_1^{\nu_1}$ where $\nu_n \ldots \nu_1$ is the expression of $i - 1$ in base 3.

The polynomials $\{R_i\}$ returned in SIadd belong to $\{H_i\}_{i=1,\ldots,3^n}$. Moreover, if r_i^n appears in $A'(n)$, then its corresponding polynomial in the vector of Sturm–Habicht queries is precisely H_i.

Given $H_i = Q_n^{\nu_n} \cdots Q_1^{\nu_1}$ and $H_j = Q_n^{\delta_n} \cdots Q_1^{\delta_1}$, we define $H_i * H_j = Q_n^{\nu_n * \delta_n} \cdots Q_1^{\nu_1 * \delta_1}$ where

$$\nu * \delta = \begin{cases} 0 & \text{if } \nu = \delta = 0 \\ 1 & \text{if } \nu + \delta \text{ is odd} \\ 2 & \text{otherwise.} \end{cases}$$

In particular, if $\nu + \delta \leq 2$ then $\nu * \delta = \nu + \delta$.

Lemma A.. *For* $1 \le i, j, k \le 3^n$, $H_k = H_i * H_j$ *if and only if* $r_k^n = r_i^n r_j^n$, *where this last product denotes the common one, performed component by component.*

Proof: Easy, by induction on n.   Q. E. D.

Remark: As a particular case, we have that if $H_k = H_i \cdot H_j$ then $r_k^n = r_i^n r_j^n$.

We shall say that a row in a matrix is *superfluous* when it depends linearly on the previous ones.

Lemma B. *Let* $1 \le i, j \le 3^n$. *Assume that the row* r_i^n *is superfluous in* $A'(n)$, *and that* H_i *divides* H_j. *Then* r_j^n *is also superfluous in* $M'(n)$.

Proof: By hypothesis we have an equality

$$r_i^n = \sum_{s < i} \lambda_s r_s^n$$

Let H_u be such that $H_j = H_i H_u$. By the preceding Lemma we have that

$$r_j^n = r_i^n r_u^n = \sum_{s < i} \lambda_s (r_s^n r_u^n)$$

Now, we only have to prove that the polynomials $H_s * H_u$ are previous to H_j.

So, let $H_s = Q_n^{\nu_n} \cdots Q_1^{\nu_1}$, $H_u = Q_n^{\delta_n} \cdots Q_1^{\delta_1}$ and $H_i = Q_n^{\gamma_n} \cdots Q_1^{\gamma_1}$. From $H_j = H_i H_u$ we deduce that $H_j = Q_n^{\gamma_n + \delta_n} \cdots Q_1^{\gamma_1 + \delta_1}$. On the other side, for every $s < i$ we have $H_s * H_u = Q_n^{\nu_n * \delta_n} \cdots Q_1^{\nu_1 * \delta_1}$.

Since $s < i$, there is an $m < n$ such that $\nu_n = \gamma_n, \ldots, \nu_{m+1} = \gamma_{m+1}$ and $\nu_m < \gamma_m$.

Then, on the one hand, for every $m + 1 \le l \le n$ it turns out that $\nu_l * \delta_l = \gamma_l * \delta_l = \gamma_l + \delta_l$ and, on the other hand, since $\nu_m < \gamma_m$, only the following two cases are possible:

$$\nu_m = 0 \Rightarrow \nu_m * \delta_m = \delta_m < \gamma_m + \delta_m$$
$$\nu_m = 1 \Rightarrow \gamma_m = 2 \text{ and } \delta_m = 0 \Rightarrow \nu_m * \delta_m = 1 < 2 = \gamma_m + \delta_m$$

which proves that $H_s * H_u$ is previous to H_j.   Q. E. D.

Lemma C. *For* $1 \le i \le 3^n$, *if* $r_i^n \notin A'(n)$ *then* r_i^n *is superfluous in* $M'(n)$.

Proof: Since $A'(1) = M'(1)$, in the case $n = 1$ there is nothing to prove.

Let us suppose then that $n > 1$. In that case, $i = s + \nu 3^{n-1}$ with $s \le 3^{n-1}$, and $r_s^{n-1} \notin A(n-1)$.

If furthermore $r_s^{n-1} \notin A'(n-1)$, then by induction hypothesis it is superfluous in $M'(n-1)$. From this fact we clearly deduce that r_s^n is superfluous in $M'(n)$. Let

$$\sum_{l \leq s} \lambda_l r_l^n = 0$$

be the equation expressing that dependence. We then have that

$$\sum_{\substack{l \leq s \\ \delta \leq \nu}} \lambda_l r_{l+\delta 3^{n-1}}^n = 0$$

and so, r_i^n is superfluous in $M'(n)$.

On the other side, if $r_s^{n-1} \in A'(n-1)$ then the row r_s^{n-1} is not conserved in the $(n-1)^{\text{th}}$ call of SIadd and that means that it is superfluous in $A'(n-1)$. This implies that it is also superfluous in $M'(n-1)$ and we can proceed as before. Q. E. D.

Lemma D. *For $1 \leq i \leq 3^n$, if r_i^n belongs to $A'(n)$ and is superfluous in $M'(n)$, then it is also superfluous in $A'(n)$.*

Proof: Let $r_i^n = \sum_{s<i} \lambda_s r_s^n$ be an expression of r_i^n as a linear combination of some previous rows. Every r_s^n appearing in this sum either belongs to $A'(n)$ or not. In this last case, such a r_s^n is superfluous in $M'(n)$ by Lemma C, and then we can replace it by a linear combination of previous (to it) rows. Iterating this process a finite number of times, we get an expression for r_i^n as a linear combination of previous rows belonging to $A'(n)$. Q. E. D.

Proof of the Theorem: It is enough to see that if a polynomial H_j is kept after the nth call to SIadd, then every H_i dividing H_j is also kept.

So, assume that H_i is not kept in SIadd, i.e. that r_i^n is superfluous in $A'(n)$. By Lemma B, r_j^n is superfluous in $M'(n)$ and by lemma D it is also superfluous in $A'(n)$. On its turn, it implies that r_j^n is not kept in $A(n)$ and consequently, nor is H_j. Q. E. D.

2. The Algorithms

We fix the following notations for the rest of this paper: Let $F \in \mathbf{Z}[X,Y]$ be a square–free polynomial, monic as a polynomial in Y, and let D denote the discriminant of F with respect to Y. Let p be the degree of F and let $\xi_1 < \xi_2 < \ldots < \xi_r$ denote the real roots of D. Set $A_0 =]-\infty, \xi_1[$, $A_r =]\xi_r, +\infty[$ and, for $i = 1, \ldots, r-1$, $A_i =]\xi_i, \xi_{i+1}[$. (If $r = 0$, we shall write $A_0 = \mathbf{R}$.)

Let $\mathcal{C} \subset \mathbf{R}^2$ be the algebraic set defined by the equation $F(X,Y) = 0$.

Given $x \in \mathbf{R}$, we shall denote by $\rho(x)$ the number of real roots of $F(x,Y)$. Let us recall that $\rho(x)$ remains constant as long as x lies in a (fixed) A_i. Moreover, the corresponding roots are given by continuous semi–algebraic functions

$$\varphi_{i,1} < \varphi_{i,2} < \ldots < \varphi_{i,\rho_i} : A_i \longrightarrow \mathbf{R}, \quad i = 0, \ldots, r.$$

One calls the graphs of such functions $\varphi_{i,j}$ the *real branches* of C over A_i.

Every such a real branch $\varphi_{i,j}$ gives rise to a half–branch of C over ξ_{i+1}^- (if $i \leq r-1$) and to a half–branch over ξ_i^+ (if $i \geq 1$). We shall denote these half–branches, as well as the corresponding infinitesimals, by $\varphi_{i,j}^-$ and $\varphi_{i,j}^+$, respectively.

2.1. Roy's algorithms

As we have said in the Introduction, Roy gives in [9] an algorithm, which we shall refer to as TOP_1, for the computation of the topology of C using Thom codes to deal with real algebraic numbers. In fact, TOP_1 computes an F–invariant, proper cylindrical algebraic decomposition of $\mathbf{R}^2$, in the notations of [1]. In the same paper Roy also sketches an algorithm, which we shall refer to as CAD_1, for the computation of a (no longer proper) F–invariant c.a.d. of $\mathbf{R}^2$ (see also [11]).

Notice that from such an F–invariant c.a.d. we can deduce whether C is empty, a finite set of points, or an algebraic curve. If such a c.a.d. is furthermore proper, we can deduce from it the topological type of C: the number of connected components, the number of singularities, the number of algebraic branches through each singularity, etc ... We can even "draw" a planar graph homeomorphic to C (cf. [3], [9]).

Since our algorithms are inspired in Roy's, for the convenience of the reader we devote this paragraph to sketch them.

Roy's algorithm CAD_1:
Specifically, CAD_1 produces the following information:
i) It characterizes the real roots $\{\xi_i\}$ of D,
ii) It finds the number $\rho(\xi_i)$ of real points of C over each ξ_i, and
iii) It finds the number ρ_i of real branches of C over each A_i.

In order to get the quoted information, CAD_1 consists of the following steps:
1) Compute the Sturm–Habicht sequence of F regarded as a polynomial in Y. This computation gives D, up to a constant.
2) Code the real roots ξ_i ($i = 1, \ldots, r$) of D by means of $RAN(D)$.
3) Compute $\rho(\xi_i)$ ($i = 1, \ldots, r$) by means of $St_1(D, F)$.
4) Compute the numbers $\rho(\xi_i^-)$ and $\rho(\xi_i^+)$ of real half–branches of C over ξ_i^- and ξ_i^+ ($i = 1, \ldots, r$) by means of $St_2(D, F)$. Notice that, for $i = 1, \ldots, r-1$, $\rho_i = \rho(\xi_i^+) = \rho(\xi_{i+1}^-)$, $\rho_0 = \rho(\xi_1^-)$ and $\rho_r = \rho(\xi_r^+)$.

Proposition 2.1. *Let n be the first integer greater than p and $|F|$. The algorithm CAD_1 runs in time $O(n^{17}(\log n)^3)$, being (4) its dominant step.*

Roy's algorithm TOP_1:

Specifically, TOP_1 produces the following information:

i) It characterizes the real roots $\{\xi_i\}$ of D,

ii) It characterizes the real roots $\{\zeta_{i,j}\}$ of $F(\xi_i, Y)$, for every i,

iii) It characterizes the real branches $\{\varphi_{i,j}\}$ of C over each A_i, and

iv) It determines the glueing relations between the points $\{(\xi_i, \zeta_{i,j})\}$ and the branches $\{\varphi_{i',j'}\}$.

As far as points (iii) and (iv) goes, one has the following result:

Proposition 2.2. ([9], Prop. 1.) *Let F, D and C be as in the beginning. Let's fix a root ξ_i of D and a half-branch $\varphi_{i-1,j}^-$ (resp. $\varphi_{i,j}^+$) of C over ξ_i^- (resp. over ξ_i^+). Let $\epsilon = (\epsilon_1, \ldots, \epsilon_p)$ be the p-tuple of signs taken by $F_Y', \ldots, F_Y^{(p)}$ on the infinitesimal corresponding to this half-branch.*

i) The sign condition ϵ characterizes the half-branch $\varphi_{i-1,j}^-$ (resp. $\varphi_{i,j}^+$) over ξ_i^- (resp. over ξ_i^+).

ii) There exists one and only one root $\zeta_{i,j}$ of $F(\xi_i, Y)$ such that its Thom code is compatible with ϵ. Moreover, the half-branch glues to such $(\xi_i, \zeta_{i,j})$.

Notice that point (i) is nothing but Proposition 1.4 (i) applied to $\mathbf{R}(\xi^+)$ and $\mathbf{R}(\xi^-)$, while point (ii) is a consequence of Thom's Lemma.

In order to get the quoted information, TOP_1 consists of the following steps:

1) Compute the Sturm–Habicht sequence of F regarded as a polynomial in Y. As in the first step of CAD_1, this computation gives D, up to a constant.

2) Code the real roots $\{\xi_i\}$ of D by means of $\mathrm{RAN}(D)$.

3) Code the real roots $\{\zeta_{i,j}\}$ of $F(\xi_i, Y)$ by means of $\mathrm{RAN}_1((D, \xi_i), F)$ $(1 \le i \le r)$.

4) Compute the n-tuples of sign conditions satisfied by $F_Y', \ldots, F_Y^{(n)}$ on the infinitesimals corresponding to half-branches of C over ξ_i^- and ξ_i^+, by means of $\mathrm{RAN}_2((D, \xi_i), F)$ $(1 \le i \le r)$.

5) Finally, from the information obtained in (3) and (4) and applying Proposition 2.2, decide how do the half-branches $\varphi_{i',j'}^+$'s and $\varphi_{i',j'}^-$'s glue to the $(\xi_i, \zeta_{i,j})$'s, and thus how the real branches $\varphi_{i',j'}$'s glue to them.

Proposition 2.3. *Let n be the first integer greater than p and $|F|$. The algorithm TOP_1 runs in time $O(n^{21}(\log n)^4)$, being (4) its dominant step.*

Remark: We skip the computations of CAD_1's and TOP_1's running times, because they can be found in [11] with respect to the upper bounds for the auxiliary procedures (RAN, St_1, and so on) given therein, and thus the interested reader can very easily rework them using the new upper bounds given in this note. Moreover, we include in the sequel full–detailed computations of the complexities of the algorithms which appear in this note for the first time.

Anyway, we want to attract the reader's attention on the fact that both CAD_1's and TOP_1's dominant steps are those using infinitesimals.

2.2. A faster c.a.d. algorithm

In this paragraph we propose a new algorithm, CAD_0, for the computation of an F–invariant c.a.d. of $\mathbf{R}^2$. The difference between Roy's CAD_1 and our CAD_0 lies in the way of determining the number ρ_i of branches of C over A_i $(i = 0, \ldots, r)$. While CAD_1 gets it as the number of half–branches over a suitable infinitesimal, CAD_0 constructs, with no extra cost, a sample point η_i for A_i and then computes ρ_i as $\rho(\eta_i)$. In this way, CAD_0 avoids CAD_1's dominant step and turns out to have a lower complexity.

Algorithm CAD_0:
The algorithm CAD_0 consists of the following steps:
1) Compute the Sturm–Habicht sequence of F as a polynomial in Y. As before, this computation yields D up to a constant.
2) Compute the codes of the real roots ξ_i of D and η_j of D' $(i = 1, \ldots, r,$ $j = 1, \ldots, s)$. To do this, apply $\mathrm{RAN}(D)$ and $\mathrm{RAN}(D')$ respectively.
   Next, sort all numbers $\{\xi_i\}$ and $\{\eta_j\}$, by using their respective codes and Proposition 1.4 (ii) (see the Remark thereafter). In particular, and because of Rolle's Theorem, for every i, $1 \le i \le r - 1$, one finds an index $j_i \in \{1, \ldots, s\}$ such that $\eta_{j_i} \in A_i$.
3) Compute $\rho(\xi_i)$ and $\rho(\eta_j)$ $(i = 1, \ldots, r, j = 1, \ldots, s)$. To do this, apply $St_1(D, F)$ and $St_1(D', F)$ respectively.
   Notice that $\rho(\eta_{j_i}) = \rho_i$, for $i = 1, \ldots, r-1$. Moreover, if there is some j_0 (resp. j_r) such that $\eta_{j_0} < \xi_1$ (resp. $\eta_{j_r} > \xi_r$) then $\rho(\eta_{j_0}) = \rho_0$ (resp. $\rho(\eta_{j_r}) = \rho_r$).
4) If there does not exist such a root $\eta_{j_0} < \xi_1$ (resp. $\eta_{j_r} > \xi_r$), find the number ρ_0 (resp. ρ_r) as the number of roots of $F(x, Y)$ when $x = -\infty$ (resp. when $x = +\infty$). To get this number, one only has to check the degrees and the signs of the leading coefficients of the Sturm–Habicht coefficients of F and to apply Theorem 1.1.

Proposition 2.4.. *Let n be the first integer greater than p and $|F|$. The algorithm CAD_0 runs in time $O(n^{16}(\log n)^3)$.*

Proof: We only have to check the complexities of steps (1) to (4).
   1) From Theorem 1.1 we deduce that the complexity of the computation of the Sturm–Habicht sequence of F is in $O(n^{10})$.

2) The degrees and the sizes of both D and D' are bounded by $O(n^2)$. Then, by Proposition 1.5 (i), the complexity of $\mathrm{RAN}(D)$ and $\mathrm{RAN}(D')$ is in $O(n^{16}(\log n)^3)$.

3) By Proposition 1.6 (i), the complexity of $\mathrm{St}_1(D, F)$ and $\mathrm{St}_1(D', F)$ is in $O(n^{15}(\log n)^3)$.

Finally, the complexity of step (4) is negligible. Q. E. D.

2.3. A faster topological type algorithm

In this paragraph we give a new algorithm, TOP_0, for the computation of an F–invariant proper c.a.d. of $\mathbf{R}^2$, and thus for the computation of the topology of $\mathcal{C}$. The main difference between this algorithm and Roy's TOP_1 consists in the way of deciding the glueing relations between the points $\{(\xi_i, \zeta_{i,j})\}$ and the branches $\{\varphi_{i',j'}\}$. While TOP_1 uses infinitesimals lying on the branches, we use real algebraic points on them which are sufficiently close to the edges. Our key observation is the following one:

Lemma 2.5.. *Let D_j be the resultant w.r.t. Y of F and $F_Y^{(j)}$, $1 \leq j \leq p-1$ (so that $D = D_1$). For every $i = 1, \ldots, r$ let $\nu_{i,+}$ denote the nearest root to ξ_i among all the roots at its right of all D_j's (if, in the case $i = r$, such a root does not exist, we take $\xi_r^+ = \infty$). In the same way, let $\nu_{i,-}$ denote the nearest root to ξ_i among all the roots at its left of all D_j's (if, in the case $i = 1$, such a root does not exist, we take $\xi_1^- = -\infty$). Then the p-tuple of signs satisfied by $F_Y', \ldots, F_Y^{(p)}$ on a the infinitesimal corresponding to a half–branch of $\mathcal{C}$ over ξ_i^+ (resp. over ξ_i^-) agrees with the code of the real root of $F(x, Y)$ corresponding to the same half–branch, for any $x \in]\xi_i, \nu_{i,+}[$ (resp. for $x \in]\nu_{i,-}, \xi_i[$).*

We construct then sample points $\eta_{i,-} \in]\nu_{i,-}, \xi_i[$ and $\eta_{i,+} \in]\xi_i, \nu_{i,+}[$ with no extra cost, using Rolle's Theorem. In this way, we avoid the use of infinitesimals and we decrease the complexity of the algorithm.

The algorithm TOP_0:

Our algorithm TOP_0 consists of the following steps:

1) Compute the Sturm–Habicht sequence of F regarded as a polynomial in Y. As always, this computation gives D $(= D_1)$, up to a constant. Compute also $D_j = \mathrm{res}_Y(F, F_Y^{(j)})$, $j = 2, \ldots, p-1$.

2) Compute the codes of the real roots of D_j $(j = 1, \ldots, p-1)$, of $(DD_j)'$ $(j = 2, \ldots, p-1)$, and of D'. To do this, apply $\mathrm{RAN}(D_j)$ $(j = 1, \ldots, p-1)$, $\mathrm{RAN}((DD_j)')$ $(j = 2, \ldots, p-1)$, and $\mathrm{RAN}(D')$, respectively.

3) Apply $\mathrm{RANI}(D_i, D_j^{(k)})$ $(1 \leq i < j \leq p-1, 0 \leq k \leq \deg D_j)$. By means of this information, applying Proposition 1.4 (ii), sort the roots of all the D_j's.

Perform also $\text{RANI}(D, (DD_j)^{(k)})$ and $\text{RANI}(D_j, (DD_j)^{(k)})$ ($2 \leq j \leq p-1$, $0 \leq k \leq \deg DD_j$). By means of this information, and again by Proposition 1.4 (ii), sort, for each $j \geq 2$, the roots of D, D_j and $(DD_j)'$.

Finally, sort also the roots of D and D' by means of their codes.

In this way, for all couples (i, ϵ) with $1 \leq i \leq r$ and $\epsilon \in \{+, -\}$ (except, possibly, for $(1, -)$ and $(r, +)$) one gets an $i_\epsilon \in \{1, \dots, p-1\}$, a root $\nu_{i,\epsilon}$ of D_{i_ϵ} and, when $i_\epsilon > 1$, a root $\eta_{i,\epsilon}$ of $(DD_{i_\epsilon})'$, while if $i_\epsilon = 1$ then one gets a root $\eta_{i,\epsilon}$ of D', in such a way that

    i) $\nu_{i,-} < \eta_{i,-} < \xi_i$ and there is no root of any D_j in $]\nu_{i,-}, \xi_i[$

    ii) $\xi_i < \eta_{i,+} < \nu_{i,+}$ and there is no root of any D_j in $]\xi_i, \nu_{i,+}[$.

4) Code the real roots $\{\zeta_{i,j}\}$ of $F(\xi_i, Y)$, by means of $\text{RAN}_1((D, \xi_i), F)$ ($1 \leq i \leq r$).

Code also the real roots of $F(\eta_{i,\epsilon}, Y)$ ($1 \leq i \leq r$, $\epsilon \in \{-, +\}$). To do that, when $i_\epsilon = 1$ apply $\text{RAN}_1((D', \eta_{i,\epsilon}), F)$, and when $i_\epsilon > 1$ apply $\text{RAN}_1(((DD_{i_\epsilon})', \eta_{i,\epsilon}), F)$.

From all these codes, Proposition 2.2 and Lemma 2.5, one can determine how the different real branches of $\mathcal{C}$ on A_k (at least for $k = 2, \dots, r-1$) glue to the $(\xi_i, \zeta_{i,j})$'s. Moreover, if in step (3) one has found an element $\eta_{1,-}$ (resp. $\eta_{r,+}$) then one can also determine how the different real branches of $\mathcal{C}$ on A_0 (resp. A_r) glue to the $(\xi_1, \zeta_{1,j})$'s (resp. $(\xi_r, \zeta_{r,j})$'s).

5) If there does not exist such a $\eta_{1,-}$ (resp. $\eta_{r,+}$), this implies that there does not exist either a real root $\nu_{1,-}$ of some D_{1_-}, smaller than ξ_1 (resp. a real root $\nu_{l,+}$ of some D_{r_+}, greater than ξ_r).

In this case, compute the signs of $F_Y', \dots, F_Y^{(p-1)}$ on the roots of $F(-\infty, Y)$ (resp. $F(+\infty, Y)$), and, again by Proposition 2.2 and Lemma 2.5, deduce from them how the different real branches of $\mathcal{C}$ on A_0 (resp. A_r) glue to the $(\xi_1, \zeta_{1,j})$'s (resp. $(\xi_r, \zeta_{r,j})$'s).

As far as the complexity of TOP_0 goes, we have the following result:

Proposition 2.6. *Let n be the first integer greater than p and $|F|$. The algorithm TOP_0 runs in $O(n^{19}(\log n)^3)$.*

Proof: We only have to check the complexities of steps (1) to (5).

1) The complexity of the computation of the Sturm–Habicht sequence of F and the complexity of the computation of D_j ($j = 2, \dots, p-1$) are in $O(n^{10})$. So the total complexity of this step is in $O(n^{11})$.

2) We apply $O(n)$ times RAN to polynomials whose degrees and sizes are bounded by $O(n^2)$. Since the complexity of such an application of RAN is in $O(n^{16}(\log n)^3)$, the total complexity of this step is in $O(n^{17}(\log n)^3)$.

3) We apply $O(n^4)$ times RANI to couples of polynomials whose degrees and sizes are bounded by $O(n^2)$. By Proposition 1.5 (ii) the complexity of such an application of RANI is in $O(n^{14}(\log n)^3)$. Thus, the total complexity of this step is in $O(n^{18}(\log n)^3)$.

4) We apply $O(n^2)$ times $\text{RAN}_1((Q, \theta), F)$ to roots θ of polynomials Q whose degrees and sizes are bounded by $O(n^2)$. Since the complexity

of such an application of $\mathrm{RAN}_1((Q,\theta),F)$ is in $O(n^{17}(\log n)^3)$, the total complexity of this step is in $O(n^{19}(\log n)^3)$.

Finally, the complexity of step (5) is negligible.   Q. E. D.

2.4. A third topological type algorithm

In this paragraph we give another algorithm, TOP_2, for the computation of an F-invariant proper c.a.d. of $\mathbf{R}^2$. Like Roy's TOP_1, it uses infinitesimals to obtain the "adjacency" information. But, while TOP_1 (and TOP_0) determines how the branches $\{\varphi_{i',j'}\}$ glue to the points $\{(\xi_i,\zeta_{i,})\}$ by associating a suitable p-tuple of signs to every branch and then comparing it with the codes of the roots $\{\zeta_{i,j}\}$ of $F(\xi_i,Y)$, TOP_2 just counts up how many branches glue to every $(\xi_i,\zeta_{i,j})$ using Theorem 1.1 (iii) on a suitable ordered domain. Notice that, since the real branches over every A_i can be characterized by their "sorting" (they do not cross each other) it turns out that this information is enough for our purposes. In particular, TOP_2 does not make any use of Proposition 2.2. This feature has made it easy to implement (see section 3).

First of all, we introduce some new (2-dimensional) infinitesimals. Given $\xi \in \mathbf{R}$, let

i) $(\xi^+)^+ \in \mathrm{Spec}_r \mathbf{R}[X_1,X_2]$ be the point $((0),<)$ where the ordering $<$ is defined as follows: a) $0 < X_2 < \varepsilon$ for all real positive numbers ε; b) $\xi < X_1 < \xi + P(X_2)$ for every positive polynomial $P(X_2) \in \mathbf{R}[X_2]$ (with respect to (a)).

ii) $(\xi^-)^+ \in \mathrm{Spec}_r \mathbf{R}[X_1,X_2]$ be the point $((0),<)$, where the ordering $<$ is defined as follows: a) $0 < X_2 < \varepsilon$ for all real positive numbers ε; b) $\xi - P(X_2) < X_1 < \xi$ for every positive polynomial $P(X_2) \in \mathbf{R}[X_2]$ (with respect to (a)).

Given $\epsilon \in \{+,-\}$, we shall denote by $\mathbf{R}((\xi^\epsilon)^+)$ the real closure of $\mathbf{R}(X_1,X_2)$ with respect to the ordering $<$ corresponding to $(\xi^\epsilon)^+$. Given a polynomial $P \in \mathbf{R}[X_1,X_2]$, we shall denote $\mathrm{sign}\,(P((\xi^\epsilon)^+))$ simply by $\mathrm{sign}^\epsilon P$. The next Proposition shows us how to compute this signs.

Proposition 2.7. *Let $P \in \mathbf{R}[X_1,X_2]$, let $\xi \in \mathbf{R}$, and let $(u,v) \in \mathbf{N}^2$ the first pair, with respect to the inverse lexicographical order on $\mathbf{N}^2$, such that*

$$\frac{\partial^{u+v}P}{\partial^u X_1 \partial^v X_2}(\xi,0) \neq 0.$$

Then

$$\mathrm{sign}^+ P = \mathrm{sign}\left(\frac{\partial^{u+v}P}{\partial^u X_1 \partial^v X_2}(\xi,0)\right)$$

and

$$\mathrm{sign}^- P = (-1)^u \mathrm{sign}\left(\frac{\partial^{u+v}P}{\partial^u X_1 \partial^v X_2}(\xi,0)\right)$$

Proof: Just develop P in powers of $(X_1 - \xi)$ and X_2 in increasing inverse lexicographical order and apply the definitions.   Q. E. D.

Proposition 2.8. *Let F, D and C as in the beginning of §2. Let ξ be a root of D, $(\xi, \zeta) \in C$ and $\epsilon \in \{+, -\}$. The number of half–branches of C over ξ^ϵ which glue to (ξ, ζ) is equal to*

$$Sth_{\zeta - X_2, \zeta + X_2}^{\mathbb{R}((\xi^\epsilon)^+)}(F(\xi, Y), 1)$$

Proof: From the good specialization properties of the Sturm–Habicht sequence, and the fact that the rules given in Proposition 2.7 for computing $\text{sign}^+ P$ give also $\text{sign}\, P(\xi + \varepsilon, \delta)$ for $\delta > \varepsilon > 0$ very, very small, one gets that

$$Sth_{\zeta - X_2, \zeta + X_2}^{\mathbb{R}((\xi^+)^+)}(F(\xi, Y), 1) = Sth_{\zeta - \delta, \zeta + \delta}^{\mathbb{R}}(F(\xi + \varepsilon, Y), 1)$$

with δ and ε as before.

Now, taking δ and ε small enough, and applying Theorem 1.1, one easily sees that this last number is equal to the number of half–branches of C through (ξ, ζ) over ξ^+.

A similar reasoning proves the statement over ξ^-.   Q. E. D.

Algorithm TOP_2:

The algorithm TOP_2 consists of the following steps:

(1), (2) and (3) as in TOP_1.

4) Decide how many half–branches of C over ξ_i^+ and over ξ_i^- glue to every $(\xi_i, \zeta_{i,j})$ $(i = 1, \dots, r, \ j = 1, \dots, \rho(\xi_i))$. To do that, use Propositions 2.7 and 2.8 in the following way:

4.1) Let $F_p = F, F_{p-1} = F_Y', F_{p-2}, \dots, F_0 = D$ be the Sturm–Habicht sequence of F which we have computed in (1). To simplify the notations, let's call

$$F_{s,u,v} = \frac{\partial^{u+v} F_s}{\partial^u X \partial^v Y}$$

For every point $(\xi_i, \zeta_{i,j})$ and every polynomial F_s $(i = 1, \dots, r, \ j = 1, \dots, \rho(\xi_i), \ s = 0, \dots, p)$, find the first pair of indices $(u_{i,j,s}, v_{i,j,s})$ (with respect to the inverse lexicographical order) such that

$$F_{s,u_{i,j,s},v_{i,j,s}}(\xi_i, \zeta_{i,j}) \neq 0$$

and compute the sign of this number. This can be done applying the algorithm $RANI_1((D, \xi_i), F; F_{s,u,v})$, for every $i = 1, \dots, r, \ s = 1, \dots, p$, in increasing (inverse lexicographical) order of (u, v) and until one has found for every $j = 1, \dots, \rho(\xi_i)$ the pair $(u_{i,j,s}, v_{i,j,s})$.

4.2) Compute the signs of

$$F_{s,u_{i,j,s},v_{i,j,s}}(\xi_i, \zeta_{i,j} - X_2) \quad \text{and} \quad F_{s,u_{i,j,s},v_{i,j,s}}(\xi_i, \zeta_{i,j} + X_2)$$

in $\mathbf{R}((\xi^+)^+)$ and $\mathbf{R}((\xi^-)^+)$ by means of Proposition 2.7.

   4.3) Apply Proposition 2.8.

Remark: In fact, in step (4) one does not need to study all points $(\xi_i, \zeta_{i,j})$, or both "sides" of each such a point. Notice for instance that if $F'_Y(\xi_i, \zeta_{i,j}) \neq 0$, then we know a priori that there is a single half–branch through $(\xi_i, \zeta_{i,j})$ over ξ_i^+ as well as over ξ_i^-. In a similar way, if $F'_Y(\xi_i, \zeta_{i,j}) = 0$ but $F'_X(\xi_i, \zeta_{i,j}) \neq 0$, then we know that there are only two half–branches through $(\xi_i, \zeta_{i,j})$, and that they are both either on ξ_i^+ or on ξ_i^-, and thus once it is known what happens on one side, the behaviour on the other side is automatically determined.

   As far as the complexity of TOP_2 goes, we have the following result:

Proposition 2.9. *Let n be the first integer greater than p and $|F|$. The running time of TOP_2 is bounded by $O(n^{25})$.*

Proof: As we have seen in Proposition 2.6, the complexities of (1), (2) and (3) are in $O(n^{10})$, $O(n^{16}(\log n)^3)$ and $O(n^{19}(\log n)^3)$ respectively.

   As far as the complexity of step (4) goes, in the worst case we compute $RANI_1((D, \xi_i), F; F_{s,u,v})$ for $1 \leq i \leq O(n^2)$, $0 \leq s \leq O(n)$, $0 \leq u \leq O(n^2)$ and $0 \leq v \leq O(n)$. Moreover, the degree and the size of $F_{s,u,v}$ are bounded by $O(n^2)$. Applying Proposition 1.6 (iii) we have that the complexity of this step is bounded by $O(n^{25})$.   Q. E. D.

3. Implementation and running times

   When implementing RAN (or $RAN_1, \dots$) we do not need necessarilly all the derivatives. In fact, we can stop the computations once we have separated all the roots. Thus, since we sort the coded roots of a polynomial beggining with the linear derivative, we can spare running time in the RAN executions by adding the derivatives from lower to higher degrees until all the roots have been separated. We can still go a bit further. If we squarefree a polynomial before coding its roots, once we have separated them in groups of at most two roots, it is sure that the first derivative will finish the task. In the sequel we shall call these two modifications as first and second simplification of RAN.

   Note that both simplifications of RAN preserve its main feature: we can code and sort the roots of a polynomial. However, in order to compare roots of different polynomials or to check a compatibility relation, we can face a lack of information because the needing of more signs of derivatives.

   Now, the RAN algorithm has been implemented in REDUCE and MAPLE, using the first and second simplifications of RAN respectively. Over this ground, the CAD algorithm has been implemented in MAPLE and TOP_2 in REDUCE.

It is not surprising the fact of not having an implementation of TOP or TOP_1. Both algorithms decide the adherence relations between points and half–branches by checking compatibility relations between codes, and moreover, TOP_1 makes several sortings of roots of differents polynomials. Thus, either we try to add the lacking information as needed, either we use RAN with no simplifications.

The first option leads to an involved programming task, and the second to unnecessary higher running times.

It is our behalf that a central virtue of TOP_2 is that it can be easily implemented with any simplified RAN. This fact can rest importance to his asymptotically more expensive behavior and this is more clear if we consider that, the in any case high complexity of the problem, limits our inputs to low degree polynomials.

A closer comparison of those algorithms should be done on the basis of programming them in the same language, and then run the resulting programs on a wide set of exemples.

We give now some running times. All them are expressed in seconds and all the executions were done on a Sun 3/60 running the UNIX operating system.

Input polynomial	CAD	TOP_2
$-x^3 + x^2y - x^2 - xy^2 + x + y^3 - y^2 - y + 1$ 6.0 line and circumference		468
$x^4 + 2x^2y^2 - x^2 + y^4 - y^2$ circumference and isolated point	7.3	760
$x^4 + 2x^2y^2 - x^2 + y^4 + y^2$ Bernoulli's lemniscata	8.2	637
$x^4 - 4x^3 + 2y^2x^2 + 4xy^2 + y^4$ trifolium	15.2	2463
$24x^3 - 4y^2x^2 + x^2 - 6xy^2 + y^4$ Ruíz Castizo's quartic	27.5	7938
$2x^4 - 3x^2y + y^4 - 2y^3 + y^2$ tacnode	68.0	12995

We close this section remarking that the implementation of TOP_2 has been done in the algebraic mode of REDUCE. An implementation in symbolic mode (and using the second simplification of RAN) is being carried on.

Acknowledgements: Theorem 1.2 relies on an idea of J. Canny. We are indebted to him as well as to M.-F. Roy, who communicated to us the central ideas from which our proof steams.

REFERENCES

[1] D. Arnon, G. Collins and S. MacCallum, *Cylindrical algebraic decomposition I and II*, SIAM J. of Computing **13** (1984).

[2] S. Arnborg and H. Fen, *Algebraic decomposition of regular curves*, J. of Symb. Comp. **5** (1988).

[3] D. Arnon and S. MacCallum, *A polynomial-time algorithm for the topological type of a real algebraic curve*, J. of Symb. Comp. **5** (1988).

[4] J. Bochnak, M. Coste and M. F. Roy, "Géométrie Algébrique Réelle," Ergebnisse, Springer–Verlag, 1987.

[5] M. Coste and M.-F. Roy, *Thom's lemma, the coding of real algebraic numbers and the computation of the topology of semi-algebraic sets*, J. of Symb. Comp. **5** (1988).

[6] F. Cucker and M.-F. Roy, *A theorem on random polynomials and some consequences in average complexity*, J. of Symb. Comp. (to appear).

[7] P. Gianni and C. Traverso, *Shape determination of real curves and surfaces*, Ann. Univ. Ferrara, Sez. VII, Sec. Math. **XXIX** (1983).

[8] L. González, H. Lombardi, T. Recio and M.-F. Roy, *Spécialisation de la suite de Sturm et sous-résultants*, RAIRO Inf. Théor. et Appl. (to appear).

[9] M.-F. Roy, *Computation of the topology of a real algebraic curve*, in "Proceedings of the Conference on Computational geometry and topology," Sevilla 1987 (to appear).

[10] M.-F. Roy and A. Szpirglas, *Complexity of computations on real algebraic numbers*, J. of Symb. Comp. (to appear).

[11] M.-F. Roy and A. Szpirglas, *Complexity of computations of cylindrical decomposition and topology of real algebraic curves using Thom's lemma*, in "Proceedings of the Conference on Real Algebraic Geometry," Trento 1988 (to appear).

Felipe Cucker
Dept. L.S.I., Facultat d'Informàtica
U. P. C.
Barcelona 08028, SPAIN

Laureano González Vega
Dept. Matem. Estad. y C.C.
Universidad de Cantabria
Santander 39071, SPAIN

Francesc Rosselló
Dept. Àlgebra i Geometria
Universitat de Barcelona
Barcelona 08007, SPAIN

Duality methods for the membership problem

ALICIA M. DICKENSTEIN   CARMEN SESSA

Introduction

The classical problem of deciding membership to arbitrary polynomial ideals is EXPSPACE complete. Moreover, the problem of finding a representation of a polynomial by generators of a given ideal may involve doubly exponential (in the number of variables) degrees ([16]). The same difficulty arises when computing Gröebner bases of arbitrary polynomial ideals ([11]). This means that all known techniques to decide membership and to find representations of polynomials with respect to a given ideal lead to doubly exponential (sequential time) worst case complexities. However, if the geometry of the underlying algebraic variety is particularly simple, e.g. if the given ideal is zero dimensional or complete intersection, algorithms of considerably lower complexity can be found (see e.g. [7], [9]). The improvements are due to recent progress concerning affine versions of the effective Nullstellensatz (compare [18] and the references given there).

Using methods from residual duality theory, we show that some of the algorithmical problems arising frequently in computational algebraic geometry can be solved in simply exponential sequential and polynomial parallel time complexity.

Let K be a subfield of the complex numbers (e.g. $K = Q$) and let $z_1, \ldots, z_n$ be indeterminates over C. Let be given a finite set of polynomials $f_1, \ldots, f_r \in K[z_1, \ldots, z_n]$ with degrees bounded by $d \geq 3$, which generate an ideal $I := I(f_1, \ldots, f_r)$.

Suppose first that $f_1, \ldots, f_r$ form a regular sequence. In this case we associated in [9] to the ideal I and to any natural number $k \in N$, an $(O(k^n) + d^{O(n^2)}) \times O(k^n)$ matrix $S_{I,k}$ with entries from K which characterizes membership to I up to degree k in the following sense: a polynomial $p \in C[z_1, \ldots, z_n]$ of degree bounded by k belongs to I iff its coefficients vector is a solution of the homogeneous linear equations system corresponding to $S_{I,k}$.

Research supported by CONICET and UBACYT, Argentina.

The matrix $S_{I,k}$ can be computed from the inputs $f_1, \ldots, f_r$ by means of an arithmetic network over K of size $k^{O(n)} + d^{O(n^2)}$ and of depth $O(n^4 \log^2(k \cdot d))$ (see [10] for the notion of arithmetical network). In a more down to earth language we can say that $S_{I,k}$ is computable in sequential time $k^{O(n)} + d^{O(n^2)}$ and (simultaneously) in parallel time $O(n^4 \log^2(k \cdot d))$. We shall call an algorithm admissible if it is realizable by an arithmetical network within these time bounds. In the same sense we shall also speak about problems solvable and functions computable in admissible time (if necessary, we allow the sequential - and parallel - complexity to depend polynomially - or polylog - on the number r of generators).

In this paper, we compute in admissible time a matrix with entries from K which characterizes membership to the radical of I up to degree k. As a by-product of our method, we are able to compute in admissible time generators for the radical of I if an a priori degree bound of type $d^{O(n)}$ for them is given (this is the case if $r = n$, i.e. if I is a zero dimensional ideal. See Proposition 2.3 below).

Suppose now that I is a zero dimensional ideal (we don't need any more the hypothesis that $f_1, \ldots, f_r$ is a regular sequence). In this case, we are able to compute in admissible time matrices with entries from K which characterize membership to I and to its radical up to a given degree bound $k \in \mathbf{N}$. As mentioned, we are able to compute in admissible time generators for the radical of I.

Our algorithms are based on admissible time computations of ideal quotients of certain zero dimensional ideals. This is combined with ideas of "Zariski-Samuel duality" (see § 2 below) in order to transfer our results on complete intersection ideals to arbitrary zero dimensional ideals.

We want to put special emphasis on the method used in [9] and in this paper, which has its origins in the analytical theory of residues developed in [6], [8], [12].

The tools from residual duality theory we need for our computational applications are outlined in section 1. For residues and duality in the context of algebraic geometry we refer to [13], [15], and also to [1], where some applications are given.

There exists already considerably work with respect to simply exponential complexity bounds in computational algebraic geometry. Pioneering work was done e.g. in [14] (non-emptiness testing and dimension determination for projective varieties), in [5] (a sequential algorithm for decomposition of algebraic varieties into irreducible components) and in [4] (non-emptiness testing, dimension determination and zero dimensional equations solving for affine varieties).

In the case of computational commutative algebra much less is done with respect to non-homogeneous ideals and modules. Our aim is to demonstrate the power of residue theory and future possibilities of its application in Computer Algebra by the presentation of some new algorithms in computational commutative algebra obtained by means of this tool.

Let us also point out that the analytical theory of residues is closely connected to the first proof of an affine effective Nullstellensatz (cf. [3]. See also [2] for further applications of this technique).

We are grateful to Joos Heintz for his generosity to share his knowledge with us.

0. Notations

- $C[z]$ denotes always the polynomial ring in n variables $C[z_1, \ldots, z_n]$ and for any $k \in N_0$, $C[z]_k := \{f \in C[z]/f = 0 \text{ or } \deg(f) \leq k\}$.
- n being fixed, we'll denote $dz = dz_1 \wedge \cdots \wedge dz_n$.
- Let $x = (x_1, \ldots, x_n) \in C^n$ and $\alpha = (\alpha_1, \ldots, \alpha_n) \in N_0^n$. We denote
$$|\alpha| := \sum_{i=1}^{n} \alpha_i \text{ and } (z - x)^\alpha := \prod_{i=1}^{n} (z_i - x_i)^{\alpha_i}.$$
- Unless otherwise stated, all the ideals considered lie in $C[z]$. For any ideal I, we denote $Z(I) := \{x \in C^n/f(x) = 0 \ \forall f \in I\}$.
- Let $f_1, \ldots, f_r \in C[z]$; $I(f_1, \ldots, f_r)$ denotes the generated ideal
$$\left\{\sum_{i=1}^{r} g_i \cdot f_i \ / \ g_i \in C[z], \quad i = 1, \ldots, r\right\}.$$
- For any ideal I, we denote $\text{rad}(I) := \{f \in C[z]/f^m \in I \text{ for some } m \in N\}$.
- Given $k \in N$, an ideal I and a matrix S such that "$P \in I \cap C[z]_k \iff$ The vector of coefficients of P is a solution of $S \cdot \mathbf{X} = 0$", we will say that $S \cdot \mathbf{X} = 0$ is a system of linear equations for $I \cap C[z]_k$.

1. Residual duality

1.1. Local analytic residues.

The local residue is a generalization of the Cauchy formula in several complex variables. It can be given the following analytic definition:

Let $x \in C^n$ (or in an n-dimensional complex manifold), $f_{1_x}, \ldots, f_{n_x} \in \mathcal{O}_x$ and let U be a ball centered at x such that there exist representatives $f_1, \ldots, f_n \in \mathcal{O}(\overline{U})$ verifying $\bigcap_{i=1}^{n}(f_i = 0) = \{x\}$. For any meromorphic n-form ω in U having its poles on $\bigcup_{i=1}^{n}(f_i = 0)$, the residue of ω at x is the complex number

$$(1.1.0) \qquad \text{Res}_x(\omega) = \frac{1}{(2\pi i)^n} \int_{\{z \in U/|f_j(z)| = \delta_j, \, 1 \leq j \leq n\}} \omega$$

for any $\delta_1 > 0, \ldots, \delta_n > 0$ sufficiently small (cf [6]; [12], ch. 5).

This definition is of course independent of the neighborhood U chosen and of the special sequence $\mathbf{f} = \{f_1, \ldots, f_n\}$ verifying the above hypotheses.

We can also think the residue of ω as an operator which assigns to any holomorphic germ $g \in \mathcal{O}_x$, the complex number $R_{\mathbf{f},x}[\omega](g) := \operatorname{Res}_x(g \cdot \omega)$; that is, we have a $\mathbb{C}$-linear operator $R_{\mathbf{f},x} : \mathcal{O}_x \to \mathbb{C}$.

Suppose $\omega = \dfrac{h \, dz}{f_1 \ldots f_n}$, $h_x \in \mathcal{O}_x$, and let $\{f'_{1_x}, \ldots, f'_{n_x}\}$ be any regular sequence in the generated ideal $I_x := I(f_{1_x}, \ldots, f_{n_x})$. The operator $R_{\mathbf{f},x}[\omega]$ can be expressed in terms of a sequence $\mathbf{f}' := \{f'_1, \ldots, f'_n\}$ of representatives by means of the following Transformation Law:

$$(1.1.1) \qquad R_{\mathbf{f},x}[\omega] = R_{\mathbf{f}',x}\left[\frac{\det A \cdot h \, dz}{f'_1 \ldots f'_n} \right]$$

where $A = (a_{ij}) \in \mathcal{O}_x^{n \times n}$ and $f'_i = \sum_{j=1}^{n} a_{ij} f_j \ \forall i = 1, \ldots, n$ (for a proof, see [12], ch. 5).

The operator $R_{\mathbf{f},x}[\omega]$ is in fact a linear differential operator acting on $\mathcal{O}_x$; more precisely, there exist $n_x \in \mathbb{N}_0$ and complex constants $(c_{\alpha,x}, \alpha \in \mathbb{N}_0^n, 0 \le |\alpha| \le n_x)$ such that for every $g \in \mathcal{O}_x$,

$$(1.1.2) \qquad R_{\mathbf{f},x}[\omega](g) = \sum_{0 \le |\alpha| \le n_x} c_{\alpha,x} \cdot \frac{\partial^{|\alpha|}(g)}{\partial z_1^{\alpha_1} \ldots \partial z_n^{\alpha_n}}(x).$$

This assertion can be easily deduced from the n-variable Cauchy formula and (1.1.1) by means of the local analytic Nullstellensatz, as follows: there exist $r = (r_1, \ldots, r_n) \in \mathbb{N}^n$ and $A = (a_{ij}) \in \mathcal{O}_x^{n \times n}$ such that

$$(z_i - x_i)^{r_i} = \sum_{j=1}^{n} a_{ij} f_j \text{ for all } i = 1, \ldots, n. \text{ Then,}$$

$$R_{\mathbf{f},x}\left[\frac{h \, dz}{f_1 \ldots f_n} \right](g) = R_{\mathbf{z}-\mathbf{x},x}\left[\frac{\det A \cdot h \, dz}{(z_1 - x_1)^{r_1} \ldots (z_n - x_n)^{r_n}} \right](g) =$$

$$= \frac{1}{\prod_j (r_j - 1)!} \frac{\partial^{|r|-n}(\det A \cdot h \cdot g)}{\partial z_1^{r_1 - 1} \ldots \partial z_n^{r_n - 1}}(x) \,.$$

And so, the coefficients $c_{\alpha,x}$ can be precisely described in terms of the derivatives of $\det A \cdot h$ at x.

Suppose that $h = h_1 f_1$, $h_{1_x} \in \mathcal{O}_x$. Then, $\operatorname{Res}_x\left(\dfrac{h \, dz}{f_1 \ldots f_n} \right) =$ $\operatorname{Res}_x\left(\dfrac{h_1 \, dz}{f_2 \ldots f_n} \right) = 0$ because the path of integration in (1.1.0) may, without crossing a singularity, be shrunk to a lower dimensional cycle by letting $\delta_1 \to 0$. Thus, by linearity, $\operatorname{Res}_x\left(\dfrac{h \, dz}{f_1 \ldots f_n} \right) = 0$ if $h_x \in I_x$. As an easy consequence, the operator $R_{\mathbf{f},x}\left[\dfrac{h \, dz}{f_1 \ldots f_n} \right]$ is identically zero when $h_x \in I_x$. In fact, the converse to the latter statement is also true:

Local duality: Let $f_{1_x}, \ldots, f_{n_x}$ be a regular sequence in $\mathcal{O}_x$. With the above notations, we have the equivalence (cf [8], [12]):

(1.1.3)
$$h \in I_x \Longleftrightarrow R_{f,x}\left[\frac{h\, dz}{f_1 \cdots f_n}\right] \equiv 0.$$

Denote Ω^n the sheaf of holomorphic differential n-forms. One can compute the $\mathcal{O}_x$-module $\mathrm{Ext}^n_{\mathcal{O}_x}(\mathcal{O}_x/I_x, \Omega^n_x)$ by means of the Koszul projective resolution of $\mathcal{O}_x/I_x$ given by the sequence $f_{1_x}, \ldots, f_{n_x}$. So, one verifies that $\mathrm{Ext}^n_{\mathcal{O}_x}(\mathcal{O}_x/I_x, \Omega^n_x) \simeq \Omega^n_x/I_x \cdot \Omega^n_x \simeq \mathcal{O}_x/I_x$.

This gives a more intrinsec formulation of the equivalence (1.1.3):
(1.1.4) The pairing

$$\mathrm{res} : \mathcal{O}_x/I_x \otimes \mathrm{Ext}^n_{\mathcal{O}_x}(\mathcal{O}_x/I_x, \Omega^n_x) \to \mathbf{C}$$

induced by

$$\mathrm{res}(g, h) = R_{f,x}\left[\frac{h\, dz}{f_1 \cdots f_n}\right](g)$$

is non-degenerated, and it is independent of the choice of the regular sequence of generators of I_x.

In the case of holomorphic regular sequences of any codimension p, one still has an explicit definition of a residual operator acting on $\mathcal{C}^\infty$ compactly supported forms (i.e. a residual current) which generalizes the punctual residue ([6]). By computing $\mathrm{Ext}^p_{\mathcal{O}_x}(\mathcal{O}_x/I_x, \Omega^n_x)$ via the injective resolution of the fibers Ω^n_x by means of the $\bar{\partial}$-complex of currents $'D^n_x$ for any x, these residual currents provide an explicit generalized local duality (cf. [8]).

1.2. Polynomial residues.

a) The 0-dimensional case.

Suppose we are given a regular sequence of n polynomials $\mathbf{q} = \{q_1, \ldots, q_n\}$ in $\mathbf{C}[z]$. Denote Q the generated ideal $I(q_1, \ldots, q_n)$ with finite zero set $Z(Q)$ in $\mathbf{C}^n$.

Grothendieck (see Hartshorne [13]) isolated the functorial aspects of the notion of local analytic residue, axiomatizing them and giving a definition of residues in a purely algebraic context. The residue is interpreted (in this particular case) as a morphism

$$\mathrm{res} : \mathrm{Ext}^n_{\mathbf{C}[z]}(\mathbf{C}[z]/Q, \Omega^n_{\mathbf{C}[z]/\mathbf{C}}) \to \mathbf{C}$$

(where $\Omega^n_{\mathbf{C}[z]/\mathbf{C}}$ denotes the Kähler module of relative differentials). To any $\omega \in \Omega^n_{\mathbf{C}[z]/\mathbf{C}}$ one may associate its Grothendieck residue symbol as follows: by means of the Koszul resolution of $\mathbf{C}[z]/Q$ given by the regular sequence $\{q_1, \ldots, q_n\}$, we may identify $\mathrm{Ext}^n_{\mathbf{C}[z]}(\mathbf{C}[z]/Q, \Omega^n_{\mathbf{C}[z]/\mathbf{C}})$ with $\Omega^n_{\mathbf{C}[z]/\mathbf{C}}/Q \cdot \Omega^n_{\mathbf{C}[z]/\mathbf{C}}$. The symbol $\left[\dfrac{\omega}{q_1 \cdots q_n}\right]$ is just the image of ω in this last quotient.

Now we are going to explain how the local analytic residue can be combined with the Grothendieck approach to give an explicit polynomial residual duality.

(1.2.1) Given $p \in C[z]$, denote $R_q[p] : C[z] \to C$ the C-linear operator

$$R_q[p](g) := \sum_{x \in Z(Q)} R_{q,x}\left[\frac{p\,dz}{q_1 \ldots q_n}\right](g) \, .$$

Then, the residue morphism is explicitely given by

$$\text{res}\left(\left[\frac{p\,dz}{q_1 \ldots q_n}\right]\right) = R_q[p](1) \, .$$

Moreover, the pairing

$$\text{Res} : C[z]/Q \otimes \text{Ext}^n_{C[z]}\left(C[z]/Q, \Omega^n_{Q[z]/C}\right) \longrightarrow C$$

induced by

$$\left(g, \left[\frac{p\,dz}{q_1 \ldots q_n}\right]\right) \longmapsto R_q[p](g)$$

is non-degenerated and independent of the choice of generators $q_1, \ldots, q_n$ of Q.

The non-degeneracy of the bilinear map associated to Res can be deduced from the local analytic duality theory as follows:

Taking into account the description (1.1.2), it is easy to see that $R_q[p] \equiv 0$ is equivalent to $R_{q,x}[p] \equiv 0 \;\; \forall x \in Z(Q)$. By (1.1.3), this condition is equivalent to the local analytic membership $p_x \in Q_x \;\; \forall x \in Z(Q)$, which by [17] is in turn equivalent to the local algebraic membership for any $x \in Z(Q)$. This is easily seen to be equivalent to the condition $p \in Q$ in $C[z]$.

We know from (1.2.1) that $p \in Q$ iff $R_q[p] \equiv 0$. Now, this latter condition can be effectively verified taking into account the following facts:

i) Given a polynomial $g \in C[z]$, $R_q[p][g]$ is effectively computable. (In fact, the resulting sequential time bounds are admissible, in the sense explained in the introduction). Moreover, if the coefficients of p, g and $q_1, \ldots, q_n$ lie in some subfield of C, so does $R_q[p](g)$ (cf. [9]).

ii) $R_q[p] \equiv 0$ iff $R_q[p](g) = 0$ for every $g \in C[z]$ with $\deg(g) \le \prod_{i=1}^n \deg(q_i) - 1$ (This can be seen considering the order of the differential operators $R_{q,x}[p]$ (cf.[9])).

Thus, the membership $p \in Q$ is equivalent to the condition that the vector of coefficients of p is a solution of a certain "residual" homogeneous linear system, which can be effectively computed in admissible time. More precisely, the following two conditions are equivalent for $p = \sum_{|\beta| \le k} c_\beta z^\beta$:

i) $p \in Q$

ii) The vector of coefficients $(c_\beta, \; \beta \in N_0^n, |\beta| \le k)$ verifies the following linear system:

$$\sum_\beta c_\beta R_q[z^\beta](z^\alpha) = 0, \qquad |\alpha| \le \prod_{i=1}^n \deg(q_i) - 1.$$

Notice that the system has at most $\left(\prod_{i=1}^{n} \deg(q_i) \right)^n$ equations.

b) Regular sequences of arbitrary dimension.

Let $\mathbf{q} = \{q_1, \ldots, q_r\}$, $r \leq n$, be a regular sequence of complex polynomials defining a complete intersection ideal $Q := I(q_1, \ldots, q_r)$ in $\mathbf{C}[z]$.

Suppose that the linear projection $\pi : \mathbf{C}^n \to \mathbf{C}^{n-r}$, $\pi(z) = (z_{r+1}, \ldots, z_n)$, restricted to $Z(Q)$ is proper and with finite fibers. Let $z' := (z_1, \ldots, z_r)$ and $z'' := (z_{r+1}, \ldots, z_n)$ be the vectors of the first r and of the last $n - r$ variables of $z = (z_1, \ldots, z_n)$. Let x'' be a point of $\mathbf{C}^{n-r}$. The specialized polynomials $q_1(z', x''), \ldots, q_r(z', x'')$ of $\mathbf{C}[z']$ define a complete intersection in $\pi^{-1}(x'')$, which is isomorphic to $\mathbf{C}^r$. So, we can apply to this context the punctual residue machinery of section a).

For $p \in \mathbf{C}[z]$, we denote by $R_{\mathbf{q}(z', x'')}[p(z', x'')]$ the residual operator acting on the fiber $\pi^{-1}(x'')$. Now let x'' vary over $\mathbf{C}^{n-r}$. One can show that for any $g \in \mathbf{C}[z']$, the mapping

$$R_{\pi, \mathbf{q}}[p](g) : \mathbf{C}^{n-r} \longrightarrow \mathbf{C}$$
$$x'' \longmapsto R_{\mathbf{q}(z', x'')}[p(z', x'')](g)$$

is a polynomial function (depending on x'') which can be effectively computed in admissible time (cf. [9]).

If we think now $R_{\pi, \mathbf{q}}[p](g)$ as a polynomial in $\mathbf{C}[z'']$, the Grothendieck residue morphism

$$\operatorname{Ext}^r_{\mathbf{C}[z]}(\mathbf{C}[z]/Q, \Omega^r_{\mathbf{C}[z]/\mathbf{C}[z'']}) \longrightarrow \mathbf{C}[z'']$$

is described by

$$\left[\frac{p(z)\, dz'}{q_1 \ldots q_r} \right] \longmapsto R_{\pi, \mathbf{q}}[p](1)$$

and the explicit pairing

$$\left(g, \left[\frac{p(z)\, dz'}{q_1 \ldots q_r} \right] \right) \longmapsto R_{\pi, \mathbf{q}}[p](g)$$

induces a residual pairing

$$R_\pi : \mathbf{C}[z]/Q \otimes \operatorname{Ext}^r_{\mathbf{C}[z]}(\mathbf{C}[z]/Q, \Omega^r_{\mathbf{C}[z]/\mathbf{C}[z'']}) \longrightarrow \mathbf{C}[z''].$$

The task of determining whether $p \in Q$ is equivalent to the membership problem arising when we restrict all the polynomials to the fibers of π. Thus, the punctual duality (1.1.4) implies that this pairing R_π is also non-degenerated.

Due to the polynomial behaviour of the fibered residue, one can effectively construct, for any $k \in \mathbf{N}$, a system of linear equations for $Q \cap \mathbf{C}[z]_k$ (in the sense of § 0) in admissible time. In fact, the degree of the image polynomial $R_{\pi, \mathbf{q}}[p](g)$ is bounded by $\deg(g) + \deg(p) + rd^n(d^r + 1)$, where $d \in \mathbf{N}_{\geq 3}$ and $d \geq \deg(q_i)$ $\forall i = 1, \ldots, r$. Detailed proofs are given in [9].

2. Zariski-Samuel duality

Let $\mathcal{I}, \mathcal{Q}$ be ideals in a ring A. We denote, as usual, $(\mathcal{Q} : \mathcal{I})$ the quotient ideal $\{f \in A \,/\, f \cdot h \in \mathcal{Q} \ \forall h \in \mathcal{I}\}$.

A careful reading of [19], ch.4, §16, reveals the following result: In case A is a local noetherian ring and $\mathcal{Q}$ is an irreducible ideal belonging to the maximal ideal of A, for any ideal $\mathcal{I}$ containing $\mathcal{Q}$ it holds:

$$\mathcal{I} = (\mathcal{Q} : (\mathcal{Q} : \mathcal{I})).$$

We will refer to the above property as Zariski-Samuel local duality.

Proposition 2.1 (N. Coleff). *Let $x \in \mathbf{C}^n$ and let $\mathcal{Q}$ be a complete intersection ideal in the local ring $\mathcal{O}_x$ such that $rad(\mathcal{Q})$ is the maximal ideal. Then,*

$$(\mathcal{Q} : (\mathcal{Q} : \mathcal{I})) = \mathcal{I}$$

for any ideal $\mathcal{I}$ in $\mathcal{O}_x$ such that $\mathcal{I} \supsetneq \mathcal{Q}$.

Proof: In view of the Zariski-Samuel local duality, it will be sufficient to show that $\mathcal{Q}$ is an irreducible ideal in $\mathcal{O}_x$.

Let $\mathcal{J}ac := \det\left(\dfrac{\partial Q_i}{\partial z_j}\right)$ be the jacobian associated to some system of n generators $\{Q_1, \ldots, Q_n\}$ of $\mathcal{Q}$ and denote $\mathcal{J} := \langle \mathcal{Q}, \mathcal{J}ac \rangle$. Then, $\mathcal{Q}$ is irreducible as an easy consequence of the following two statements:

i) $\mathcal{J} \neq \mathcal{Q}$

ii) For any ideal $\mathcal{J}' \supseteq \mathcal{Q}$, $\mathcal{J}' \supseteq \mathcal{J}$.

In order to prove i) and ii), we will use the local duality law (1.1.3): "$\forall \varphi \in \mathcal{O}_x, \varphi \in \mathcal{Q} \Longleftrightarrow R_{\mathcal{Q},x}[\varphi] \equiv 0$".

For any $h \in \mathcal{O}_x$, $R_{\mathcal{Q},x}[\mathcal{J}ac](h) = c \cdot (2\pi i)^n h(x)$, where $c \in \mathbf{N}$ is the (non-vanishing) intersection number of $(Q_1 = 0), \ldots, (Q_n = 0)$ at x (cf. [6], [12]). As $R_{\mathcal{Q},x}[\mathcal{J}ac] \not\equiv 0$, we deduce that $\mathcal{J}ac \notin \mathcal{Q}$, proving i).

In order to see ii), let $f \in \mathcal{J}' - \mathcal{Q}$ and $A = \{g \in \mathcal{O}_x \,/\, g \cdot f \notin \mathcal{Q}\}$. Then, A is not empty (because $1 \in A$) and there exists $m \in \mathbf{N}$ such that $(z - x)^\alpha \notin A$, $\forall |\alpha| \geq m$ (because, by the local Hilbert Nullstellensatz, there exists $m \in \mathbf{N}$ such that $(z - x)^\alpha \in \mathcal{Q}$, $\forall |\alpha| \geq m$). Hence, there is an element $g \in A$ such that $(z_i - x_i) \cdot g \notin A$, for all $i = 1, \ldots, n$ (i.e. $g \cdot f \notin \mathcal{Q}$ and $(z_i - x_i) \cdot g \cdot f \in \mathcal{Q}$ for all i).

The assumption $g \cdot f \notin \mathcal{Q}$ implies that $R_{\mathcal{Q},x}[g \cdot f] \not\equiv 0$. For any $h \in \mathcal{O}_x$, we have

$$R_{\mathcal{Q},x}[g \cdot f](h) = R_{\mathcal{Q},x}[g \cdot f]\left(h(x) + \sum_{i=1}^n (z_i - x_i) \cdot h_i(z)\right) =$$

$$= h(x) \cdot R_{\mathcal{Q},x}[g \cdot f](1), \quad \text{and so} \quad c_1 := R_{\mathcal{Q},x}[g \cdot f](1) \neq 0 \,.$$

Let's define $\psi := \dfrac{c}{c_1} \cdot g \cdot f - \mathcal{J}ac$; we deduce, by duality again, that $\psi \in \mathcal{Q}$ since $R_{\mathcal{Q},x}[\psi] \equiv 0$. In particular, $\psi \in \mathcal{J}'$, and so $\mathcal{J}ac \in \mathcal{J}'$.Q. E. D.

Theorem 2.2. *Given a punctual complete intersection ideal $Q \in \mathbf{C}[z]$,*

$$(Q : (Q : I)) = I$$

for any ideal I such that $I \supseteq Q$.

Proof: By the previous proposition, we only need to show that $(Q : I)_x = (Q_x : I_x)$, $\forall x \in \mathbf{C}^n$ (where for any polynomial ideal J, J_x denotes its image in the local ring $\mathcal{O}_x$).

The inclusion $\subseteq$ is trivial. Suppose now $I = I(g_1, \ldots, g_r)$ and $f \in (Q_x : I_x)$. For each $i = 1, \ldots, r$, there exists $p_i \in \mathbf{C}[z]$ such that $p_i(x) \neq 0$ and $p_i \cdot f \cdot g_i \in Q$ (because $f \cdot g_i \in Q_x$). Therefore, $\prod_{i=1}^{r} p_i \cdot f \in (Q : I)$, i.e. $f \in (Q : I)_x$.   Q. E. D.

Proposition 2.3 and Remark 2.4 below will be useful in order to obtain an effective membership test for zero dimensional ideals:

Proposition 2.3. *Let Q be a zero dimensional ideal in $\mathbf{C}[z]$ and suppose $\deg(Q) \leq M$. Then, for any ideal J containing Q, there is a system of generators $f_1, \ldots, f_r$ of J such that $\deg(f_i) \leq M$, $\forall i = 1, \ldots, r$.*

Proof: To avoid cumbersome notation, let's suppose that $Z(Q) = \{a, b\}$ and $\deg_a(Q) = n_a$, $\deg_b(Q) = n_b$ ($n_a + n_b \leq M$). Then, $(z-a)^\alpha \cdot (z-b)^\beta \in Q$, for any pair $\alpha, \beta \in \mathbf{N}_0^n$ such that $|\alpha| \geq n_a$ and $|\beta| \geq n_\beta$.

Given a polynomial $P \in \mathbf{C}[z]$, by iterated Taylor expansions around a and b, there exists constants c_α, $c_{\alpha\beta}$ and polynomials $P_{\alpha\beta}$ such that

$$P = \sum_{|\alpha| < n_a} c_\alpha (z-a)^\alpha + \sum_{\substack{|\alpha| = n_a \\ |\beta| < n_b}} c_{\alpha\beta}(z-a)^\alpha \cdot (z-b)^\beta +$$

$$+ \sum_{\substack{|\alpha| = n_a \\ |\beta| = n_b}} P_{\alpha\beta} \cdot (z-a)^\alpha \cdot (z-b)^\beta \ .$$

That is, $P = P_1 + P_2$ with $\deg(P_1) < M$ and $P_2 \in I((z-a)^\alpha \cdot (z-b)^\beta$, $|\alpha| = n_a$ and $|\beta| = n_b) \subseteq Q \subseteq J$. Then, one can find a (finite) system of generators for J belonging to the set $\{f \in J \ / \ \deg(f) < M\} \cup \{(z-a)^\alpha \cdot (z-b)^\beta, \ |\alpha| = n_a \text{ and } |\beta| = n_b\} \subseteq \{f \in J \ / \ \deg f \leq M\}$.   Q. E. D.

Remark 2.4: Let $Q \subseteq \mathbf{C}[z]$ be an ideal. Suppose that for any $m \in \mathbf{N}$ we have a system of linear equations for $Q \cap \mathbf{C}[z]_m$ (in the sense of § 0), given by a matrix S_m.

Let $I = I(f_1, \ldots, f_r)$ with $\deg(f_i) \leq D$ $\forall i = 1, \ldots, r$. As $(Q : I) = \bigcap_{i=1}^{r}(Q : f_i)$, for any fixed $m \in \mathbf{N}$, $h \in (Q : I) \cap \mathbf{C}[z]_m \iff$ The vector of coefficients of $h \cdot f_i$ is a solution of $S_{m+D} \cdot \mathbf{X} = 0$ $\forall i = 1, \ldots, r$.

Now, let's call M_i the associated matrix (in the canonical basis of monomials) to the C-linear mapping $C[z]_m \longrightarrow C[z]_{m+D}$ given by $g \longmapsto g \cdot f_i$. Therefore, $h \in (Q : I) \cap C[z]_m \Longleftrightarrow$ The vector of coefficients of h is a solution of $S_{m+D} \cdot M_i \cdot \mathbf{X} = 0 \quad \forall i = 1, \ldots, r$.

Thus, we have a system of linear equations for $(Q : I) \cap C[z]_m$. Clearly, the number of equations of this system is r times the number of equations of S_{m+D}.

3. The membership problem in the case of a zero dimensional ideal

In this section, the duality properties of § 1 and § 2 will be our tools to design an admissible algorithm for zero dimensional ideals.

The *inputs* of the algorithm are:

– A set $\{f_1, \ldots, f_r\}$ of polynomials in $C[z]$ such that the generated ideal $I(f_1, \ldots, f_r)$ is zero dimensional.

– $d \in N_{\geq 3}, \ d \geq \deg(f_i) \quad \forall i = 1, \ldots, r$.

– $k \in N$.

The *output* of the algorithm is a matrix S with $\binom{n+k}{n}$ columns and at most $(d+1)^{2n^2}$ rows, satisfying:

– For any complex polynomial $P = \sum_{|\beta| \leq k} c_\beta z^\beta, \ P \in I$ iff the vector of coefficients (c_β) is a solution of the linear system $S \cdot \mathbf{X} = 0$.

– The entries of S can be computed in simply exponential in n (and polynomial in d and r) time.

– If $f_1, \ldots, f_r \in K[X_1, \ldots, X_n]$ for some subfield K of C, the entries of S also belong to K.

The sketch of the algorithm is as follows:

First step: Find $q_1, \ldots, q_n \in C[z]$ such that:

i) $Q := I(q_1, \ldots, q_n) \subseteq I$

ii) $\{q_1, \ldots, q_n\}$ is a regular sequence

iii) $\deg(q_i) \leq d \quad \forall i = 1, \ldots, n$.

The polynomials $q_1, \ldots, q_n$ can be effectively found in admissible time, by taking linear combinations of the data $f_1, \ldots, f_r$ of the form $f_1 + \gamma f_2 + \gamma^2 f_3 + \cdots + \gamma^{r-1} f_r$, with γ varying in any finite set $\Gamma \subseteq C$ with at least $r \cdot d^n$ elements. A complete proof can be given combining the following three ingredients: a) the proof of Prop. 3 in: Heintz, J.: Definability and Fast Quantifier Elimination in Algebraically Closed Fields. Theoretical Computer Science 24 (1983), 239-277; b) lemma 2.42 in: Giusti, M. and Heintz, J.: Algorithmes - disons rapides - pour la décomposition d'une variété algébrique en composantes irréductibles et équidimensionnelles. Submitted to MEGA 90; and c) Corollary (1.9.1) in [7] (for the bounds in computing the dimension of the ideal generated by a given sequence of linear combinations).

Second step: Find the matrix S' of a system of linear equations for $Q \cap \mathbf{C}[z]_{d^n + \max\{k,d\}}$.

The matrix S' can be effectively constructed (see § 1.2 a)) since Q is a complete intersection ideal.

Third step: Find the matrix S'' of a system of linear equations for $(Q : I) \cap \mathbf{C}[z]_{d^n}$.

Following Remark 2.4, one can obtain S'' from S' since $Q \subseteq (Q : I)$.

Fourth step: Find a system of generators $h_1, \ldots, h_s$ of $(Q : I)$.

By Proposition 2.3, it is enough to find a $\mathbf{C}$-basis $\{h_1, \ldots, h_s\}$ of the space of solutions of the system $S'' \cdot \mathbf{X} = 0$ constructed in the third step.

Notice that $\deg(h_i) \leq d^n \ \forall i = 1, \ldots, n$ and $s \leq \dim \mathbf{C}[z]_{d^n} = \binom{d^n + n}{n}$.

Fifth step: Find the matrix S of a system of linear equations for $I \cap \mathbf{C}[z]_k$.

By Theorem 2.2, we know that $I = (Q : (Q : I))$. Moreover, by the fourth step, we have generators $\{h_1, \ldots, h_s\}$ of $(Q : I)$ with degrees bounded by d^n. So, we can construct S from S' (obtained in the second step), following again Remark 2.4.

4. Dealing with radicals

4.1. Complete intersections.

Given a complete intersection ideal Q in $\mathbf{C}[z]$ and $k \in \mathbf{N}$, we will show how to construct a linear system for $\mathrm{rad}(Q) \cap \mathbf{C}[z]_k$ in admissible time.

We first need the following result:

Theorem 4.1.1. *Let* $Q = I(Q_1, \ldots, Q_p)$ *be a complete intersection ideal* $(\dim Q = n - p)$. *Let* $\mathcal{A}$ *denote the set of all increasing sequences A of p indexes $1 \leq A_1 < A_2 < \cdots < A_p \leq n$, and for each $A \in \mathcal{A}$, let's call*

$$M_A := \det\left(\frac{\partial Q_i}{\partial z_{A_j}}\right)_{1 \leq i,j \leq p} . \quad \text{Then, for any } f \in \mathbf{C}[z],$$

$$f \in \mathrm{rad}(Q) \Longleftrightarrow f \cdot M_A \in Q , \qquad \forall A \in \mathcal{A} .$$

In the particular case $\dim Q = 0$,

$$f \in \mathrm{rad}(Q) \Longleftrightarrow f \cdot Jac \in Q$$

(where $Jac := \det\left(\dfrac{\partial Q_i}{\partial z_j}\right)_{1 \leq i,j \leq n}$).

Proof: The proof is based on the two following facts:

i) There is a well defined residual current acting on $(n - p, n - p)$ compactly supported $\mathcal{C}^\infty$ forms

$$\mathrm{Res}\left[\frac{dQ_1 \wedge \cdots \wedge dQ_p}{Q_1 \ldots Q_p}\right] \equiv \frac{1}{(2\pi i)^p} \int_{[Q^{-1}(0)]} \qquad \text{(cf. [6]),}$$

where $[Q^{-1}(0)]$ denotes the intersection cycle (with integer multiplicities along each irreducible component).

ii) The generalized duality results ([8]) give in particular:

$$f \cdot \operatorname{Res} \left[\frac{dQ_1 \wedge \cdots \wedge dQ_p}{Q_1 \dots Q_p} \right] \equiv 0 \Longleftrightarrow f \cdot dQ_1 \wedge \cdots \wedge dQ_p \in Q \cdot \Omega^p(\mathbf{C}^n) \,.$$

Then,

$$f \cdot M_A \in Q, \quad \forall A \in \mathcal{A} \Longleftrightarrow f \cdot dQ_1 \wedge \cdots \wedge dQ_p \in Q \cdot \Omega^p(\mathbf{C}^n)$$

$$\Longleftrightarrow f \cdot \operatorname{Res} \left[\frac{dQ_1 \wedge \cdots \wedge dQ_p}{Q_1 \dots Q_p} \right] \equiv \frac{1}{(2\pi i)^p} \int_{[Q^{-1}(0)]} f \wedge \cdot \equiv 0 \Longleftrightarrow$$

$$f|_{Z(Q)} \equiv 0 \Longleftrightarrow f \in \operatorname{rad}(Q) \,.$$

Q. E. D.

Remark 4.1.2: Suppose that the linear projection $\pi : Z(Q) \longrightarrow \mathbf{C}^{n-p}$, $\pi(z) = (z_1, \dots, z_{n-p})$ has finite fibers, and let $A_0 := (n - p + 1, n - p + 2, \dots, n)$. Then, the conditions $f \cdot M_A \in Q, \ \forall A \in \mathcal{A}$ (in the above theorem) can be replaced by the single condition $f \cdot M_{A_0} \in Q$ (cf. [8], §4). In fact, one can always effectively find a system of coordinates x in "Noether position" for Q (for which $\pi' : Z(Q) \longrightarrow \mathbf{C}^{n-p}$, $\pi'(x) = (x_1, \dots, x_{n-p})$ has finite fibers) (cf. [7], §1).

4.1.3. The algorithm.

INPUTS:

- $Q_1, \dots, Q_p \in \mathbf{C}[z]$ $(p \le n)$ defining a complete intersection ideal Q.
- $d_1, \dots, d_p \in \mathbf{N}$ verifying $d_i = \deg(Q_i) \ \forall i = 1, \dots, p$.
- $k \in \mathbf{N}$.

OUTPUTS:

A matrix S verifying:

i) S is the matrix of a system of linear equations for $\operatorname{rad}(Q) \cap \mathbf{C}[z]_k$.

ii) S has at most $\left(k + \prod_{i=1}^{p} d_i + p \cdot d^n (d^p + 1) + \sum_{i=1}^{p} (d_i - 1) \right)^{n-p} \cdot \left(\prod_{i=1}^{p} d_i \right)^n$ rows (and $\binom{k+n}{n}$ columns), where $d \in \mathbf{N}_{\ge 3}$ and $d \ge \deg(Q_i) \ \forall i = 1, \dots, p$.

First step: Find a system of linear equations $S' \cdot \mathbf{X} = 0$ for $Q \cap \mathbf{C}[z]_{k+\sum_{i=1}^{p}(d_i-1)}$.

Second step: Find a system of linear equations $S \cdot \mathbf{X} = 0$ for $\operatorname{rad}(Q) \cap \mathbf{C}[z]_k$.

S can be obtained from S' following the Remark 2.4, because by theorem 4.1.1 $\operatorname{rad}(Q) = (Q : I(M_A, A \in \mathcal{A}))$.

We refer to [9] for the construction of S' and the bounds in ii).

4.2. The zero dimensional case.

Given an arbitrary zero dimensional ideal $I = I(f_1, \ldots, f_r)$ and $k \in \mathbf{N}$, we will show in 4.2.3 how to construct a linear system for $\mathrm{rad}(Q) \cap \mathbf{C}[z]_k$ in admissible time. As a consequence, we will have an effective method to find a system of generators of $\mathrm{rad}(I)$ in admissible time.

Proposition 4.2.1. *Let J be a zero dimensional radical ideal. For any ideal I verifying $J \subseteq \mathrm{rad}(I)$,*

$$\mathrm{rad}(I) = (J : (J : I)) .$$

Proof: Let $x \in \mathbf{C}^n$ and denote $\mathcal{M}_x$ the maximal ideal in $\mathcal{O}_x$.

In case $x \in Z(I)$, $I_x \subseteq \mathcal{M}_x$ and by the assumptions on J, $J_x = \mathcal{M}_x$. So

$$(J : (J : I))_x = (J_x : (J_x : I_x)) = (\mathcal{M}_x : \mathcal{O}_x) = \mathcal{M}_x = \mathrm{rad}(I)_x .$$

In case $x \notin Z(I)$, $I_x = \mathcal{O}_x$ and so

$$(J_x : (J_x : I_x)) = (J_x : J_x) = \mathcal{O}_x = \mathrm{rad}(I)_x .$$

Q. E. D.

Remark 4.2.2: Let $Q = I(Q_1, \ldots, Q_n)$ be a zero dimensional complete intersection ideal. By Theorem 4.1.1, $\mathrm{rad}(Q) = (Q : \mathrm{Jac})$. Then, for any ideal $I = I(f_1, \ldots, f_r)$, $(\mathrm{rad}(Q) : I) = (Q : I(f_1 \cdot \mathrm{Jac}, \ldots, f_r \cdot \mathrm{Jac}))$.

4.2.3. The algorithm.

INPUTS:
- $f_1, \ldots, f_r \in \mathbf{C}[z]$ defining a zero dimensional ideal I.
- $d \in \mathbf{N}_{\geq 3}$, $d \geq \deg(f_i)$ $\forall i = 1, \ldots, r$.
- $k \in \mathbf{N}$.

OUTPUT:

A matrix S verifying:

i) S is the matrix of a system of linear equations for $\mathrm{rad}(I) \cap \mathbf{C}[z]_k$.

ii) S has at most $(d+1)^{2n^2}$ rows (and $\binom{k+n}{n}$ columns).

First step: Same as first step in the algorithm in § 3.

Second step: Find the matrix S' of a system of linear equations for $Q \cap \mathbf{C}[z]_{d^n + n(d-1) + \max\{k,d\}}$.

Third step: Find the matrix S'' of a system of linear equations for $(\mathrm{rad}(Q) : I) \cap \mathbf{C}[z]_{d^n}$.

Following Remark 2.4, one can obtain S'' from S' since $(\mathrm{rad}(Q) : I) = (Q : I(f_1 \cdot \mathrm{Jac}, \ldots, f_r \cdot \mathrm{Jac}))$ by Remark 4.2.2.

Fourth step: Find a system of generators $h_1, \ldots, h_s$ of $(\mathrm{rad}(Q) : I)$.

As $Q \subseteq (\mathrm{rad}(Q) : I)$ and $\deg(Q) \leq d^n$, it is enough to find a C-basis $h_1, \ldots, h_s$ of solutions of the system $S'' \cdot \mathbf{X} = 0$ constructed in the third step (by Proposition 2.3).

Notice that $\deg(h_i) \leq d^n$ and $s \leq \binom{d^n+n}{n}$.

Fifth step: Find the matrix S of a system of linear equations for $\mathrm{rad}(I) \cap \mathbf{C}[z]_k$.

By Proposition 4.2.1, $\mathrm{rad}(I) = (\mathrm{rad}(Q) : (\mathrm{rad}(Q) : I))$. Moreover, by Remark 4.2.2,

$$\mathrm{rad}(I) = (Q : I(h_1 \cdot \mathrm{Jac}, \ldots, h_s \cdot \mathrm{Jac})) .$$

Consequently, S can be constructed from the matrix S' found in the second step, following Remark 2.4.

4.2.4. Generators of $\mathrm{rad}(I)$.

In case $I = I(f_1, \ldots, f_r)$ is a zero dimensional ideal with $\deg(f_i) \leq d$ $\forall i = 1, \ldots, r$, there exists (by Proposition 2.3) a system of generators of $\mathrm{rad}(I)$ with degrees bounded by d^n.

Let $k = d^n$ and S the corresponding output of the algorithm described in 4.2.3. Then, a C-basis of solutions of $S \cdot \mathbf{X} = 0$ provides a system of generators of $\mathrm{rad}(I)$.

REFERENCES

[1] Angeniol, B., *Résidus et Effectivité*, Preprint (1983).

[2] Berenstein, C., Yger, A., *Bounds for the degrees in the division problem*, Preprint Univ. of Maryland (1989).

[3] Brownawell, W.D., *Bounds for the degrees in the Nullstellensatz*, Annals of Math. **126** (1987), 577–591.

[4] Caniglia, L., Galligo, A., Heintz, J., *Some new effectivity bounds in Computational Geometry*, in "Applied Algebra, Algebraic Algorithms and Error Correcting Codes. Proc. 6th Int'l Conf., Rome 1988," (Ed. T.Mora), Springer LN Comput. Sci. 357, 1989, pp. 131–151.

[5] Chistov, A.L., Grigor'ev, D.Yu., *Subexponential time solving systems of algebraic equations*, LOMI preprints E-9-83 E-10-83, Leningrad (1983).

[6] Coleff, N., Herrera, M., "Les Courants Residuels Associés à une Forme Meromorphe," Springer LN Math. 633, 1978.

[7] Dickenstein, A., Fitchas, N., Giusti, M., Sessa, C., *The membership problem for unmixed polynomial ideals is solvable in single exponential time*, in "Discrete Applied Algebra, Proc. AAECC-7, Toulouse 1989" (to appear).

[8] Dickenstein, A., Sessa, C., *Canonical Representatives in Moderate Cohomology*, Invent. Math. **80** (1985), 417–434.

[9] Dickenstein, A., Sessa, C., *An Effective Residual Criterion for the Membership Problem in* $C[z_1, \ldots, z_n]$, J. Pure Appl. Algebra (to appear).

[10] von zur Gathen, J., *Parallel arithmetic computations. A survey*, in "Proc. 13th Symp. MFCS 1986," Springer LN Comput. Sci. 233, 1986, pp. 93–112.

[11] Giusti, M., *Complexity of standard bases in projective dimension zero*, Preprint Ecole Polytechnique Paris (1987).

[12] Griffiths, P., Harris, J., "Principles of Algebraic Geometry," John Wiley & Sons, 1978.

[13] Hartshorne, R., "Residues and Duality," Springer L.N. Math. 20, 1966.

[14] Lazard, D., *Algèbre linéaire sur* $K[x_1, \ldots, x_n]$ *et élimination*, Bull. Soc. Math. France **105** (1977), 165–190.

[15] Lipman, J., *Dualizing sheaves, differentials and residues on algebraic varieties*, Astérisque **117** (1984).

[16] Mayr, E., Meyer, A., *The complexity of the word problem for commutative semigroups and polynomial ideals*, Advances in Math. **46** (1982), 305–329.

[17] Serre, J.P., *Géometrie Algébrique et Géometrie Analytique*, (G.A.G.A.), Annales de l'Institut Fourier **VI** (1956), 1–42.

[18] Teissier, B., *Résultats récents d'algèbre commutative effective*, in "Séminaire Bourbaki, 42ème année, 1989-90," n° 718, pp. 1–19.

[19] Zariski, O., Samuel, P., "Commutative Algebra," Vol. 1, Van Nostrand, New York, 1958.

Alicia Dickenstein - Carmen Sessa
Departamento de Matemática - FCEyN
Universidad de Buenos Aires
Ciudad Universitaria - Pabellón I
(1428) Buenos Aires, Argentina.

Exemples d'ensembles de Points en Position Uniforme

ANDRÉ GALLIGO

Introduction

Un ensemble E de n points d'un espace affine, dont les équations forment un idéal de polynômes de colongueur n, est en position uniforme si pour tout sous-ensemble de n' points E' de E la fonction de Hilbert de E' ne dépend que de n'.

Cette condition exprime une grande symétrie des points de E, elle apparait dans les problèmes de classification de courbes lorsqu'on considère des sections hyperplanes générique.

Bien qu'on démontre de jolies propriétés sur de tels ensembles, notamment que presque tout ensemble de points est en position uniforme, on dispose de peu d'exemples non triviaux donnés explicitement.

L'une des possibilités du Calcul Formel (Computer Algebra) étant de fournir des exemples à la Géométrie Algébrique, on a essayé de répondre à cette préoccupation. Le mode de calcul proposé ne découle pas directement des définitions mais nécessite des développements auxiliaires.

On introduit la notion plus restrictive d'ensemble de points G-symétrique qui, bien que moins facile à visualiser, se prête mieux aux calculs d'exemples. Plus précisemment, si k est le plus petit corps contenant les coefficients d'équations définissant E, on demande que l'équation de la projection de E sur une droite générale soit irréductible sur k et ait pour groupe de Galois le groupe symétrique.

On compare ensuite les deux notions et on est conduit à utiliser le principe de Dedekind, cf. Van der Warden [25] ou les exercices de Bourbaki [1], qui permet de construire explicitement de tels polynômes. Cette procédure a été é galement étudiée par J. Heintz [17], en vue d'applications à la théorie de la complexité de calcul.

On produit ainsi une grande famille d'exemples qui permet l'expérimentation sur ordinateur. Dans cet article, pour simplifier notre propos, nous nous plaçons en caractéristique zéro et on ne traite essentiellement que le cas de points du plan. La généralisation à des ensembles de points dans des espaces de dimension plus grande est abordée au dernier paragraphe. La caractéristique positive ainsi qu'une présentation plus générale de notre approche sera traitée dans un prochain travail.

On démontre deux théorèmes d'existence sur les ensembles G-symétriques ayant une fonction de Hilbert fixée ou certains degrés de générateurs fixés. Enfin, on discute brièvement d'algorithmes adaptés pour le calcul de la fonction de Hilbert et du nombre minimum de générateurs de l'idéal homogène définissant un ensemble de points.

1   Définitions et rappels

On note k un corps, K sa cloture algébrique, $A = k[x, y]$ l'anneau des polynômes, $B = K[x, y]$. I étant un idéal de A, on note I_m le sous espace vectoriel des polynômes de degré $\leq m$. La fonction de Hilbert de A/I est $H_{A/I}(m) = \dim_k(A_m) - \dim_k(I_m)$.

Un idéal I de A de colongueur $n = \dim_k(A/I)$ définit une famille E de n points simples dans l'espace affine $\mathbf{A}_2(K)$ si IB se décompose en une intersection d'idéaux maximaux. Ces points sont dits en *position uniforme* ssi pour tout idéal J de B contenant IB (donc définissant une sous famille de E), la fonction de Hilbert $H_{B/J}$ ne dépend que de la colongueur $\dim_K(B/J)$.

On munit les monômes de l'ordre "diagonal": $x^i y^j < x^a y^b$ ssi $i + j < a + b$ ou ($i + j = a + b$ et $j < b$), tout P de A s'écrit $P = cx^a y^b +$des termes inferieurs, ce qui permet de définir l'exposant et le terme principal d'un polynôme, puis l'escalier $E(I)$:

$$(a, b) = \exp(P), \ cx^a y^b = lt(P), \ E(I) = \{\exp(P) \mid P \in I\}.$$

A partir d'un système de générateur d'un idéal I de A et par pseudo-divisions successives dans A, on peut par l'algorithme de Buchberger calculer une base standard (on dit aussi une base de Groebner réduite) de I: c'est une famille de polynômes P_i ($1 \leq i \leq s$) de I telle que $E(I) = \cup_{1 \leq i \leq s}(\exp(P_i) + \mathbf{N}^2)$ et qui est minimale pour cette propriété.

Un algorithme de division généralisé permet alors pour tout polynôme P de calculer l'unique polynôme $R = red(P)$ égal à P modulo I et dont aucun exposant de monôme n'est dans $E(I)$.

Comme la graduation par l'ordre "diagonal" raffine la graduation par le degré, la connaissance de $E(I)$ implique celle de la fonction de Hilbert de A/I. (cf. par exemple [10], [4], [22]). D'où:

Lemme 1 *La fonction de Hilbert de A/I se calcule à partir d' un système de générateurs de l'idéal I par un algorithmique qui n'effectue que des opérations rationnelles sur le corps k.*

Si k est assez grand (notamment si k est de caractéristique 0 donc infini) il existe une projection de E sur une autre droite qui évite l' ensemble fini des droites passant par deux points de E, donc qui envoie E sur n points distincts. D'où:

Lemme 2 *Après changement linéaire de coordonnées, tout idéal I de colongueur n de A peut être engendré par deux polynômes $\{g, y-h\}$ où g et h sont des polynôme de $k[x]$, g de degré n et h de degré strictement inférieur à n. De plus g et h peuvent être calculés par un algorithmique qui n'effectue que des opérations rationnelles sur le corps k.*

Preuve: Il suffit de calculer une base standard relativement à l'ordre lexicographique.

Par ailleurs, on a le résultat "classique" suivant sur les fonctions de Hilbert.

Commençons par remarquer que la fonction de Hilbert H d'un idéal I du type précédent est croissante, par contre son saut $H^1(l) = H(l) - H(l-1)$ commence par croitre puis éventuellement stationne, puis décroît jusqu'à s'annuler. Si lors de cette décroissance, elle décroit toujours strictement, on dira qu'elle est *de type décroissant* (on dit aussi que le caractère numérique correspondant est connexe ou sans lacune). Si ce n'est pas le cas, on aura $H^1(l+1) = H^1(l)$ et on dira qu'elle est lacunaire en degré $l+1$.

Lemme 3 (de décomposition): *Si la fonction de Hilbert de I est lacunaire en degré l, alors il existe un polynôme non constant qui divise tous les générateurs de degré ≤ 1 d'une base standard (de Groebner) de I. De plus ce polynôme est à coefficients dans le corps k.*

Preuve: Cf. par exemple [6], Th 4.1, ou [14]. Briévement: Pour faciliter l'argument on se place dans des coordonnées génériques. $E(I)$ a des "marches de hauteur 1" donc les relations élémentaires entre les générateurs de degré $\leq l$ de la base standard réduite (qui s'obtiennent à partir des S-polynômes) ne font pas intervenir les générateurs de degré $> l$. En effet, ces relations ont un degré total $2l + 1$ et l'hypothèse implique qu'il n'y a pas de générateur de degré $l + 1$. La matrice des relations se découpe ainsi en quatre blocs dont un formé de zéros. Comme les générateurs de la base standard sont les mineurs maximaux de cette matrice, on en déduit le résultat annoncé.

Remarque 4 Dans la thèse de Coppo [5], on trouvera une version plus géométrique de ce lemme et des applications à l'étude des schémas de Hilbert ponctuels.

2   Ensembles de points G-symétriques.

Définition 5 Soit E un ensemble de n points du plan A_2 définis par des équations $\{g, y - h\}$, où g et h sont des polynôme de $k[x]$, g de degré n et

h de degré strictement inférieur à n. On dira que E est G-symétrique ssi g est irréductible dans $k[x]$ et si son groupe de Galois sur k est le groupe symétrique $\mathbf{S}_n$.

Proposition 6 *Si E est G-symétrique, tout sous-ensemble de E est G-symétrique.*

Preuve: En effet, si E est défini par $\{g, y - h\}$, notons L le corps de décomposition de g sur k, et soit $g = c(x - a_1) \ldots (x - a_n)$ une décomposition de g dans $L[x]$. Un sous-ensemble à $n - n'$ points de E est défini par $\{g_1, y - h_1\}$ où g_1 est de la forme g/g_2 avec $g_2 = (x - a_{\sigma(1)}) \ldots (x - a_{\sigma(n')})$, σ étant une permutation de $\mathbf{S}_n$. On en déduit que g_1 appartient à l'extension algébrique k' de k, $k' = k(a_{\sigma(1)}, \ldots, a_{\sigma(n')})$. Le groupe de Galois de g étant le groupe symétrique, cela implique que g_1 est irréductible sur k' et que son groupe de Galois est le groupe symétrique $\mathbf{S}_{n-n'}$.

Proposition 7 *Si un ensemble E est défini par l'idéal I engendré par les équations $\{g, y - h\}$ et si g est irréductible, les générateurs d'une base standard réduite de I sont irréductibles.*

Preuve: En effet, l'anneau quotient $k[x, y]/I = k[x]/(g)$ est alors intègre. Si un générateur de la base standard de I se décomposait $f = f_1 f_2$, comme $\exp(f) = \exp(f_1) + \exp(f_2)$ on aurait $\exp(f_1) \notin E(I)$ par minimalité de $\exp(f)$ relativement à l'ordre produit de $\mathbf{N}^n$, donc $f_1 \notin I$.

Du Lemme 3 de décomposition on déduit:

Proposition 8 *Si E est G-symétrique, défini par un idéal I, la fonction de Hilbert $H_{A/I}$ est de type décroissant.*

Remarque 9 La réciproque de la proposition 7 est fausse.

Il suffit de considérer l'exemple de trois points du plan d'abscisses -1, 0, 1, et d'ordonnées α, 0, β dans une extension algébrique k de $\mathbf{Q}$. Par ces points passent trois coniques irréductibles dont les équations forment une base standard: $y - ax^2 - bx$, $x - a'y^2 - b'y$, $(x - a'')(y - b'') - 1$, les exposants principaux sont $(2,0)$, $(0,2)$, $(1,1)$. Pourtant on a $g = x(x - 1)(x + 1)$.

3   G-symétrique implique position uniforme.

Théorème 10 *Un ensemble de points E qui est G-symétrique est en position uniforme.*

Preuve: Reprenons les notations précédentes: E est défini par $\{g, y - h\}$ à coefficients dans k, L le corps de décomposition de g sur k, et soit $g = c(x - a_1) \ldots (x - a_n)$ une décomposition de g dans $L[x]$.

Un sous-ensemble à $n - n'$ points de E est défini par un idéal J_σ de $L[x]$ engendré par $\{g_\sigma, y - h_1\}$ où g_σ est de la forme g/g_2 avec $g_2 =$

$(x - a_{\sigma(1)}) \ldots (x - a_{\sigma(n')})$, σ étant une permutation de $\mathbf{S}_n$. Le polynôme h_1 est le reste de la division de h par g_σ.

Soient σ et τ deux permutations de $\mathbf{S}_n$. Le groupe de Galois de L sur k étant le groupe symétrique $\mathbf{S}_n$, il existe un automorphisme ϕ de L sur k qui échange les générateurs de J_σ et J_τ.

L'automorphisme ϕ conjugue aussi les polynômes obtenus par des opérations rationnelles, donc les deux bases standard (de Groebner) de J_σ et J_τ. Deux polynômes conjugués ayant les mêmes exposants, on a $E(J_\sigma) = E(J_\tau)$ donc les mêmes fonctions de Hilbert.

Remarque 11 La réciproque est fausse.

En effet, il suffit de considérer un ensemble E formé par quatre points du plan d'abscisses entières, disons 0, 1, 2, 3, et d'ordonnées entières $0, b_1, b_2, b_3$. On remarque que $g = x(x-1)(x-2)(x-3)$ n'est pas irréductible sur $\mathbf{Q}$. Cependant si E ne contient pas trois points alignés, il sera en position uniforme, puisqu'il n'y a que deux fonctions de Hilbert possibles pour un ensemble de trois points. On choisit par exemple $b_1 = 1$, $b_2 = 3$, $b_3 = 7$, les pentes des droites sont alors 1, 2, 3, 4, 6, 7 qui sont distinctes.

Remarque 12 On n'a même pas une réciproque faible en imposant des conditions supplémentaires draconniènes du type suivant:

Assertion fausse: Soit E un ensemble de n points en position uniforme défini par un idéal I dont une base standard pour l'ordre diagonal est formée de polynômes irréductibles sur le corps k de ses coefficients. Si cette propriété est encore vraie pour une famille de sous-ensembles de E de cardinal compris entre 1 et n, alors E est G-symétrique.

En effet, il suffit de reprendre l'exemple des trois points du paragraphe 2.

4   Exemples d'ensembles G-symétriques

Nous proposons de prendre pour corps k des coefficients: $\mathbf{Q}$ ou une extension algébrique de $\mathbf{Q}$ et d'exploiter une remarque du manuel d'Algèbre de van der Wærden [25], p 200, qui fait référence à un travail de van der Warden de 1934, [26] et qui affirme qu'on peut construire un grand nombre de polynôme irréductible de degré n à coefficients entiers bornés par une constante $N(n)$.

On trouve également dans les exercices du chapitre 5 d'Algèbre de Bourbaki, (exercices 11 et 12, 412, p 159 de [1]) la description du même "critère de Dedekind":

Soit f un polynôme de $\mathbf{Z}[x]$ de degré n, et soient p_1, p_2, p_3 trois nombres premiers distincts ne divisant pas le coefficient dominant de f; soient f_i, $i = 1, 2, 3$ les images canoniques de f dans $\mathbf{Z}/p_i\mathbf{Z}[x]$. On suppose que f_1 est *produit d'un facteur linéaire et d'un facteur irréductible de degré $n-1$, que f_2 est produit d'un facteur du second degré et de facteurs irréductibles de degré impairs, enfin que f_3 est irréductible. Alors f est irréductible dans $\mathbf{Q}[x]$ et son groupe de Galois sur $\mathbf{Q}$ est le groupe symétrique $\mathbf{S}_n$.*

Par ailleurs, on trouve dans les exercices de Bourbaki d'autres critères dans le cas où le degré n de g est premier. Ceci laisse supposer qu'il existe une littérature "constructiviste" abondante sur le sujet.

Enfin, J. Heintz dans [17] a étudié la complexité de calcul d'une famille de polynômes construits suivant un principe analogue et "faciles à calculer", il en a tiré deux applications sur l'estimation de bornes inférieures.

Ceci implique en particulier qu'il y a une infinité de g vérifiant le théorème.

On peut alors faire varier les coefficients de h dans $\mathbf{Q}$ ou dans une extension algébrique k de $\mathbf{Q}$ sur lequel le groupe de g est encore $\mathbf{S}_n$ pour obtenir une grande famille d'exemples.

Pour illustrer, traitons "à la main" le cas bien connu $g(x) = x^5 - x - 1$, $h(x) = bx^4$ on obtient les 2 fonctions de Hilbert distinctes qui sont de type décroissant:

$$b = 0 \quad H(1) = 2 \quad H(2) = 3 \quad H(3) = 4 \quad H(4) = 5$$
$$b = 1 \quad H(1) = 3 \quad H(2) = 5$$

On donne aussi une famille d'exemples obtenus en utilisant le système de calcul formel SISYPHE développé par l'INRIA et l'Université de Nice [12].

On prend $g = x^{12} + 13x^{11} - 15x^{10} + 12x^9 + 6x^8 + 6x^7 + 24x^6 + 22x^5 + 9x^4 + 6x^3 + 18x^2 + 34x + 3$, qui est calculé par le procédé décrit plus haut.

Pour n=12, il y a 5 fonctions de Hilbert de type décroissant H_1, H_2, H_3, H_4, H, qu'on représente par le dessin de leurs escaliers génériques: il suffit de compter le nombre de points sous l'escalier. On vérifie aisément qu'ils sont obtenus pour les idéaux $I_i = (g, y - h_i)$, avec $h_1 = 0$; $h_2 = x^2$; $h_3 = x^3$; $h_4 = x^3 + ax^2$, $a = 13/4$; $h_5 = x^4$.

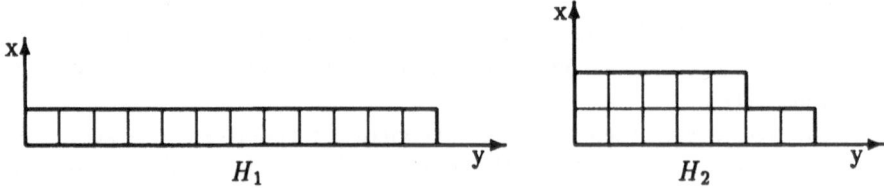

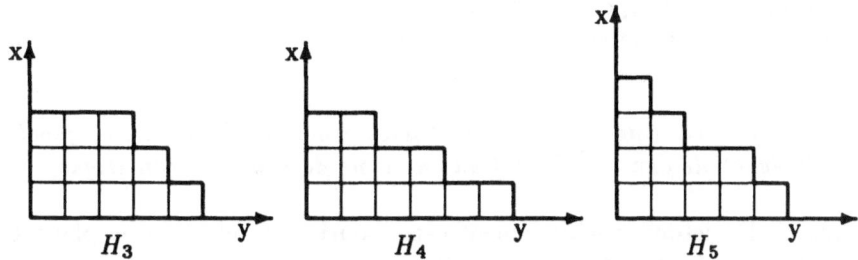

5   Quelques questions d'existence.

On vient de voir qu'on a une infinité d'exemples d'ensembles de points G-symétriques, mais la première question naturelle est:

Q1) Toute fonction de Hilbert de type décroissant est elle la fonction de Hilbert d'un ensemble de points G-symétriques?

Dans [19], il est démontré que cette propriété est vraie pour les points en position uniforme. Ce résultat est aussi une conséquence de [15], th 2.5 et du lemme de position uniforme de [16].

Les questions suivantes sont également pertinentes:

Q2) Toute fonction de Hilbert de type décroissant peut elle être obtenue avec $k = \mathbf{Q}$? ou avec une "petite" extension algébrique de $\mathbf{Q}$?

Q3) Peut elle être obtenue pour toute ou "presque toute" g du bon degré et G-symétrique?, y a -t- il des différences de comportement entre les différentes g? Comment le lire sur l'expression de g?

Q4) Toute fonction de Hilbert de type décroissant est elle la fonction de Hilbert d'un ensemble de points en position uniformes mais non G-symétriques?

Une autre voie qui pourrait être suivie est celle d'explorer les ensembles de points G-symétriques qui sont contenus dans des ensembles G-symétriques déja construits, notamment dans des intersections complètes, ainsi que leurs complémentaires (on dit résiduel ou en liaison). Il faut bien remarquer que lorsqu'on considère de tels sous-ensembles le corps des coefficients a tendance à augmenter.

Enfin, il serait intéressant d'affiner ces questions en tenant compte du nombre minimaux de générateurs de ces ensembles dans P^2, cf. par exemple le travail de T. Sauer [23]. On donnera un algorithme de calcul de ce nombre au paragraphe 7.

Un système de calcul formel aidera à affiner notre intuition sur ce type de questions. Pour cela, il sera utile de constituer des tables de fonctions de Hilbert "atteignables".

Nous pouvons répondre affirmativement à Q1) en reformulant le résultat de Harris et en le couplant avec le résultat de Gruson et Peskine (ou [19]).

Proposition 13 *Une section plane "assez générale" d'une courbe gauche lisse irréductible dans $P^3(C)$ est un ensemble de points G-symétrique.*

Preuve: Projetons génériquement cette courbe sur une courbe plane C puis sur une droite. C est un revêtement ramifié de la droite, on peut toujours supposer que le revêtement est étale à l'infini et se placer dans le cas affine. On obtient ainsi pour la restriction C' de C, une équation $g(x,y)$ irréductible dans $C[x,y]$ de degré n et monique en y, on considère qu'on projète sur l'axe des x. D'après Harris [16], le groupe de monodromie du revêtement qui agit sur une fibre générique (donc comportant n points) est le groupe symétrique S_n. On peut interpréter ce résultat en terme d'extensions de corps: $g(x,y)$ est irréductible dans $C(x)[y]$ et son groupe de Galois est S_n (voir [9], 1.8). Considérons maintenant la plus petite extension k de Q qui contient les coefficients de g et sa cloture algébrique K. Si x_0 est un élément transcendant sur k et K la cloture algébrique $k(x_0)$, alors $g(x_0,y)$ est clairement irréductible dans $K[y]$ et son groupe de Galois est nécessairement S_n.

En effet le groupe de Galois ne peut que diminuer par extension de corps. Donc la fibre de C au dessus de tout x_0 de C transcendant sur K est un ensemble de points G-symétrique. Plus précisément, on peut voir que l'ensemble des points x_0 tels que $g(x_0,y)$ est réductible dans $K[y]$ ou tels que son groupe de Galois n'est pas S_n, forme un fermé algébrique. On en déduit le:

Théorème 14 *Toute fonction de Hilbert de type décroissant est la fonction de Hilbert d'un ensemble de points G-symétriques*

6 Calcul de la fonction de Hilbert

On considère comme entrées un polynôme g irréductible de degré n à coefficients entiers et un polynôme h de degré $n-1$ à coefficients rationnels (dans une première étape). Le polynôme g pourra être choisi irréductible et à groupe de Galois symétrique, mais cela n'intervient pas dans l'algorithme que nous allons décrire.

Pour effectuer de l'expérimentation, il faudra lancer souvent l'algorithme de calcul de la fonction de Hilbert de $Q[x,y]/I$. Il est donc utile de le rendre plus efficace que le calcul habituel de base de Groebner.

On propose d'exploiter deux faits particuliers:

- I est engendré par deux générateurs $g(x)$, $h(x) - y$
- On connait a priori la colongueur n de I.

On va donc constituer un pré-processeur, puis utiliser un algorithme de complétion où le test de colongueur remplacera le test de fin de calcul par la nullité des réduction des S-polynômes, cf. par exemple [13]. On pourrait utiliser aussi l'algorithme de changement d'ordre de [8].

On doit disposer des fonctions auxiliaires:

Reste(A, B, x) qui calcule le reste de la division de A par B, B est supposé monique en tant que polynômes de la variable x, la lettre y n'intervenant pas dans les termes principaux.

Reduc$(A, B, diag)$ qui calcule le résultat de la réduction de A par B: si le terme principal de B est $lt(B) = cx^a y^b$, on itère le remplacement (dans les monômes multiples de $x^a y^b$) de $lt(B)$ par $B' = B - lt(B)$, ce qui abaisse la graduation défini par l'ordre diagonal.

Red$(A; listeDiv)$ avec $listeDiv = \{B_1, \ldots, B_t\}$ calcule le résultat de l'application itérée et dans l'ordre indiqué des fonctions *Reduc*$(A, B_i, diag)$.

:= désigne l'affectation.

Range prend en argument une collection d'éléments de $\mathbf{N}^2$ et rend les générateurs minimaux du monoïde qu'ils engendrent $\{(a_0, b_0), \ldots, (a_s, b_s)\}$, $a_0 > \ldots > a_s$ et met à jour la variable $Min := a_s$;

colong? est un teste qui vérifie alors si l'on a: $b_0 = 0$ et $a_s = 0$ et $n = (a_0 - a_1)b_1 + (a_1 - a_2)b_2 + \ldots + a_s - 1b_s$.

Ajour est un procédure qui met à jour la liste des diviseurs *listeDiv* en éliminant ceux dont les exposants principaux ne sont pas les générateurs minimaux du monoïde.

PaireC est une procédure qui met à jour et ordonne la liste des paires critiques PC qui vont fournir des S-polynômes, en tenant compte des critères usuels et d'éventuelles stratégies; elle fournit le polynôme *Spoly* qui est le S-polynôme construit avec le premier élément de *PaireC*, puis suprime de la liste PC la paire critique correspondante. Rappelons que le S-polynôme de f et g est $Sp(f, g) = (f \cdot lt(g) - g \cdot lt(f))/\mathrm{pgcd}(lt(f), lt(g))$.

Hilb est une fonction qui prend en argument une liste d'éléments de $\mathbf{N}^2$: $(a_0, b_0), \ldots, (a_s, b_s)$ avec $a_0 > \ldots > a_s$; $a_s = 0$; $b_0 = 0$ et à l'aide d'une formule, rend la fonction de Hilbert correspondant à l'idéal monomial engendré par les monômes ayant ces exposants. Pour une généralisation de ce calcul voir [21].

Algorithme 15 (Preprocessing)
$A := g$; $B := h - y$;

```
    Repeat      B := Reste(A, B, x); A := B
    Until       deg(B) > deg_x(B);
a_0 := deg(A); b_0 := 0; a_1 := deg_x(B); b_1 := deg(B) - deg_x(B);
            { autrement dit exp(A) := (a_0, 0); exp(B) := (a_1, b_1) }
```

(Initialisation)

$listeDiv := \{A, B\}$; $Exp := \{\exp(A), \exp(B)\}$; $PaireC := \{(A, B)\}$;
(Boucle)
    While $(Min \neq 0)$ **or** $(Min = 0 \text{ and } colong? = \textbf{false})$ **Do**
    $PaireC$; $C := Red(Spoly, listeDiv)$
    If $C \geq 0$ **then**
        $listeDiv := listeDiv U \{C\}$;
        $Exp := Range(Exp \cup \{\exp(C)\})$
        $Ajour$;
(Sortie)
    $Hilb(Exp)$.

7 Degrés des générateurs minimums.

Lorsqu'on dispose d'une base standard (de Groebner) de l'idéal I de $k[x, y]$ pour l'ordre diagonal, en homogénéisant les générateurs on obtient des générateurs de l'ideal homogénéisé J de $k[x, y, T]$.

Le nombre de générateurs minimum de J est $m = \dim_k(J/MJ)$, M étant l'idéal maximal. En considérant les escaliers de J et de MJ et le fait que, par construction, aucun terme principal d'un générateur de J ne dépend de T, on déduit que $m = dim k(I/(x, y)I)$.

Ce nombre peut être calculé facilement à partir de la matrice B des relations élémentaires entre les générateurs de la base standard de I : $s + 1 - m$ est égal au rang de la matrice $(B_{j,i}(0, 0))$.

En voici une explication qui reprend une illustration de [24] et qui donne également les degrés de ces générateurs. Il s'agit de calculer le cardinal de $E(I) - E((x, y)I)$, donc de "combler" éventuellement les trous en pointillé s dans le dessin suivant

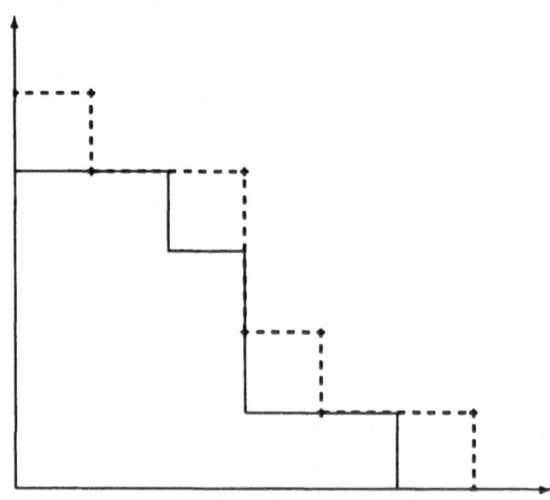

les coins sortants représentent les exposants principaux des générateurs de la base de Groebner de $I : f_0, \ldots, f_s$, les coins rentrants représentent les exposants des S-polynômes qui définissent les s relations élémentaires (qu'on suppose avoir déja calculé):

$$x^{a_j - a_{j-1}} f_j - y^{b_{j-1} - b_j} f_{j-1} = \sum_i A_{j,i}(x, y) f_i,$$

les pointillés représentent l'escalier intermédiaire défini par les $2(s + 1)$ générateurs f_i et y de $(x, y)I$ qu'il faut compléter en une base de Groebner, mais dont l'escalier $E((x, y)I)$ est bien sûr limité par $E(I)$; on remarque que les $2s + 1$ S-polynômes correspondants à ces générateurs sont soit nuls, soit égaux à des S-polynômes entre les f.

Il suffit donc de lire sur l'expression de la réduction de ces S-polynômes si celle-ci ne fait intervenir que des xf_i et des yf_i auquel cas le "nouveau reste" est nul, sinon elle produit un élément non nul de $I/(x, y)I$.

A l'aide de cette construction, on retrouve un résultat de Campanella repris dans [14], qui explicite les degrés possibles des générateurs minimums compatibles avec une fonction de Hilbert fixée. L'auteur l' explicite plus particulièrement dans l'hypothèse où les deux premiers générateurs n'ont pas de facteurs commun, ce qui est le cas pour les ensembles G-symétriques d' après la proposition 7.On dira que ces degrés sont admissibles.

Notre proposition 13 nous permet, comme au §5, de reprendre le théorème en question de [14], dont la démonstration est obtenue en juxtaposant les articles [23] et [14], afin de déduire le:

Théorème 16 *Etant donnée une fonction de Hilbert de type décroissant et des degrés minimum admissibles, il existe un ensemble de points G-symétrique qui les réalise.*

8   Ensembles de points dans $\mathbf{P}^m$, $m > 2$

Tout le début de notre étude: les paragraphes 1 (mis à part le lemme de décomposition), 2, 3, 4, 6, et le début du 7, se généralise immédiatement pour les ensembles de points de $\mathbf{P}^m$, $m > 2$.

Par contre, on ne connait pas encore les propriétés des fonctions de Hilbert des ensembles de points G-symétriques (ni en position uniforme). Les idéaux correspondants sont plus compliqués, par exemple ils ne sont plus determinantiels et la donnée de la fonction de Hilbert n'est plus équivalente à celle de l'escalier générique [11]. Il est donc important de disposer d'un outil d'expérimentation pour défricher ce sujet.

Conclusion

Pour calculer des exemples d'ensembles de points en position uniformes, nous avons introduit la notion d'ensemble de points G-symétrique. En nous appuyant sur des résultats de Gruson-Peskine, Harris, Maggioni-Ragusa, nous avons démontré que toute fonction de Hilbert de type décroissant est obtenue pour un ensemble G-symétrique.

Dans un prochain travail, nous reprendrons de façon constructive les preuves des résultats précédents afin de réaliser ces fonctions de Hilbert avec des "petits" corps de coefficients, on examinera aussi comment ces arguments s'adaptent en caractéristique positive. Enfin, en suivant la même idée et en reprenant les études faite dans [2] et [18], nous présenterons un nouveau théorème de structure sur les "schémas de dimension 0" de $\mathbf{P}^2$

Par ailleurs, nous comptons implémenter efficacement les différents algorithmes esquissés ici afin de réaliser des tables des différentes configurations d'ensembles de n points, pour des "petites" valeurs de n. Avec cet appareil calculatoire on espère pouvoir d'une part deviner (puis valider) les fonctions de Hilbert des ensembles de points en position uniforme de $\mathbf{P}^3$ voire $\mathbf{P}^m$ et dans une deuxième étape, examiner "expérimentalement" des relations d'incidence entre différentes strates de $Hilb^n(\mathbf{P}^2)$ considérées dans [3] et dans [5].

Remerciements: Je remercie A. Hirschowitz et M. A. Coppo qui m'ont mis en contact avec ce sujet et m'ont indiqué des références bibliographiques sur les ensembles de points en position uniforme.

REFERENCES

[1] N. Bourbaki, "Algèbre," ch 5, Hermann, 1981.

[2] J. Briançon, A. Galligo, *Déformations distinguées de points*, Astérisque n. 7 et 8 (1973), 129–138.

[3] J. Brun, A. Hirschowitz, *Le problème de Brill-Noether pour les idéaux de $\mathbf{P}^2$*, Ann. Sc. E.N.S. **20** (1987), 171–200.

[4] B. Buchberger, *Groebner bases: An algorithmic method in Polynomial Ideal Theory*, in "Multidimensional System Theory," D.Reidel Publishing Co., 1985, pp. 184–232.

[5] M.A. Coppo, "Familles maximales de points surabondants dans la plan projectif," Thèse, 1989. Université de Nice

[6] E. Davis, *Complete intersections . . . revisited*, Rend. Sem. Mat. Torino **43, 2** (1985), 333–353.

[7] Dubreil, *Sur quelques propriétés des systèmes de points dans le plan et des courbes gauches algébriques*, Bull. Soc. Math. France **61** (1933), 258–283.

[8] J. C. Faugère, P. Gianni, D. Lazard, T. Mora, *Efficient Computation of Zero-dimensional Gröbner Bases by Change of Ordering*, (submitted), J. Symb. Comp. (1989). Technical Report LITP 89-52

[9] O. Forster, "Lectures on Riemann Surfaces," Graduate Texts in Math. 81, Spinger, 1981.

[10] A. Galligo, "Algorithmes de calcul de bases standard," Pre-print Université de Nice, 1983.

[11] A. Galligo, *A propos du théorème de préparation de Weierstrass*, in "Séminaire Norguet," Springer Lect. Notes in Math. vol 409, 1974, pp. 543–579.

[12] A. Galligo, J. Grimm, L. Pottier, *The design of SISYPHE*, in "DISCO 1990," Springer Lect.N. in C.S. 429, 1990, pp. 30–39.

[13] A. Galligo, L. Pottier, C. Traverso, *GECD and standard basis completion algorithm*, in "ISSAC88," Springer Lect.N. in C.S. vol 358, 1989, pp. 162–176.

[14] A. V. Geramita, J. C. Migliore, "Hyperplane sections of smooth curves," Queen's Papers in Pure and Applied Math. vol 80, 1988.

[15] L. Gruson, Ch. Peskine, "Genres de courbes de l'espace projectif," Springer Lect. Notes in Math. 687, 1978.

[16] J. Harris, *The genus of space curves*, Math. Ann. **249** (1980), 191–204.

[17] J. Heintz, *On polynomials with symmetric Galois group which are easy to compute*, T. C. S. **47** (1986), 99–105.

[18] D. Lazard, *Ideal basis and Primary decomposition: case of two variables*, J. S. C **1** (1985), 261–270.

[19] R. Maggioni, A. Ragusa, *The hilbert function of generic plane sections of curves in P^3*, Invent. Math. **91** (1988), 253–255.

[20] Matzat, "Konstruktive Galoistheorie," Springer Lect.N. in Math, 1987.

[21] M. Moeller, T. Mora, *The computation of the Hilbert function*, in "Proc. Eurocal '83," Springer Lect Notes in C.S., 1983, pp. 157–167.

[22] L. Robbiano (ed.), "Computational aspect of commutative Algebra," Academic Press, 1989.

[23] T. Sauer, *The number of equations defining points in general position*, Pacific J. of Math. **120** (1985), 199–213.

[24] C. Traverso, *Groebner trace algorithms*, in "Proceedings ISSAC 88," Springer Lect. Notes in C.S. vol 358, 1989, pp. 125–138.

[25] B.L. van der Wærden, "Algebra," volume 1, F. Ungar, 1970.

[26] B.L. van der Wærden, *Zur Algebraischen Geometrie*, V, Math. Ann. **109** (1934), 128–133.

André Galligo
Département de Mathématiques
Université de Nice
Parc Valrose
F-06034 NICE cedex
galligo@mirsa.inria.fr
aussi:
équipe SAFIR à l'INRIA, Sophia Antipolis.

Efficient Algorithms and Bounds for Wu-Ritt Characteristic Sets

GIOVANNI GALLO* BHUBANESWAR MISHRA[†]

Abstract. The concept of a characteristic set of an ideal was originally introduced by J.F. Ritt, in the late forties, and later, independently rediscovered by Wu Wen-Tsün, in the late seventies. Since then Wu-Ritt Characteristic Sets have found wide applications in Symbolic Computational Algebra, Automated Theorem Proving in Elementary Geometries and Computer Vision. The original algorithm of Ritt, and subsequent modifications by Wu, has a non-elementary worst-case time complexity, and could be used for computing only an extended characteristic set. In this paper, we present optimal algorithms for computing a characteristic set with simple-exponential sequential and polynomial parallel time complexities. These algorithms are derived, via linear algebra, from simple-exponential degree bounds for a characteristic set. The degree bounds are obtained by using the recent effective version of Hilbert's Nullstellensatz, due to D. Brownawell and J. Kollár, and a version of Bezout's Inequality, due to J. Heintz.

1 Introduction.

In the late forties, J.F. Ritt, in his now classic book *Differential Algebra* [28], introduced an effective process to construct a triangular set of equations from a set of differential equations, with various interesting properties. The techniques of Ritt, when specialized to commutative algebra, have provided various interesting tools in Symbolic Computational Algebra. Recently, such triangular sets, derived from a set of polynomial equations, were independently rediscovered by Wu Wen-Tsün [29,30,31] in China, and have found applications in Mechanical Theorem Provers, e.g. *China Prover*. These triangular sets have come to be known as *Wu-Ritt Characteristic Sets* of the ideal generated by the given polynomials [8,15,19,28].

---

*supported by Italian Government Graduate Fellowship
[†]Supported by NSF Grant #DMS-87-03458, ONR Grant #N00014-89-J3042

With its increasing popularity, characteristic sets have begun to receive closer scrutiny, especially in its power in solving algebraic problems symbolically.

The original *Wu-Ritt Process*, first devised by Ritt, and subsequently modified by Wu, has a non-elementary worst-case time complexity, and could be used for computing only an extended characteristic set. To the best of our knowledge, even the constructivity of the more powerful characteristic set had not been explicitly demonstrated.

On the other hand, considerable amounts of experimental data are available, comparing the time-complexity of various implementations of Wu-Ritt Process with Gröbner Bases algorithms [7,8,9,21,24,25]. While the worst-case complexity of Gröbner Bases algorithms are fairly well-understood at this point [2,3,4,13,14,16,18,22,23,26,32], the computational complexity of Wu-Ritt Characteristic Sets has remained open.

In this paper, we resolve the complexity question for Wu-Ritt characteristic sets, as follows:

- We show that the degrees of the polynomials in a characteristic set of an ideal I in $k[x_1, \ldots, x_n]$ (where k is an arbitrary field) is bounded from both above and below, by a simple exponential function of n. More precisely, if d bounds the total degrees of the generators of a zero-dimensional ideal $I \subseteq k[x_1, \ldots, x_n]$, then we give a lower bound of $\Omega(d^n)$ and an upper bound of $O(d^{O(n)})$, for the degrees of polynomials in a characteristic set of I with respect to arbitrary ordering of the indeterminates. For non-zero-dimensional ideals, the corresponding lower and upper bounds are, respectively, $\Omega(d^{(n-\dim I)})$ and $O(sd^{O((n-\dim I)^2)})$, where s is the number of given generators of the ideal; in this case, the characteristic sets are assumed to be with respect to certain special orderings of the indeterminates.

- Using these bounds and via linear algebra, we derive several efficient parallel and sequential algorithms to compute characteristic sets. The most general classes of our algorithms have $O(s^{O(n)}d^{O(n^3)})$ sequential and $O(n^7 \log^2(s+d))$ parallel complexity. In particular, if the number of variables n is assumed to be a constant, our algorithm is in NC_2.

- Lastly, we present degree bounds and algorithms for the process of *generalized pseudo-division*, which plays a crucial role in Wu's China Prover. We give an $O(d^{O(n)})$ degree bound for g_0, the *generalized pseudo-remainder* of a polynomial g with respect to an ordered sequence of polynomials $\langle f_1, \ldots, f_r \rangle$, (the sequence of f_i's satisfy certain technical conditions), where all the polynomials have their degrees bounded by d.

- As a corollary of the above results, we see that a geometric theorem (involving no inequalities) can be verified in $O(2^{O(C(P+C)^3)})$ sequential time and $O(C^2(P + C)^7)$ parallel time, where it is assumed that the geometric theorem is presented by P points in the plane, and C constructions involving straight lines and circles.

Also, these results may be of independent interest to the algebraists, as it explores the power of the techniques based on some recent results of Brownawell and Kollár on the *Effective Hilbert's Nullstellensatz* [6,20].

The paper is organized as follows: In section 2, we provide a quick introduction to the concepts, relevant to characteristic sets, extended characteristic sets and Wu-Ritt Process. In section 3, we derive the degree bounds for characteristic sets. In this section, we also prove a single-exponential lower bound for the degree of an extended characteristic set. In section 4, we present efficient sequential and parallel algorithms for computing a characteristic set of an ideal. In section 5, we present quantitative versions of pseudo-division lemmas, with algorithms for computing the pseudo-remainders. We conclude the paper, with some discussion of the open problems.

2   Preliminaries.

Let $k[x_1, \ldots, x_n]$ denote the ring of polynomials in n indeterminates, with coefficients in a field k. Consider a fixed ordering on the set of indeterminates; without loss of generality, we may assume that the given ordering is the following:

$$x_1 \prec x_2 \prec \cdots \prec x_n.$$

Definition 2.1 Class and Class Degree. Let $f \in k[x_1, \ldots, x_n]$ be a multivariate polynomial with coefficients in k. An indeterminate x_j is said to be *effectively present in f*, if some monomial in f with nonzero coefficient contains a (strictly) positive power of x_j. For $1 \leq j \leq n$, *degree of f with respect to x_j*, $\deg_{x_j}(f)$, is defined to be the maximum degree of the indeterminate x_j in f. The *degree* [1] *of f* is defined to be $\deg(f) = \sum_{j=1}^{n} \deg_{x_j}(f)$.

The *class* and the *class degree* (cdeg) of a polynomial $f \in k[x_1, \ldots, x_n]$ with respect to a given ordering is defined as follows:

1. If *no indeterminate x_j is effectively present in f*, (i.e., $f \in k$) then, by convention,
   class$(f) = 0$ and cdeg$(f) = 0$; otherwise,

2. If *x_j is effectively present in f, and no $x_i \succ x_j$ is effectively present in f*
   (i.e., $f \in k[x_1, \ldots, x_j] \setminus k[x_1, \ldots, x_{j-1}]$) then class$(f) = j$ and cdeg$(f) = \deg_{x_j}(f)$.

---

[1] Our definition of *degree* is somewhat nonstandard; however, it makes the subsequent discussion considerably simpler.

Thus, with each polynomial $f \in k[x_1, \ldots, x_n]$, we can associate a pair of integers, its *type*:

$$\text{type} \; : \; k[x_1, \ldots, x_n] \to \mathbf{N} \times \mathbf{N}$$
$$: \; f \mapsto \langle \text{class}(f), \text{cdeg}(f) \rangle.$$

Definition 2.2 Ordering on the Polynomials. Given two polynomials f_1 and $f_2 \in k[x_1, \ldots, x_n]$, we say f_1 is of *lower rank than* f_2, $f_1 \prec f_2$, if either (i) $\text{class}(f_1) < \text{class}(f_2)$, or (ii) $\text{class}(f_1) = \text{class}(f_2)$ and $\text{cdeg}(f_1) < \text{cdeg}(f_2)$. This is equivalent to saying that the polynomials are ordered according to the lexicographic order on their types:

$$f_1 \prec f_2 \quad \text{if and only if} \quad \text{type}(f_1) <_{\text{lex}} \text{type}(f_2).$$

Note that there are distinct polynomials f_1 and f_2 that are not comparable under the preceding order. In this case, $\text{type}(f_1) = \text{type}(f_2)$, and f_1 and f_2 are said to be of the *same rank*, $f_1 \sim f_2$.

Thus, a polynomial f of class j and class degree d can be written as

(1)
$$f = I_d(x_1, \ldots, x_{j-1})x_j^d +$$
$$I_{d-1}(x_1, \ldots, x_{j-1})x_j^{d-1} + \cdots + I_0(x_1, \ldots, x_{j-1}),$$

where $I_l(x_1, \ldots, x_{j-1}) \in k[x_1, \ldots, x_{j-1}]$, for $l = 0, 1, \ldots, d$.

Definition 2.3 Initial Polynomial. Given a polynomial f of class j and class degree d, its *initial polynomial*, $\text{in}(f)$, is defined to be the polynomial $I_d(x_1, \ldots, x_{j-1})$ as in equation (1).

Lemma 2.1 Pseudo-Division Lemma.
Consider two polynomials f and $g \in k[x_1, \ldots x_n]$, with $\text{class}(f) = j$. Then there exist two polynomials q and r, and an integer α such that the following equation holds:
(2)
$$\text{in}(f)^\alpha g = qf + r,$$
where $\deg_{x_j}(r) < \deg_{x_j}(f)$ and $\alpha \leq \deg_{x_j}(g) - \deg_{x_j}(f) + 1$. If α is assumed to be the smallest possible power satisfying equation (2) then q and r are uniquely determined. In this case, the polynomial r is said to be the pseudo-remainder *of g with respect to f, and denoted by, $\text{prem}(g, f)$. We say a polynomial g is* reduced *with respect to f if $g = \text{prem}(g, f)$.*

Definition 2.4 Ascending Set.
A sequence of polynomials $\mathcal{F} = \langle f_1, f_2, \ldots, f_r \rangle \subseteq k[x_1, \ldots, x_n]$ is said to be an *ascending set* (or *chain*), if one of the following two conditions holds:

1. $r = 1$ and f_1 is not identically zero;

2. $r > 1$, and $0 < \text{class}(f_1) < \text{class}(f_2) < \cdots < \text{class}(f_r) \leq n$, and each f_i is reduced with respect to the preceding polynomials, f_j's $(1 \leq j < i)$.

Every ascending set is finite and has at most n elements. The *dimension* of an ascending set $\mathcal{F} = \langle f_1, f_2, \ldots, f_r \rangle$, $\dim \mathcal{F}$, is defined to be $(n - r)$.

Thus, with each ascending set $\mathcal{F}$ we can associate an $(n+1)$-vector, its *type*,

$$\text{type: Family of Ascending Sets} \rightarrow (\mathbf{N} \cup \{\infty\})^{n+1},$$

where ∞ is assumed to be greater than any integer. For all $0 \le i \le n$, the ith component of the vector is

$$\text{type}\,(\mathcal{F})\,[i] = \begin{cases} \text{cdeg}(g), & \text{if } (\exists!\ g \in \mathcal{F})\ \text{class}(g) = i; \\ \infty, & \text{otherwise.} \end{cases}$$

Definition 2.5 Ordering on the Ascending Sets. Given two ascending sets

$$\mathcal{F} = \langle f_1, \ldots, f_r \rangle \quad \text{and} \quad \mathcal{G} = \langle g_1, \ldots, g_s \rangle,$$

we say $\mathcal{F}$ is of *lower rank* than $\mathcal{G}$, $\mathcal{F} \prec \mathcal{G}$, if one of the following two conditions is satisfied,

1. There exists an index $i \le \min\{r, s\}$ such that

$$(\forall\ 1 \le j < i)\ f_j \sim g_j, \quad \text{and} \quad f_i \prec g_i;$$

2. $r > s$ and $(\forall\ 1 \le j \le s)\ f_j \sim g_j.$

Note that there are distinct ascending sets $\mathcal{F}$ and $\mathcal{G}$ that are not comparable under the preceding order. In this case $r = s$, and $(\forall\ 1 \le j \le s)\ f_j \sim g_j$, and $\mathcal{F}$ and $\mathcal{G}$ are said to be of the *same rank*, $\mathcal{F} \sim \mathcal{G}$.

Hence,

$$\mathcal{F} \prec \mathcal{G} \quad \textit{if and only if} \quad \text{type}\,(\mathcal{F}) <_{\text{lex}} \text{type}\,(\mathcal{G}).$$

Thus the map, *type*, is a partially ordered homomorphism from the family of ascending sets to $(\mathbf{N} \cup \{\infty\})^{n+1}$, where $(\mathbf{N} \cup \{\infty\})^{n+1}$ is ordered by the lexicographic order. Hence, the family of ascending sets endowed with the ordering '$\prec$' is a well ordered set.

Definition 2.6 Characteristic Set. Let I be an ideal in $k[x_1, \ldots, x_n]$. Consider the family of all ascending sets, each of whose component is in I,

$$\mathbf{S}_I = \Big\{ \mathcal{F} = \langle f_1, \ldots, f_r \rangle : \mathcal{F} \text{ is an ascending set and } f_i \in I, 1 \le i \le r \Big\}.$$

A minimal element in $\mathbf{S}_I$ (with respect to the $\prec$ order on ascending sets) is said to be a *characteristic set* of the ideal I.

It can be easily shown that if $\mathcal{G}$ is a characteristic set of I then

$$n \ge |\mathcal{G}| \ge n - \dim I.$$

For a given ordering of the indeterminates, characteristic set of an ideal is not necessarily unique. However, any two characteristic sets of an ideal must be of the same rank.

Lemma 2.2 Generalized Pseudo-Division Lemma.
Consider an ascending set $\mathcal{F} = \langle f_1, f_2, \ldots, f_r \rangle \subseteq k[x_1, \ldots, x_n]$, and a polynomial $g \in k[x_1, \ldots, x_n]$. Then there exists a sequence of polynomials

(called a pseudo-remainder sequence*), g_0, g_1, ..., $g_r = g$, such that for each $1 \leq i \leq r$, the following equation holds,*

$$(\exists! \, q_i') \, (\exists! \, \alpha_i) \, \left[\mathrm{in}(f_i)^{\alpha_i} g_i = q_i' f_i + g_{i-1} \right]$$

where g_{i-1} is reduced with respect to f_i and α_i assumes the smallest possible power, achievable. Thus, the pseudo-remainder sequence is uniquely determined. Moreover, each g_{i-1} is reduced with respect to f_i, f_{i+1}, ..., f_r.

$$(3) \qquad \mathrm{in}(f_r)^{\alpha_r} \mathrm{in}(f_{r-1})^{\alpha_{r-1}} \cdots \mathrm{in}(f_1)^{\alpha_1} g = \sum_{i=1}^{r} q_i f_i + g_0.$$

The polynomial $g_0 \in k[x_1, ..., x_n]$ is said to be the *(generalized) pseudo-remainder* of g with respect to the ascending set $\mathcal{F}$,

$$g_0 = \mathrm{prem}\,(g, \mathcal{F}).$$

By the earlier observations, g_0 is uniquely determined, and reduced with respect to f_1, f_2, ..., f_r. We say a polynomial g is *reduced with respect to* an ascending set $\mathcal{F}$ if $g = \mathrm{prem}\,(g, \mathcal{F})$.

For an ascending set $\mathcal{F}$, we describe the set of all polynomials that are *pseudo-divisible* by $\mathcal{F}$, by the following notation:

$$\mathcal{M}(\mathcal{F}) = \left\{ g \in k[x_1, ..., x_n] \colon \mathrm{prem}(g, \mathcal{F}) = 0 \right\}.$$

Proposition 2.1 *Let I be an ideal in $k[x_1, ..., x_n]$. Then the ascending set $\mathcal{G} = \langle g_1, ..., g_r \rangle$ is a characteristic set of I if and only if*

$$(\forall \, f \in I) \, \left[\mathrm{prem}(f, \mathcal{G}) = 0 \right].$$

For any set of polynomials $F \subseteq k[x_1, ..., x_n]$, we write $\mathcal{Z}(F)$, to denote its *zero set*:

$$\mathcal{Z}(F) = \left\{ \langle p_1, ..., p_n \rangle \in k^n \colon \left(\forall \, f \in F \right) f(p_1, ..., p_n) = 0 \right\}.$$

Proposition 2.2 *Let I be an ideal in $k[x_1, ..., x_n]$ with a characteristic set $\mathcal{G} = \langle g_1, ..., g_r \rangle$. Then*

$$\mathcal{Z}(\mathcal{G}) \setminus \left(\bigcup_{i=1}^{r} \mathcal{Z}\left(\mathrm{in}(g_i) \right) \right) \subseteq \mathcal{Z}(I) \subseteq \mathcal{Z}(\mathcal{G}).$$

The following triangulation process, due to J.F. Ritt and Wu Wen-Tsün, computes a so-called *extended characteristic set* of an ideal by repeated applications of the generalized pseudo-division.

Lemma 2.3 Ritt's Principle.
Let $F = \{f_1, ..., f_s\} \subseteq k[x_1, ..., x_n]$ be a finite nonempty set of polynomials, and $I = (F)$ be the ideal generated by F. The Wu-Ritt Process described below obtains an ascending set $\mathcal{G}$ such that either

    1. $\mathcal{G}$ consists of a polynomial in $k \cap I$, or

2. $\mathcal{G} = \langle g_1, \ldots, g_r \rangle$ with class(g_1) > 0 and such that

$$g_i \ \in \ I, \quad \text{for all } i = 1, \ldots, r,$$
$$\text{prem}(f_j, \mathcal{G}) \ = \ 0, \quad \text{for all } j = 1, \ldots, s.$$

Ascending chain $\mathcal{G}$ is called an extended characteristic set of F.

Algorithm 2.1

Wu-Ritt Process:
Input: $F = \{f_1, \ldots, f_s\} \subseteq k[x_1, \ldots, x_n]$.
Output: $\mathcal{G}$, an extended characteristic set of F.

$\mathcal{G} := \emptyset; \quad R := \emptyset;$
loop

$\quad\quad F := F \cup R; \quad F' := F; \quad R := \emptyset;$
$\quad\quad$**while** $F' \neq \emptyset$ **loop**
$\quad\quad\quad\quad$ Choose a polynomial $f \in F'$ of minimal rank;
$\quad\quad\quad\quad F' := F' \backslash \ \{g: \text{class}(g) = \text{class}(f) \ and$
$\quad\quad\quad\quad\quad\quad\quad\quad g \ is \ not \ reduced \ with \ respect \ to \ f\};$
$\quad\quad\quad\quad \mathcal{G} := \mathcal{G} \cup \{f\};$
$\quad\quad$**end{loop};**
$\quad\quad$**return** $f \in F \backslash \mathcal{G}$ **loop**
$\quad\quad\quad\quad$**if** $r := \text{prem}(f, \mathcal{G}) \neq 0$ **then**
$\quad\quad\quad\quad\quad\quad R := R \cup \{r\};$
$\quad\quad\quad\quad$**end{if};**
$\quad\quad$**end{loop};**
until $R = \emptyset;$
return$\mathcal{G};$
end{Wu-Ritt Process}.

In general, an extended characteristic set of an ideal is not a characteristic set of the ideal. However, an extended characteristic set *does* satisfy the following property, in a manner similar to a characteristic set.

Proposition 2.3 Let $F \subseteq k[x_1, \ldots, x_n]$ be a basis of an ideal I, with an extended characteristic set $\mathcal{G} = \langle g_1, \ldots, g_r \rangle$. Then

$$\mathcal{Z}(\mathcal{G}) \backslash \left(\bigcup_{i=1}^{r} \mathcal{Z}\left(\text{in}(g_i)\right) \right) \subseteq \mathcal{Z}(I) \subseteq \mathcal{Z}(\mathcal{G}).$$

We conclude this section, with the following technical lemma:

Lemma 2.4 Let $H = \{h_1, h_2, \ldots, h_n\} \subseteq k[x_1, x_2, \ldots, x_n]$ be a set of univariate monic polynomials, one in each variable:

$$h_j(x_j) \in k[x_j] \backslash k, \quad and \quad \text{in}(h_j) = 1.$$

Then $\mathcal{H} = \langle h_1, h_2, \ldots, h_n \rangle$ is a characteristic set of (H) with respect to the following ordering of the indeterminates:

$$x_1 \prec x_2 \prec \cdots \prec x_n.$$

Proof:[2] First, note that $\mathcal{H}$ is an extended characteristic set of H (with respect to the ordering $x_1 \prec x_2 \prec \cdots \prec x_n$); this is easily seen by an application of the *Wu-Ritt Process* to the set H. Thus

$$H \subseteq \mathcal{M}(\mathcal{H}).$$

We claim that the set $\mathcal{M}(\mathcal{H})$ is an ideal. Thus

$$(H) \subseteq \left(\mathcal{M}(\mathcal{H}) \right) = \mathcal{M}(\mathcal{H}),$$

and $\mathcal{H}$ is a characteristic set of (H), since $\left(\forall f \in (H) \right) \operatorname{prem}(f, \mathcal{H}) = 0$.

In order to prove the claim, we make the following simple observations:

1. Since $\operatorname{in}(h_j) = 1$, for all $1 \le j \le n$,

$$\left[\operatorname{prem}(g', \mathcal{H}) = g_0' \ \wedge \ \operatorname{prem}(g'', \mathcal{H}) = g_0'' \right] \Rightarrow$$
$$(A) \ \operatorname{prem}(g' + g'', \mathcal{H}) = g_0' + g_0''$$
$$(B) \ \operatorname{prem}(cg', \mathcal{H}) = cg_0', \quad \left(\forall c \in k \right).$$

2. Since $h_j \in k[x_j] \setminus k$, for all $1 \le j \le n$,

$$\left[\operatorname{prem}(g', \mathcal{H}) = 0 \right] \Rightarrow \operatorname{prem}(x_i g', \mathcal{H}) = 0, \quad \left(\forall x_i \right).$$

Assume that $h_i = x_i^d + c_{d-1} x_i^{d-1} + \cdots + c_0$, and let the following sequence

$$g_0'(= 0), g_1', \ldots, g_{i-1}', \ldots, g_n'(= g')$$

denote the sequence of pseudo-remainders, obtained by dividing g' by $\mathcal{H}$. If $\deg_{x_i} g_{i-1}' < (d-1)$ then, clearly

$$\operatorname{prem}(x_i g', \mathcal{H}) = \operatorname{prem}\left(x_i g', \langle h_{i-1}, \ldots, h_1 \rangle \right) = 0.$$

---

[2] A somewhat indirect proof of the above lemma may be obtained by first observing that H is a Gröbner basis of the ideal (H) (with respect to any admissible ordering), which immediately implies that

$$\left(\forall g \in (H) \right) \left(\forall 1 \le j \le n \right) \left[\deg_{x_j}(g) \ge \operatorname{cdeg}(h_j) \right].$$

Since, $\operatorname{class}(h_j) = j$, and since it has the minimal possible class degree, indeed, $\mathcal{H}$ is a characteristic set of (H).

Otherwise, $\deg_{x_i} g'_{i-1} = (d-1)$, and

$$
\begin{aligned}
x_i g'_{i-1} = {} & \alpha_d(x_1, \ldots, x_{i-1}, x_{i+1}, \ldots, x_n) x_i^d \\
& + \alpha_{d-1}(x_1, \ldots, x_{i-1}, x_{i+1}, \ldots, x_n) x_i^{d-1} + \cdots \\
& + \alpha_1(x_1, \ldots, x_{i-1}, x_{i+1}, \ldots, x_n) x_i,
\end{aligned}
$$

where $\operatorname{prem}(\alpha_k, \langle h_{i-1}, \ldots, h_1 \rangle) = 0$. But

$$
\operatorname{prem}(x_i g'_{i-1}, h_i) = (\alpha_{d-1} - c_{d-1} \alpha_d) x_i^{d-1} + \cdots + (\alpha_1 - c_1 \alpha_d) x_i - c_0 \alpha_d,
$$

and, by (1), $\operatorname{prem}((\alpha_k - c_k \alpha_d), \langle h_{i-1}, \ldots, h_1 \rangle) = 0$.
Hence

$$
\operatorname{prem}(x_i g', \mathcal{H}) = \operatorname{prem}\Big(\operatorname{prem}(x_i g'_{i-1}, h_i), \langle h_{i-1}, \ldots, h_1 \rangle\Big) = 0.
$$

Thus if g', $g'' \in \mathcal{M}(\mathcal{H})$ and $f \in k[x_1, \ldots, x_n]$ then, by observation (1A), $g' + g'' \in \mathcal{M}(\mathcal{H})$ and, by observations (1) and (2), $fg' \in \mathcal{M}(\mathcal{H})$. Q. E. D.

3   Degree Bounds

Let $I \subseteq k[x_1, \ldots, x_n]$ be an ideal generated by a set of s generators, $f_1, \ldots, f_s$, in the ring of polynomials in n indeterminates over the field k. Further, assume that each of the polynomial f_i in the given set of generators has its degree bounded by d:

$$
\Big(\forall\, 1 \le i \le s\Big)\ \deg(f_i) \le d.
$$

Let $\mathcal{G} = \langle g_1, \ldots, g_r \rangle$ be a characteristic set of the ideal, and let $D = \max\{\deg(g_i) : 1 \le i \le r\}$ be the degree of this characteristic set. In the next two subsections, we shall give upper and lower bounds for D, as function of s, d and n, for a large class of orderings on the indeterminates.

3.1   Lower Bounds

Theorem 3.1 Lower Bound Theorem.
   Let k be an algebraically closed field, and let I be an ideal in $k[x_1, \ldots, x_n]$, generated by the following n polynomials, $F = \{f_1, \ldots, f_n\}$, each of degree d:

$$
\begin{aligned}
f_1 &= x_1 - x_n^d, \\
f_2 &= x_2 - x_1^d, \\
&\ \ \vdots \\
f_{n-1} &= x_{n-1} - x_{n-2}^d, \\
f_n &= x_n - x_{n-1}^d.
\end{aligned}
$$

Then, independent of the ordering on the indeterminates,

(1)   Every characteristic set of F is of degree $D \geq d^n$;

(2)   Every extended characteristic set of F is of degree $D \geq d^n$.

proof: Let $\pi \in S_n$ be a permutation of $[1..n]$, and the arbitrary but fixed ordering on the indeterminates be the following

$$x_i = x_{\pi(1)} \prec x_{\pi(2)} \prec \cdots \prec x_{\pi(n)}.$$

Note that $I \cap \left(k[x_i] \setminus k \right)$ contains a nonzero polynomial $x_i - x_i^{d^n}$ of *minimal possible degree*, d^n.

(1)   Let $\mathcal{G} = \langle g_1, \ldots, g_r \rangle$ be a characteristic set of F, with respect to the chosen ordering. Since $\dim I = 0$,

$$|\mathcal{G}| = n, \quad \text{and} \quad g_1 \in I \cap \left(k[x_i] \setminus k \right).$$

Thus, $D \geq \mathrm{cdeg}(g_1) = \min\left\{ \deg(f) : f \in I \cap (k[x_i] \setminus k) \right\} = d^n.$

(2)   Let $\mathcal{G}' = \langle g_1', \ldots, g_r' \rangle$ be an extended characteristic set of F, with respect to the chosen ordering. By an examination of the Wu-Ritt process, we see that $\mathrm{in}(g_j') = 1$, for all $1 \leq j \leq r$, and by Lemma 2.3,

$$\mathcal{Z}(\mathcal{G}') = \mathcal{Z}(I) = \text{a finite set.}$$

Hence

$$|\mathcal{G}'| = n, \quad \text{and} \quad g_1 \in I \cap \left(k[x_i] \setminus k \right).$$

Thus, $D \geq \mathrm{cdeg}(g_1) = \min\left\{ \deg(f) : f \in I \cap (k[x_i] \setminus k) \right\} \geq d^n.$ Q. E. D.

In the subsequent discussion, we shall focus only on the characteristic sets of an ideal, since it is dubious whether an extended characteristic set improves the computational power of our algorithms. We remark that if we choose the following ordering on the indeterminates:

$$x_1 \prec x_2 \prec \cdots \prec x_n,$$

then a characteristic set of I of theorem 3.1 is given by,

$$\mathcal{G} = \left\langle \begin{array}{rcl} g_1 &=& x_1 - x_1^{d^n}, \\ g_2 &=& x_2 - x_1^d, \\ &\vdots& \\ g_{n-1} &=& x_{n-1} - x_1^{d^{n-2}}, \\ g_n &=& x_n - x_1^{d^{n-1}} \end{array} \right\rangle.$$

A slightly better bound of

$$\deg(f_1) \deg(f_2) \cdots \deg(f_n),$$

can be obtained, by a simple modification of the construction in theorem 3.1.

We note that the ideal I of theorem 3.1 is a zero-dimensional ideal; the previous example can be modified to show that, in general, the degree of a characteristic set is $\Omega(d^{(n-\dim I)})$. It is interesting to ask if a better lower bound can be constructed for more general cases.

3.2 Upper Bounds

We start by recalling the following version of *effective Hilbert's Nullstellensatz*, originally due to D. Brownawell [6] and later sharpened by J. Kollár [20].

Theorem 3.2 Effective Hilbert's Nullstellensatz.
Let $I = (f_1, \ldots, f_s)$ be an ideal in $k[x_1, \ldots, x_n]$, where k is an arbitrary field, and $\deg(f_i) \leq d$, $1 \leq i \leq s$. Let $h \in K[x_1, \ldots, x_n]$ be a polynomial with $\deg(h) = \delta$. Then

$$h \in \sqrt{I}$$

if and only if

$$\left(\exists\, b_1, \ldots, b_s \in k[x_1, \ldots, x_n] \right) h^M = \sum_{i=1}^{s} b_i f_i,$$

where $M \leq 2(d+1)^n$ and $\deg(b_i f_i) \leq 2(\delta+1)(d+1)^n, 1 \leq i \leq s$.

Using this version of Nullstellensatz, we shall first derive a degree bound for a characteristic set of a zero-dimensional ideal, and then use a 'lifting' procedure to obtain a bound for the more general cases.

3.2.1 Zero-Dimensional Case

The following proposition shows that every zero dimensional ideal I contains a univariate polynomial $h_j(x_j)$, in each variable x_j, with the additional property that $\deg(h_j) \leq 2(d+1)^{2n}$. Since the sequence $\langle h_1, \ldots, h_n \rangle$ is clearly an ascending set in $\mathbf{S}_I$, we get a similar bound on the class degrees of the polynomials in a characteristic set of I. This leads to a degree bound of $2n(d+1)^{2n}$ for a characteristic set of I.

Lemma 3.1 *Let $I = (f_1, \ldots, f_s)$ be a zero-dimensional ideal in $k[x_1, \ldots, x_n]$, where k is an arbitrary field, and $\deg(f_i) \leq d$, $1 \leq i \leq s$. Then for every indeterminate x_j, there exists a univariate polynomial $h_j \in I \cap k[x_j]$ such that*

$$\left(\exists\, b_{j,1}, \ldots, b_{j,s} \in k[x_1, \ldots, x_n] \right) h_j = \sum_{i=1}^{s} b_{j,i} f_i,$$

where $\deg(h_j) \leq 2(d+1)^{2n}$, and $\deg(b_{j,i} f_i) \leq 4(d+1)^{2n}, 1 \leq i \leq s$.

Proof: Suppose first that k is algebraically closed. Assume that the finite set of T zeroes of I are given by

$$\mathcal{Z}(I) = \left\{ P_i = \langle p_{i,1}, \ldots, p_{i,j}, \ldots, p_{i,n} \rangle \in k^n : 1 \le i \le T \right\} \ne \emptyset.$$

By the Bezout's inequality of J. Heintz [17], we have $T \le (d+1)^n$. Now consider the following monic polynomial $h'_j(x_j) \in k[x_j]$,

$$h'_j(x_j) = \prod_{i=1}^{t} (x_j - p_{i,j}),$$

where $p_{i,j}$ is the jth coordinate of the ith zero of I. As h'_j vanishes at all the zeros of I, by Hilbert's Nullstellensatz, it must be in the *radical* of I,

$$h'_j \in \sqrt{I}, \quad \text{that is,} \quad h_j = (h'_j)^M \in I.$$

But by construction h_j is a monic polynomial of degree TM, in $k[x_j]$. Moreover, by theorem 3.2, we see that $M \le 2(d+1)^n$, and

$$\deg(h_j) \le 2(d+1)^{2n}, \quad 1 \le j \le n.$$

The bound on the representation of h_j comes directly from theorem 3.2,

$$\deg(b_{j,i} f_i) \le 2(T+1)(d+1)^n \le 4(d+1)^{2n}.$$

If k is not algebraically closed the above construction can be done in its algebraic closure. Observe now that, since the degree of h_j is bounded, its coefficients are determined by a system of linear equations over k, and thus, $h_j \in k[x_j]$.   Q. E. D.

Corollary 3.1 *Let $I = (f_1, \ldots, f_s)$ be a zero-dimensional ideal in $k[x_1, \ldots, x_n]$, where k is an arbitrary field, and $\deg(f_i) \le d$, $1 \le i \le s$. Let $g \in k[x_1, \ldots, x_n]$ be a polynomial with $\deg(g) = \Delta$. Then*

$$g \in I$$

if and only if

$$\left(\exists\, a_1, \ldots, a_s \in k[x_1, \ldots, x_n] \right) g = \sum_{i=1}^{s} a_i f_i,$$

where $\deg(a_i f_i) \le \max\{d, \Delta\} + 6n(d+1)^{2n}, 1 \le i \le s$.

Proof: Assume that $g \in I$. Then, using the univariate, monic polynomials h_j's (obtained in lemma 3.1) as rewriting rules modulo $J = (h_1, \ldots, h_n) \subseteq I$, g may be expressed as follows:

$$g = h + \sum_{i=1}^{s} a'_i f_i$$

where

$$\deg(a_i' f_i) \leq \max_{1 \leq i \leq s} \Big\{ \deg(f_i) \Big\} + \sum_{j=1}^{n} \deg_{x_j}(h_j) \leq d + 2n(d+1)^{2n}.$$

Thus, we have

$$h = g - \sum_{i=1}^{s} a_i' f_i \in J, \quad \text{and} \quad \deg(h) \leq \max\Big\{ \Delta, d + 2n(d+1)^{2n} \Big\}.$$

But, by Lemma 2.4, $\mathcal{H} = \langle h_1, \ldots, h_n \rangle$ is a characteristic set of J. Since $h \in J$, we see that $\mathrm{prem}(h, \mathcal{H}) = 0$. But since $\mathrm{in}(h_j) = 1$, for all $1 \leq j \leq n$, we have,

$$h = \sum_{j=1}^{n} q_j h_j = \sum_{i=1}^{s} \left(\sum_{j=1}^{n} q_j b_{j,i} \right) f_i = \sum_{i=1}^{s} a_i'' f_i,$$

where

$$\begin{aligned}
\deg(a_i'' f_i) &\leq \max_{j} \Big\{ \deg(q_j) \Big\} + \max_{i,j} \Big\{ \deg(b_{j,i} f_i) \Big\} \\
&\leq \deg(h) + 4(d+1)^{2n} \leq \max\{d, \Delta\} + 6n(d+1)^{2n}.
\end{aligned}$$

Thus g may be expressed as

$$g = \sum_{i=1}^{s} (a_i'' + a_i') f_i = \sum_{i=1}^{s} a_i f_i,$$

where

$$\deg(a_i f_i) \leq \max_{1 \leq i \leq s} \Big\{ \deg(a_i'' f_i), \deg(a_i' f_i) \Big\} \leq \max\{d, \Delta\} + 6n(d+1)^{2n}.$$

Q. E. D.

Theorem 3.3 Zero-Dimensional Upper Bound Theorem.

Let $I = (f_1, \ldots, f_s)$ be a zero-dimensional ideal in $k[x_1, \ldots, x_n]$, where k is an arbitrary field, and $\deg(f_i) \leq d$, $1 \leq i \leq s$. Then I has a characteristic set $\mathcal{G} = \langle g_1, \ldots, g_n \rangle$ with respect to the ordering,

$$x_1 \prec x_2 \prec \cdots \prec x_n,$$

where for all $1 \leq j \leq n$,

1. $\mathrm{class}(g_j) = j$.

2. $\deg(g_j) \leq 2n(d+1)^{2n}$.

3.

$$\left(\exists \, a_{j,1}, \ldots, a_{j,s} \in k[x_1, \ldots, x_n]\right) \, \left[g_j = \sum_{i=1}^{s} a_{j,i} f_i,\right],$$

and $\deg(a_{j,i} f_i) \leq 8n(d+1)^{2n}, 1 \leq i \leq s.$

Proof: Note that $\mathcal{H} = \langle h_1, \ldots, h_n \rangle$ of lemma 3.1, is an ascending set in S_I. The conditions (1) and (2) follows from the facts that, for all $1 \leq j \leq n$.
1. $\mathrm{class}(g_j) = \mathrm{class}(h_j) = j$, and
$\mathrm{cdeg}(g_j) \leq \mathrm{cdeg}(h_j) \leq \deg(h_j) \leq 2(d+1)^{2n}.$
2. For all $1 \leq i < j$, $\deg_{x_i}(g_j) < \deg_{x_i}(g_i)$, as g_j is reduced with respect to the preceding g_i's. Thus

$$\deg(g_j) \leq \left(\sum_{i=1}^{j} \mathrm{cdeg}(g_i)\right) - j + 1 \ \leq \ 2j(d+1)^{2n} - j + 1 \ \leq \ 2n(d+1)^{2n}.$$

Since each g_j is in I, the condition (3) is an immediate consequence of the corollary 3.1.   Q. E. D.

3.2.2   General Cases

The results on the bounds for a characteristic set of a zero-dimensional ideal can be extended to the more general classes of ideals, by the following 'lifting' process. However, in this case, we need to restrict the class of orderings on the indeterminates that may be chosen.
   Let $I = (f_1, \ldots, f_s)$ be an ideal in the ring, $k[x_1, \ldots, x_n]$, where k is an arbitrary field, $\deg(f_i) \leq d$. Assume that the indeterminates, $x_{\pi(1)} = u_1, \ldots, x_{\pi(l)} = u_l$, are the *independent* variables with respect to I. That is, these independent variables form the largest subset of $\{x_1, \ldots, x_n\}$ such that

$$I \cap k[u_1, \ldots, u_l] = (0).$$

Thus, $\dim(I) = l > 0$. The remaining $r = (n-l)$ indeterminates, $x_{\pi(l+1)} = v_1, \ldots, x_{\pi(n)} = v_r$, form the set of *dependent* variables.
   We write $k[u_1, \ldots, u_l, v_1, \ldots, v_r]$ for the ring of polynomials, $k[x_1, \ldots, x_n]$.
   The ordering on the indeterminates is assumed to be so chosen that

$$(\forall \, 1 \leq i \leq l) \ (\forall \, 1 \leq j \leq r) \ u_i \prec v_j.$$

For a polynomial $f \in k[u_1, \ldots, u_l, v_1, \ldots, v_r]$, we shall use the notation, $\deg_U(f)$ and $\deg_V(f)$, to imply

$$\deg_U(f) = \sum_{i=1}^{l} \deg_{u_i}(f) \leq \deg(f)$$

and

$$\deg_V(f) = \sum_{j=1}^{r} \deg_{v_j}(f) \le \deg(f).$$

Let I' be the *extended ideal* of I in the ring $k(u_1, \ldots, u_l)[v_1, \ldots, v_r]$, i.e.

$$I' = I\Big\{k(u_1, \ldots, u_l)[v_1, \ldots, v_r]\Big\}.$$

Then I' is a zero-dimensional ideal in $k(u_1, \ldots, u_l)[v_1, \ldots, v_r]$, and by the results of the previous subsection, I' has a characteristic set $\mathcal{G}' = \langle g_1', \ldots, g_r' \rangle$, where

$$\text{class}(g_j') = \pi(l + j),$$

and

$$\deg_V(g_j') \le 2r(d+1)^{2r}.$$

Additionally, by the results of the previous section, each g_j' can be expressed as

$$(4) \qquad\qquad g_j' = \sum_{i=1}^{s} a_{j,i}' f_i,$$

where

$$\deg_V(a_{j,i}' f_i) \le 8r(d+1)^{2r}.$$

Thus the system of linear equations, defined by (4) consists of at most

$$\Gamma = \binom{8r(d+1)^{2r} + r}{r}$$

equations in at most $(s+1)\Gamma$ unknowns, where the known coefficients are polynomials of degree at most d in $k[u_1, \ldots, u_l]$. Note that there is one equation per each power-product in r variables of degree $\le 8r(d+1)^{2r}$, and one unknown per coefficient of the polynomials g_j' and $a_{j,i}'$'s. Thus the coefficients of g_j' and $a_{i,j}'$'s are rational functions in $k(u_1, \ldots, u_d)$, with the numerators and denominators being determined by determinants of polynomial matrices of order $\le \Gamma$ and with entries of degree $\le d$ in $k[u_1, \ldots, u_l]$. Hence, these numerators and denominators, themselves, have their degrees bounded by $d\Gamma$.

Now, the left and right hand sides of the equation (4) can be multiplied by an appropriate polynomial, $m \in k[u_1, \ldots, u_l]$, to clear out the denominators, and yield the following equation,

$$(5) \qquad\qquad g_j = \sum_{i=1}^{s} a_{j,i} f_i,$$

where all the coefficients are in $k[u_1, \ldots, u_l]$. Note that m can be chosen to be the *least common multiplier* of all the denominators, and since there are at most $(s+1)\Gamma$ such denominators, we have

$$\begin{aligned} \deg_U(m) &\leq d(s+1)\Gamma^2, \\ \deg_U(g_j) &\leq d\Gamma + d(s+1)\Gamma^2, \quad \text{and} \\ \deg_U(a_{j,i}f_i) &\leq d + d\Gamma + d(s+1)\Gamma^2. \end{aligned}$$

In the following theorem, we show that $\mathcal{G} = \langle g_1, \ldots, g_r \rangle$ is, in fact, a characteristic set of the ideal I.

Theorem 3.4 General Upper Bound Theorem.

Let $I = (f_1, \ldots, f_s)$ be an ideal in $k[x_1, \ldots, x_n]$, where k is an arbitrary field, and $\deg(f_i) \leq d$, $1 \leq i \leq s$. Assume that $x_1, \ldots, x_l$ are the independent variables with respect to I. Let $r = n - \dim I = n - l$. Then I has a characteristic set $\mathcal{G} = \langle g_1, \ldots, g_r \rangle$ with respect to the ordering

$$x_1 \prec x_2 \prec \cdots \prec x_n,$$

where for all $1 \leq j \leq r$,

1. $\text{class}(g_j) = j + l$.

2. $\deg(g_j) \leq 4(s+1)(9r)^{2r}d(d+1)^{4r^2}$.

3.

$$\left(\exists\, a_{j,1}, \ldots, a_{j,s} \in k[x_1, \ldots, x_n] \right)\, g_j = \sum_{i=1}^{s} a_{j,i} f_i,$$

and $\deg(a_{j,i}f_i) \leq 11(s+1)(9r)^{2r}d(d+1)^{4r^2}, 1 \leq i \leq s$.

Proof: As before, let us assume that the independent variables with respect to I are $u_1 = x_1, \ldots,$ and $u_l = x_l$, and we write $k[u_1, \ldots, u_l, v_1, \ldots, v_r]$ for the ring $k[x_1, \ldots, x_n]$. Let $\mathcal{G} = \langle g_1, \ldots, g_r \rangle$ be the sequence of polynomials defined in the preceding paragraph. As g_j's were derived from an ascending set $\mathcal{G}'$ of $\mathbf{S}_{I'}$ in $k(u_1, \ldots, u_l)[v_1, \ldots, v_r]$, by multiplication with a polynomial in $k[u_1, \ldots, u_l]$, and since $g_j \in I$ (by construction), clearly $\mathcal{G}$ is an ascending set in $\mathbf{S}_I$.

Furthermore, $\mathcal{G}$ is a minimal set in $\mathbf{S}_I$, since, if there were another ascending set $\mathcal{H} \prec \mathcal{G}$ in $\mathbf{S}_I$, then $\mathcal{H}$ itself would be an ascending set in $\mathbf{S}_{I'}$ and of smaller rank than $\mathcal{G}'$, when the polynomials in $\mathcal{H}$ are treated as polynomials in $k(u_1, \ldots, u_l)[v_1, \ldots, v_r]$; this, however, would contradict the assumption that $\mathcal{G}'$ is a characteristic set of the extended ideal I'.

By construction, $\text{class}(g_j) = j + l$. The rest involves some simple calculations.   Q. E. D.

4   Algorithms.

In this section, we show how one can compute a characteristic set of an ideal, by using the degree bounds of the *General Upper Bound Theorem* and fairly simple ideas from linear algebra. In particular, we shall use, as subroutines, fast sequential and parallel algorithms for computing the *rank* and *determinants* of matrices of order N over arbitrary field. For this purpose, we assume that the sequential algorithms have time complexity $O(N^{2.376})$ [10,1] and that the parallel algorithms have time complexity $O(\log^2 N)$, (and polynomial size) [11,5,27]. Our complexity model assumes a $O(1)$-complexity cost for all field operations over k.

Let I be an ideal given by a set of generators $\{f_1, \ldots, f_s\} \in k[x_1, \ldots, x_n]$, where k is an arbitrary field, $\deg(f_i) \leq d$. Assume that after some reordering of the indeterminates, the indeterminates $x_1, \ldots, x_n$ are so arranged that the first l of them are *independent* with respect to I, and the remaining $(n - l)$ variables, *dependent*. The ordering on the indeterminates is

$$x_1 \prec x_2 \prec \cdots \prec x_n.$$

These conditions can be verified in $O(s^7 d^{O(n^2)})$ sequential time, or equivalently, in $O(n^4 \log^2(sd))$ parallel time by an algorithm of Dickenstein et. al. [12].

Assume, inductively, that the first $(j - 1)$ elements $g_1, \ldots, g_{j-1}$, of a characteristic set, $\mathcal{G}$, of I have been computed, and we wish to compute the jth element g_j of $\mathcal{G}$. By the theorem 3.4, we know that class$(g_1) = (l + 1)$, ..., class$(g_{j-1}) = (l + j - 1)$ and class$(g_j) = (l + j)$. Let

$$\text{cdeg}(g_1) = d_{l+1}, \ldots, \text{cdeg}(g_{j-1}) = d_{l+j-1}.$$

Thus, the polynomial g_j, sought, must be a nonzero polynomial of least degree in x_{l+j}, in $I \cap k[x_1, \ldots, x_{l+j}]$ such that

$$\deg_{x_{l+1}}(g_1) < d_{l+1}, \ldots, \deg_{x_{l+j-1}} < d_{l+j-1}.$$

Furthermore, we know, from the *General Upper Bound Theorem*, that

(6)
$$\left(\exists\, a_{j,1}, \ldots, a_{j,s} \in k[x_1, \ldots, x_n] \right)\, g_j = \sum_{i=1}^{s} a_{j,i} f_i,$$

and $\deg(g_j), \deg(a_{j,i} f_i) \leq 11(s + 1)(9r)^{2r} d(d + 1)^{4r^2}, \quad 1 \leq i \leq s$.
Thus g_j satisfying all the properties can be determined by solving an appropriate system of linear equations.

Let $M_1, M_2, \ldots, M_\rho$ be an enumeration of all the power products in $x_1, \ldots, x_n$ (called, PP$(x_1, \ldots, x_n)$) of degree less than $11(s+1)(9r)^{2r} d(d+1)^{4r^2}$; thus ρ satisfies the following bound,

$$\rho = \binom{11(s + 1)(9r)^{2r} d(d + 1)^{4r^2} + n}{n} = O\left(s^{O(n)}(d + 1)^{O(n^3)} \right).$$

The enumeration on the power products is assumed to be so chosen that the indices $\lambda < \mu$, *only if* one of the following three conditions is satisfied:

1. $M_\lambda \in \mathrm{PP}(x_1, \ldots, x_{l+j})$ and $M_\mu \in \mathrm{PP}(x_1, \ldots, x_n) \setminus \mathrm{PP}(x_1, \ldots, x_{l+j})$.

2. $M_\lambda, M_\mu \in \mathrm{PP}(x_1, \ldots, x_{l+j})$ and

$$(\forall\, l < i < l + j)\ \deg_{x_i}(M_\lambda) < d_i$$

and

$$(\exists\, l < i < l + j)\ \deg_{x_i}(M_\mu) \geq d_i.$$

3. $M_\lambda, M_\mu \in \mathrm{PP}(x_1, \ldots, x_{l+j})$,

$$(\forall\, l < i < l + j)\ \deg_{x_i}(M_\lambda), \deg_{x_i}(M_\mu) < d_i,$$

and $\deg_{x_{l+j}}(M_\lambda) < \deg_{x_{l+j}}(M_\mu)$.

Let us express $a_{i,j}$ and g_j symbolically as follows:

$$a_{i,j} \;=\; \sum_{\lambda=1}^{\rho} \alpha_{(i-1)+\lambda} M_\lambda, \quad 1 \leq i \leq s,$$

$$g_j \;=\; \sum_{\mu=1}^{\rho} \beta_\mu M_\mu.$$

By equating the the terms of the same monomials on both sides of the equation 6, we get at most

$$\binom{d + l + j}{l + j} \cdot \rho \leq (d+1)^n \rho$$

equations, in $(s+1)\rho$ unknowns, α's and β's. We represent the homogeneous system as follows:

$$\begin{bmatrix} A_1 \cdots A_{s\rho} \vdots B_1 \cdots B_\rho \end{bmatrix} \begin{bmatrix} \alpha_1 \\ \vdots \\ \alpha_{s\rho} \\ \cdots \\ \beta_1 \\ \vdots \\ \beta_\rho \end{bmatrix} = 0.$$

Any solution of this system that minimizes the index of the last nonzero entry of the β's will correspond to the desired g_j. The existence of a solution of the desired nature follows from the theorem 3.4, and the rest follows simply from the choice of the ordering on the power products.

Let $\lambda_1 (= 1)$, λ_2, ..., λ_l and m, be a sequence of indices such that, for $1 \leq i < l$,

$$0 < \text{rank}\left[A_1 \cdots A_{\lambda_i}\right] =$$

$$= \text{rank}\left[A_1 \cdots A_{\lambda_i + 1}\right] = \cdots =$$

$$\text{rank}\left[A_1 \cdots A_{\lambda_{i+1} - 1}\right] < \text{rank}\left[A_1 \cdots A_{\lambda_{i+1}}\right],$$

and

$$\text{rank}\left[A_1 \cdots A_{s\rho}\right] < \text{rank}\left[A_1 \cdots A_{s\rho} \vdots B_1\right] < \cdots <$$

$$\text{rank}\left[A_1 \cdots A_{s\rho} \vdots B_1 \cdots B_{m-1}\right] = \text{rank}\left[A_1 \cdots A_{s\rho} B_1 \cdots B_m\right].$$

Then, we know that in the desired solution all the α's and β's, not including α_{λ_1}, ..., α_{λ_l} and β_1, ..., β_m, must be zero. Thus, we need to solve the following system of linear equations

$$\left[A_{\lambda_1} \cdots A_{\lambda_l} \vdots B_1 \cdots B_{m-1}\right] \begin{bmatrix} \alpha_{\lambda_1} \\ \vdots \\ \alpha_{\lambda_l} \\ \cdots \\ \beta_1 \\ \vdots \\ \beta_{m-1} \end{bmatrix} = B_m.$$

Again by a sequence of rank computations (while adding one row at a time), all but $(l + m - 1)$ rows can be eliminated, to yield a square matrix in the left. This system can be solved by Cramer's rule, which only requires computations of the appropriate set of determinants. Then g_j is a monic polynomial given by:

$$g_j = M_m - \beta_{m-1} M_{m-1} - \cdots - \beta_1 M_1.$$

If $\Sigma = ((d+1)^n + s + 1)\rho$, then in the computation, we require, $O(\Sigma)$-many rank and determinant computations, each requiring $O(\Sigma^{2.376})$ sequential time and $O(\log^2 \Sigma)$ parallel time. Since Σ is $O(s^{O(n)}(d+1)^{O(n^3)})$, we see that the sequential and parallel complexity of the algorithm to compute the jth element of a characteristic set are given, respectively, by $O(s^{O(n)}(d+1)^{O(n^3)})$ and $O(n^6 \log^2(s + d + 1))$. Thus, a characteristic set of the given ideal I, can be computed by at most n-many iterations in $O(s^{O(n)}(d+1)^{O(n^3)})$ sequential time and $O(n^7 \log^2(s + d + 1))$ parallel time.

138   GIOVANNI GALLO   BHUBANESWAR MISHRA

Theorem 4.1 Complexity of Characteristic Sets.

Let $I = (f_1, \ldots, f_s)$ be an ideal in $k[x_1, \ldots, x_n]$, where k is an arbitrary field, and $\deg(f_i) \leq d$, $1 \leq i \leq s$. Then with respect to any ordering on the indeterminates:

$$x_1 \prec x_2 \prec \cdots \prec x_n,$$

where the first $\dim(I)$-many variables are independent, it is possible to compute a characteristic set of I, in $O(s^{O(n)}(d+1)^{O(n^3)})$ sequential time or $O(n^7 \log^2(s+d+1))$ parallel time. The polynomials in the computed characteristic set are of degree $O(s(d+1)^{O(n^2)})$.

If the ideal is zero-dimensional, then the arguments in this section can be specialized to yield better bounds.

Theorem 4.2 Complexity of Characteristic Sets of a Zero-Dimensional Ideal.

Let $I = (f_1, \ldots, f_s)$ be a zero-dimensional ideal in $k[x_1, \ldots, x_n]$, where k is an arbitrary field, and $\deg(f_i) \leq d$, $1 \leq i \leq s$. Then with respect to any ordering on the indeterminates:

$$x_1 \prec x_2 \prec \cdots \prec x_n,$$

one can compute a characteristic set of I, in $O(s^{3.376}(d+1)^{O(n^2)})$ sequential time or $O(n^5 \log^2(s+d+1))$ parallel time. The polynomials in the computed characteristic set are of degree $O((d+1)^{O(n)})$.

5   Generalized Pseudo-Division.

The techniques used above can be adapted to develop a parallelizable procedure for computing the pseudo-remainder of a given polynomial f.

Lemma 5.1 Quantitative Pseudo-Division Lemma.

Consider two polynomials f and $g \in k[x_1, \ldots x_n]$, where k is an arbitrary field, $\deg(f) \leq d$ and $\deg(g) \leq \delta$. If $r = \text{prem}(g, f)$ is the pseudo-remainder of g with respect to f then

$$\deg(r) \leq (d+1)(\delta+1) - 1.$$

Also, the pseudo-remainder r can be computed in sequential time $O((d\delta+1)^{O(n)})$ and parallel time $O(n^2 \log^2(d\delta+1))$.

Proof: Since r is a pseudo-remainder of g with respect to f then there is a quotient q and a minimal possible integer $\alpha \leq \delta + 1$ such that

$$\text{in}(f)^\alpha g = qf + r.$$

Thus $\deg(\mathrm{in}(f)^\alpha g) \leq (d+1)(\delta+1) - 1$, which implies that

$$\deg(qf), \deg(r) \leq (d+1)(\delta+1) - 1.$$

Now, it is easily seen that r can be determined by solving at most $(\delta+1)$ linear systems, each consisting of at most $(d+1)^n \rho$ equations in 2ρ unknowns, where

$$\rho = \binom{(d+1)(\delta+1)+n}{n} = O\left((d\delta+1)^{O(n)}\right).$$

Thus as before, if $\Sigma = ((d+1)^n+2)\rho$, then this involves at most $O(\Sigma)$-many rank and determinant computations, each requiring $O(\Sigma^{2.376})$ sequential time and $O(\log^2 \Sigma)$ parallel time. Since Σ is $O((d\delta+1)^{O(n)})$, the complexity given in the statement of the lemma is immediate.   Q. E. D.

Lemma 5.2 Quantitative Generalized Pseudo-Division Lemma.
Consider an ascending set $\mathcal{F} = \langle f_1, \ldots, f_r \rangle \subseteq k[x_1, \ldots, x_n]$ and a polynomial $g \in k[x_1, \ldots x_n]$, where k is an arbitrary field, $\deg(f_i) \leq d$, for all $1 \leq i \leq s$, and $\deg(g) \leq \delta$. If $g_0 = \mathrm{prem}(g, \mathcal{F})$ is the generalized pseudo-remainder of g with respect to $\mathcal{F}$ then

$$\deg(g_0) \leq (d+1)^r(\delta+1).$$

Also, the generalized pseudo-remainder g_0 can be computed in sequential time $O(\delta^{O(n)}(d+1)^{O(nr)})$ and parallel time $O(n^2 r^3 \log^2(d\delta+1))$.

Proof: Let $g_0, g_1, \ldots, g_r = g$ be the pseudo-remainder sequence of g with respect to $\mathcal{F}$, satisfying the following equation

$$\mathrm{in}(f)^{\alpha_i} g_i = q_i' f_i + g_{i-1},$$

where

$$\alpha_i \leq \deg(g_i) + 1$$
$$\deg(q_i' f_i), \deg(g_{i-1}) \leq d\alpha_i + \deg(g_i) \leq (\deg(g_i)+1)(d+1) - 1.$$

Thus, we see that $\deg(g_i)$ satisfies the following recurrence equation

$$\deg(g_r) + 1 = \delta + 1$$
$$\deg(g_{i-1}) + 1 \leq (d+1)(\deg(g_i)+1).$$

Thus, for all $1 \leq i \leq r$, $\deg(g_i) \leq (d+1)^{r-i}(\delta+1)$.

Hence, if we assume that g_{i+1} is available then g_i can be computed in $O(\delta^{O(n)}(d+1)^{O(n(r-i))})$ sequential time, and in $O(n^2(r-i)^2 \log^2(d\delta+1))$ parallel time.

Thus g_0 can be computed in sequential time $O(\delta^{O(n)}(d+1)^{O(nr)})$ and parallel time $O(n^2 r^3 \log^2(d\delta+1))$. Since $r \leq n$, r may be eliminated in the above bounds to yield a simpler formula.   Q. E. D.

6   Conclusion.

We have seen that the degrees of a characteristic set of an ideal is bounded from both above and below by a simple-exponential function in the number of variables. Also, if the ideal is zero-dimensional, this is true, irrespective of the ordering on the indeterminates that may be chosen. However, for the general ideals, the upper-bound is true, only for certain special orderings on the indeterminates; such orderings are abundant $(O(2^{O(n \log n)}))$, and can be selected efficiently (i.e., simple exponential sequential, or polynomial parallel time). Thus, in all practical applications, this will not be a cause of a problem. However, the question of finding optimal degree bounds for any particular ordering of indeterminates, remains an intriguing and important open problem.

Acknowledgement: We are grateful to Joos Heintz for introducing us to the results of Dickenstein et. al.[12] and to Pina Carrá, Teo Mora and an anonymous referee for many useful discussions and comments on an earlier draft.

REFERENCES

[1] A. V. Aho, J. E. Hopcroft, J. D. Ullman, "The Design and Analysis of Computer Algorithms," Addison-Wesley Publishing Company, Reading, Massachusetts, 1974.

[2] D. Bayer, *The Division Algorithm and the Hilbert Scheme*, PhD. Harvard University, 1982.

[3] D. Bayer and M. Stillman, *A Criterion for Detecting m-Regularity*, Inventiones Mathematicae **87** (1987), 1–11.

[4] D. Bayer and M. Stillman, *On the Complexity of Computing Syzygies*, J. of Symbolic Computation 6 (1988), 135–147.

[5] S. J. Berkowitz, *On Computing the Determinant in Small Parallel Time using a Small Number of Processors*, Information Processing Letters **18** (1984), 147–150.

[6] D. Brownawell, *Bounds for the Degree in the Nullstellensatz*, Annals of Mathematics **126** (1987), 577–591.

[7] B. Buchberger, *Gröbner Bases: An Algorithmic Method in Polynomial Ideal Theory*, in "Recent Trends in Multidimensional System Theory," (Ed. Bose), Reidel, 1985.

[8] Chou Shang-Ching, "Mechanical Geometry Theorem Proving," D. Reidel Publishing Company, Kluwer Academic Publishers Group, Dordrecht, Boston, 1988.

[9] Chou Shang-Ching, *Proving and Discovering Theorems in Elementary Geometries Using Wu's Method*, PhD.Thesis, Department of Mathematics, University of Texas at Austin, 1985.

[10] D. Coppersmith and S. Winograd, *Matrix Multiplication via Arithmetic Progressions*, in "Proceedings of the Nineteenth Annual ACM Symposium on Theory of Computing," 1987, pp. 1–6.

[11] L. Csanky, *Fast Parallel Matrix Inversion Algorithms*, SIAM Journal of Computing **5** (1976), 618–623.

[12] A. Dickenstein, N. Fitchas, M. Giusti and C. Sessa, *The Membership Problem for Unmixed Polynomial Ideals is Solvable in Subexponential Time*, 1989, (Unpublished Manuscript).

[13] T. Dubé, B. Mishra and Chee-keng Yap, *Complexity of Buchberger's algorithm for Grobner bases*, Extended Abstract, 1986.

[14] T. Dubé, *Quantitative Analysis Problems in Computer Algebra: Grobner Bases and the Nullstellensatz*, PhD. Courant Institute of Mathematical Sciences, New York University", 1989.

[15] G. Gallo, *La Dimostrazione Automatica in Geometria e Questioni di Complessita' Correlate*, Tesi di Dottorato di Ricerca, Catania, 1989.

[16] M. Giusti, *Some Effectivity Problems in Polynomial Ideal Theory*, in "EUROSAM 84," Lecture Notes in Computer Science, 174,, Springer-Verlag,, 1984, pp. 159–171.

[17] J. Heintz, *Definability and Fast Quantifier Elimination over Algebraically Closed Fields*, Theoretical Computer Science **24** (1983), 239–277.

[18] D. T. Huynh, *A Superexponential Lower Bound for Gröbner Bases and Church-Rosser Commutative Thue Systems*, Information and Control **86** (1986), 196–206.

[19] E. R. Kolchin, "Differential Algebra and Algebraic Groups," Academic Press, New York, 1973.

[20] J. Kollár, *Sharp Effective Nullstellensatz* , Unpublished Manuscript, 1988.

[21] B. Kutzler and S. Stifter, *Automated Geometry Theorem Proving Using Buchberger's Algorithm*, in "Proceedings of the 1986 Symposium on Symbolic and Algebraic Computation," 1986, pp. 209–214.

[22] D. Lazard, *Gröbner Bases, Gaussian Elimination and Resolution of Systems of Algebraic Equations*, in "Computer Algebra Proceedings," Lecture Notes in Computer Science, 162, Springer-Verlag, 1983, pp. 146–157.

[23] E. W. Mayr and A. R. Meyer, *The Complexity of the Word Problems for Commutative Semigroups and Polynomial Ideals*, Advances in Mathematics **46** (1982), 305–329.

[24] B. Mishra, "Algorithmic Algebra," Accepted for publication by Springer-Verlag in the Texts and Monographs in Computer Science Series, 1989.

[25] B. Mishra and Chee-keng Yap Notes on Gröbner Bases, Information Sciences **48** (1989), 219–252.

[26] H. M. Möller and F. Mora, *Upper and Lower Bounds for the Degree of Gröbner Bases*, in "EUROSAM 84," Lecture Notes in Computer Science, 174, Springer-Verlag, 1984, pp. 172–183.

[27] K. Mulmuley, *A Fast Parallel Algorithm to Compute the Rank of a Matrix over an Arbitrary Field*, Combinatorica **7** (1987), 101–104.

[28] J. F. Ritt, "Differential Algebra," American Mathematical Society, New York, 1950.

[29] Wu Wen-Tsün, *On the Decision Problem and the Mechanization of Theorem Proving in Elementary Geometry*, Scientia Sinica **21** (1978), 157–179.

[30] Wu Wen-Tsün, *Basic principles of Mechanical Theorem Proving in Geometries*, Journal of Sys. Sci. and Math. Sci. **4** (1984), 207–235; Also in, Journal of Automated Reasoning **2** (1986), 221–252.

[31] Wu Wen-Tsün, *Some Recent Advances in Mechanical Theorem-Proving of Geometries*, in "Automated Theorem Proving: After 25 Years," Contemporary Mathematics, 1984, pp. 235–242 American Methematical Society.

[32] Chee-keng Yap, *A Double-Exponential Lower Bound for Degree-Compatible Gröbner Bases*, NYU-Courant Robotics Laboratory Report , n.181, 1988.

Giovanni Gallo
Courant Institute of Mathematical Sciences
New York University, New York (USA)
and Universitá di Catania (ITALY)

Bhubaneswar Mishra
Courant Institute of Mathematical Sciences
New York University, New York (USA)

Noetherian Properties and Growth
of some Associative Algebras

TATIANA GATEVA-IVANOVA

Abstract. We consider finitely generated associative algebras over a fixed K of arbitrary characteristic. For such an algebra A we impose some structural restrictions (we call A strictly ordered). We are interested in the implication of strict order on A for its noetherian properties and type of growth. In particular, we prove that if A is a graded standard finitely presented strictly ordered algebra, then A is right (left) noetherian if and only if A has polynomial growth. In this case A is almost commutative. It follows from this that the conjecture we made in [9] is true.

1. Introduction

In this paper K denotes a fixed field of arbitrary characteristic and the term *K-algebra* is used to denote an associative algebra with unit over K. Given a non-empty set $X = \{x_1, \ldots, x_n\}$, $\langle X \rangle$ will denote the free monoid with unit, generated by X and $K\langle X \rangle$ will denote the free associative algebra (with 1) generated by X. In what follows we shall always consider $\langle X \rangle$ ordered by the *degree-lexicographic order* $<_O$ (we set $x_1 <_O \cdots <_O x_n$).

For any $g = \sum_{i=1}^{m} \alpha_i u_i$, $\alpha_i \in K \setminus \{0\}$, $u_i \in \langle X \rangle$ we denote by $HM(g)$ the highest monomial of g, i.e. $HM(g) = u_i$ if $u_j <_O u_i$, for all $j \neq i$. We say $HM(0) = 0$.

Let F be a set of polynomials in $K\langle X \rangle \setminus \{0\}$. The monomial u is *normal* (*modulo F*) if it does not contain any of the monomials $HM(f)$, $f \in F$, as a subword. $N(F)$ will denote *the set of all normal* (mod F) *monomials*.

Let I be a nontrivial ideal in $K\langle X \rangle$, $A = K\langle X \rangle / I$. We shall always assume, without loss of generality, that the presentation $A = K\langle X \rangle / I$ is minimal, that means that I does not contain polynomials of degree 1. Let $N(I)$ (sometimes we write $N = N(I) = N(A)$) be the set of all normal (mod I) monomials. It is clear that $X \subseteq N(I)$ and there is an equality

Partially supported by grant 11 04/88 by Ministry of Science and High Education of Bulgaria.

$K\langle X \rangle = \text{Span } N(I) \oplus I$ as vector spaces. For any $f \in K\langle X \rangle$ one has $f = \tilde{f} + g$, where $\tilde{f} \in \text{Span } N(I)$ and $g \in I$ are uniquely determined. The polynomial $\tilde{f}$ is called the *normal form of f*. Obviously, there is an isomorphism of vector spaces $A \approx \text{Span } N(I)$, $N(I)$ projects to a K-basis of A. We give here two of the well known equivalent definitions of a standard basis of an ideal I in in $K\langle X \rangle$ (with respect to the degree-lexicographic order $<_O$). For more details cf. [6], [8], [7], [12], [13].

(1.1) *Definition:* A set F of polynomials in $K\langle X \rangle \setminus \{0\}$, generating I as two-sided ideal is a *standard* (or *Gröbner*) basis of I if one of the following equivalent conditions hold : a) For any $h \in I$ there is $f \in F$ and $a, b \in \langle X \rangle$ such that $HM(h) = aHM(f)b$; b) There is an equality $N(F) = N(I)$.

A standard basis

$$F = \{f_p = w_p - g_p \mid p \in P, \ w_p \in \langle X \rangle, \ g_p \in K\langle X \rangle, \ HM(g_p) <_O w_p\}$$

is *reduced* if for any $p \in P$ the polynomial f_p is in normal form modulo $F \setminus \{f_p\}$.

It is well known that if I is an ideal of $K\langle X \rangle$ then there exists a uniquely determined reduced standard basis F of I (we also say that F is the reduced standard basis of $A = K\langle X \rangle / I$). The algebra $A = K\langle X \rangle / I$ is *standard finitely presented* (or s. f. p.) if the ideal I has a finite standard basis. Obviously in this cases the reduced standard basis of A is finite as well.

As is well known finitely generated commutative K-algebras are s. f. p. and noetherian. We are interested in the problem of when a finitely generated associative algebra is noetherian. Various positive results are given in [2], [3], [11], [13], [14] and [15]. In contrast with [2], [13], [14], [15], in [9] we impose some structural restrictions on the algebra A but none on the shape of the relations. For convenience of the reader we recall some definitions and notation from [9].

We fix an ideal I of $K\langle X \rangle$.

(1.2) Let $\bar{f}$ denote the highest monomial of the normal form of f, $f \in K\langle X \rangle \setminus I$, i.e. $\bar{f} = HM(\tilde{f})$. Obviously $\bar{f} = 0$ iff $f \in I$.

(1.3) *Definition,[9]:* The algebra $A = K\langle X \rangle / I$ is *strictly ordered* if
 i) for any two polynomials $f, g \in K\langle X \rangle \setminus \{0\}$ in normal form one has
    $\overline{fg} = \overline{\bar{f}\bar{g}} \neq 0$;
 ii) $\deg \overline{x_p x_q} = 2$, for $1 \leq p, q \leq n$.

(1.4) **Remark,[9].** *Let $A = K\langle X \rangle / I$ be a strictly ordered algebra, and let $N = N(A)$ be the set of normal monomials. Define an operation $\star$ on N by*

$$u \star v = \overline{uv}, \qquad \text{for } u, v \in N,$$

and take the restriction of the degree-lexicographic order $<_O$ on N. Then $(N; \star; <_O)$ is a well-ordered semigroup.

One can extend the notion of standard (Groebner) basis for ideals in A as it is done in [12], see also [9,1.12]. It is interesting to describe the class of strictly ordered algebras A with the following properties:

(1.4): Any two-sided (resp.: left, or right) ideal in A has a finite (resp.: left, or right) Groebner basis.

A wide class of strictly ordered algebras are the algebras of solvable type (w.r.t. $<_O$), cf. (2.6), defined and studied by Kandri-Rody and Weispfenning in [11]. As shown in [11] any algebra of solvable type satisfies (1.4). In Theorem II below we give necessary and sufficient conditions for (1.4) in case A is a graded strictly ordered standard finitely presented algebra.

(1.5) *Definition*: For $A = K\langle X\rangle/I$ we say that
  i) A is *almost commutative* iff $(N, \star)$ is a commutative semigroup;
  ii) A is *graded* (by length) *quadratic* if the ideal I is generated by homogeneous relations of degree 2. (In [5] such algebras are called 2-related.)

The main results of this paper are:

(1.6) **Theorem I.** *Let $A = K\langle x_1, \ldots, x_n\rangle/I$ be a strictly ordered algebra and suppose the monomial $x_n x_{n-1}$ is not normal mod I. Then A is almost commutative.*

(1.7) **Theorem II.** *Let I be an ideal of $K\langle X\rangle$ generated by homogeneous polynomials. Suppose $A = K\langle X\rangle/I$ be a strictly ordered standard finitely presented algebra. Then the following conditions are equivalent:*
  1) A is almost commutative;
  2) A has polynomial growth;
  3) A is right noetherian;
  4) A is left noetherian;
  5) Any right ideal in A has a finite right standard basis;
  6) Any left ideal of A has a finite left standard basis;
  7) The reduced standard basis F of I has the following property: for any pair of integers i, j, $1 \le i \le j \le n$, either i) there is a polynomial f_{ji} in F such that $f_{ji} = x_j x_i - c_{ji} x_i x_j - p_{ji}$, where $c_{ji} \in K \setminus \{0\}$, $p_{ji} \in Span\ N$, and (if $p_{ji} \ne 0$) $HM(p_{ji}) <_O x_i x_j$; or ii) there are two polynomials f_{ji}, f_{ij} in F such that $f_{ji} = x_j x_i - c_{ji} x_s x_t - p_{ji}$, $f_{ij} = x_i x_j - c_{ij} x_s x_t - p_{ij}$, where $s < i, t > j, c_{ji}, c_{ij} \in K \setminus \{0\}$, $p_{ji}, p_{ij} \in Span\ N$, and $HM(p_{ji} <_O x_s x_t, HM(p_{ij}) <_O x_s x_t$.

In particular, if this is the case for A then any two-sided ideal in A has a finite standard basis.

(1.8) **Theorem III.** *Let $A = K\langle X\rangle/I$ be a graded (by lenght) strictly ordered algebra. Then the following two conditions are equivalent:*
 1) A is a quadratic (resp.: s. f. p.) algebra with Hilbert series $H_A(z) = 1/(1-z)^n$;
 2) A is an algebra of solvable type (w.r.t. $<_O$);
  If the ideal I is generated by binomials (i.e. by relations of the form u − cv, where $u, v \in \langle X\rangle$, $c \in K$) then the above conditions are also equivalent to
 3) A is s. f. p. (resp.: quadratic) with polynomial growth and finite global dimension.

2. Almost Commutative strictly ordered Algebras

For convenience we recall some more definitions and results from [9].

(2.1): For $f, g \in K\langle X\rangle$ we set $f \sim g$ iff $\bar{f} = \bar{g}$ (cf. (1.2)).

This, obviously, is an equivalence relation.
Define also a partial order on $K\langle X\rangle$ by

(2.2): $f < g$ iff $\bar{f} <_O \bar{g}$, for $\bar{f}, \bar{g} \in N$; $0 < f$, for $\bar{f} \in N$

Both relations depend on I and on the order $<_O$ on $\langle X\rangle$. We write $f \lesssim g$ iff $f < g$ or $f \sim g$.
It is proved in [9, L.2.5] that

(2.3) **Lemma.** *Let $A = K\langle X\rangle/I$ be a strictly ordered algebra and let $f_1, f_2, h \in K\langle X\rangle$, $h \notin I$. Then*
 i) $f_1 \sim f_2$ iff $f_1 h \sim f_2 h$ (resp.: $h f_1 \sim h f_2$);
 ii) $f_1 < f_2$ iff $f_1 h < f_2 h$ (resp.: $h f_1 < h f_2$).

It follows from here that $(N, \star, <)$ is a well ordered semigroup with cancelation.
The following lemma is a straightforward consequence from (2.2) and (2.3.ii).

(2.4) **Lemma.** *Let $A = K\langle X\rangle/I$ be a strictly ordered algebra and let $x_p x_q \notin N$, for some p, q, $1 \le p, q \le n$. Then $\overline{x_p x_q} = x_s x_t$, where $s < p$, $t > q$. In particular, the monomials $x_1 x_i$ and $x_i x_n$ are normal for all i, $1 \le i \le n$.*

Lemma 2.4 is generalised in Section 3, cf. 3.1.

(2.5) *Definition, [3]:* A monomial of the form $x_1^{k_1}, \ldots, x_n^{k_n}$, $k_i \geq 0$, $1 \leq i \leq n$, is called an *ordered monomial.* T will denote the subset of $\langle X \rangle$ consisting of all ordered monomials.

(2.6) *Definition, [11]:* The algebra $A = K\langle X \rangle / I$ is an *algebra of solvable type* (w.r.t. $<_O$) if the reduced standard basis F of I is of the form:

$$(2.7) \qquad F = \{x_j x_i - c_{ij} x_i x_j - p_{ij} \mid 1 \leq i \leq j \leq n\},$$

where for $1 \leq i < j \leq n$, $c_{ij} \in K \setminus \{0\}$, $p_{ij} \in \text{Span } T$, and (if $P_{ij} \neq 0$) $HM(p_{ij}) <_O x_i x_j$.

Obviously, if A is of solvable type, then $\overline{x_j x_i} = x_i x_j \in N$, hence $x_j \star x_i = x_i \star x_j$, for all i, j, $1 \leq i, j \leq n$, and A is almost commutative. In [9, Theorem I] we prove that a strictly ordered algebra A for wich $N(A) = T$ is solvable (w.r.t. $<_O$).

(2.8) **Lemma.** *Let $A = K\langle X \rangle / I$ be a strictly ordered algebra, $(N, \star)$ be the semigroup of normal monomials. Then the conditions (1.7.1) and (1.7.7) are equivalent. In particular, $N \subseteq T$.*

Proof: Since the elements $x_1, \ldots, x_n$ generate N as a semigroup, it is clear that N is commutative iff $x_j \star x_i = x_i \star x_j$, or equivalently, $x_j x_i \sim x_i x_j$, for all i, j, $1 \leq i, j \leq n$ (cf. (1.4) and (2.1)). The lemma follows then from [9, Remark 4.19].

It follows directly from Lemma 2.4 that

(2.9) **Remark.** *Let $A = K\langle X \rangle / I$ be a strictly ordered algebra. Let i, j be integers, $1 \leq i, j \leq n$. The following then holds:*
  i) If $x_i x_j \in N$, then $x_i x_k \in N$, for all k, $j \leq k \leq n$.
  ii) If $x_i x_j \notin N$, then $x_i x_k \notin N <$ for all k, $1 \leq k \leq j$.

(2.10) **Lemma.** *Let A be a strictly ordered algebra and let the monomial $x_i x_j$ be normal for some i, j, $1 \leq i \leq j \leq n$. Then:*
  i) If $x_j x_i \sim x_i x_j$, then $x_k x_i \gtrsim x_i x_k \in N$, for all k, $j \leq k \leq n$;
  ii) If $x_n x_i \sim x_i x_n$, then $x_k x_i \lesssim x_i x_k \in N$, for all k, $j \leq k \leq n$.

Proof: i). It follows from (2.9.i) that all the monomials $x_i x_k$, $j \leq k \leq n$ are normal. Now we shall use induction on k. There is nothing to prove for $k = j$. Assume we have proved that $x_k x_i \gtrsim x_i x_k$ for some $k \geq j$. This together with the obvious inequality

$$x_{k+1} x_i > x_k x_i$$

gives

$$(2.11) \qquad x_{k+1} x_i > x_i x_k$$

Since $x_i x_{k+1}$ is the minimal among the normal monomials u with the property $u > x_i x_k$, it follows from (2.11) that $x_{k+1} x_i \gtrsim x_i x_{k+1}$.
  ii) can be proved by similar arguments.

(2.12) **Remark, [9, Cor. 2.21].** *Assume that for some integers i, k, $1 \leq i \leq k \leq n$, one has $\overline{x_k x_i} = x_i x_k$ and $\overline{x_n x_i} = x_i x_n$. Then for all j with $k \leq j \leq n$, one has $\overline{x_j x_i} = x_i x_j$.*

(2.13) **Lemma.** *Let $A = \langle X \rangle / I$ be a strictly ordered algebra and let the monomial $x_n x_{n-1}$ be not normal. Then $x_n x_i \lesssim x_i x_n$, for $1 \leq i \leq n$.*

Proof: We shall use decreasing induction on i. Since the monomial $x_n x_{n-1}$ is not normal, by Lemma 2.4 one has

$$(2.14) \qquad\qquad x_n x_{n-1} \lesssim x_{n-1} x_n.$$

This gives the base for the induction. Assume we have proved that

$$x_n x_j \lesssim x_j x_n \ , \ \text{for all } j,\ n > j > i.$$

We shall prove that $x_n x_i \lesssim x_i x_n$. Assume the contrary, namely:

$$(2.15) \qquad\qquad x_n x_i > x_i x_n.$$

Then one has

$$(2.16) \qquad x_n x_i \sim x_{i+1} x_s \in N, \quad \text{for some } s,\ n > s > i.$$

This follows from Lemma 2.4, (2,15) and from the inequalities

$$x_n x_i < x_n x_{i+1} \qquad \text{(since } A \text{ is strictly ordered)}$$
$$\lesssim x_{i+1} x_n \qquad \text{(by the inductive assumption)}.$$

Consider now the relations:

$$x_n x_i x_n \sim x_{i+1} x_s x_n \qquad : \text{by (2.16) and (2.3.i)}$$
$$\left.\begin{array}{l} \gtrsim x_{i+1} x_n x_s \\ \gtrsim x_n x_{i+1} x_s \end{array}\right\} \quad : \text{by the inductive assumption and (2.3.ii)}$$

Hence, by (2.3.ii) one has

$$x_i x_n > x_{i+1} x_s$$

which is impossible, since by (2.16) $x_{i+1} x_s \in N$. Thus the assumption (2.15) gives a contradiction and we have proved that $x_n x_i \lesssim x_i x_n$ for all i, $1 \leq i \leq n$.

(2.17) **Lemma.** *Let $A = K\langle X \rangle / I$ be a strictly ordered algebra, $x_n x_{n-1} \notin N$. Let i be an integer, $1 \leq i \leq n$, such that for any pair j, k, $1 \leq j < i$, $1 \leq k \leq n$, one has $x_k x_j \sim x_j x_k$. Then $x_n x_i \sim x_i x_n$.*

Proof: We shall first find an integer t such that $\overline{x_i x_i} = x_i x_t \in N$. If $x_i x_i \in N$ we set $t = i$, otherwise $t = t_v$, where t_v is the last element of the sequence

$$i = t_0 < t_1 < t_2 < \ldots < t_{v-1} < t_v$$

defined as follows:

$$\begin{aligned}
\overline{x_i x_i} &= x_{s_1} x_{t_1} && \in N, && s_1 < i, \ i > t_1 \\
\overline{x_i x_{t_1}} &= x_{s_2} x_{t_2} && \in N, && s_2 < i, \ t_1 > t_2
\end{aligned}$$

(2.18)

$$\cdots$$

$$\begin{aligned}
\overline{x_i x_{t_{v-1}}} &= x_{s_v} x_{t_v} && \in N, && s_v < i, \ t_{v-1} > t_v \\
x_i x_{t_v} &&& \in N.
\end{aligned}$$

Note that by Lemma 2.4 such a sequence always exists. Using induction on q we shall prove that $x_{t_q} x_i \sim x_i x_{t_q}$, for all q, $0 \le q \le v$.

The obvious equality $x_i x_{t_0} = x_i x_i = x_{t_0} x_i$ gives the base for the induction. Assume we have proved that

(2.19) $$x_{t_{q-1}} x_i \sim x_i x_{t_{q-1}} \quad \text{for some } q > 1.$$

The following relations hold:

$$\begin{aligned}
x_{s_q} x_{t_q} x_i &\sim x_i x_{t_{q-1}} x_i && : \text{ by (2.13) and Lemma 2.3} \\
&\sim x_i x_i x_{t_{q-1}} && : \text{ by (2.19) and Lemma 2.3} \\
&\sim x_i x_{s_q} x_{t_q} && : \text{ by (2.18) and Lemma 2.3} \\
&\sim x_{s_q} x_i x_{t_q} && : \text{ by the assumption of Lemma, since } s_q < i.
\end{aligned}$$

Thus, by Lemma 2.3.i, one has

$$x_{t_q} x_i \sim x_i x_{t_q}.$$

In particular, for $t = t_v$ one has

$$x_t x_i \sim x_i x_t \in N.$$

Now one can apply Lemma 2.10.i to obtain

$$x_n x_i \gtrsim x_i x_n.$$

Since the opposite inequality holds by Lemma 2.13 one gets

$$x_n x_i \sim x_i x_n.$$

(2.20) **Proof of the Theorem I:** Assume that $A = K\langle X \rangle / I$ is a strictly ordered algebra and $x_n x_{n-1} \notin N$. We have to prove that the semigroup $(N, \star)$ is commutative. It will be enough for that purpose to show that $x_j \star x_i = x_i \star x_j$, or equivalently

(2.21) $$x_j x_i \sim x_i x_j \ , \quad \text{for all } i, j, \ 1 \le i, j \le n.$$

We shall do this using induction on i.

Step 1: The base for the induction:

(2.22) $$x_k x_1 \sim x_1 x_k \ , \quad \text{for all } k, \ 1 \le k \le n.$$

Note first that

(2.23) $$x_n x_1 \sim x_1 x_n.$$

Indeed, by Lemma (2.4) the monomial $x_1 x_1$ is normal, thus applying Lemma 2.10.i one obtains $x_n x_1 \gtrsim x_1 x_n$. This together with the inequality $x_n x_1 \lesssim x_1 x_n$ proved in Lemma 2.13 gives (2.23)

Now one has

$$\overline{x_1 x_1} = x_1 x_1 \quad : \quad \text{by L.2.4 and (1.2)}$$

$$\overline{x_n x_1} = x_1 x_n \quad : \quad \text{by (2.23) and by the fact that } x_1 x_n \in N.$$

It follows then from Remark 2.12 that $\overline{x_k x_1} = x_1 x_k$, for $1 \leq k \leq n$, which gives (2.22).

Step 2: Assume we have already proved that for all j, $1 \leq j \leq i$, one has

$$x_k x_j \sim x_j x_k \ , \text{ for all } k, \ 1 \leq k \leq n.$$

We shall prove that

(2.24) $x_k x_i \sim x_i x_k \ , \text{ for all } k, \ 1 \leq k \leq n.$

Note first that we are under the assumption of Lemma 2.17, hence

(2.25) $x_n x_i \sim x_i x_n.$

Let $q = q(i)$ be the minimal integer such that $x_i x_q \in N$. We shall show that

(2.26) $x_q x_i \sim x_i x_q.$

It follows from (2.25) and Lemma 2.10.ii that $x_q x_i \lesssim x_i x_q$. Assume that $x_q x_i < x_i x_q$. Since $x_{i-1} x_n$ is the maximal normal monomial which is less that $x_i x_q$, one has

(2.27) $x_q x_i \lesssim x_{i-1} x_n.$

Now consider the relations

$$x_i x_q x_i \lesssim x_i x_{i-1} x_n \quad : \quad \text{by (2.27) and Lemma 2.3.}$$

$$\sim x_{i-1} x_i x_n \quad : \quad \text{by the inductive assumption and Lemma 2.3}$$

$$\sim x_{i-1} x_n x_i \quad : \quad \text{by (2.25) and Lemma 2.3.}$$

Applying Lemma 2.3 again one obtains

$$x_i x_q \lesssim x_{i-1} x_n.$$

This is impossible, since by the choice of q, $x_i x_q$ is normal, a contradiction. Thus, $x_q x_i \sim x_i x_q$.

By (2.25) and (2.26) we are under the assumptions of Remark 2.12, thus

(2.28) $x_k x_i \sim x_i x_k \in N \ \text{ for all } k, \ q \leq k \leq n.$

Assume that there is an integer k, $1 \leq k \leq n$, such that

(2.29) $x_k x_i \not\sim x_i x_k,$

and let k be the maximal with this property. It follows from (2.28) and from the inductive assumption that $1 < i < k \leq q - 1$. Then $x_i x_k \notin N$ and, by Lemma 2.4 one has

(2.30) $x_i x_k \sim x_s x_t \in N$ for some s, t, $s < i$, $t > k$.

With s and t as above, the following equivalences hold:

$$x_i x_k x_i \sim x_s x_t x_i \quad : \quad \text{by (2.30) and Lemma 2.3.i}$$
$$\sim x_s x_i x_t \quad : \quad \text{since, by (2.30), } t > k$$
$$\sim x_i x_s x_t \quad : \quad \text{by the inductive assumption, since } s < i$$

Thus applying Lemma 2.3.i again one gets

$$x_k x_i \sim x_s x_t \sim x_i x_k,$$

a contradiction with the choice of k. We have proved that (2.24) holds.   Q. E. D.

3. Noetherian properties and Growth of strictly ordered Algebras

We keep the notation of the previous sections. We shall assume that A is a strictly ordered algebra given by a fixed presentation $A = K < x_1, \dots, x_n > /I$, and N is the set of normal modulo I monomials.

The following lemma is a generalization of [9, L.2.6] which we need in case that the algebra A is not graded. It can easily be proved if one applies Lemma 2.3 and the definitions of the orders $<_O$ and $<$.

(3.1) **Lemma.** *Let $u, v \in N$ and $uv \notin N$. Then either*
1) $\deg \overline{uv} = \deg uv$ and $\overline{uv} = u_1 v_1$, where $u_1 v_1 \in N$, $\deg u_1 = \deg u$, $\deg v_1 = \deg v$ and $u_1 <_O u$, $v <_O v_1$; or
2) $\max\{\deg u, \deg v\} \leq \deg \overline{uv} < \deg uv$ and there exist normal monomials u_1, v_1, u_2, v_2 such that
   i) $\overline{uv} = u_1 v_1$, $\deg v_1 = \deg v$, $v_1 >_O v$, $\deg u_1 < \deg u$ (and obviously $u_1 <_O u$);
   ii) $uv = u_2 v_2$, $\deg u_2 = \deg u$, $u_2 >_O u$, $\deg v_2 < \deg v$ ($v_2 <_O v$).

As an immediate consequence from (3.1.1) and (3.1.2.i) one obtains

(3.2) **Corollary.** *1) Let $u, v \in N$, $uv \notin N$ and let v_1 be the right segment of uv of lenght $\deg v$. Then $v_1 > v$. 2) If $u \in N$ then $ux_n^k \in N$ for all $k \geq 0$. In particular, $x_n^k \in N$ for all $k \geq 1$.*

Corollary 3.2.2 was proved in [9,Cor.2.9.2] in case that A is graded. Here we do not need this restriction on A.

(3.3) **Corollary [9,2.11].** *Let $u, v, w \in N$, and let $v < w$, $\deg v = \deg w$. Then: i) from $uv \in N$ it follows that $uw \in N$; ii) from $uw \notin N$ it follows that $uv \notin N$.*

(Here, in difference with [9, Cor.2.11] it is not necessary that $\deg \overline{uv} = \deg u + \deg v$.)

(3.4) **Lemma.** *Let $a \in N$ and let $x_n a \in N$. Then $x_n^s a x_n^k \in N$ for all $s, k \geq 0$.*

Proof: There is nothing to prove in case that $a = x_n^p$ for some p, one can just apply Cor. 3.2.2. Assume now that

$$a < x_n^{\deg a}.$$

Then for the monomials $a x_n$, $x_n a$ one has

(3.5) $a x_n <_O x_n a.$

Clearly,

(3.6) $a x_n, \ x_n a \in N, \ \deg a x_n = \deg \ x_n a,$

hence, by (3.5),

(3.7) $a x_n < x_n a.$

Using induction on s we shall prove that $x_n^s a \in N$ for all $s \geq 1$. The hypothesis of the Lemma gives the base for the induction. Assume $x_n^s a \in N$ for some $s \geq 1$. Then by (3.2.2) one has

(3.8) $x_n^s a x_n \in N.$

It follows from (3.6) and (3.7) that for $u = x_n^s$, $v = a x_n$, $w = x_n a$ one has $u, v, w \in N, v < w, \deg v = \deg w$, thus we are under the assumption of Cor. 3.3. By (3.8) the monomial $uv = x_n^s a x_n$ is normal, hence by (3.3.1) the monomial $uw = x_n^{s+1} \cdot a$ is normal as well. We have proved that $x_n^s a \in N$ for all $s \geq 1$. It follows then from Cor 3.2.2 that $x_n^s a x_n^k \in N$ for all $s, k \geq 0$.

(3.9) **Corollary.** *If $x_n x_{n-1} \in N$ then $x_n^s x_{n-1} x_n^k \in N$ for all $k, s \geq 0$.*

(3.10) **Lemma.** *Let I be an ideal in $K\langle X \rangle$ generated by homogeneous polynomials. Assume $A = K\langle X \rangle / I$ is a strictly ordered right noetherian algebra. Then $x_n x_{n-1} \notin N$.*

Proof: Assume the contrary, $x_n x_{n-1} \in N$. By Lemma 3.4 all the monomials $v_i = x_n^i x_{n-1}$ are normal for all $i \geq 1$. It follows from the right noetherian property of A that there exists an integer $k > 1$ such that

(3.11) $f = v_k - \displaystyle\sum_{\substack{i,j \\ i < k}} c_{ij} v_i b_{ij} \equiv 0 \pmod{I},$

where $c_{ij} \in K \setminus \{0\}$, $b_{ij} \in N$, $1 \leq i < k$, $1 \leq j \leq s_i$. Since the ideal I is graded we may assume that $\deg b_{ij} = k - i$, thus one has

(3.12) $b_{ij} \leq x_n^{k-i}.$

Consider now the relations:

$$v_i b_{ij} = x_n^i x_{n-1} b_{ij}$$
$$\leq x_n^i x_{n-1} x_n^{k-i} \quad \text{(by 3.12)}$$
$$< x_n^i x_n^{k-i} x_{n-1}$$
$$= v_k.$$

Thus v_k is the highest monomial of f, which is impossible since $f \in I$ and v_k is normal. We have proved that the assumption $x_n x_{n-1} \in N$ leads to a contradiction.

(3.13) Corollary. *Let I be an ideal in $K\langle X \rangle$ generated by homogeneous polynomials, let $A = K\langle X \rangle / I$ be a strictly ordered algebra. If A is right noetherian, then A is almost commutative.*

Proof: This follows from Lemma 3.12 and Theorem I.

(3.14) Remark, [9, (4.3)]. *Let $A = \langle X \rangle / I$ be a standard finitely presented algebra. Let F be the reduced standard basis of A. Define an integer $m = m(F)$ as*

$$(3.15) \qquad m = \max\{\deg f \mid f \in F\} - 1.$$

Then for any three monomials $u, v, w \in N$ such that $\deg v \geq m$, uv, $vw \in N$, the monomial uvw is normal as well.

The following Lemma is proved in [9], cf. L. 4.6 in case that A is graded. Here such a restriction on A is not necessary, since one can apply Cor. 3.9 and an argument similar to that of [9,l. 4.6].

(3.16) Lemma. *Let $A = \langle X \rangle / I$ be a standard finitely presented strictly ordered algebra. If A is left noetherian then $x_n x_{n-1} \in N$.*

(3.17) Corollary. *Let $A = K\langle X \rangle / I$ be a standard finitely presented strictly ordered algebra. If A is left noetherian then A is almost commutative.*

(3.18) Lemma. *Let $A = K\langle X \rangle / I$ be a graded standard finitely presented strictly ordered algebra. If A has polynomial growth then $x_n x_{n-1} \notin N$. (Recall that a graded algebra $A = \bigoplus_{s \geq 0} A_s$ has polynomial growth if there is a polynomial with integer coefficients $p(x) \in z[x]$ such that $\dim_K A_s \leq p(s)$ for s large.)*

Proof: Assume the contrary, that $x_n x_{n-1} \in N$. For m as in (3.15) it follows from Lemma 3.2. that the monomials $u = x_n^m$ and $v = x_{n-1} x_n^m$ are normal. Applying now Lemmas 3.2.2, 3.4 and Remark 3.14 one obtains that all the monomials of the type

$$u^{a_1} \cdot v^{b_1} \cdot u^{a_2} \cdot v^{b_2} \ldots u^{a_s} v^{b_s} \ , \ a_i, b_i \geq 0 \ , \ 1 \leq i \leq s,$$

are normal as well. Hence A contains the free algebra B on generators u and v and has exponential growth. This gives a contradiction with the hypothesis of the lemma. Thus the assumption $x_n x_{n-1} \in N$ leads to a contradiction.

(3.19) Corollary. *Let $A = K\langle X \rangle / I$ be a graded standard finitely presented strictly ordered algebra. If A has polynomial growth then it is almost commutative.*

We end this section with the proof of Theorem II. Recall first the following fact proved in [9].

(3.20) **Fact.**[9, Proposition 3.18]. *Let $A = K\langle X \rangle / I$ be an almost commutative strictly ordered algebra. Then: 1) Any left (right) ideal of A has a finite left (right) standard basis. In particular, A is left and right noetherian. 2) Any two-sided ideal of A has a finite standard basis.*

(3.21) **Proof of Theorem II:** The implications 1) $\Longrightarrow$ 3),4),5),6) follow from (3.20). 1) $\Longrightarrow$ 2),7) follows from Lemma 2.8. 2) $\Longrightarrow$ 1) by Corollary 3.19. 3) $\Longrightarrow$ 1) follows from Corollary 3.13. Corollary 3.17 gives 4) $\Longrightarrow$ 1). 5) $\Longrightarrow$ 3) and 6) $\Longrightarrow$ 4) are obvious. 7) $\Longrightarrow$ 1) by Lemma 2.8.   Q. E. D.

Strictly ordered Algebras with Polynomial Growth

In this section we shall prove Theorem III.

(4.1) **Remark,** [9, Lemma 2.14]. *Let $A = K\langle X \rangle / I$ be a strictly ordered algebra. Let $x_j x_{n-1} \notin N$, for some j, $2 \leq j \leq n$. Then $x_j x_{n-1} = x_{j-1} x_n$.*

(4.2) **Lemma.** *Let $A = K\langle X \rangle / I$ be a strictly ordered algebra and let $x_{n-1} x_{n-1} \notin N$. Then A is almost commutative.*

Proof: We shall prove that $x_n x_{n-1} \notin N$. By the assumption of the lemma $x_{n-1} x_{n-1} \notin N$, thus by (4.1)

(4.3) $$x_{n-1} x_{n-1} \sim x_{n-2} x_n \in N.$$

Note that by Remark 2.9.ii, $x_{n-1} x_j \notin N$ for all $j \leq n-1$. We shall prove that

(4.4) $$x_{n-1} x_{n-2} \leq x_{n-2} x_{n-1}.$$

It will be enough for that purpose to show that

(4.5) The monomial $\overline{x_{n-2} x_{n-1}}$ is the largest among the normal monomials v such that $v < x_{n-2} x_n$. Two cases are possible. 1) $x_{n-2} x_{n-1} \in N$. Then $\overline{x_{n-2} x_{n-1}} = x_{n-2} x_{n-1}$ and (4.5) is obvious. 2) $x_{n-2} x_{n-1} \notin N$. Then by (4.1)

(4.6) $$\overline{x_{n-2} x_{n-1}} = x_{n-3} x_n$$

and, since by 2.9.ii, $x_{n-2} x_j \notin N$, for all $j \leq n-1$, the monomial $x_{n-3} x_n$ is obviously the largest among the normal monomials v with the property $v < x_{n-2} x_n$. This together with (4.6) gives (4.5). Obviously, it follows from here that (4.4) holds. Consider now the relations

$$x_{n-2} x_n x_{n-1} \sim x_{n-1} x_{n-1} x_{n-1} \quad \text{by lemma 2.3 and (4.3)}$$

$$\sim x_{n-1} x_{n-2} x_n$$

$$\lesssim x_{n-2} x_{n-1} x_n \quad \text{by (4.4) and Lemma 2.3.}$$

Applying Lemma 2.3 again (this time to cancel x_{n-2} from the left) one obtains $x_n x_{n-1} \lesssim x_{n-1} x_n$, hence the monomial $x_n x_{n-1}$ is not normal. It follows then from Theorem I that A is almost commutative.

Recall that the algebra $A = K\langle X \rangle / I$ is quadratic graded (by length) iff the ideal I is generated by homogeneous relations of degree 2, cf. (1.5.ii)

(4.7) Remark. Let $A = K\langle X \rangle / I$ be a quadratic graded algebra. Let v be a monomial in $\langle X \rangle$ such that $\deg v = 3$, $v \notin N$. Then either i) there is a subword u of v, $\deg u = 2$, such that $u \notin N$, or ii) v is the highest monomial of a polynomial $f \in I$, obtained in the process of solving some ambiguity by means of reductions, cf. [6]. Following Bergman, one can easily see that in this case there are some integers p, q, r, $1 \leq p, q, r \leq n$, such that

$$(4.8) \qquad v <_O x_p x_q x_r \quad \text{and} \quad x_p x_q \notin N, \ x_q x_r \notin N.$$

(4.9) Lemma. Let A be a graded quadratic strictly ordered algebra. Assume $x_n x_{n-1} \in N$. Then $x_n x_{n-1} x_{n-1} \in N$.

Proof: It follows from Lemmas 4.2 and 2.4 that under the assumption of the lemma one has

$$(4.10) \qquad \text{If } x_{n-1} x_{n-1} \lesssim_O x_p x_q \text{ (as monomials in } \langle X \rangle) \text{ then } x_p x_q \in N.$$

In particular, $x_{n-1} x_{n-1} \in N$. We shall prove that $x_n x_{n-1} x_{n-1} \in N$. Assume the contrary,

$$(4.11) \qquad v = x_n x_{n-1} x_{n-1} \notin N.$$

Clearly, by (4.10), any subword of v of degree 2 is normal. Thus (4.7.ii) and (4.8) hold. One can easily see that (4.8) and (4.9) give a contradiction, due to the assumption (4.10). We have proved that $x_n x_{n-1} x_{n-1} \in N$.

(4.12) Lemma. Let A be a quadratic graded strictly ordered algebra and let $x_n x_{n-1} \in N$. Then A contains a free subalgebra on two generators. In particular, A has exponential growth.

Proof: We shall prove that the subalgebra B of A generated by the elements $a = x_n$, $b = x_{n-1} x_n$ is free with a and b as free generators. It will be enough for that purpose to prove that any monomial c of the type

$$(4.13) \quad c = x_n^{t_0} x_{n-1} x_n^{t_1} \ldots x_n^{t_{s-1}} x_{n-1} x_n^{t_s}, \quad \text{where } t_0 \geq 0, \ t_i \geq 1, \ 1 \leq i \leq s,$$

is normal. Assume the contrary and take a c, $c \notin N$, of the type (4.13) with minimal length. Note that in this case for s as in (4.13) one has $s > 1$, since any monomial of the type $x_n^t x_{n-1}$, $t \geq 0$, is normal. (This follows from Cor. 3.11.) Consider the monomials

$$w = x_n^{t_1} x_{n-1} \ldots x_n^{t_{s-1}} x_{n-1} x_n^{t_s}, \quad \text{with } s \text{ and } t_i \text{ as in (4.13)}$$

$$v = x_{n-1} x_n^{(\deg\ w - 1)}.$$

Obviously, w and v are of the type (4.13), $v <_O w$, $c = x_n^{t_0} x_{n-1} w$, $1 \leq \deg v = \deg w < \deg c$. Thus, by the choice of c, $v \in N$, $w \in N$. Clearly, $v < w$.

For $u = x_n^{t_0} x_{n-1} \in N$, v and w as above, one has $uw = c \notin N$. It follows then from Corollary 3.3.ii that

$$uv = x_n^{t_0} x_{n-1} x_{n-1} x_n^{(\deg w - 1)} \notin N.$$

Thus, by Corollary 3.2.2, $x_n^{t_0} x_{n-1} x_{n-1} \in N$. This is impossible, since, by the assumption of the Lemma $x_n x_{n-1} \in N$, hence, by Lemma 4.9, $x_n x_{n-1} x_{n-1} \in N$, which gives (by L.3.4) that $x_n^t x_{n-1} x_{n-1} \in N$ for all $t \geq 0$. We have proved that any monomial c of the type (4.13) is normal. This shows that the subalgebra B is free with free generators a and b as above. Thus A has exponential growth.

(4.14) **Corollary.** *Let A be a quadratic graded strictly ordered algebra. If A has polynomial growth then A is almost commutative.*

(4.15) Recall that for a graded algebra $A = \bigoplus_{i \geq 0} A_i$ the *Hilbert series* $H_A(z)$ is defined as $H_A(z) = \sum_{i \geq 0} \dim_K A_i z^i$.

(4.16) **Remark.** *Let $A = K\langle X \rangle / I$ be an algebra, N be the set of normal mod I monomials. Let $N \subseteq T$, where T is the set of ordered monomials (cf. (2.5)). Then 1) There is an equality $H_A(z) = 1/(1-z)^n$ iff $N = T$. 2) If $N = T$ then A is standard finitely presented.*

The following fact is a straightforward consequence from [9, Propositon 2.22].

(4.17) **Remark.** *Let $A = K\langle X \rangle / I$ be a strictly ordered algebra and let $N = T$. Then A is an algebra of solvable type. In particular, A is quadratic and standard finitely presented.*

(4.18) Let I be an ideal of $K\langle X \rangle$, F be the reduced standard basis of I. Denote by V the set

(4.19) $$V = \{HM(f) \mid f \in F\}.$$

It is clear that V is the set of obstructions for I (cf. [1]).
    One can easily see that

(4.19) **Remark.** *Let $A = K\langle X \rangle / I$ be a graded (by lenght) algebra and assume the reduced standard basis of I contains only polynomials of degree 2 (or equivalently, $\deg v = 2$ for all $v \in V$). Then Anick's resolution [1] is minimal.*

(4.20) **Proof of the Theorem III:**
    1) $\implies$ 2). Assume that A is a quadratic (resp.: standard) finitely presented algebra with Hilbert series $H_A(z) = 1/(1-z)^n$. Then, obviously, A has polynomial growth and applying Cor. 4.14 (resp.: Cor 3.19) one obtains that A is almost commutative. This gives $N \subseteq T$. It follows now from remark 4.16.1 that $N = T$. Thus, by (4.17), A is an algebra of solvable type.
    2) $\implies$ 1) is obvious.
    2) $\implies$ 3). It is known (cf. 3, Cor. 6.7) that for any algebra $A = K < x_1, \ldots, x_n > /I$ such that $N = T$ one has gl.dim $A = n$. (This obviously follows also from Remark 4.18)

3) $\implies$ 2). (Sketch of the proof). Assume that A is a graded standard finitely presented resp.: quadratic) strictly ordered algenra with finite global dimension and polynomial growth. Assume that the ideal I is generated by binomials. We shall prove that A is of solvable type.

Note first that in this case the reduced standard basis F of I consists also of binomials, i.e.

$$F = \{u_p - c_p v_p \mid 1 \le 0 \le P, \ c_p \in K, \ u_p, v_p \in \langle X \rangle, \ v_p <_O u_p\}.$$

Let V be the set of obstructions for I, that is $V = \{u_p \mid 1 \le p \le P\}$. Let

$$V_0 = \{x_j x_i \mid 1 \le i < j \le n\}.$$

Since A has polynomial growth it follws from Cor. 3.4 (resp.: Cor.4 14) that A is almost commutative. Hence $N \subseteq T$, or equivalently $V_0 \subseteq V$. We shall prove that $V_0 = V$. Assume the contrary, $V_0 \subsetneq V$. Obviously, in this case any $v \in V \setminus V_0$ is an ordered monomial, i.e. $v = x_1^{k_1} \ldots x_n^{k_n}$, $k_i \ge 0$. Applying Corollary 3.3 one can see that if

$$v = x_i x_{j_1} \ldots x_{j_s} \in V, \ i \le j_1 \le j_2 \le \ldots j_s,$$

then the monomial $x_i^{\deg v}$ is not normal. Thus $x_i^j \in V$, for some $j \le \deg v$. Take the minimal i such that $x_i^j \in V$, for some $j > 1$. Now we shall use Anick's resolution [1, Thm. 1.4]. It can be easily seen that for any integer $d \ge 1$, and for r_d defined as

$$r_d = \begin{cases} jk, & \text{if } d = 2k - 1, \\ jk + 1, & \text{if } d = 2k, \end{cases}$$

the monomial $x_j^{r_d}$ is a d-chain. (The definition of a d-chain is given in [1].). More detailed argument shows that

$$x_i^{r_d} \in \operatorname{Tor}_d^A(K_A, {}_A K), \quad \text{for all } d \ge 1.$$

This shows that A has infinite global dimension, a contradiction due to the assumption that $V_0 \subsetneq V$. We have proved that $V_0 = V$, that is $N = T$. It follows then from Remark 4.17 that A is an algebra of solvable type. Q. E. D.

Acknowledgement: I should like to thank Teo Mora with whom I had some useful and constructive discussions. I also thank E. S. Golod and V. N. Latyshev for their attention to my work and many valuable discussions. Finally I should like to express my gratitude to the Chair of Algebra of Moscow State University and especially to the participants in the specialized seminars on Ring theory and Computer algebra for the cooperation.

REFERENCES

[1] D. Anick, *On the homology of associative algebras*, Trans. AMS **296** (1986), 641–659.

[2] J. Apel, *Gröbnerbasen in nichtkommutativen Algebren und ihre Anwendung*, Dissertation A, Karl-Marx-Universitat, Leipzig.

[3] M. Artin and W. Schelter, *Graded algebras of global dimension 3*, Advances in Mathematics **66** (1987), 171–216.

[4] M. Artin, J. Tate, and M. Van Den Bergh, *Some algebras associated to automorphisms of elliptic curves*, in "The Grothendieck Festschrift," Vol. I, eds P. Cartier et al., Birkhäuser, 1990, pp. 33–85.

[5] J. Backelin, R. Fröberg, *Koszul algebras, Veronese subrings and r rings with linear resolutions*, Rev. Roumaine Math. Pures Appl. **30** (1985), 85–97.

[6] G. M. Bergman, *The diamond lemma of ring theory*, Advances in Math. **29** (1978), 178–218.

[7] B. Buchberger, *An algorithm for finding a basis for the residue class ring of a zero-dimensional polynomial ideal*, (German). Ph D Thesis, Univ. of Innsbruck (Austria), Math. Inst., 1965, Aequationes Mathematicae **4/3** (1970), 374–383.

[8] T. Gateva-Ivanova, V. Latyshev, *On recognisable properties of associative algebras*, J. Symb. Comp. **6** (1988), 371–388.

[9] T. Gateva-Ivanova, *On the noetherianity of some associative finitely presented algebras*, J. Algebra (to appear).

[10] E. S. Golod, *Standard bases and homology*, in "Algebra — some current trends," Lect. notes in Math. **1352**, 1988, pp. 88–95.

[11] A. Kandri-Rody, V. Weispfenning, *Non-commutative Groebner bases in algebras of solvable type*, J. Symb. Comp..

[12] T. Mora, *Seven variations on standard bases*, Univ. di Genova, Dip. di Matematica, N.45 Preprint, 1988.

[13] T. Mora, *Groebner bases for non-commutative algebras,*, in "ISSAC 88," L. N. C. S., Springer Verlag.

[14] J. Okninski, *On monimial algebras*, Arch. Math. **50** (1988), 417–423.

[15] S. P. Smith and J. T. Stafford, *Regularity of the four dimensional Sklyanin algebra*, preprint, 1990.

Tatiana Gateva-Ivanova
Institute of Mathematics
Bulgarian Academy of Sciences
Sofia 1090, Bulgaria

Codes and Elliptic Curves

GERARD VAN DER GEER

Introduction

In this paper we discuss recent results ([1], [2], [3], [4], [5], [9]) on codes and algebraic curves. We are not concerned with algebraic geometric Goppa codes but rather with another link between coding theory and algebraic geometry. In our case the codes correspond to *families* of algebraic curves over a finite field, whereas Goppa codes come from a fixed algebraic curve. We use algebraic geometry to determine the weight distributions of certain codes but also we show that results from coding theory may be used to obtain results about the variation of the number of points in families of algebraic curves over a finite field. We do not assume that the reader has a knowledge of coding theory.

2. Codes

Let p be a prime and let $q = p^m$ for some natural number m. Very often $p = 2$ for applications but here this is a matter of convenience rather than necessity. Codes are used to transfer messages without errors. For this we choose an alphabet, say with q symbols. There is no objection to take the finite field F_q as alphabet. We also fix a word length : n. A (linear) *code* is a very simple mathematical object : it is just a linear subspace C of the F_q-vector space F_q^n. The space C acts as the "dictionary" of allowable words while F_q^n is the set of all words. An obvious invariant of C is its dimension $k = k(C) = \dim_{\mathsf{F}_q}(C)$. Under the assumption that not many mistakes are made during the transfer of our message we expect that the received words are "near" the original words. We try to retrieve the original message by replacing a received word which does not belong to C by a word from C which is near the received word. Here near is to be understood with respect to the metric defined by the *Hamming distance* on F_q^n

$$d(x, y) = \#\{i \in \mathbb{Z} : 1 \le i \le n, x_i \ne y_i\}.$$

We define the *weight* of a word $x \in \mathsf{F}_q^n$ as

$$w(x) = \#\{i : x_i \ne 0\},$$

and we have

$$w(x) = d(x, 0).$$

A little reflection will show that we would like to have large distances between the words of the codes. For example, if we know the *minimum distance* of the code C defined by

$$d = d(C) = \min_{x,y \in C, x \neq y} d(x, y)$$

then we know that we can detect $d - 1$ mistakes and that we can correct $[(d - 1)/2]$ mistakes and simultaneously detect $[d/2]$ mistakes, since a ball with radius $[(d - 1)/2]$ contains at most one vector from C. So the minimum distance d is obviously another important invariant and we would like to maximize it. On the other hand we would also like to have a large dimension (w.r.t. the word length n). These requirements on k and on d are conflicting. Coding theory consists to some extent of finding a good compromise.

A refinement of the invariant d is the *weight distribution* of the code C, i.e. the set of non-negative integers

$$A_i = \#\{c \in C : w(c) = i\}$$

for $0 \leq i \leq n$. An object that gathers the information of the weight distribution is the *weight enumerator*

$$A = \sum_{i=0}^{n} A_i z^i \qquad \in \mathbf{Z}[z].$$

The weight enumerator determines the so-called error probability of the code, cf. [8]. We introduce the *dual code* :

$$C^{\perp} = \{x \in \mathbf{F}_q^n : \sum_{i=1}^{n} x_i y_i = 0 \text{ for all } y \in C\}.$$

Often a code is specified by a *generator matrix*, i.e. a $k \times n$-matrix the rows of which form a basis of C. A *parity check matrix* of C is a $(n - k) \times n$-matrix H which is a generator matrix of the dual code $C^{\perp}$. Obviously,

$$x \in C \iff H^t x = 0.$$

Therefore a code can be specified by a parity check matrix.

(2.1) *Example*: The $(n = 2^r - 1, k = 2^r - r - 1)$-*binary Hamming code*. This is a code over $\mathbf{F}_2$ given by a parity check matrix H the colums of which are the binary representations of $1, 2, 3, ..., 2^r - 1$. For example, the $(7, 4)$-binary Hamming code has as parity check matrix

$$H = \begin{pmatrix} 0 & 0 & 0 & 1 & 1 & 1 & 1 \\ 0 & 1 & 1 & 0 & 0 & 1 & 1 \\ 1 & 0 & 1 & 0 & 1 & 0 & 1 \end{pmatrix}.$$

This is a code with minimum distance $d = 3$. Suppose now that we receive a word $x = c + e$ with $c \in C$ and e the error vector. If we suppose that at most one mistake is made, i.e. $w(e) \leq 1$ then either $H^t x = 0$ or $H^t x =$ binary representation of the place of the mistake. So decoding is extremely easy here. Usually it is not.

A central (but not very deep) result of coding theory is the relation between the weight enumerator of C and that of $C^{\perp}$.

(2.2) **The MacWilliams Identities.** *If C is an (n, k) code over $\mathbf{F}_q$ with weight enumerator $A(z)$ and $C^{\perp}$ is the dual $(n, n - k)$-code with weight enumerator $B(z)$ then*

$$B(z) = q^{-k}(1 + (q - 1)z)^n A\left(\frac{1 - z}{1 + (q - 1)z}\right).$$

Let us give some more examples of codes.

(2.3) *Reed-Muller Codes*: Let

$$P_r = \{f \in \mathbf{F}_2[X_1, ..., X_m] : \text{total degree of } f \leq r\}$$

and consider the evaluation map

$$\beta: P_r \to (\mathbf{F}_2)^n \qquad \text{with } n = 2^m$$

defined by evaluating f in the points of $\mathbf{F}_2^m$:

$$\beta(f) = (f(v))_{v \in \mathbf{F}_2^m}.$$

We define the *Reed-Muller code* with parameters r, m over $\mathbf{F}_2$ by

$$R(r, m) = \text{Im}(\beta).$$

This is a classical code. The weight distribution is not known except for special cases, e.g. when $r \leq 2$.

(2.4) *Cyclic Codes*: A code is called *cyclic* if with every word c of C, $c = (a_0, ..., a_{n-1})$, also the word $(a_{n-1}, a_0, ..., a_{n-2})$ belongs to C.

We now use the identification of vector spaces over $\mathbf{F}_q$

$$\mathbf{F}_q^n \cong \mathbf{F}_q[X]/(X^n - 1)$$

given by

$$(a_0, ..., a_{n-1}) \to \sum_{i=0}^{n-1} a_i X^i.$$

Then $C \subset \mathbf{F}_q^n$ corresponds to an additive subgroup I of $\mathbf{F}_q[X]/(X^n - 1)$, but C is cyclic if and only if I is an ideal (i.e. stable under multiplication by X). Such an ideal is principal, hence can be written as (g) . The generator polynomial g can be chosen to be a monic divisor of $X^n - 1$ and is given by its zeroes in an extension field of $\mathbf{F}_q$. So for example, let α be a generator of the cyclic group $\mathbf{F}_q^*$ with $q = 2^m$. We define a code M' as the cyclic code of length $n = q - 1$ over $\mathbf{F}_q$ with $g = (X - \alpha)(X - \alpha^{-1})$. In order to obtain a code over $\mathbf{F}_2$ we take the restriction :

$$\text{the Melas Code } M(q) = \text{Res}_{\mathbf{F}_2}(M'),$$

where Res means taking those vectors of M' which happen to have coordinates $0, 1$ only. This is a very classical code.

It is easy to see that the dual code of the cyclic code M' corresponds to the ideal generated by $(X^n - 1)/X^{\deg(g)}g(1/X)$, hence has the complementary zeroes $1, \alpha^2, \alpha^3, ..., \alpha^{q-3}$. But these are exactly the zeroes of the polynomials

$$\sum_{i=0}^{q-2}(a\alpha^{-i} + b\alpha^i)X^i \in \mathbf{F}_q[X]/(X^{q-1} - 1) \qquad \text{with } a, b \in \mathbf{F}_q.$$

Therefore we find

$$(M')^{\perp} = \{(ax + \frac{b}{x})_{x \in \mathbf{F}_q^*} : a, b \in \mathbf{F}_q\}.$$

There are two ways of constructing a code over the prime field $\mathbf{F}_p$ from a code over $\mathbf{F}_q$; one is restriction as explained above defined by restricting oneself to those vectors that have entries from $\mathbf{F}_p$, the other is the trace, by applying Tr to the entries of a vector, where Tr is the usual trace map

$$Tr: \mathbf{F}_q \to \mathbf{F}_p, \qquad x \to x + x^p + ... + x^{p^{m-1}}.$$

The relation between these is given by

(2.4) Delsarte's Theorem. *If C is a code over $\mathbf{F}_q$ then*

$$(\text{Res}|_{\mathbf{F}_p} C)^{\perp} = Tr(C^{\perp}).$$

Hence the dual of the classical Melas Code is given by

$$M(q)^{\perp} = \{((\text{Tr}(ax + \frac{b}{x}))_{x \in \mathbf{F}_q^*} : a, b \in \mathbf{F}_q\}.$$

It is a code of dimension $2m$ over $\mathbf{F}_2$.

We refer to [6] for an introduction to coding theory. The bible of coding theory is [8] , but it does not cover the recent developments linking coding theory and algebraic geometry.

3. Elliptic Curves

An *elliptic curve* over a field k is a non-singular projective curve of genus 1 with a k-rational point P. Suitable sections of $L(3P)$ define an embedding of the curve into $\mathbf{P}^2$ such that an affine equation is given by

$$(1) \qquad y^2 + a_1xy + a_3y = x^3 + a_2x^2 + a_4x + a_6 \qquad \text{with } a_i \in k.$$

The point P goes to $(0 : 1 : 0) \in \mathbf{P}^2$. The elliptic curve can be identified with its jacobian and this defines a group law on it, i.e. for each extension field K/k the K-rational points $E(K)$ of E form a group. It is well-known that the isomorphism class of E over an algebraic closure $\overline{k}$ of k is uniquely determined by the *so-called* j-invariant of E , which is a rational expression in the coefficients a_i of (1).

Suppose that k is a field of characteristic p. The multiplication morphism $[p]: E \to E$ is inseparable and has degree p^2. The kernel of multiplication by p on $E(\bar{k})$ is a subgroup $E[p](\bar{k})$ of $E(\bar{k})$ and it is isomorphic to $\mathbf{Z}/p\mathbf{Z}$ or (0) (the latter if $[p]$ is purely inseparable). The elliptic curve is called *ordinary* or *supersingular* accordingly.

Two elliptic curves E_1 and E_2 are called *isogenous* if there exists a finite morphism $f: E_1 \to E_2$. If k is a finite field then E_1 and E_2 are isogenous over k if and only if they have the same number of k-rational points.

4. The Melas Code and its Dual for $p = 2$ and $p = 3$

As we saw in Section 2 the dual $M(q)^\perp$ of the Melas code $M(q)$ consists of the words

$$c(a, b) = (\mathrm{Tr}(ax + \frac{b}{x}))_{x \in \mathbf{F}_q^*} \qquad (a, b \in \mathbf{F}_q).$$

There is an $\mathbf{F}_q^*$-action on the code defined by

$$c(a, b) \to c(\lambda a, \lambda^{-1} b) \qquad \text{for } \lambda \in \mathbf{F}_q^*.$$

The weight is invariant under this action. Therefore if we are interested in the weight distribution we may assume that $b = 0$ or 1 in $\mathbf{F}_p$.

(4.1) **Lemma.** *The weight of a word $c(a, b)$ with $b = 0, 1$ is as follows :*

$$w(c(0, 0)) = 0,$$
$$w(c(0, b)) = q(1 - 1/p) \text{ for } b = 1,$$
$$w(c(a, 0)) = q(1 - 1/p) \text{ for } \neq 0,$$

while

$$w(c(a, 1)) = q - 1 + \frac{2}{p} - \frac{1}{p} \# C_a(\mathbf{F}_q) \qquad \text{for } a \neq 0,$$

where C_a is the non-singular projective curve with affine equation

(2) $$y^p - y = ax + \frac{1}{x}.$$

Proof: We treat the last case. We have $q - 1 - w(c(a, 1)) = \#\{x \in \mathbf{F}_q^* : \mathrm{Tr}(ax + \frac{1}{x}) = 0\}$. Since the kernel of Tr equals $\{y^p - y : y \in \mathbf{F}_q\}$ we find

$$q - 1 - w(c(a, 1)) = \frac{1}{p}\{\# C_a(\mathbf{F}_q) - 2\},$$

where the 2 comes from the points lying over 0 and ∞.   Q. E. D.

This Lemma tells us that we have to study the family of algebraic curves given by (2). Their genus is $p - 1$ for $a \neq 0$. Let us start with $p = 2$ and assume $q = 2^m$ with $m \geq 3$, cf [9]. We then have a family of elliptic curves whose equation can be written in the Weierstrass form :

(3) $$y^2 + xy = x^3 + ax \qquad (a \in \mathbf{F}_q^*).$$

Such a curve has j-invariant $j = a^{-2}$. Over $\overline{\mathbf{F}}_q$ there is exactly one iso-morphism class of elliptic curves with this j-invariant, but over the finite field $\mathbf{F}_q$ we have to take into account Galois cohomology. Over $\mathbf{F}_q$ there are exactly two isomorphism classes of elliptic curves defined over $\mathbf{F}_q$ with a given $j \neq 0$. They can be distinguished by their number of points over $\mathbf{F}_q$. If one of them has $q + 1 - t$ points defined over $\mathbf{F}_q$, then the other has $q + 1 + t$ such points. If a runs through $\mathbf{F}_q^*$ then j runs through $\mathbf{F}_q^*$ once; so our family contains out of these two exactly one. To see which one, note that our elliptic curve E_a defined by (3) has a $\mathbf{F}_q$-rational point of order $4 : (a^{q/2}, 0)$. Hence 4 divides $E_a(\mathbf{F}_q)$; so if $\#E_a(\mathbf{F}_q) = q + 1 - t$ then $t \equiv 1 \pmod 4$. But then $q + 1 + t \not\equiv 0 \pmod 4$.

But we are not so much interested in the isomorphism classes but rather in the isogeny classes. It is well known that the number of isomorphism classes in an isogeny class of elliptic curves with $q + 1 - t$ points is given by a class number (for t odd):

$$H(t^2 - 4q).$$

Here $H(\Delta) = $ *Kronecker class number* $= \#\mathrm{SL}_2(\mathbf{Z})$ -equivalence classes of positive definite binary integral quadratic forms $\phi = aX^2 + bXY + cY^2$ with discriminant $\Delta = b^2 - 4ac$, each class $[\phi]$ being counted with weight $2/\#\mathrm{Aut}(\phi)$, where Aut is the number of orientation preserving automor-phisms.

Combination of these ingredients yields the result (cf. [9])

(4.2) Theorem. *The non-zero weights in the dual $M(q)^\perp$ of the Melas code $M(q)$ are $w_t = \frac{q-1+t}{2}$ with $t \in \mathbf{Z}, t^2 < 4q, t \equiv 1 \pmod 4$. The frequency of w_t for $t \neq 1$ equals $(q-1)H(t^2 - 4q)$, while for $t = 1$ the frequency of w_1 is $(q-1)(H(1 - 4q) + 2)$.*

The possible weights were determined by Lachaud and Wolfmann, [5].

Of course, an application of the MacWilliams Identities yields the weight distribution of the Melas Code itself. The result is a somewhat messy expression in terms of class numbers. However, somewhat similar expressions occur in the Selberg Trace Formula which describes the trace of the Hecke operators on spaces of modular forms. Rewriting the result with the help of this gives the following theorem.

(4.3) Theorem. *Let*

$$A_i = \# \text{ words of weight } i \text{ in } M(q).$$

Then

$$q^2 A_i = \binom{q-1}{i} + 2(-1)^{[(i+1)/2]}(q-1)\binom{\frac{q}{2}-1}{[i/2]} +$$

$$-(q-1) \sum_{j=0, j \equiv i(2)}^{i} W_{i,j}(q)(1 + \tau_{j+2}(q));$$

here $W_{i,j}(q) \in \mathbf{Z}[q]$ *for* $0 \leq j \leq i, j \equiv i(mod\ 2)$, *with* $W_{0,0} = 1, W_{1,1} = -1$ *and the other ones defined by*

$$(i+1)W_{i+1,j+1} = -qW_{i,j+2} - W_{i,j} - (q-i)W_{i-1,j+1}.$$

and otherwise $W_{i,j} = 0$. *Moreover, for* $k \geq 3$

$$\tau_k(q) = trace\ of\ Hecke\ operator\ T(q)$$

$$on\ the\ space\ of\ modular\ forms\ S_k(\Gamma_1(4))$$

and $\tau_k(q) = -q$ *for* $k = 2$.

In case $p = 3$ we find a family of curves of genus 2, but their jacobians are isogenous to a product of elliptic curves. Therefore, we can prove similar results in that case. Here we get elliptic curves with a point of order three and the group $\Gamma_1(3)$, see [2]. However, in case $p > 3$ it is not clear what to do.

In [2],[9] and [1] we use these results to derive the weight distributions for the so-called *Zetterberg* codes.

5. Reed-Muller Codes and Supersingular Elliptic Curves

For convenience we set $q = 2^m, m \geq 3$. In the preceding section we used ordinary elliptic curves; here we establish a link between *supersingular* elliptic curves and the Reed-Muller codes. Recall

$$R(r,m) = \{(f(v))_{v \in \mathbf{F}_2^m} : f \in \mathbf{F}_2[X_1, ..., X_m], \deg(f) \leq r\}.$$

Let R be a *linearized polynomial*, i.e.

$$R = \sum_{i=0}^{h} a_i X^{2^i}, \qquad a_i \in \mathbf{F}_q.$$

It is called linearized because $R(a + b) = R(a) + R(b)$ and the zeroes of R in $\mathbf{F}_q$ form a $\mathbf{F}_2$-linear subspace of $\mathbf{F}_q$. The expression

$$Q(x) = \mathrm{Tr}[xR(x)],$$

with $\mathrm{Tr}: \mathbf{F}_q \to \mathbf{F}_2$ the usual trace map, is a quadratic form on $(\mathbf{F}_2)^m \cong \mathbf{F}_q$ since $\mathrm{Tr}[xR(y) + yR(x)]$ is bilinear.

Consider the space

$$R_h = \{R = \sum_{i=0}^{h} a_i X^{2^i} : a_i \in \mathbf{F}_q\} \qquad \text{for } h \leq [\tfrac{m}{2}].$$

Then

$$C_h = \{(\mathrm{Tr}[xR(x)]_{x \in \mathbf{F}_q}) : R \in R_h\}$$

is a subcode of $R(2, m)$. The weight of a word $(\mathrm{Tr}[xR(x)]_{x \in \mathbf{F}_q})$ equals

$$q - \frac{1}{2}\{\# \text{ of rat. points on the affine curve } y^2 + y = xR(x)\}.$$

and thus coding theory leads to the study of the family of curves

$$C_R : y^2 + y = xR(x), \qquad R \in R_h.$$

These curves have a large automorphism group over an algebraic closure of $\mathbf{F}_q$.

(5.1) **Theorem.** *The automorphism group of C_R over $\overline{\mathbf{F}}_q$ is the semi-direct product of a normal subgroup $\overline{G}$ which is an extra-special 2-group of order 2^{2h+1} and a cyclic group of order $g.c.d.\{2^i+1 : i \geq 1, a_i \neq 0\}$. This group $\overline{G}$ is the central product of $h-1$ dihedral groups of order 8 and one quaternion group of order 8 with identified centres.*

What happens over $\mathbf{F}_q$? We define

$$W_R = \{x \in \mathbf{F}_q : \text{Tr}[xR(y) + yR(x)] = 0 \text{ for all } y \in \mathbf{F}_q\}.$$

We have

$$m \equiv \dim_{\mathbf{F}_2}(W_R) \pmod{2} .$$

Define

$$E_{h,R}(x) = (R(x))^{2^h} + \sum_{i=0}^{h}(a_i x)^{2^{h-i}}.$$

Then

$$W_R = \{x \in \mathbf{F}_q : E_{h,R}(x) = 0\}.$$

(Note that here both the vector space structure as well as the structure of a field is used for $\mathbf{F}_q$.) Let

$$\overline{W}_R = \{x \in \overline{\mathbf{F}}_q : E_{h,R}(x) = 0\}.$$

Then we find an isomorphism

(5) $\overline{G}/Z \to \overline{W}$

where Z is the center of $\overline{G}$ generated by the hyperelliptic involution ϕ with $\phi(x) = x, \phi(y) = y + 1$. The isomorphism is induced by sending an automorphism ψ to $\psi(x) - x \in \overline{\mathbf{F}}_q$. The commutator on the Heisenberg group $\overline{G}$ gives rise to a symplectic form on $\overline{W}$ and W. This form is non-degenerate on $\overline{W}$. This symplectic form on the kernel of another symplectic form (namely $\text{Tr}[xR(x)]$) plays a crucial role in the study of these curves. An important remark is that the isomorphism (5) is equivariant for the action of $\text{Gal}(\overline{\mathbf{F}}_q/\mathbf{F}_q)$. This explains what happens over $\mathbf{F}_q$.

We are interested in the variation of the number of points on the curves C_R as R varies through R_h. We define

$$V = \{x \in W : \text{Tr}[xR(x)] = 0\}.$$

We have $V = W$ or $\dim(V) = \dim(W) - 1$. The following proposition gives the number of points (up to a sign).

(5.2) **Proposition.** *If $V \neq W$ then $\#C_R(\mathbf{F}_q) = q + 1$, while if $V = W$ we have $\#C_R(\mathbf{F}_q) = q + 1 \pm \sqrt{q2^w}$, where $w = \dim_{\mathbf{F}_2} W$.*

We now first study families $\mathcal{F}$ given by $y^2 + y = x(R' + a_0 x)$, where R' is a fixed linearized polynomial and where a_0 varies in $\mathbf{F}_q$. For this family the equation $E_h = 0$ is independent of a_0.

(5.3) Proposition. *For the curves C_R in the family $\mathcal{F}$ the frequency of the trace of Frobenius $t = 0$ (resp. $t = \pm\sqrt{q2^w}$) is*

$$n_0 = \frac{2^w - 1}{2^w} q \qquad \left(resp.\ n_\pm = \frac{q \mp \sqrt{q2^w}}{2^{w+1}}\right).$$

The next thing we do is studying the families $y^2 + y = xR(x)$, where $R = \sum_{i=1}^h a_i X^{2^i}$ and the a_i for $i > 0$ vary over $\mathbf{F}_q$. Here it turns out coding theory can help algebraic geometry. The results on subcodes of binary Reed-Muller codes $R(2, m)$ give explicit formulas for the numbers

$$n_w^{(h)} = \#\{R \in R_h : a_0 = 0, a_h \neq 0, \dim(W_R) = w\}.$$

We refer to [3] for the formulas. But algebraic geometry provides also new relations between these numbers:

(5.4) Theorem. *We have*

$$\sum_w (2^w - 1)n_w^{(h)} = 2(q - 1)^2 q^{h-2} \qquad for\ h \geq 2,$$

and

$$\sum_w (2^w - 1)n_w^{(1)} = q - 1.$$

Using the formulas for the $n_w^{(h)}$ we find infinitely many *maximal* and *minimal* curves, i.e. curves for which the Hasse-Weil bound is assumed.

In case $h = 1$ we find elliptic curves. These curves are supersingular as one easily checks. This holds more generally.

(5.5) Theorem. *The curve C_R is supersingular over $\overline{\mathbf{F}}_q$.*

This means that over $\overline{\mathbf{F}}_q$ the jacobian of C_R splits up to isogeny as a product of supersingular elliptic curves. This is proved by induction : one takes a non-hyperelliptic involution ϕ in $\overline{G}$. Then one shows that the curve $C_R/ <\phi>$ is again of the form

$$y^2 + y = xS(x),$$

where S is linearized and $\deg(S) = 2^{h-1}$. However, over $\mathbf{F}_q$ the situation is much more complex, cf [3].

The methods employed here work as well for characteristic $p > 2$. For other families of curves related to codes we refer to [4].

The link between the Reed-Muller codes and supersingular elliptic curves shows that the relationship between coding theory and algebraic geometry can be beneficial for both.

REFERENCES

[1] van der Geer, G., Schoof, R., van der Vlugt, M., *Weight Formulas for Ternary Melas Codes*, Preprint 1990.

[2] van der Geer, G., van der Vlugt, M., *Artin-Schreier Curves and Codes*, Preprint 1989, Journal of Algebra (to appear).

[3] van der Geer, G., van der Vlugt, M., *Reed-Muller Codes and Supersingular Curves. I.*, Preprint 1990.

[4] van der Geer, G., van der Vlugt, M., *Families of algebraic curves and codes*, In preparation.

[5] Lachaud, G., Wolfmann, J., *Sommes de Kloosterman, courbes elliptiques et codes cycliques en charactéristique 2*, Comptes Rendus Acad. Sci. Paris **305** (1987), 881–883.

[6] van Lint, J. H., "Introduction to Coding Theory," Grad. Texts in Math., Springer Verlag, 1982.

[7] van Lint, J. H., van der Geer, G., "Introduction to Coding Theory and Algebraic Geometry," Birkhäuser, 1988.

[8] MacWilliams, F. J., Sloane, N. J. A., "The Theory of Error Correcting Codes," North Holland, Amsterdam, 1983.

[9] Schoof, R., van der Vlugt, M., *Hecke operators and the weight distribution of certain codes*, Preprint 1989, Journal of Comb. Theory (to appear).

Gerard van der Geer
Faculteit Wiskunde en Informatica
Universiteit van Amsterdam
Plantage Muidergracht 24
1018 TV Amsterdam
The Netherlands

Algorithmes - disons rapides - pour la décomposition d'une variété algébrique en composantes irréductibles et équidimensionnelles

MARC GIUSTI* JOOS HEINTZ

Résumé. Nous décrivons dans cet article deux algorithmes qui construisent les décompositions irréductible et équidimensionnelle d'une variété algébrique affine (ou projective), définie par un ensemble fini de polynômes. Le premier calcule les composantes irréductibles, donc dépend d'un algorithme de factorisation, et en conséquence ne peut se paralléliser que partiellement, du moins à notre connaissance. Le second correspond aux composantes équidimensionelles, et est susceptible d'une parallélisation totale. Comme applications, le lecteur trouvera en appendice un calcul de la forme de Chow d'une variété projective quelconque, dû à T. Krick et P. Solerno, et un calcul du degré d'une variété affine.

Ces algorithmes sont décrits par des réseaux arithmétiques, ce qui permet un déroulement séquentiel ou parallèle (avec la restriction concernant la factorisation). En séquentiel, leurs complexités admettent une borne supérieure simplement exponentielle en la taille des polynômes d'entrée (quantité, degré, nombre de variables et éventuellement taille arithmétique). En parallèle, la complexité du deuxième algorithme devient polynomiale en cette taille.

1 Introduction

Pour l'ensemble de ce travail, nous allons fixer les notations.

1.1 Notations de base

Soient k un corps commutatif, parfait et infini, et $\bar{k}$ une clôture algébrique. Afin de traiter simultanément les situations affine et projective, nous allons considérer des espaces ambiants de même dimension n: l'espace affine $\mathbf{A}_{\bar{k}}^n$ (resp. l'espace projectif $\mathbf{P}_{\bar{k}}^n$), munis de la topologie de Zariski, et notés plus simplement $\mathbf{A}^n$ et $\mathbf{P}^n$.

*Avec un soutien du GDR G0060 "Calcul Formel, Algorithmes, Langages et Systèmes" et du PRC "Mathématiques et Informatique".

Nous voulons considérer des sous-variétés vivant dans ces espaces ambiants. Il y a deux points de vue: le premier est *intrinsèque*, et consiste à parler d'une sous-variété sans avoir précisé coordonnées ou équations. Mais ce deuxième point de vue *non intrinsèque* est souvent inévitable et il faut donc en passer par là: soient donc $x_1, \ldots, x_n$ (resp. $x_0, x_1, \ldots, x_n$) des indéterminées sur k, et f un système fini de polynômes $f_1, f_2, \ldots, f_s$ de $k[x_1, \ldots, x_n]$ (resp. des polynômes homogènes de $k[x_0, x_1, \ldots, x_n]$). L'idéal engendré sera noté $I(f_1, \ldots, f_s)$ ou plus simplement $(f_1, \ldots, f_s)$, voire I, et la variété algébrique, au sens classique, affine (resp. projective) définie par f ou I sera notée $V_{\bar{k}}(I)$, ou plus simplement $V(I)$, $V(f)$ ou V quand aucun doute n'est possible.

1.2 Considérations de complexité

Les algorithmes considérés ci-dessous vont admettre comme entrées et sorties des ensembles finis de polynômes. Il nous faut donc d'abord quantifier et mesurer la taille d'un tel ensemble, puis préciser le type d'algorithmes qui vont le manipuler.

1.2.1 Taille des entrées-sorties

Soit $f = f_1, \ldots, f_s$ un ensemble fini de polynômes à n variables (resp. homogènes à $n + 1$ variables). Nous supposerons qu'ils sont rangés par degré total décroissant $d_1 \geq \cdots \geq d_s$ et, dans le cas projectif, nous noterons $d(f)$ (ou d pour simplifier) le plus grand d'entre eux, à savoir d_1. Dans le cas affine, nous noterons $d(f)$ le maximum de d_1 et de 3; cette restriction technique est due à l'utilisation d'un Nullstellensatz affine effectif, où elle semble être pour le moment inévitable. Sans restriction de généralité, nous supposerons aussi que la dimension de l'espace ambiant est au moins 2.

Il existe un cas particulier important de corps de base: celui des rationnels. En fait comme aucun des algorithmes n'utilise la division dans le corps de base, nous supposerons les coefficients entiers. Dans ce cas la complexité de notre ensemble doit inclure la *taille arithmétique*, c'est-à-dire le nombre maximal $t(f)$ (ou plus simplement t) de bits qu'il faut pour écrire n'importe quel coefficient des polynômes de f.

Autrement dit, la hauteur de chaque f_i ($i = 1, \ldots, s$), au sens classique des arithméticiens, qui est le maximum des valeurs absolues de tous ses coefficients, est majoré par 2^t.

Une bonne mesure de la complexité du système f semble donc être le triplet (d, n, s) ou le quadruplet (d, n, s, t) dans le cas entier, ce que nous appellerons la *taille* de f. En effet nous pouvons aisément majorer la place mémoire nécessaire pour stocker f à partir de sa taille. Le nombre de monômes sur n lettres de degré au plus d (resp. sur $n + 1$ lettres de degré d) est le coefficient binomial $(d + n)!/d!n!$, quantité majorée par $(d + 1)^n$. Si nous convenons de coder un polynôme par le vecteur de ses coefficients dans la représentation dense, il faut stocker au plus $s(d+1)^n$ coefficients, et s'ils sont entiers, cela demande au plus $s(d+1)^n t$ bits.

Par ailleurs, remarquons qu'au prix d'un algorithme standard d'algèbre linéaire nous pouvons supposer que les polynômes $f_1, \ldots, f_s$ sont linéairement indépendants sur le corps de base, et donc que s est majoré par $(d+1)^n$. Le coût de cette préparation n'est pas très élevé, et nous le préciserons au paragraphe suivant. Nous la supposerons donc dorénavant faite systématiquement. Avec cette remarque, il suffirait donc de nous restreindre au couple (d,n) ou au triplet (d,n,t) pour caractériser la complexité de f. Mais nous conserverons s dans les données, car posséder un petit nombre de générateurs constitue pour un idéal une propriété intéressante dont il faut garder trace.

Par contre, nous aurons aussi envie de considérer des quantités intrinsèques caractérisant la complexité de V: ce seront essentiellement sa dimension $r(V)$, majorée par n, et son degré $deg(V)$ majoré par d^n (voir la définition et l'inégalité de Bézout correspondante dans Heintz (1983)).

1.2.2   Algorithmes et leur complexité

Les algorithmes que nous allons utiliser ou introduire seront en principe décrits par un *réseau arithmétique*, représenté par un graphe orienté acyclique (von zur Gathen, 1986). A chaque sommet correspond un processeur élémentaire qui effectue une opération de base du corps de base k, et chaque arête indique l'envoi d'une sortie d'un processeur comme entrée du second. Les opérations de base sont l'addition, la soustraction, la multiplication et un sélecteur associé au test d'égalité. Si la caractéristique est positive, il convient d'y ajouter l'extraction des racines $p^{ièmes}$. Nous parlerons alors de *réseau arithmétique étendu*.

Chaque polynôme est codé par le vecteur de ses coefficients dans la représentation dense. Un algorithme travaille donc sur un vecteur d'éléments du corps de base, et admet un déroulement séquentiel ou parallèle. La *complexité séquentielle* est la taille du réseau, c'est-à-dire le nombre de processeurs ou sommets du graphe. La *complexité parallèle* est la profondeur du réseau, c'est-à-dire la longueur du plus long chemin dans le graphe orienté.

Dans le cas entier chaque processeur arithmétique devient lui-même un réseau booléen dont les processeurs manipulent maintenant les bits. Comme nous ne voulons pas entrer dans des considérations de parallélisation à ce niveau booléen, nous nous restreindrons à l'aspect séquentiel en interprétant le déroulement dans le modèle des machines de Turing. Il faut alors tenir compte de la croissance éventuelle des coefficients des polynômes intermédiaires.

Pour une discussion plus approndie de ce modèle de complexité, nous renvoyons à von zur Gathen (1986) et Fitchas-Galligo-Morgenstern (1987).

Dans ce cadre, nous dirons qu'un algorithme est de complexité *admissible* si pour traiter une entrée de taille (d, n, s) le réseau correspondant admet une complexité séquentielle $(sd)^{n^{O(1)}}$ et parallèle en $(n \log(sd))^{O(1)}$. Le cas échéant nous préciserons la constante $O(1)$ en parlant d'un algorithme *admissible d'exposant a*. Dans le cas entier, nous parlerons de complexité admissible si nous pouvons contrôler la taille des entiers qui apparaissent dans les calculs intermédiaires par des bornes du type $(tsd)^{n^{O(1)}}$. Observons que dans le cas où le résultat est constitué de polynômes, la somme de leurs degrés est au plus $(sd)^{n^{O(1)}}$.

Par exemple, l'algorithme de réduction de s, mentionné à la fin du paragraphe précédent, est admissible. C'est une conséquence immédiate de l'algorithme de Berkowitz (1984), et la borne dépend linéairement de s en séquentiel (resp. de $\log(s)$ en parallèle).

1.3   Enoncé des problèmes et des résultats

Un problème fondamental pour la description de V consiste en la détermination de l'ensemble $\mathcal{C}(V)$ des composantes irréductibles de V. Du point de vue algébrique, ceci peut être évidemment résolu par une décomposition primaire irredondante de I. Mais qu'on utilise l'algorithme originel de Hermann (1926), repris par Seidenberg (1974) ou celui plus récent de Gianni-Trager-Zacharias (1988), on se heurte à l'utilisation massive de résolutions de systèmes linéaires à coefficients polynomiaux ou de constructions de bases standard, ce qui entraîne a fortiori une complexité au moins doublement exponentielle.

Mais si nous ne nous intéressons pas au schéma correspondant à l'idéal donné, mais seulement à l'espace topologique sous-jacent, c'est-à-dire à l'ensemble algébrique constitué par les points à coordonnées dans $\bar{k}$ qui annulent tous les polynômes de l'idéal, la situation s'améliore très nettement: ceci constitue la contribution fondamentale de Chistov-Grigoriev (1983), montrant que la détermination ensembliste des composantes irréductibles d'une variété projective peut se calculer par un algorithme séquentiel admissible d'exposant 2. Malheureusement ce dernier dépend d'un algorithme de factorisation, et se parallélise donc mal (il n'en reste pas moins que le reste de leur algorithme peut, lui, fort bien se paralléliser).

Nous nous proposons dans cet article de nous libérer de cette contrainte en exposant un algorithme qui calcule la décomposition équidimensionnelle d'une variété V, c'est-à-dire partant des équations définissantes rend, pour chaque dimension intermédiaire i, $(0 \leq i \leq r(V))$ un ensemble fini de polynômes définissant la réunion des composantes irréductibles ensemblistes de dimension i. Cet algorithme se parallélise totalement car il n'utilise que des sous-algorithmes d'algèbre linéaire sur k.

Chemin faisant, nous redémontrerons dans le cas entier par une méthode originale une version un peu plus faible du théorème de Chistov-

Grigoriev, en calculant les composantes irréductibles sur Q sans entrer dans des questions d'extensions algébriques, en fait en renonçant à la factorisation absolue.

Du point de vue du coût, les complexités sont admissibles, et nous pouvons préciser l'exposant pour la détermination des composantes irréductibles et équidimensionnelles (2, comme chez Chistov-Grigoriev): en séquentiel $s^5 d^{O(n^2)}$ et en parallèle $O(n^4 \log^2 sd)$).

Nous avons appris pendant la préparation de cet article que P. Hintenhaus (1989) avait également traité d'une version du résultat de Chistov-Grigoriev.

L'algorithme de décomposition équidimensionnelle permet de ramener la considération d'une variété arbitraire au cas d'une variété équidimensionnelle. Par exemple, signalons le calcul en temps admissible de la forme de Cayley (ou forme de Chow) d'une variété projective en utilisant l'algorithme d'élimination des quantificateurs de la théorie élémentaire de $\bar{k}$ (Fitchas-Galligo-Morgenstern, 1989, ou Krick, 1990). Ce travail est rassemblé dans un appendice.

Une méthode plus directe pour calculer la forme de Chow d'un idéal équidimensionnel est donné dans Caniglia (1989). Remarquons aussi que découle du même algorithme le calcul du degré de la variété (au sens donné par Heintz (1983), à savoir la somme des degrés de toutes les composantes irréductibles ensemblistes). Ceci découle de Caniglia-Galligo-Heintz (1989, Theorem 19) ou de notre méthode de calcul du polynôme de Chow.

Terminons enfin cette introduction par une remarque technique. Ce qui suit utilise de manière essentielle et constante les divers Nullstellensätze effectifs. Dans le cas projectif, il s'agit d'un résultat de D. Lazard (1977), bien plus facile à établir que son homologue affine. La première version de ce résultat affine a été obtenue par Brownawell (1987), et a été suivie par des extensions de Caniglia-Galligo-Heintz, Kollár, Berenstein-Yger, Philippon, et bien d'autres ...Une bibliographie sur ce sujet se trouve dans la revue de Teissier (1989) au séminaire Bourbaki. Pour ce qui nous concerne ici, nous emploierons principalement la version "à la Hentzelt" (voir par exemple Fitchas-Galligo, 1988 Théorème 10) ou la version locale de Caniglia-Guccione-Guccione (1990).

Une partie de la rédaction de cet article a été effectuée pendant la tenue à Bonn d'un "workshop on algebraic and geometric complexity". Il nous est agréable de remercier ici Marek Karpinski de son invitation.

2   Mise en position générique

Il se peut que les données sur la variété V à décomposer soient très mal conditionnées. Comment est définie V au départ? Dans le cas affine par exemple, comme fibre au dessus de 0 d'une application polynomiale de $\mathbf{A}^n$ dans $\mathbf{A}^s$. Nous allons être amenés à opérer des changements linéaires de coordonnées à la source. Dans le cas projectif, il faudra également changer de coordonnées à la source, mais comme f ne donne pas naturellement naissance à une application projective, la préparation sur les générateurs de l'idéal est un peu plus compliqué qu'un changement de coordonnées.

2.1 Définition des variables normales et générales

Le but est de préparer notre situation, au besoin à l'aide d'un changement linéaire des variables, de manière à ce que la variété étudiée se projette de la façon la plus agréable possible, sur une sous-variété linéaire de même dimension, puis de dimension un de plus.

Le premier pas correspond au lemme algébrique de normalisation de Noether, le second est plus subtil car de nature géométrique: après projection, il s'agit de pouvoir continuer à séparer les composantes irréductibles.

Afin d'alléger les notations, nous allons noter r la dimension $r(V)$ de V.

Précisons ce qui a été indiqué ci-dessus. Etant donné un système de coordonnées $x_1, \ldots, x_n$ de l'espace affine $\mathbf{A}^n$, nous considérerons les projections $\pi_0 : \mathbf{A}^n \to \mathbf{A}^r$ et $\pi : \mathbf{A}^n \to \mathbf{A}^{r+1}$ définies par $\pi_0(x_1, \ldots, x_n) = (x_1, \ldots, x_r)$ et $\pi(x_1, \ldots, x_n) = (x_1, \ldots, x_{r+1})$.

Dans la situation projective, il faut éviter les sous-variétés linéaires $V(x_0, x_1, \ldots, x_r)$ ou $V(x_0, x_1, \ldots, x_{r+1})$ centres de projection. Etant donné un système de coordonnées $x_0, x_1, \ldots, x_n$ de l'espace projectif $\mathbf{P}^n$, nous considérerons les projections $\pi_0 : \mathbf{P}^n - \mathbf{P}^{n-r-1} \to \mathbf{P}^r$ et $\pi : \mathbf{P}^n - \mathbf{P}^{n-r-2} \to \mathbf{P}^{r+1}$ définies par $\pi_0(x_0, x_1, \ldots, x_n) = (x_0, x_1, \ldots, x_r)$ et $\pi(x_0, x_1, \ldots, x_n) = (x_0, x_1, \ldots, x_{r+1})$.

En l'honneur de Max Noether et de son lemme de normalisation, nous dirons que les variables $x_1, \ldots, x_n$ sont en position *normale* par rapport à V si la restriction $\pi_0|_V$ de π_0 à V est un morphisme fini et surjectif de variétés. Ceci revient à dire que l'homorphisme canonique $k[x_1, \ldots, x_r] \to k[x_1, \ldots, x_n]/(f_1, \ldots, f_s)$ est injectif et correspond à une extension entière d'anneaux.

Nous dirons que les variables $x_1, \ldots, x_n$ sont en position *générale* par rapport à V si elles sont en position normale par rapport à la variété et si de plus π sépare les composantes de V, c'est-à-dire que $\pi(C)$ n'est pas contenu dans $\pi(C')$ pour tout couple de composantes distintes C et C' de $\mathcal{C}(V)$.

2.2 Méthode brutale

La mise en position générique peut s'obtenir par une extension trancendante brutale du corps de base k, obtenue en adjoignant un certain nombre de lettres. Décrivons la dans le cas affine.

Soient A la $n \times n$-matrice carrée à coefficients génériques a_{ij} $(i, j = 1, \ldots, n)$, et B la $(n+1) \times s$-matrice à coefficients génériques b_{ij} $(i = 1, \ldots, n+1,$ $j = 1, \ldots, s)$. Si x et f sont respectivement les vecteurs colonnes des

$x_1, \ldots, x_n$ et des $f_1, \ldots, f_s$, on change linéairement de coordonnées à la source et au but en introduisant $y = Ax$ et $g = Bf$.

Dans le cas projectif, la matrice carrée de changement de base à la source doit être d'ordre $n + 1$. En ce qui concerne le but, afin de conserver le caractère homogène des générateurs, il faut d'abord les multiplier par des polynômes ad hoc de degré $d_1 - d_i$ avant d'effectuer des combinaisons linéaires à coefficients constants.

En considérant que le nouveau corps de base est $K = k(A, B)$, et que k est infini, les lemmes d'évitement classiques assurent la mise en position générique. Le lecteur peut également se référer au paragraphe suivant.

Il est clair que la méthode précédente n'est pas très bonne du point de vue de la complexité. Cependant si un algorithme est de complexité admissible sur K, il le reste sur k, essentiellement parce que le nombre de monômes de degré $d^{n^{(O(1))}}$ sur n lettres reste du même ordre. [S'il faut changer de générateurs, il est nécessaire d'éviter que s grimpe en exposant, ce que nous pourrons faire en utilisant Heintz-Schnorr (1980, Theorem 4.4), mais nous n'en aurons pas besoin ici.]

2.3   Méthodes raffinées de mise en position générique

Cependant si nous voulons de meilleures bornes, il faut un peu plus travailler.

2.3.1   Mise en position normale

Dans le cas affine, comme k est infini nous pouvons appliquer l'algorithme de normalisation décrit dans Dickenstein-Fitchas-Giusti-Sessa (1989, 1.15) ou dans Logar (1988, A.4). Celui-ci calcule un changement k-linéaire de variables, telles que les nouvelles variables soient en position normale par rapport à V, avec une complexité séquentielle $s^5 d^{O(n^2)}$ et parallèle $O(n^4 \log^2(sd))$. L'algorithme mentionné se base sur une version effective du Nullstellensatz affine (Fitchas-Galligo, 1988, Théorème 10). Dans le cas projectif, nous pouvons nous servir de l'algorithme de normalisation homogène donné par Giusti (1988, Théorème 5.6.2), qui utilise la version effective du Nullstellensatz projectif démontrée par Lazard (1977). Ces deux algorithmes incluent le calcul de la dimension.

Il nous faut maintenant passer à la mise en position générale, que nous allons traiter dans le cas affine. L'extension au cas projectif est laissée (pour le moment !) au lecteur. La clef est de disposer d'un lemme d'évitement très précis, et en fait la dimension 0 suffit. L'idée de ce lemme est due à Chistov-Grigoriev (1983) et repose sur la propriété bien connue des déterminants de Vandermonde.

2.3.2   Définition: les formes x_γ et les projections associées

Si γ est un élément du corps de base, notons x_γ la forme linéaire sur $\mathbf{A}^n$ (resp. $\mathbf{P}^n$) définie par: $x_\gamma(x_1,\ldots,x_n) = x_1 + \gamma x_2 + \cdots + \gamma^{n-1} x_n$ (resp. $x_\gamma(x_0,\ldots,x_n) = x_0 + \gamma x_1 + \cdots + \gamma^n x_n$). Remarquons qu'un ensemble d'éléments distincts deux à deux donne naissance à des formes linéairement indépendantes, pourvu bien sûr qu'elles ne soient pas en nombre trop grand.

Nous aurons aussi à considérer le morphisme d'espaces affines défini par $\pi_\gamma(x) := (x_1,\ldots,x_r,x_\gamma(x))$ où $x = (x_1,\ldots,x_n)$ est un point de $\mathbf{A}^n$.

2.3.3   Un lemme d'évitement en dimension zéro affine

Lemme. *Soit Γ un ensemble fini de c éléments du corps de base k. Soit m un entier vérifiant l'inégalité $(n-1)m < c$. Pour tout sous-ensemble fini M de $\mathbf{A}^n$, ne contenant pas l'origine, et de cardinal au plus m, il existe un élément γ dans Γ tel que la forme linéaire x_γ ne s'annule pas sur M.*

Preuve: Soit t une nouvelle indéterminée, et considérons le polynôme non identiquement nul $w := \prod_{(x_1,\ldots,x_n)\in M}(x_1 + \cdots + x_n t^{n-1})$. Le degré de w ne dépasse pas $(n-1)m$. Comme le cardinal de Γ est strictement supérieur à cette quantité, w ne peut pas s'annuler sur tous les points de Γ: soit γ un point le contredisant. Mais $w(\gamma)$, non nul, n'est pas autre chose que $\prod_{x\in N} x_\gamma(x)$. Il s'ensuit que la forme x_γ ne s'annule pas sur M.

Corollaire 2.3.4 *Soit Γ un ensemble fini de c éléments du corps de base k. Soit m un entier vérifiant l'inégalité $(n-1)m(m-1)/2 < c$. Pour tout sous-ensemble fini M de $\mathbf{A}^n$, de cardinal au plus m, il existe un élément γ dans Γ tel que la forme linéaire x_γ sépare les points de M, c'est-à-dire prenne des valeurs distinctes $x_\gamma(x) \neq x_\gamma(x')$ pour tout couple de points distincts $x \neq x'$ de M.*

Preuve: Formons l'ensemble $N = \{x - x' \mid x,x' \in M, x \neq x'\}$, de cardinal au plus $m(m-1)/2$, et qui ne contient pas zéro par construction. Il suffit de démontrer qu'il existe un point γ de Γ tel que la forme x_γ ne s'annule pas sur l'ensemble N. Mais c'est clair par 2.3.3.

2.3.5   Algorithme non déterministe de calcul de variables générales

Lemme. *Supposons les variables $x_1,\ldots,x_n$ en position normale par rapport à V et la dimension $r(V)$ non triviale, c'est-à-dire $r(V) < n$. Alors si Γ est un ensemble de valeurs de k, de cardinal au moins $n(deg(V))^4$,* *il existe un point γ de Γ tel que les variables $x_1, \ldots, x_r, x_\gamma, x_{r+2}, \ldots, x_n$ soient en position générale par rapport à V.*

Preuve: Choisissons pour chaque composante C de V un point x_C qui n'appartienne à aucune autre composante C' du moment qu'elle soit distincte de C. La définition du degré que nous avons choisie (Heintz, 1983) implique que le nombre de composantes est majoré par le degré de la variété. Comme $\pi_0 \mid_V$ est un morphisme fini, la fibre $\pi_0^{-1}(\pi_0(x_C)) \cap V$ n'est constituée que de points en nombre fini majoré par $deg(V)$, d'après l'inégalité de Bézout affine (Heintz, 1983, Théorème 1). Le cardinal de la réunion de tous ces points, $M := \cup_{C \in \mathcal{C}(V)} \pi_0^{-1}(\pi_0(x_C)) \cap V$ est donc au plus $(deg(V))^2$. Pour chaque élément γ de Γ, considérons le morphisme π_γ introduit en 2.3.2. Observons que $\pi_\gamma \mid_C$ est aussi un morphisme fini pour toute composante C de V. D'après le corollaire 2.3.4, il existe un γ tel que π_γ sépare les points de M. Soient C et C' deux composantes distinctes de V. Alors $\pi_\gamma(x_C)$ ne peut pas appartenir à la projection de C' par π_γ (voir Heintz, 1983, preuve de Lemma 3, pour détails). Ceci implique évidemment que $\pi_\gamma(C)$ n'est pas contenue dans $\pi_\gamma(C')$.

Certes il est désagréable de ne pouvoir déterminer cet élément γ algorithmiquement par un réseau arithmétique, mais la cardinalité admissible de l'ensemble d'où nous le tirons suffira pour nos calculs de complexité.

Une dernière remarque de convention: par abus de notation, nous considérerons toujours être revenu dans les anciennes variables $x_1, \ldots, x_r$, $x_{r+1}, \ldots, x_n$ en n'écrivant jamais le changement de variables correspondant.

3   Trouver une équation de petit degré pour la projection

Dorénavant, nous allons supposer que les variables $x_1, \ldots, x_n$ sont normales pour la variété V à décomposer, et que sa dimension est strictement inférieure à n.

Groupons les composantes de V en deux sous-variétés: d'abord la réunion $T(V)$ des composantes de plus grande dimension. Nous nous référerons de manière imagée à la *tête* de la variété. Il reste éventuellement la réunion $Q(V)$ des autres composantes de plus petite dimension, que nous appellerons le *croupion* de la variété. Nous remercions Jacques Morgenstern de nous avoir dissuadé d'utiliser un autre mot (qui semblait pourtant s'imposer phonétiquement), et donc de sa contribution décisive en faveur de cette terminologie. [1]

---

[1] " ... we take the occasion to stress that English is strongly recommended for the final version (at least the Program Committee suggests that an English translation of "croupion" is provided to help readers not very familiar with French!)" *The MEGA 90 Program Committee.*

D'après le *Petit Robert* (Paul Robert, 1970, traduction anglaise de notre

3.1 Cas affine

Nous allons définir de deux manières différentes la projection de la tête de la variété sur une sous-variété linéaire de dimension 1 de plus en position générale.

3.1.1 Le polynôme minimal

Observons que comme le système de coordonnées est en position normale, la restriction de π à la variété est un morphisme fini de V sur son image. La tête $T(V)$ étant une variété équidimensionnelle définissable sur k, $\pi(T(V))$ l'est aussi. Donc cette dernière est une hypersurface de $\mathbf{A}^{r+1}$, et soit $h(V)$ un polynôme de degré minimal qui la définit. Comme k est parfait par hypothèse, nous en déduisons que $h(V)$ est sans carrés ("quadratfrei !", comme aime à le rappeler le théoricien des nombres Maurice Mignotte, qui possède une solide culture germanique), et que son degré est majoré par celui de la tête (voir Heintz, 1983, Lemma 2), donc par celui de la variété, et en dernier ressort par d^n d'après l'inégalité de Bézout.

3.1.2 Le polynôme calculable

Posons maintenant $D = D(V) := d^n(d^n + 1)$, et analysons le k-espace vectoriel $E(V)$ formé des polynômes de I ne dépendant que des $r + 1$ premières variables, et tels que leur écriture sur la base des générateurs donnés $f_1, \ldots, f_s$ se réalise en degré au plus D:

$$E(V) = \{f \in k[x_1, \ldots, x_{r+1}] \mid \exists\, p_1, \ldots, p_s \in k[x_1, \ldots, x_n]$$
$$avec \; f = \sum_{1 \le i \le s} p_i f_i \; et \; Max\, \{deg(p_i f_i) \le D \mid i = 1, \ldots s\}\}$$

Comme D a été choisi suffisamment grand, $E(V)$ n'est pas réduit à 0 (voir Dickenstein-Fitchas-Giusti-Sessa, 1989, Proposition 1.11), et sa dimension est au plus $(D+1)^n$, lui-même majoré par $d^{O(n^2)}$. Soit $g(V)$ (ou g pour simplifier) le plus grand commun diviseur d'une base de $E(V)$: il est évidemment de degré au plus D.

Nous pouvons décrire les coefficients de f et des p_i qui interviennent dans la définition de $E(V)$ par un système d'équations linéaires, donné par

cru):

CROUPION. n. m. (v. 1460); de *croupe*. 1° Extrémité postérieure du corps de l'oiseau, composée des dernières vertèbres dorsales et supportant les plumes de la queue (V. Uropygial) (**rump**). *Morceau délicat au-dessus du croupion d'une volaille.* V. Sot-l'y-laisse. (**parson's, pope's nose**). 2° *Plaisant.* Le derrière humain. *V. Croupe.* (**rump**) 3° *Hist.* (trad. angl.). *Le Parlement Croupion:* convoqué par Charles Ier en 1640, dissous par Cromwell en 1653 et rappelé à deux reprises (**The Rump**).

une matrice d'ordre au plus $(D+1)^n \times (s+1)(D+1)^n$ dont les entrées sont formées des coefficients des f_i. Une base de cet espace vectoriel se calcule par un réseau arithmétique, en temps séquentiel $s^5 d^{O(n^2)}$ et parallèle $O(n^2 \log^2 sD) = O(n^4 \log^2 sd)$. Le calcul du pgcd se ramène à des des techniques d'algèbre linéaire sur k et s'effectue avec les mêmes bornes de complexité (voir Brown, 1971 et von zur Gathen, 1986).

Lemme 3.1.3 *Les variétés définies dans* $\mathbf{A}^n$ *par* h *et* g *coïncident.*

Quoique les polynômes h et g ne dépendent que des $r+1$ premières variables, nous les considérerons comme éléments de $k[x_1, \ldots, x_n]$ sauf mention explicite du contraire. En ce sens la tête de la variété est bien contenue dans le lieu des zéros de h et de g.

Preuve: Comme tout polynôme de E appartient à $I \cap k[x_1, \ldots, x_{r+1}]$, il s'annule évidemment sur $\pi(T(V))$. Mais le polynôme minimal h qui définit cette dernière variété dans l'espace ambiant $\mathbf{A}^{r+1}$ étant libre de carrés, il divise tous les polynômes de E, donc leur pgcd g. Ceci implique donc que $V(h)$ est contenu dans $V(g)$.

Montrons maintenant l'inclusion dans l'autre sens. Comme $\pi(Q(V))$ est une variété définissable sur k, de dimension strictement inférieure à celle de V et de degré au plus celui du croupion de V, nous en déduisons par des arguments analogues à ceux utilisés dans Heintz (1983, Proposition 3) qu'il existe des polynômes q_1 et q_2 de $k[x_1, \ldots, x_{r+1}]$, sans facteurs communs et de degré au plus $deg(Q(V))$, qui s'annulent sur $\pi(Q(V))$ donc sur $Q(V)$. Comme par ailleurs h s'annule sur la tête, hq_1 et hq_2 s'annulent sur la variété V. Maintenant le degré de chaque hq_i est majoré par $deg(V)$:

$$deg(hq_i) = deg(h) + deg(q_i) \leq deg(T(V)) + deg(Q(V)) = deg(V)$$

Appliquons à cette situation Dickenstein-Fitchas-Giusti-Sessa 1989, Remark 1.7: hq_1 et hq_2, élevés à une puissance convenable (d^n) tombent dans $E(V)$. Donc $g(V)$ divise une puissance de h, et $V(g)$ est contenue dans $V(h)$.

3.1.4 Définition: les polynômes minimaux $h_\gamma(V)$ et les polynômes calculables $g_\gamma(V)$

Il est clair d'après ce qui précède que les polynômes minimaux $h(V)$ et calculables $g(V)$ dépendent fortement des coordonnées. Afin de rendre cette dépendance plus explicite, dans le cas où la necessité de travailler avec des variables générales conduit à introduire des formes x_γ (voir 2.3.2), nous noterons $h_\gamma(V)$ le polynôme minimal défini en 3.1.1 et $g_\gamma(V)$ le polynôme calculable défini en 3.1.2 quand les coordonnées sont constituées de $x_1, \ldots, x_r, x_\gamma, x_{r+2}, \ldots, x_n$. Conformément à ce que nous avons dit dans 2.3.5, ces polynômes sont considérés comme éléments de $k[x_1, \ldots, x_n]$.

3.2   Cas projectif

Nous pourrions évidemment brutalement appliquer ce que nous avons dit
dans le cas affine ci-dessus. Mais il n'est pas inintéressant de traiter à part
le cas projectif dans la mesure où nous pouvons remplacer le Nullstellensatz
affine effectif par sa version projective beaucoup plus facile à établir. Nous
allons donc seulement en esquisser les grandes lignes.

Comme nous sommes en position normale, l'idéal $I \cap k[x_0, \ldots, x_r]$ se
réduit à (0), alors que $I \cap k[x_0, \ldots, x_{r+1}]$ n'est pas trivial et définit $\pi(V)$
(voir par exemple Giusti, 1988, 2.12). De plus tous ses éléments sont divisi-
bles par un polynôme non trivial, qui définit $\pi(T(V))$ puisque nous sommes
en position générale. L'idée consiste à trouver dans $I \cap k[x_0, \ldots, x_{r+1}]$ un
polynôme $p(V)$ de degré $d^{O(n)}$, en étudiant une certaine base standard.

3.2.1   Le morphisme fondamental

Nous définissons l'application ϕ_0 suivante:

$$
\begin{aligned}
k[x_0, \ldots, x_n] &\longrightarrow K[t, x_{r+1}, \ldots, x_n] \\
f(x_0, \ldots, x_n) &\longmapsto f(tx_0, \ldots, tx_r, x_{r+1}, \ldots, x_n)
\end{aligned}
$$

où K est le corps $k(x_0, \ldots, x_r)$.

C'est un morphisme de k-algèbres qui conserve le caractère homogène
des polynômes, et le degré. Autrement dit, il s'agit d'un morphisme d'algèbres graduées de poids 0.

Maintenant si I est un idéal homogène de $k[x_0, \ldots, x_n]$, son image par
ϕ_0 n'est malheureusement plus un idéal de $K[t, x_{r+1}, \ldots, x_n]$, et ce pour
une raison stupide de non-surjectivité (x_0 n'admet pas de contre-image !).
Nous allons donc caractériser l'image de ϕ_0.

3.2.2   L'image du morphisme fondamental

La sous-algèbre $k[x_0, \ldots, x_r][t, x_{r+1}, \ldots, x_n]$ de $K[t, x_{r+1}, \ldots, x_n]$ constituée des éléments à coefficients dans $k[x_0, \ldots, x_r]$ admet une double graduation par le degré partiel en $t, x_{r+1}, \ldots, x_n$ et le degré partiel en $x_0, \ldots, x_r$,
$x_{r+1}, \ldots, x_n$. L'image de ϕ_0 est évidemment composée des polynômes dont
les deux degrés partiels ci-dessus sont égaux.

Qui plus est, il n'est pas difficile de voir en utilisant cette bigraduation
que l'image de l'idéal I par ϕ_0 est exactement l'intersection de l'idéal J
engendré par $\phi_0(I)$ et de l'image de ϕ_0.

3.2.3   La fibre générique

L'idéal J définit une variété de dimension zéro, qui n'est autre que la fibre
générique de $\pi_0 \mid_{V(I)}$, de degré égal à celui de la tête de V. Nous pouvons

alors utiliser les résultats connus sur les bases standard en dimension 0 (voir Giusti, 1989 et 1990). En choisissant un ordre total à la Bayer-Stillman sur les monômes de $K[t, x_{r+1}, \ldots, x_n]$, séparant t et x_{r+1} des autres variables, nous voyons qu'il existe dans cette base standard un polynôme en t et x_{r+1} de degré au plus $H(J)^{O(n)} Sup(deg(V(J)), H(J))$, où $H(J)$ est la régularité de la fonction de Hilbert de J. De plus, comme les variables sont normales, son monôme dominant est une puissance pure de x_{r+1}.

Proposition 3.2.4 *Il existe dans* $I \cap k[x_0, \ldots, x_{r+1}]$ *un polynôme* p *de degré majoré par une expression ne faisant intervenir que des quantités intrinsèques, au plus* $d^{O(n)}$ *au vu des majorations standard.*

Notons pour simplifier $\delta(J) = Sup(deg(V(J)), H(J))$. Une fois obtenu le polynôme du paragraphe précédent, le problème vient de la remontée. Mais ses coefficients, qui sont des polynômes en $x_0, \ldots, x_r$, s'obtiennent en triangulant une matrice d'ordre au plus $(H(J + (t, x_{r+1})^n \delta(J))^{O(1)}$, ce qui donne la puissance maximale de t par laquelle multiplier le dit polynôme pour tomber dans l'image de ϕ_0.

3.2.5 Corollaire: une version faible du théorème de Chistov-Grigoriev

Là encore nous nous contenterons d'esquisser l'algorithme, le résultat de complexité étant connu.

L'intérêt de la proposition ci-dessus réside en ce que la majoration du degré de $p(V)$ n'implique que des quantités intrinsèques. C'est ce qui va permettre d'itérer la construction sans aboutir à l'explosion doublement exponentielle de la complexité.

Plaçons nous dans le cas entier, et factorisons $p(V)$ sur $\mathbf{Q}$: $p = \prod_i p_i$, ce qui peut se faire polynomialement en taille et degré (algorithme de Lenstra-Lenstra-Lovascz) (répétons que nous ne voulons pas entrer dans des considérations d'extensions algébriques, et donc nous nous contenterons d'irréductibilité sur $\mathbf{Q}$). Il s'agit de réitérer la construction précédente sur $I + (p_i)$, un simple calcul de dimension séparant composantes de la tête et du croupion.

4   Décomposition équidimensionnelle

Nous sommes maintenant arrivés au point suivant: après mise en position normale, et application de 2.3.5, il existe un point γ dans un ensemble de cardinal au moins $n(deg(V))^4$ tel que les variables $x_1, \ldots, x_r, x_\gamma, x_{r+2}, \ldots, x_n$ soient en position générale par rapport à V. Pour simplifier les notations dans les raisonnements non algorithmiques qui suivent, nous poserons $x_{r+1} := x_\gamma$.

4.1   Détermination de la tête

Supposons que V ne soit pas équidimensionnelle, c'est-à dire que dans la décomposition $V = T(V) \cup Q(V)$ le croupion $Q(V)$ ne soit pas vide. Rappelons que nous avons introduit un polynôme $h = h(V)$ qui définit la projection de la tête de V. Pour récupérer la tête, une idée naturelle consiste alors à couper la décomposition tête-croupion par le lieu des zéros de h: comme h s'annule sur $T(V)$, la tête ne bouge pas, par contre comme les variables sont générales, le croupion doit être coupé proprement. C'est ce que va confirmer le lemme suivant.

Définissons donc $V' := V \cap V(h)$. D'après l'inégalité de Bézout, nous avons $deg(V') \leq deg(V)deg(h) \leq (deg(V))^2$.

Lemme 4.1.1 *La dimension de $Q(V')$ est strictement inférieure à celle de $Q(V)$.*

Preuve: Soit C' une composante de $Q(V')$. Choisissons une composante C de V tel que C' soit une composante irréductible de $C \cap V(h)$: C est nécessairement une composante du croupion, par définition de h. Nous disons que C est coupée proprement par $V(h)$, car les variables étant en position générale, $\pi(C)$ n'est pas contenue dans $\pi(T(V))$ et par conséquent h ne s'annule pas sur $\pi(C)$. Par conséquent C' a une dimension inférieure à celle de C, cqfd.

Choisissons un ensemble Γ de valeurs de k de cardinal au moins nd^{4n^2}. D'après 2.3.5, il existe alors un point γ' dedans, tel que les variables $x_1' := x_1, \ldots, x_r' := x_r, x_{r+1}' := x_{\gamma'}, x_{r+2}' := x_{r+2}, \ldots, x_n' := x_n$ soient en position générale par rapport à V', et nous pouvons définir les polynômes $h(V')$ et $g(V')$ d'après 3.1.

Théorème 4.1.2 (algorithme de calcul de la tête et sa complexité) *Soit V une variété définie par $f_1, \ldots, f_s$, avec les notations de base 1.1, et Γ un ensemble de valeurs de k de cardinal au moins nd^{4n^2}. Pour chaque point γ de Γ, suivant 3.1.4 calculons le polynôme $g_\gamma(V)$ de $k[x_1, \ldots, x_r, x_\gamma]$. Alors $T(V)$ est défini par l'idéal $(f_1, \ldots, f_s, g_\gamma; \gamma \in \Gamma)$. La complexité séquentielle est d'ordre $s^5 d^{O(n^2)}$ et la complexité parallèle d'ordre $O(n^4 \log^2 sd)$.*

Démonstration: Comme g_γ s'annule sur la tête de V, nous obtenons une inclusion dans un sens. Pour démontrer l'inclusion dans l'autre sens, l'idée consiste à introduire une suite de variétés démarrant par V et aboutissant à $T(V)$, définie par récurrence comme suit:

Soit $V_0 = V$; il existe alors par 2.3.5 un élément γ_0 dans Γ autorisant la mise en position générale de V, et soit h_{γ_0} le polynôme minimal correspondant (3.1.1). Nous définissons par récurrence $V_i = V_{i-1} \cap V(h_{\gamma_{i-1}}(V_{i-1}))$ $(i = 1, \ldots, q)$. Ces variétés ont toutes la même tête que V (donc les degrés des polynômes $h(V_i)$ restent majorés par $deg(V)$) et les dimensions de leurs croupions forment une suite strictement décroissante d'après le lemme 4.1.1, aboutissant au croupion de V_q qui est vide. Par suite V_q est équidimensionnelle et égale à la tête de V, et q est majoré par $r(V)$. Remarquons que le degré de chaque V_i reste majoré par d^{n^2} puisque $deg(V_i) \leq deg(V_0)deg(h(V_0))\ldots deg(h(V_{i-1})) \leq (deg(V))^{q+1} \leq d^{n^2}$. En conséquence nous pouvons utiliser le même ensemble Γ de cardinal au moins nd^{4n^2} pour les mises en positions générales successives. Enfin pour chaque $i = 0,\ldots,q-1$ calculons par 3.1.4 le polynôme g_{γ_i} correspondant. Le lemme 3.1.3 implique que g_{γ_i} et h_{γ_i} définissent la même variété, et donc:

$$T(V) = V_q = V \cap V(g_{\gamma_0})\ldots \cap V(g_{\gamma_{q-1}}) = V(f_1,\ldots,f_s,g_{\gamma_0},\ldots,g_{\gamma_{q-1}})$$

Le degré de chaque g_γ est majoré par $D(V) = d^{O(n)}$, et nous pouvons définir la tête avec au plus $s+ \mid \Gamma \mid = s + d^{O(n^2)}$ équations. Par conséquent du point de vue coût, il faut répéter $\mid \Gamma \mid = d^{O(n^2)}$ fois l'algorithme décrit dans 3.1.2, ce qui conserve le même ordre de borne que pour en calculer un seul.

4.2   Détermination des parties équidimensionnelles de dimension non maximale

Pour la détermination de la tête $T(V)$ en temps séquentiel $s^5 d^{O(n^2)}$ et parallèle $O(n^4 \log^2 sd)$ nous avons pu nous contenter des polynômes calculables $g(V)$ pour définir les projections de la tête. Ces polynômes ont un degré d'ordre $d^{O(n)}$: en itérant naïvement ce processus du calcul de la tête pour décomposer V en parties équidimensionelles, cette majoration du degré conduirait à une complexité doublement exponentielle.

En étant plus subtil, il est possible de conserver une complexité admissible et de borner par $d^{O(n^2)}$ le degré de tous les polynômes qui interviennent dans les calculs intermédiaires.

Mais pour conserver le même type de bornes, c'est-à-dire une complexité séquentielle d'ordre $s^5 d^{O(n^2)}$, parallèle $O(n^4 \log^2 sd)$ et des polynômes d'ordre $d^{O(n)}$ en degré, nous serons amenés à calculer le polynôme $h(V)$, dont le degré est majoré par la quantité intrinsèque $deg(V)$. Pour ce faire, nous demanderons donc à disposer d'une opération arithmétique additionelle (qui ne sera en fait utile que dans le cas où le corps de base k est de caractéristique positive p): l'extraction des racines $p^{\`emes}$. Nous appellerons les réseaux arithmétiques qui disposent d'une telle opération des *réseaux arithmétiques étendus.*

Lemme 4.2.1 (calcul du polynôme $h(V)$). *Le polynôme $h(V)$ se calcule à partir de $f_1,\ldots,f_s$ par un réseau arithmétique étendu avec une complexité admissible d'exposant 2.*

Reprenons la démonstration du lemme 3.1.3: g est à une constante près une puissance de h, et ce dernier est sans facteur carré.

En caractéristique nulle, comme les variables sont en position normale, le polynôme g dépend effectivement de la variable x_{r+1}, et donc $\frac{\partial g}{\partial x_{r+1}} \neq 0$. Pour récupérer le polynôme sans facteur carré h, il suffit de prendre $g / \frac{\partial g}{\partial x_{r+1}}$. Et ce calcul s'effectue avec les mêmes bornes de complexité qu'en 3.1.3.

Supposons maintenant la caractéristique p positive. Comme k est parfait et h sans carré, il existe une variable, disons x_{r+1}, qui apparait à une puissance non divisible par p dans h, et $\frac{\partial h}{\partial x_{r+1}} \neq 0$. Sans restriction de généralité on peut supposer $g = h^{p^a b}$, avec a et b entiers, et b non divisible par la caractéristique. Des inégalités $p^a \leq deg(g) \leq D = d^n(d^n + 1)$ nous déduisons que a est un $O(n \log d)$. Par ailleurs il se détermine aisément par inspection des monômes qui apparaissent dans g.

Nous commençons donc à calculer h^b à partir de g en extrayant consécutivement a fois la racine $p^{ième}$ des coefficients de g et en réduisant les monômes correspondants, puis h à partir de h^b sans problème par la formule:

$$ h = \frac{h^b}{pgcd(h^b, \frac{\partial h^b}{\partial x_n})} $$

tout à fait licite puisque b non divisible par la caractéristique implique que $\frac{\partial h^b}{\partial x_n}$ est non nul.

4.2.2   Une amélioration de l'algorithme de calcul de la tête

Comme nous l'avions annoncé, voici une version du calcul de la tête qui consiste à remplacer les polynômes calculables g_γ par les polynômes minimaux h_γ.

Proposition. *Soit V une variété définie par $f_1, \ldots, f_s$, avec les notations de base 1.1, et Γ un ensemble de valeurs de k de cardinal au moins nd^{4n^2}. Pour chaque point γ de Γ, suivant 3.1.4 et 4.2.1 calculons le polynôme $h_\gamma(V)$ de $k[x_1, \ldots, x_r, x_\gamma]$. Alors $T(V)$ est défini par l'idéal $(h_\gamma; \gamma \in \Gamma)$. La complexité séquentielle est d'ordre $s^5 d^{O(n^2)}$ et la complexité parallèle d'ordre $O(n^4 \log^2 sd)$.*

Démonstration: il nous suffit de reprendre celle de 4.1.2 en utilisant le calcul des h_γ effectuée en 4.2.1. Enfin il est inutile de conserver les générateurs initiaux, en modifiant légèrement les démonstrations données dans (Heintz, 1983, Lemma 3 et Proposition 3), voir aussi 2.3.5.

4.2.3   Stratégie de calcul des composantes équidimensionnelles

Pour tout entier m ($0 \leq m \leq n$), soient $T_m(V)$ la réunion des composantes de V de dimension m et $Q_m(V)$ la réunion des composantes de V de dimension strictement inférieure à m. Par exemple, si m est strictement supérieur à la dimension r de V, $T_m(V)$ est vide mais $Q_m(V)$ est V tout entier.

Observons que le résultat 4.2.2 nous donne un idéal définissant la tête $T_r(V)$: notons le $J_r(V)$ ou J_r s'il n'y a pas d'ambiguïté. Notre but va être de fabriquer successivement des idéaux J_m définissant les composantes équidimensionnelles T_m ($r \geq m$).

Une fois la tête calculée via l'idéal J_r, la première idée qui vient à l'esprit consiste à itérer le processus en recommençant avec l'idéal quotient $(I : J_r)$ qui définit ensemblistement le croupion. En effet, rappelons le résultat élémentaire classique qui affirme que étant donné deux idéaux I et J, l'idéal quotient $(I : J)$ définit la réunion des composantes irréductibles de $V(I)$ non contenues dans $V(J)$, qui est encore la clôture de Zariski de la différence $V(I) - V(J)$. Malheureusement, les degrés des générateurs d'un tel idéal quotient sont beaucoup trop grands a priori pour sauvegarder une complexité admissible. Il faut donc être un peu plus rusé.

4.2.4   Définition et notations utiles à la récurrence descendante

Soit W une variété définie par des polynômes $g_1, \ldots, g_t$ de degré borné par d^n. Nous allons maintenant introduire de manière un peu analogue à 3.1.2 un k-espace vectoriel $E(V, W)$ comme suit: choisissons un entier $D := d^n(2d^n + 1)$ (légèrement plus grand que celui de 3.1.2), et considérons:

$E(V, W) =$
$\{f \in k[x_1, \ldots, x_{r+1}] \mid \forall j,\ 1 \leq j \leq t\ \exists\, p_1{}^{(j)}, \ldots, p_s{}^{(j)} \in k[x_1, \ldots, x_n]$
$avec\ fg_j{}^{d^n} = \sum_{1 \leq i \leq s} p_i{}^{(j)} f_i\ et\ Max\,\{deg(p_i{}^{(j)} f_i)\} \leq D \mid i = 1, \ldots, s\}$

Comme $E(V)$, cet espace vectoriel dépend en fait des générateurs f et g des idéaux $I = (f_1, \ldots, f_s)$ et $J = (g_1, \ldots, g_t)$ plutôt que des idéaux ou des variétés. Cependant, le k-espace vectoriel $E(V)$ (de dimension finie) représentait une troncature (par le degré) de l'idéal définissant une projection de V, considéré comme k-espace vectoriel de dimension infinie. Cette troncature était suffisamment haute pour que le lieu des zéros d'une base de $E(V)$ définisse effectivement cette projection. De la même manière, le k-espace vectoriel de dimension finie $E(V, W)$ consistue une troncature (par le degré) d'une sorte d'idéal quotient. Cette troncature serait-elle suffisamment haute pour définir la projection de la clôture de Zariski de la différence $V - W$, autrement dit la réunion des composantes de V non contenues dans W?

Avant de parler de projection, il nous faut d'abord étudier ce qui se passe en haut. Introduisons l'espace vectoriel auxiliaire:

$E'(V, W) =$
$\{f \in k[x_1, \ldots, x_n] \mid \forall j,\ 1 \leq j \leq t\ \exists\, p_1{}^{(j)}, \ldots, p_s{}^{(j)} \in k[x_1, \ldots, x_n]$
$avec\ fg_j{}^{d^n} = \sum_{1 \leq i \leq s} p_i{}^{(j)} f_i\ et\ Max\,\{deg(p_i{}^{(j)} f_i)\} \leq D \mid i = 1, \ldots, s\}$

Proposition 4.2.5 *Les polynômes de $E'(V, W)$ définissent la clôture de Zariski de $V - W$.*

Démonstration: Soit f un élément de $E'(V, W)$ et soit C une composante irréductible de V non contenue dans W. Nous allons démontrer que f s'annule sur C. Si ce n'était pas le cas, il existerait un ouvert de Zariski non vide U de C où f ne s'annulerait pas. Mais C n'étant pas contenue dans W, il existe un polynôme g_j $(1 \leq j \leq t)$ et un ouvert de Zariski non vide U' de C tels que g_j ne s'annule pas sur U'. Donc le produit $f g_j^{d^n}$ ne s'annule pas non plus sur l'ouvert non vide $U \cap U'$ de C. Ceci contredit l'hypothèse de l'appartenance de f à E'.

Réciproquement, d'après Heintz (1983, Proposition 3 and Theorem 3) la variété $\overline{V - W}$ peut être définie par $n+1$ polynômes $f_1', \ldots, f_{n+1}'$ de degré au plus son propre degré, majoré par le degré de V, lui-même majoré par d^n. Pour tout l $(1 \leq n+1)$ et tout j $(1 \leq j \leq t)$, le produit $f_l' g_j$ s'annule sur V, donc appartient au radical de $(f_1, \ldots, f_s)$. En appliquant une version adaptée du Nullstellensatz effectif (voir par exemple Fitchas-Galligo, 1988, Théoreème 10), nous obtenons qu'il existe des polynômes $p_1, \ldots, p_s$ tels que $(f_l' g_j)^{d^n} = \sum_{i=1}^{s} p_i f_i$ et $Max\{deg(p_i f_i) \mid i = 1, \ldots, s\} \leq D$.

Corollaire 4.2.6 (fabrication descendante des composantes équidimensionnelles). *Soit m un entier $(0 \leq m < r)$. Supposons que pour tout entier $l = r, r-1, \ldots, m+1$ nous ayons construit des idéaux $J_l = (h_1^{(l)}, \ldots, h_{t_l}^{(l)})$ (définissant comme précisé en 4.2.3 $T_l(V)$), dont les générateurs sont de degré majoré par $\deg T_l$.*

Supposons de plus que nous ayons aussi construit un sous-ensemble K_{m+1} de $\{h_{j_r}^{(r)} \ldots h_{j_{m+1}}^{(m+1)} \mid 1 \leq j_r \leq t_r, \ldots, 1 \leq j_{m+1} \leq t_{m+1}\}$ de cardinal $d^{O(n^2)}$ dont les éléments engendrent l'idéal $J_r \ldots J_{m+1}$.

Alors il existe un réseau arithmétique étendu qui construit en temps séquentiel $s^5 d^{O(n^2)}$ et parallèle $O(n^4 \log^2 sd)$ un idéal $J_m = (h_1^{(m)}, \ldots, h_{t_m}^{(m)})$, définissant $T_m(V)$, engendré par des polynômes en nombre $t_m = d^{O(n^2)}$ et de degré majoré par $\deg T_m(V)$.

De plus le réseau construit un sous-ensemble K_m de $\{h_{j_r}^{(r)} \ldots h_{j_m}^{(m)} \mid 1 \leq j_r \leq t_r, \ldots, 1 \leq j_m \leq t_m\}$ de cardinal $d^{O(n^2)}$ dont les éléments engendrent l'idéal $J_r \ldots J_m$.

Esquisse de la démonstration: Reprenons les notations de 4.2.4 et appliquons la proposition 4.2.5 à la variété W définie par l'ensemble K_{m+1}, puisque les polynômes constituant celui-ci sont de degré au plus $\deg T_r + \cdots + \deg T_{m+1} \leq \deg V \leq d^n$. Les polynômes contenus dans $E'(V, W)$ définissent bien $Q_{m+1}(V)$.

Il va nous suffire maintenant de généraliser *mutatis mutandis* les sections de 2.3.1 à 2.3.5, 3.1, 4.1.2 et 4.2.2. L'idée consiste à ne faire que des opérations d'algèbre linéaire sur $E'(V, W)$, ce qui conserve une complexité séquentielle d'ordre $s^5 d^{O(n^2)}$ et parallèle d'ordre $O(n^4 \log^2 sd)$ des algorithmes.

C'est ainsi que nous pouvons modifier l'algorithme 1.15 de Dickenstein-Fitchas-Giusti-Sessa (1989) pour obtenir en temps séquentiel $s^5 d^{O(n^2)}$ et parallèle $O(n^4 \log^2 sd)$ des variables en position normale par rapport à $Q_{m+1}(V)$. L'algorithme détermine aussi sa dimension r', si bien que nous pouvons supposer sans restriction de généralité que les variables $x_1, \ldots, x_n$ sont en position normale par rapport à $Q_{m+1}(V)$.

Le reste, mise en position générale, projection et détermination de la tête de Q_{m+1} se déroule sans problèmes. Le seul point nouveau consiste en la fabrication de l'ensemble K_m.

Théorème 4.2.7 (décomposition équidimensionnelle et sa complexité).
Soit V une variété affine définie avec les notations de base 1.1. Alors le réseau arithmétique étendu décrit ci-dessus calcule pour chaque entier m ($0 \leq m \leq n$) la réunion $T_m(V)$ des composantes de V de même dimension m, c'est-à-dire des polynômes $h_1^{(m)}, \ldots, h_{t_m}^{(m)}$ en nombre $t_m = d^{O(n^2)}$, de degré majoré par $\deg T_m(V) \leq d^n$ et tels que leur lieu des zéros soit $T_m(V)$.

La complexité du réseau est en temps séquentiel $s^5 d^{O(n^2)}$ et parallèle $O(n^4 \log^2 sd)$.

Si nous nous limitons aux réseaux arithmétiques non étendus, nous pouvons obtenir en modifiant de façon ad hoc ce qui précède un algorithme de décomposition équidimensionnelle de complexité légèrement supérieure mais qui reste néanmoins admissible:

Remarque 4.2.8 *Soit V une variété affine définie avec les notations de base 1.1. Alors il existe un réseau arithmétique (non étendu) qui calcule pour chaque entier m ($0 \leq m \leq n$) la réunion $T_m(V)$ des composantes de V de même dimension m, c'est-à-dire des polynômes $h_1^{(m)}, \ldots, h_{t_m}^{(m)}$ en nombre $t_m = d^{O(n^3)}$, de degré d'ordre $d^{O(n^2)}$ et tels que leur lieu des zéros soit $T_m(V)$. La complexité du réseau est en temps séquentiel $s^5 d^{O(n^3)}$ et parallèle $O(n^6 \log^2 sd)$.*

5 Appendice: application au calcul de la forme de Chow et du degré

Nous allons montrer que la forme de Chow d'une variété équidimensionnelle se calcule par un algorithme parallélisable de complexité admissible. D'après ce qui précède, ce résultat s'étendra donc à une variété arbitraire. Comme corollaire, le degré s'obtient de la même manière. Puis nous donnerons une meilleure méthode, plus directe, pour obtenir ce dernier invariant.

5.1   Rappel de la notion de forme de Chow

Dans tout ce qui suit, nous identifierons un point du dual $\check{\mathbf{P}}^n$ de l'espace projectif avec l'hyperplan lieu des zéros de la forme linéaire qu'il définit.

Soit V une sous-variété équidimensionnelle, fermée et de dimension r de l'espace projectif $\mathbf{P}^n$. Considérons le sous-ensemble fermé Ω de $(\check{\mathbf{P}}^n)^{r+1} \times V$ défini par:

$$\Omega = \{(y^{(0)}, \ldots, y^{(r)}, x) \in (\check{\mathbf{P}}^n)^{r+1} \times V \mid x \in y^{(j)} \ 0 \le j \le r\}$$

Soit π la projection canonique $(\check{\mathbf{P}}^n)^{r+1} \times V \longrightarrow (\check{\mathbf{P}}^n)^{r+1}$. Alors $\pi(\Omega)$ est une hypersurface de $(\check{\mathbf{P}}^n)^{r+1}$ dont le polynôme minimal F_V est appelé la *forme de Chow* de V (voir par exemple comme référence Shafarevitch (1972) ou Nesterenko (1977). Elle possède les deux propriétés suivantes:

(i) $deg(F_V) = deg(V)$

(ii) $V = \bigcap_{j=0}^{r} \{y^{(j)} \mid F_V(y^{(0)}, \ldots, y^{(r)}) = 0\}$

Si V est une sous-variété fermée arbitraire de l'espace projectif, la forme de Chow F_V est définie comme le produit des formes de Chow de toutes les composantes équidimensionnelles de V:

$$F_V = \prod_{j=0}^{r(V)} \{F_{T_j(V)} \mid T_j(V) \neq \emptyset\}$$

Ces rappels effectués, nous remercions T. Krick de nous avoir communiqué l'application suivante de l'algorithme de complexité admissible de décomposition équidimensionnelle:

Théorème 5.2 (Krick, 1990) *Soit V une variété projective définie par des polynômes homogènes $f_1, \ldots, f_s$ de $k[x_0, \ldots, x_n]$, avec les notations de base 1.1. Alors la forme de Chow F_V est calculable par un réseau arithmétique étendu de complexité admissible.*

Esquisse de la démonstration: en utilisant la version projective du théorème 4.2.7, nous nous ramenons d'abord au cas où V est équidimensionnelle de dimension r.

Soient $y_0^{(0)}, \ldots, y_n^{(0)}, \ldots, y_0^{(r)}, \ldots, y_n^{(r)}$ des nouvelles indéterminées et considérons la forme de Chow F_V, que nous noterons désormais F pour simplifier, comme un polynôme de $k[y_l^{(j)} ; 0 \le j \le r , 0 \le l \le n]$. Puisque k est parfait, F est sans facteur carré. Pour tout indice j $(0 \le j \le r)$ soit $L^{(j)}$ la forme linéaire en $x_0, \ldots, x_n$, définie par $L^{(j)}(x_0, \ldots, x_n) = y_0^{(j)} x_0 + \cdots + y_n^{(j)} x_n$, et W l'hypersurface $\{w \in \mathbf{A}^{(n+1)(r+1)} \mid F(w) = 0\}$. En tenant compte de la propriété (ii) des formes de Chow, nous voyons que l'ensemble W est définissable par une formule élémentaire Φ du langage des corps algébriquement clos contenant k, et est de la forme:

$$(\exists x_0)\,(\exists x_n)\,(f_1(x_0,\ldots,x_n)=0) \wedge \ldots \wedge (f_s(x_0,\ldots,x_n)=0) \wedge$$
$$(L^{(0)}(y_0^{(0)},\ldots,y_n^{(0)},x_0,\ldots,x_n)=0) \wedge \ldots \wedge$$
$$(L^{(r)}(y_0^{(r)},\ldots,y_n^{(r)},x_0,\ldots,x_n)=0)$$

et contient $(n+1)(r+2)$ variables, dont $y_0^{(0)},\ldots,y_n^{(0)},\ldots,y_0^{(r)},\ldots,y_n^{(r)}$ sont libres et $x_0,\ldots,x_n$ liées, et $s+r+1$ polynômes de degré au plus d et un seul bloc de quantificateurs. Appliquons maintenant à Φ la procédure efficace d'élimination des quantificateurs décrite dans Fitchas-Galligo-Morgenstern (1987, 1989) ou dans Krick (1990). En temps admissible nous obtenons une formule équivalente Ψ sans quantificateurs qui est une disjonction de conjonctions où apparaissent des polynômes $G_1,\ldots,G_Q$ de $k[y_l^{(j)} \mid 0 \le j \le r,\ 0 \le l \le n]$ qui satisfont la borne suivante:

$$\sum_{q=1}^{Q} deg(G_q) = (sd)^{n^{O(1)}}$$

Sans restriction de généralité, nous pouvons supposer que Ψ est une disjonction de $M = (sd)^{n^{O(1)}}$ conjonctions Ψ_m de la forme:

$$(G_{q_1}=0) \wedge \ldots \wedge (G_{q_m}=0) \wedge (G_{q_m+1} \ne 0)$$

qui définit un sous-ensemble localement fermé non vide W_m de $\mathbf{A}^{(n+1)(r+1)}$. Nous pouvons déterminer la dimension de chaque W_m $(1 \le m \le M)$ (voir Caniglia-Galligo-Heintz (AAECC 6), 1989, Theorem 19 ou Dickenstein-Fitchas-Giusti-Sessa, 1989, Corollary 1.9.1).

Notons $N := (n+1)(r+1)$ et $S := \{m \in \mathbf{N} \mid 1 \le m \le M,\ dim(W_m) = N - 1\}$. Ce dernier ensemble n'est pas vide parce que $W = \bigcup m = 1^M W_m$ est une hypersurface de $\mathbf{A}^N$. Considérons un de ses éléments m, et la conjonction Ψ_m qui peut être supposée sans restriction de généralité de la forme $(G_q = 0) \wedge \ldots \wedge (G_q = 0) \wedge (G_{q+1} \ne 0)$. Puisque l'ensemble W_m défini par Ψ_m a la dimension $r(W_m) = N - 1$ et comme W_m est contenue dans W nous voyons que l'entier q est strictement positif et qu'il existe un facteur premier de F qui divise $G_1,\ldots,G_q$ mais pas G_{q+1}.

Soit H_m le polynôme sans facteur carré associé au pgcd de $G_1,\ldots,G_q$; d'après 4.2.1, il est possible de le calculer par un réseau arithmétique étendu en temps admissible. Nous pouvons avec la même complexité calculer le polynôme $P_m := H_m/pgcd(H_m, G_{q+1})$ et P le polynôme sans facteur carré associé à $\prod_{m \in S} P_m$.

Il suffit pour terminer la démonstration de prouver que P et F coïncident. Soit R un polynôme irréductible qui divise F dans $k[y_l^{(j)} \mid 0 \le j \le r,\ 0 \le l \le n]$. Comme l'hypersurface $V(R) = \{w \in \mathbf{A}^{(n+1)(r+1)} \mid R(w) =$

0} est contenue dans $W = V(F)$, il existe un entier m $(1 \leq m \leq M)$ tel que la dimension de $W_m \cap V(R)$ est $N - 1$, ce qui implique que m appartient à S et que R divise P_m: par suite R divise aussi P. Comme R était un facteur irréductible arbitraire de F, lui-même sans facteur carré, nous en déduisons que F divise P.

Démontrons maintenant que P divise F. Soit R un facteur irréductible de P. Choisissons un entier m dans S tel que R divise P_m; nous pouvons supposer sans restriction de généralité que Ψ_m est de la forme $(G_1 = 0) \wedge \ldots \wedge (G_q = 0) \wedge (G_{q+1} \neq 0)$. Alors R divise $G_1, \ldots, G_q$ mais pas G_{q+1}, ce qui implique que $V(R)$ est contenu dans la clôture de Zariski de W_m qui est un sous-ensemble de $W = V(P)$: R divise donc F. De la même manière qu'auparavant nous concluons donc que P divise F, c.q.f.d.

5.3   Calcul du degré d'une variété en temps admissible

Le théorème précédent implique que nous pouvons calculer le degré d'une variété projective en temps admissible. En utilisant le fait que la clôture projective d'une variété affine s'obtient en temps admissible (voir Caniglia-Galligo-Heintz (AAECC 7), 1989, Theorem 2.10), nous en déduisons le même résultat algorithmique pour les variétés affines. Ceci généralise Caniglia-Galligo-Heintz (AAECC 6), 1989, Theorem 19, (iii) où un algorithme de complexité admissible était annoncé. En fait, par une méthode plus directe nous allons produire un résultat plus précis:

Proposition 5.3.1 *Le calcul du degré d'une variété affine de V de $\mathbf{A}^n$, définie par des polynômes $f_1, \ldots, f_s$ conformément aux notations de base 1.1, peut s'effectuer par un réseau arithmétique étendu avec une complexité séquentielle en $s^5 d^{O(n^3)}$ et parallèle en $O(n^6 \log^2 sd)$.*

Démonstration: en calculant une décomposition équidimensionnelle de V au moyen du théorème 4.2.7, nous pouvons nous ramener au cas où V est une variété équidimensionnelle de dimension connue, disons r, et définie par $t = d^{O(n^2)}$ polynômes $h_1, \ldots, h_t$ de $k[x_1, \ldots, x_n]$ de degré majoré par d^n.

Introduisons de nouvelles indéterminées $y_1^{(1)}, \ldots, y_n^{(1)}, \ldots, y_1^{(r)}, \ldots, y_n^{(r)}$, et $y_1, \ldots, y_n$, en nombre $(r + 1)n$. Notons A (respectivement K) l'extension $k[y_l^{(j)}, y_j \ ; \ 1 \leq j \leq r, \ 1 \leq l \leq n]$ (respectivement l'extension $k(y_l^{(j)}, y_j \ ; \ 1 \leq j \leq r, \ 1 \leq l \leq n))$. Pour tout j $(1 \leq j \leq r)$, définissons la forme linéaire $l^{(j)}$ de $A[x_1, \ldots, x_n]$ comme suit: $l^{(j)}(x_1, \ldots, x_n) := y_j + \sum_{l=1}^n y_l^{(j)} x_l$. L'idéal $J := (h_1, \ldots, h_t, l^{(1)}, \ldots, l^{(r)})$ de $K[x_1, \ldots, x_n]$ définit une variété de dimension zéro.

Fixons un ordre total admissible sur les monômes de $K[x_1, \ldots, x_n]$, et en utilisant un réseau arithmétique sur A, calculons alors en temps séquentiel $s^5 d^{O(n^3)}$ et parallèle $O(n^6 \log^2 sd)$ l'escalier relativement à cet ordre du radical $rad(J)$ de J (voir Krick-Logar, 1990 et Dickenstein-Fitchas-Giusti-Sessa, 1989 pour détails). Sur l'escalier ainsi obtenu nous lisons la dimension du K-espace vectoriel $K[x_1, \ldots, x_n]/rad(J)$ qui n'est autre que $degV$.

Reste à voir quelle est la complexité d'un réseau arithmétique sur le corps de base k qui exécute les opérations arithmétiques sur A nécessaires au calcul de l'escalier du radical de J. En fait, pour obtenir ce dernier, nous n'avons qu'à tester s'il existe des polynômes dans $rad(J)$, de degré borné par d^n et contenant certains monômes. Pour cela, nous utilisons seulement des éléments de A qui comme polynômes en les $(r+1)n$ variables $y_1^{(1)}, \ldots, y_n^{(1)}, \ldots, y_1^{(r)}, \ldots, y_n^{(r)}$, et $y_1, \ldots, y_n$, ont un degré majoré par d^{n^2}. Il nous faut tester la nullité ou non de ces polynômes, donnés par nos calculs dans A. D'après Heintz-Schnorr (1980), cela est possible par des évaluations de ces polynômes en $s^5 d^{O(n^3)}$ points choisis à propos dans $k^{(r+1)n}$. Ces évaluations peuvent être effectuées par un réseau arithmétique sur k en en temps séquentiel $s^5 d^{O(n^3)}$ et parallèle $O(n^6 \log^2 sd)$.

REFERENCES

[1] D. Bayer, M. Stillman, *A theorem on refining division orders by the reverse lexicographic order*, Lecture Notes for the meeting on Algebraic Geometry and Computing, 1985, Trento, Italy.

[2] S. J. Berkowitz, *On computing the determinant in small parallel time using a small number of processors*, Information Processing Letters **18** (1984), 147–150.

[3] W. S. Brown, *On Euclid's algorithm and the computation of polynomial greatest common divisors*, J. ACM **18** (1971), 478–504.

[4] D. Brownawell, *Bounds for the degree in the Nullstellensatz*, Ann. Math. **126** (1987), 577–591.

[5] L. Caniglia, *Complejidad de algoritmos en geometria computational*, Thèse, Universidad de Buenos Aires 1989.

[6] L. Caniglia, *How to compute the Chow form of an unmixed polynomial ideal in simple exponential time*, (submitted), Applicable Algebra in Engeneering Communication and Computer Science (1989).

[7] L. Caniglia, A. Galligo, J. Heintz, *Borne simple exponentielle pour les degrés dans le théorème des zéros sur un corps de caractéristique quelconque*, C. R. A. S. Paris **307** (1988), 255–258.

[8] L. Caniglia, A. Galligo, J. Heintz, *Some new effectivity bounds in computational geometry,*, in "Proc. AAECC 6," Springer LNCS **357**, 1989, pp. 131–151.

[9] L. Caniglia, A. Galligo, J. Heintz, *Equations for the projective closure of an affine algebraic variety*, Discrete Appl. Math. Proc. AAECC-7, Toulouse (1989). (to appear)

[10] L. Caniglia, J. A. Guccione, J. J. Guccione, *Local membership problems for polynomial ideals,*, in "Proc. MEGA 90," 1990.

[11] A. L. Chistov, D. Yu. Grigoriev, *Subexponential-time solving systems of algebraic equations I, II*, Steklov Mathematical Institute, Lenigrad department, LOMI Preprints E-9-83, 0E-10-c83 (1983).

[12] A. Dickenstein, N. Fitchas, M. Giusti, C. Sessa, *The membership problem for unmixed polynomial ideals is solvable in subexponential time*, Discrete Appl. Math. Proc. AAECC-7, Toulouse (1989). (to appear)

[13] N. Fitchas, A. Galligo, *Nullstellensatz effectif et conjecture de Serre (Théorème de Quillen-Suslin) pour le calcul formel*, Math. Nachrichten (1988). (à paraître)

[14] Noaï Fitchas, A. Galligo, J. Morgenstern, *Algorithmes rapides en séquentiel et parallèle pour l'élimination des quantificateurs en géométrie élémentaire*, in "Séminaire Structures Algébriques Ordonnées 1986–87," Université Paris VII, 1987. (à paraître)

[15] Noaï Fitchas, A. Galligo, J. Morgenstern, *Precise sequential and parallel complexity bounds for quantifier elimination over algebraically closed fields*, Journal of Pure and Applied Algebra (1987). (to appear)

[16] J. Von zur Gathen, *Parallel arithmetic computations: a survey,*, in "Proc. 13th Symp. MFCS 1986," Springer Lecture Notes in Computer Science **233**, 1986, pp. 93–112.

[17] P. Gianni, B. Trager, G. Zacharias, *Gröbner bases and primary decomposition of polynomial ideals*, J. Symbolic Computation **6** Computational aspects of commutative algebra (1988), 249–265.

[18] M. Giusti, *Combinatorial Dimension Theory of Algebraic Varieties*, J. Symbolic Computation **6** Computational aspects of commutative algebra (1988), 249–265.

[19] M. Giusti, *Complexity of standard bases in projective dimension zero*, in "Proceedings of EUROCAL 87," (European Conference on Computer Algebra, Leipzig, RDA), Lecture Notes in Computer Science **378**, Springer Verlag, 1989, pp. 333–335.

[20] M. Giusti, *Complexity of standard bases in projective dimension zero II*, in "Proceedings of AAECC," Tokyo, 1990. (to appear)

[21] J. Heintz, *Definability and fast quantifier elimination over algebraically closed fields,*, Theoretical Computer Science **24** (1985), 239–277. Russian translation in Kyberneticeskij Sbornik, Novaja Serija, Mir Moscow **22**, pp. 113–158

[22] J. Heintz, C.P. Schnorr, *Testing polynomials which are easy to compute*, in "Proc. 12th Annual Symposium ACM on computing," 1980, pp. 262–272. also in: "Logic and Algorithmic. An international Symposium held in honour of Ernst Specker" Monographie **30** de l'Enseignement Mathématique, Genève 1982, pp. 237–254

[23] G. Hermann, *Die Frage der endlich vielen Schritte in der Theorie der Polinomideale*, Math. Ann. **95** (1926), 736–788.

[24] P. Hintenhaus, *Decomposing and Parameterizing the Solution Set of an Algebraic System*, Ph. D. Thesis, preprint RISC-Linz **89-27**.

[25] T. Krick, *Complejidad para problemas de geometriá elemental*, Thèse, Universidad de Buenos Aires 1990.

[26] T. Krick, A. Logar, *Membership problems, representation problems and the computation of the radical for one-dimensional ideals*, in "Proceedings MEGA-90," (these proceedings), 1990.

[27] T. Krick, P. Solernó, *Calcul effectif du polynôme de Chow d'un idéal homogène ensemblistement équidimensionnel*, Prépublication, Buenos Aires 1989.

[28] D. Lazard, *Algèbre linéaire sur $K[x_1, \ldots, x_n]$ et élimination*, Bull. Soc. Math. France **105** (1977), 165–190.

[29] A. Logar, *A computational proof of the Noether normalization lemma*, in "Proc. AAECC 6," (Roma), Springer LNCS **357**, 1988, pp. 259–273.

[30] Y. V. Nesterenko, *Estimates for the order of zeroes of functions of a cerain class and applications in the theory of transcendantal numbers*, Izvestija Akad. Nauk. SSR **41** (1977). Translation in: Math. USSR Izvestija **11**

[31] P. Robert, "Dictionnaire alphabétique et analogique de la langue française,," Société du nouveau Littré, Paris, 1970.

[32] A. Seidenberg, *Constructions in algebra*, Trans. Amer. Math. Soc. **197** (1974), 273–313.

[33] I. R. Shafarevich, *Basic Algebraic Geometry*, Nauka, Moscow. English version: Springer, Berlin 1974

[34] B. Teissier, *Résultats récents d'algèbre commutative effective*, in "Séminaire Bourbaki," 42ième année, **718**, 1989.

Marc Giusti
SDI CNRS 6176 "Calcul formel, Algèbre et Géométrie algorithmiques"
(associé au Centre de Mathématiques, URA CNRS D.0169)
Laboratoire d'Informatique
Ecole Polytechnique, 91128 Palaiseau Cedex, France
giusti@cmep.polytechnique.fr
cfmagi@frpoly11.BITNET

Joos Heintz
Groupe de travail Noaï Fitchas
Instituto Argentino de Matemática (CONICET)
Viamonte 1636, (1055) Buenos Aires, Argentina
atina!mate!nfitchas@uunet.uu.net
nfitchas@mate.edu.ar
atina!dcfcen!mate!nfitchas

[31] P. Robert, "Dictionnaire alphabétique et analogique de la langue française,," Société du nouveau Littré, Paris, 1970.

[32] A. Seidenberg, *Constructions in algebra*, Trans. Amer. Math. Soc. **197** (1974), 273–313.

[33] I. R. Shafarevich, *Basic Algebraic Geometry*, Nauka, Moscow. English version: Springer, Berlin 1974

[34] B. Teissier, *Résultats récents d'algèbre commutative effective*, in "Séminaire Bourbaki," 42ième année, **718**, 1989.

Marc Giusti
SDI CNRS 6176 "Calcul formel, Algèbre et Géométrie algorithmiques"
(associé au Centre de Mathématiques, URA CNRS D.0169)
Laboratoire d'Informatique
Ecole Polytechnique, 91128 Palaiseau Cedex, France
giusti@cmep.polytechnique.fr
cfmagi@frpoly11.BITNET

Joos Heintz
Groupe de travail Noaï Fitchas
Instituto Argentino de Matemática (CONICET)
Viamonte 1636, (1055) Buenos Aires, Argentina
atina!mate!nfitchas@uunet.uu.net
nfitchas@mate.edu.ar
atina!dcfcen!mate!nfitchas

Complexity of Solving Systems
of Linear Equations
over the Rings of Differential Operators

DIMITRI YU. GRIGOR'EV

Introduction

Denote by $\mathcal{A}_n = \mathcal{A}_n(F) = F[X_1, \ldots, X_n, D_1, \ldots, D_n]$ the Weyl algebra over a field F([2]) determined by the relations $X_i X_j = X_j X_i$, $D_i D_j = D_j D_i$, $X_i D_i = D_i X_i - 1$, $X_i D_j = D_j X_i$ for $i \neq j$, and by

$$\mathcal{K}_n = \mathcal{K}_n(F) = F(X_1, \ldots, X_n)[D_1, \ldots, D_n] \supset \mathcal{A}_n$$

the algebra of differential operators.

Any element $a \in \mathcal{A}_n$ can be uniquely represented in the form

$$a = \sum_{I,J} a_{I,J} D_n^{i_n} \ldots D_1^{i_1} X_n^{j_n} \ldots X_1^{j_1}$$

where $a_{I,J} \in F$, for multiindices $I = (i_n, \ldots, i_1)$, $J = (j_n, \ldots, j_1)$, in a similar way any element $f \in \mathcal{K}_n$ can be uniquely represented in the form $f = a_1 c^{-1}$ where $a_1 \in \mathcal{A}_n$, $0 \neq c \in F[X_1, \ldots, X_n]$ and the degree $\deg(c)$ is the least possible. Define the degree $\deg(D_n^{i_n} \ldots D_1^{i_1} X_n^{j_n} \ldots X_1^{j_1}) = i_n + \ldots + i_1 + j_n + \ldots + j_1$ according to the Bernstein filtration ([2]), the degree

$$\deg(a) = \max_{a_{I,J} \neq 0} \deg(D_n^{i_n} \ldots D_1^{i_1} X_n^{j_n} \ldots X_1^{j_1})$$

and lastly $\deg(f) = \max\{\deg(a_1), \deg(c)\}$.

The purpose of this paper is to estimate the complexity of solving a linear system

$$\sum_{1 \leq \ell \leq s} u_{k,\ell} V_\ell = w_k, \ 1 \leq k \leq m \tag{1}$$

over the ring $\mathcal{A}_n$ (i.e. the coefficients $u_{k,\ell}$, $w_k \in \mathcal{A}_n$ and the unknowns $V_\ell \in \mathcal{A}_n$), or respectively, over the ring $\mathcal{K}_n$. In the formulations of the theorem and the corollary $\mathcal{R}$ denotes either $\mathcal{A}_n$ or $\mathcal{K}_n$.

Theorem. *Assume that the system (1) is solvable in the ring $\mathcal{R}$, and that* $\deg(u_{k,\ell})$, $\deg(w_k) < d$, $1 \leq k \leq m$, $1 \leq \ell \leq s$. *Then there exists a solution of (1) for which* $\deg(V_\ell) < (md)^{2^{O(n)}}$, $1 \leq \ell \leq s$.

Corollary. *Let a field F be given in an effective way, e.g. as an explicitly finitely generated extension of $\mathbf{Q}$ (see [3,6,7]) and the bit-size of each coefficient of $u_{k,\ell}, w_k$ is at most M. Then one can test solvability of (1) over $\mathcal{R}$ and yield a solution (provided that it exists) in polynomial in $M, (md)^{2^{O(n)}}$, s time.*

For systems over the polynomial ring $F[X_1, \ldots, X_n]$ the bound from the theorem is well-known [8], however one cannot generalize directly the Hermann-Seidenberg construction to noncommutative rings of differential operators. Nevertheless, the general approach from [8] is used in the proof of the theorem.

We give the proof of the theorem only for the case $\mathcal{R} = \mathcal{A}_n$, the proof for $\mathcal{K}_n$ is similar with slight modifications. In section 1 we estimate the complexity of reducing a matrix over $\mathcal{A}_n$ by left elementary transformations to trapezium form. The main difficulty here is that unlike the commutative case one cannot use determinants. In the section 2 a normalization lemma for $\mathcal{A}_n$ is proved, based on the transitivity of the sympletic group $Sp_{2n}(F)$ ([1]). With the help of it we show how to eliminate D_n and X_n from the system (1) (similar to [8]).

Let us mention that in [7] a polynomial-time algorithm for finding the greatest common divisor of a family of linear ordinary differential operators is given, it entails, in particular, the theorem for the case $\mathcal{R} = \mathcal{K}_1$.

Let us mention also that in [4] for an important particular case of the problem under discussion, namely, recognizing unity of an ideal in a polynomial ring $F[X_1, \ldots, X_n]$, i.e. solvability of an equation $\sum\limits_{1 \leq \ell \leq s} u_\ell V_\ell = 1$, the sharp bound $\deg(V_\ell) \leq d^{O(n)}$ is proved. It is unknown, whether a similar result holds for the rings $\mathcal{A}_n$ and $\mathcal{K}_n$? Several algorithmical problems in the theory of ideals in the rings of differential operators were posed also in [5].

1.   Estimate of entries of quasiinverse matrix over a Weyl algebra

Let a matrix $A = (a_{i,j})_{1 \leq i \leq m-1, 1 \leq j \leq m}$ have its entries in a Weyl algebra: $a_{i,j} \in \mathcal{A}_n$ and $\deg(a_{i,j}) \leq d$.

Lemma 1. *There exists a vector $0 \neq f = (f_1, \ldots, f_m) \in (\mathcal{A}_n)^m$ such that $Af = 0$ and $\deg(f) \leq N = 4n(m-1)d$.*

Proof: Consider a linear space $\mathcal{B} \subset (\mathcal{A}_n)^m$ over F consisting of all the vectors $c = (c_1, \ldots, c_m)^T$ such that $\deg(c) \leq N$. Then $\dim \mathcal{B} = \binom{N+2n}{2n} m$.

For any vector $c \in \mathcal{B}$ we have $\deg(Ac) \leq N + d$, i.e. $Ac \in \gamma$ where the space γ consists of all the vectors $e = (e_1, \ldots, e_{m-1}) \in (\mathcal{A}_n)^{m-1}$ for which $\deg(e) \leq N + d$, henceforth $\dim \gamma = \binom{N+d+2n}{2n}(m - 1)$.

Let us prove an inequality $\binom{N+2n}{2n}m > \binom{N+d+2n}{2n}(m - 1)$, whence lemma would follow immediately. Indeed,

$$
\binom{N + d + 2n}{2n} \Big/ \binom{N + 2n}{2n} =
$$
$$
= \frac{N + d + 2n}{N + 2n} \cdot \frac{N + d + 2n - 1}{N + 2n - 1} \cdots \frac{N + d + 1}{N + 1} \leq (N + d + 1)^{2n}.
$$

It suffices to check the inequality $(\frac{N+d+1}{N+1})^{2n} < 1 + \frac{1}{m-1}$. It follows from the inequality

$$
(1 + \frac{1}{m - 1})^{1/2n} > 1 + \frac{1}{2n(m - 1)} + \frac{1}{2}\frac{1}{2n}(\frac{1}{2n} - 1)\frac{1}{(m - 1)^2} >
$$
$$
> 1 + \frac{1}{4n(m - 1)} > 1 + \frac{d}{N + 1}
$$

q. e. d. Lemma 1 implies that $\mathcal{A}_n$ is an Ore domain ([2]).

Let us call an $m \times n$ matrix $C = (c_{i,j})$ a right (resp. left) quasiinverse to an $m \times m$ matrix $B = (b_{i,j})$, if the matrix BC (resp. CB) has diagonal form with nonzero diagonal entries.

Lemma 2. *If $m \times m$ matrix B over $\mathcal{A}_n$ (suppose that $\deg(B) \leq d$) has a right quasiinverse matrix over $\mathcal{A}_n$ then B has also left quasiinverse matrix G over $\mathcal{A}_n$ such that $\deg(G) \leq N$.*

Proof: Observe that there does not exist vector $0 \neq b \in (\mathcal{A}_n)^m$ for which $bB = 0$, since $\mathcal{A}_n$ is a domain ([2]). Consider a matrix $B^{(i)}$ obtained from B by deleting its i-th column. By virtue of lemma 1 there is a vector $0 \neq g_i \in (\mathcal{A}_n)^m$ such that $g_i B^{(i)} = 0$ and $\deg(g_i) \leq N$. Then the $m \times m$ matrix G with the rows $g_1, \ldots, g_m$ is a left quasiinverse for B, q.e.d.

Notice that B has a right quasiinverse matrix iff, considered as a matrix over the skew-field of fractions $\mathcal{D}_n$ of $\mathcal{A}_n$ ([2]), B is nonsingular, i.e. has a nonvanishing Dieudonné determinant ([1]). So, one can define the rank $r = rg(A)$ of a matrix A over $\mathcal{A}_n$ as the maximal size of its nonsingular submatrices. In the following lemma, the $m_1 \times m_2$ matrix A is such that $r = rg(A)$ and its $r \times r$ submatrix A_1 in the left upper corner is nonsingular.

Lemma 3. *Let C_1 be a left quasiinverse matrix for A_1 over $\mathcal{A}_n$. Then there exists $(m_1 - r) \times r$ matrix C_2 over $\mathcal{D}_n$ such that*

$$
\begin{pmatrix} C_1 & 0 \\ C_2 & E \end{pmatrix} A = \left(
\begin{array}{ccc|c}
a_1 & & 0 & \\
 & \ddots & & * \\
0 & & a_r & \\
\hline
 & 0 & & 0
\end{array}
\right)
$$

where from now on E denotes the unit matrix.

Proof: The matrix C_2 is determined uniquely by the requirement that in the product of matrices under consideration the left lower corner is zero. Then the right lower corner is also zero by the definition of the rank. q.e.d.

Returning to system (1), apply lemma 3 to the $m \times s$ matrix $(U_{k,\ell})$ and get matrices C_1, C_2 (after suitable renumerations of columns and rows one can assume that $r \times r$ submatrix of $(U_{k,\ell})$ in the left upper corner is the nonsingular submatrix of $(U_{k,\ell})$ of maximal size). In the case when the vector $(C_2 E)(w_1, \ldots, w_m)^T$ does not vanish, the system (1) is inconsistent. If $(C_2 E)(w_1, \ldots, w_m)^T = 0$ then (1) is equivalent to a system of the following form (see Lemma 3)

$$
a_k V_k + \sum_{r+1 \leq \ell \leq s} a_{k,\ell} V_\ell = b_k, \quad 1 \leq k \leq r \tag{2}
$$

Lemma 2 entails the bounds $\deg(a_k)$, $\deg(a_{k,\ell})$, $\deg(b_k) \leq 4nrd \leq 4nmd$.

Fix a certain $r+1 \leq \ell \leq s$ for the time being. According to lemma 1 there exists $h_1^{(\ell)}, \ldots, h_r^{(\ell)} \in \mathcal{A}_n$ such that

$$
a_k h_k^{(\ell)} + a_{k,\ell} h^{(\ell)} = 0, \quad 1 \leq k \leq r \tag{3}
$$

and moreover $\deg(h_1^{(\ell)}), \ldots, \deg(h_r^{(\ell)})$, $\deg(h^{(\ell)}) \leq 16n^2 m^2 d$.

2.   Normalization in $\mathcal{A}_n$ and eliminating $\mathcal{D}_n, X_n$.

Let $g_1, \ldots, g_t \in \mathcal{A}_n$ be a certain family of elements. The next lemma is an analogy of the normalization lemma for $\mathcal{A}_n$.

Lemma 4. *There is a nonsingular linear transformation over F of $2n$-dimensional space with basis $X_1, \ldots, X_n$, $D_1, \ldots, D_n$ under which*

$$
X \to \Gamma_{X_i} = \sum_{1 \leq j \leq n} \gamma_{i,j}^{(1,1)} X_j + \sum_{1 \leq j \leq n} \gamma_{i,j}^{(1,2)} D_j;
$$

$$
D_i \to \Gamma_{D_i} = \sum_{1 \leq j \leq n} \gamma_{i,j}^{(2,1)} X_j + \sum_{1 \leq j \leq n} \gamma_{i,j}^{(2,2)} D_j
$$

such that the following relations hold:

$$\Gamma_{X_i}\Gamma_{D_i} = \Gamma_{D_i}\Gamma_{X_i} - 1; \Gamma_{X_i}\Gamma_{X_j} = \Gamma_{X_j}\Gamma_{X_i};$$
$$\Gamma_{D_i}\Gamma_{D_j} = \Gamma_{D_j}\Gamma_{D_i}; \Gamma_{D_i}\Gamma_{X_j} = \Gamma_{X_j}\Gamma_{D_i}$$

for $1 \le i \ne j \le n$ (cf. the relations defining the Weyl algebra), and furthermore for every $1 \le \ell \le t$ the leading coefficient $0 \ne lc_{D_n}(\tilde{g}_\ell)$ with respect to D_n belongs to F, where $\tilde{g}_\ell$ is obtained from g_ℓ by the indicated linear transformation, in other words $\tilde{g}_\ell = (lc_{D_n}(\tilde{g}_\ell))D_n^{\deg(g_\ell)} + \tilde{\tilde{g}}_\ell$, where $\deg_{D_n}(\tilde{\tilde{g}}_\ell) < \deg(g_\ell)$.

Proof: Consider $2n \times 2n$ matrix $\Gamma = \begin{pmatrix} \Gamma^{1,1} & \Gamma^{1,2} \\ \Gamma^{2,1} & \Gamma^{2,2} \end{pmatrix}$ with rows corresponding to the linear forms $\Gamma_{x_1},...,\Gamma_{X_n}, \Gamma_{D_1},...,\Gamma_{D_n}$. Then the relations in the lemma hold iff Γ belongs to the sympletic group $\Gamma \in Sp_{2n}(F)[1]$, i.e. satisfies the equality $\Gamma \begin{pmatrix} O & -E \\ E & O \end{pmatrix} \Gamma^T = \begin{pmatrix} O & -E \\ E & O \end{pmatrix}$. For every $1 \le \ell \le t$ the leading coefficient $lc_{D_n}(\tilde{g}_\ell)$ can be regarded as a homogeneous polynomial in the entries $\gamma_{1,n,...}^{(1,2)}, \gamma_{n,n}^{(1,2)}, \gamma_{1,n,...}^{(2,2)}, \gamma_{n,n}^{(2,2)}$ of the last column of the matrix Γ. Since the group $Sp_{2n}(F)$ is transitive on the set of the vectors $F^{2n} \setminus \{0\}[1]$, there exists $\Gamma \in Sp_{2n}(F)$ such that $lc_{D_n}(\tilde{g}_\ell) \ne 0, 1 \le \ell \le t$, q.e.d.

Apply lemma 4 to the family of elements $h^{(\ell)}, r+1 \le \ell \le s$ (see (3)). One can consider A_n as Weyl algebra over $\Gamma_{X_1},...,\Gamma_{X_n}, \Gamma_{D_1},...,\Gamma_{D_n}$. Therefore we can w.l.o.g. suppose that $0 \ne \alpha_\ell = lc_{D_n}(h^{(\ell)}) \in F; r+1 \le \ell \le s$ making (if necessary) a linear transformation of variables.

Henceforth one can divide (with remainder) V_ℓ for $r+1 \le \ell \le s$ (see (2)) from the left by $h^{(\ell)}$ (see (3)) with respect to D_n in the ring A_n, i.e. $V_\ell = h^{(\ell)}\overline{\overline{V}}_\ell + \bar{V}_\ell$ where $\overline{\overline{V}}_\ell, \bar{V}_\ell \in A_n$ and moreover $\deg_{D_n}(\bar{V}_\ell) < \deg(h^{(\ell)}) \le 16n^2m^2d$.

For every $1 \le k \le r$ add to the equation (2) the equations (3) multiplied on the right by $-\overline{\overline{V}}_\ell$ for all $r+1 \le \ell \le s$, respectively. As a result we get the system as follows, having a solution in $(A_n)^s$ iff (2) has a solution (and finally (1) has a solution):

$$a_k \bar{V}_k + \sum_{r+1 \le \ell \le s} a_{k,\ell} \bar{V}_\ell = b_k, \quad 1 \le k \le r \tag{4}$$

Since $\deg_{D_n}(b_k), \deg_{D_n}(a_{k,\ell}\bar{V}_\ell) \le N_0 = (nmd)^{0(1)}$ for $1 \le k \le r$.

One can represent $\bar{V}_j = \sum_{0 \le i \le N_0} D_n^i \bar{V}_{j,i}$ for $1 \le j \le s$, where $\bar{V}_{j,i} \in A_{n-1}[X_n]$. Then the system (4) can be replaced by an equivalent system in the unknowns $\bar{V}_{j,i}; 1 \le j \le s, 0 \le i \le N_0$, each equation from (4) being replaced by $(N_0 + 1)$ equations. This proves the following

Lemma 5. *System (1) is equivalent to the following system of linear equations*

$$\sum_{1 \leq \ell \leq N_2} f_{k,\ell} Z_\ell = g_k \ , \ 1 \leq k \leq N_3 \tag{5}$$

over the ring $\mathcal{A}_{n-1}[X_n]$, moreover $\deg(f_{k,\ell})$, $\deg(g_k)$, $N_3 \leq (nmd)^{0(1)}$; $N_2 \leq s(nmd)^{0(1)}$. Furthermore, if system (5) has a solution $Z_1, \ldots, Z_{N_2} \in (\mathcal{A}_{n-1}[X_n])^{N_2}$ satisfying the bounds $\deg(Z_\ell) \leq N_1$, $1 \leq \ell \leq N_2$ for a certain N_1 then system (1) has a solution $(V_1, \ldots, V_s) \in (\mathcal{A}_n)^s$ such that $\deg(V_\ell) \leq N_1 + N_0$, $1 \leq \ell \leq s$.

Thus, lemma 5 allows us to eliminate D_n and reduce our considerations to the systems of linear equations over the ring $\mathcal{A}_{n-1}[X_n]$. Next we eliminate X_n in a similar way. First, we observe that lemma 1 and by the same token lemma 2 are valid also for matrices over the ring $\mathcal{A}_{n-1}[X_n]$.

Applying the construcion exhibited in section 1 to the system (5), considering it as the initial system (1), one can reduce (5) to the trapezium form (cf. (2)) as follows

$$p_k Z_k + \sum_{r_1+1 \leq \ell \leq N_2} p_{k,\ell} Z_\ell = q_k \ , \ 1 \leq k \leq r_1 \tag{2'}$$

where $p_k, p_{k,\ell}, q_k \in \mathcal{A}_{n-1}[X_n]$ and r_1 equals the rank of the $N_3 \times N_2$ matrix $(f_{k,\ell})$. By virtue of lemma 2 and the above observation we have $\deg(p_k)$, $\deg(p_{k,\ell})$ $\deg(q_k) \leq (nmd)^{0(1)}$. In view of lemma 1 and the above observation, for every $r_1 + 1 \leq \ell \leq N_2$ there exist $y_1^{(\ell)}, \ldots, y_{r_1}^{(\ell)}, y^{(\ell)} \in \mathcal{A}_{n-1}[X_n]$ such that $\deg(y_k^{(\ell)})$, $\deg(y^{(\ell)}) \leq (nmd)^{0(1)}$ and the following equalities are true (cf. (3)):

$$p_k y_k^{(\ell)} + p_{k,\ell} y^{(\ell)} = 0, \ 1 \leq k \leq r_1 \tag{3'}$$

Let $g_1^{(1)}, \ldots, g_t^{(1)} \in \mathcal{A}_{n-1}[X_n]$ be a certain family of elements. The following lemma (and also its proof) is similar to lemma 4.

Lemma 4'. *There exists a nonsingular linear (over F) transformation of $(2n-1)$-dimensional space, under which $X_n \to X_n$, $X_i \Delta_{X_i} = X_i + \delta_i^{(1)} X_n$, $D_i \to \Delta_{D_i} = D_i + \delta_i^{(2)} X_n$, $1 \leq i \leq n-1$ such that $0 \neq \mathrm{lc}_{X_n}(\tilde{g}_k^{(1)}) \in F$, $1 \leq k \leq t$ where $\tilde{g}_k^{(1)}$ is obtained from $g_k^{(1)}$ by the indicated linear transformation.*

Apply lemma 4' to the family of elements $y^{(\ell)}$, $r_1 + 1 \leq \ell \leq N_2$ (see (3')). One can consider $\mathcal{A}_{n-1}$ as a Weyl algebra in $\Delta_{X_1}, \ldots, \Delta_{X_{n-1}}$, $\Delta_{D_1}, \ldots, \Delta_{D_{n-1}}$, since the linear transformation from lemma 4' keeps the defining relations of the Weyl algebra. So, we can assume w.l.o.g. that $0 \neq \mathrm{lc}_{X_n}(y^{(\ell)}) \in F$, $r_1 + 1 \leq \ell \leq N_2$ making (if necessary) a suitable linear transformation of variables.

In a similar way as above divide (with remainder) Z_ℓ, $r_1 + 1 \leq \ell \leq N_2$ (see, (2')) by $y^{(\ell)}$ (see (3')) from the left with respect to X_n in the ring $\mathcal{A}_{n-1}[X_n]$. Thus, $Z_\ell = y^{(\ell)}\overline{\overline{Z}}_\ell + \bar{Z}_\ell$ and $\deg_{X_n}(\bar{Z}_\ell) < \deg_{X_n}(y^{(\ell)}) \leq (nmd)^{0(1)}$ (cf. (3')). Now, for every $1 \leq k \leq r_1$ add to the equality (2') the equalities (3') multiplied on the right by $-\overline{\overline{Z}}_\ell$ for all $r_1 + 1 \leq \ell \leq N_2$, respectively. As a result we obtain the following system of linear equations over $\mathcal{A}_{n-1}[X_n]$

$$p_k \bar{Z}_k + \sum_{r_1+1 \leq \ell \leq N_2} p_{k,\ell} \bar{Z}_\ell = q_k, \ 1 \leq k \leq r_1 \tag{4'}$$

having a solution in $(\mathcal{A}_{n-1}[X_n])^{N_2}$ iff system (2') has a solution.

As earlier we estimate

$$\deg_{X_n}(p_k \bar{Z}_k) \leq \max_{r_1+1 \leq \ell \leq N_2} \{\deg_{X_n}(p_{k,\ell} \bar{Z}_\ell), \ \deg_{X_n}(q_k)\} \leq N_4 (nmd)^{0(1)}.$$

In a similar way we represent $\bar{Z}_j = \sum_{0 \leq i \leq N_4} X_n^i \bar{Z}_{j,i}$ where $\bar{Z}_{j,i} \in \mathcal{A}_{n-1}$ and replace each of the equations from (4') by $(N_4 + 1)$ linear equations in the variables $\bar{Z}_{j,i}$, $1 \leq j \leq N_2$, $0 \leq i \leq N_4$ over the ring $\mathcal{A}_{n-1}$ (cf. above). The following lemma can be proved in a similar way to lemma 5.

Lemma 5'. *System (5) is equivalent to appropriate system of linear equations*

$$\sum_j f'_{i,j} Y_j = h'_i \tag{5'}$$

over the ring $\mathcal{A}_{n-1}$, where the number of equations and the degrees of all $f'_{i,j}$, h'_i do not exceed $(nmd)^{0(1)}$. Furthermore, if system (5') has a solution $\{Y_j \in \mathcal{A}_{n-1}\}$ with $\deg(Y_j) \leq N_5$ for a certain N_5 and all j then system (5) has a solution $\{Z_\ell \in \mathcal{A}_{n-1}[X_n]\}$ satisfying the bounds $\deg(Z_\ell) \leq N_5 + N_4$, moreover $N_4 \leq (nmd)^{0(1)}$.

One can complete the proof of the theorem (see introduction) in the usual way (see [8]) by induction on n based on lemmas 5, 5'.

REFERENCES

[1] Artin E., "Geometric algebra," Interscience publishers, 1957.

[2] Björk J.-E., "Rings of differential operators," North-Holland, 1979.

[3] Chistov A.L., Grigor'ev D.Yu., *Subexponential-time solving systems of algebraic equations*, Preprints LOMI E-9-83, E-10-83. Leningrad, 1983.

[4] Fitchas N., Galligo A., *Nullstellensatz effectif et Conjecture de Serre (Théorème de Quillen-Suslin) pour le Calcul Formel*, Séminaire "Structures algébriques ordonnées", Paris VII, 1988 (to appear in *Mathematische Nachrichten*).

[5] Galligo A., *Some algorithmical questions on ideals of differential operators.*, in ",", Lect. Notes Comput. Sci. 204, 1985, pp. 413–421.

[6] Grigor'ev D. Yu., *Computational complexity in polynomial algebra*, in "Proc. Intern. Congr. Mathem.," Berkeley, 1986, pp. 1452–1460.

[7] Grigor'ev D. Yu., *Complexity of factoring and GCD calculating of linear ordinary differential operators*, J. Symbol. Comput. (to appear).

[8] Seidenberg A., *Constructions in algebra*, Trans. Amer. Math. Soc. 197 (1974), 273–313.

Dimitri Yu. Grigor'ev
Leningrad Department of Mathematical
V.A.Steklov Institute
Academy of Sciences of the USSR
Fontanka 27, Leningrad, 191011, USSR

Membership problem, Representation problem and the Computation of the Radical for one-dimensional Ideals

TERESA KRICK   ALESSANDRO LOGAR

1. Introduction

In this paper we consider membership problem, representation problem and also the computation of the radical for one-dimensional ideals in the polynomial ring $k[X_1, \ldots, X_n]$ from a complexity point of view. Our aim is to give bounds for the complexity of the above problems which are simply exponential in the number n of variables in the one-dimensional case. Moreover we show that in the general case the first two problems are doubly exponential only in the dimension of the ideal, and parallelizable. Many authors considered membership problem (i.e. given polynomials $F, F_1, \ldots, F_s \in k[X_1, \ldots, X_n]$, decide whether F belongs to $I := (F_1, \ldots, F_s)$) and representation problem (compute a representation $F = \sum_{i=1}^{s} A_i F_i$, $A_i \in k[X_1, \ldots, X_n]$) from this effective approach. In particular [18] gives a lower bound, doubly exponential in (a fraction of) the numbers of variables n. On the other hand, [4] shows that in "good" cases, for instance for unmixed ideals, the membership problem is simply exponential in n.

In section 2, we apply results over efficient Noether Normalization Lemma (see [4] and [17]) to convert membership and representation problems into a problem of compatibility and resolution of a linear system of equations, of dimension related with the number of variables and the dimension of the ideal. By [10] (or also [20], [15] and [9]), we can therefore show that representation problem (and hence membership problem) is at most doubly exponential (in sequential) and simply exponential (in parallel) in the dimension of the ideal considered. This implies that for any fixed dimension, these problems are simply exponential in n, and parallelizable. In section 3, we show (more generally) that it is possible to give bounds of the same order also for the computation of a large class of Gröbner bases of the ideal I. Examples presented in [19] suggest that these bounds cannot be substantially improved. Finally, section 4 deals with the problem of the computation of the radical of a polynomial ideal. The construction of the radical (and the primary decomposition) of an ideal was first solved in [10] (see also [20]). In [7], a new approach to the problem (using Gröbner bases techniques) is considered, but no explicit bounds for the complexity of the computations are given, while in [3] the problem is solved also from the complexity point of view. Here, we use effective linear algebra tecniques over $k[X_1]$ (crf. for instance [11], [12], [13] and [14]) to compute in sequential single exponential time in the number of variales the radical of a one-dimensional ideal I. The approach presented here seems to be new and completely elementary and the algorithm is easily implementable; moreover, it is promising for a generalization for higher dimensions.

In the following k will indicate an infinite field (of any characteristic, if not otherwise specified). We assume that the elementary operations on k require constant time. The notions of sequential and parallel complexities used here correspond to the size and depth of arithmetical networks introduced in [6].

2. The membership and the representation problems

Let $k[X_1, \dots, X_n]$ be a polynomial ring over the field k, which, as said above, is assumed to be infinite. Suppose that polynomials $P_{d+1}, \dots, P_n$ are given such that:

- $P_{d+i} \in k[X_1, \dots, X_d, X_{d+i}]$, $i = 1, \dots, n-d$;
- P_{d+i} is monic in X_{d+i};
- $\deg(P_{d+i}) \leq D$ (where $\deg(\cdot)$ is the total degree and $D \in \mathbf{N}$ is a given bound).

Let Σ (or $\Sigma(P_{d+1}, \dots, P_n)$) be the set of all term-orders on $k[X_1, \dots, X_n]$ such that $M_<(P_{d+i})$ is a power of X_{d+i} ($M_<(F)$ is the maximal monomial of F w.r.t. the term-order $<$ considered). It is clear that $\Sigma \neq \emptyset$ because if $<_1$ is any term-order on $k[X_1, \dots, X_d]$ and $<_2$ is any term-order on $k[X_{d+1}, \dots X_n]$, then the *product* term-order $<$ on $k[X_1, \dots, X_n]$ defined by $\phi\psi < \phi'\psi'$ if $\psi <_2 \psi'$ or $\psi = \psi'$ and $\phi <_1 \phi'$ (where ϕ, ϕ' are monomials in $k[X_1, \dots, X_d]$ and ψ, ψ' are monomials in $k[X_{d+1}, \dots, X_n]$) is in Σ. In the following we shall fix a term-order $<$ in Σ, and $M(\cdot)$ will indicate the maximal monomial w.r.t. it. It holds:

Lemma 2.1. *It is possible to construct a $k[X_1, \dots, X_d]$-isomorphism*

$$\phi : k[X_1, \dots, X_n]/(P_{d+1}, \dots, P_n) \to k[X_1, \dots X_d]^r$$

where $r := \prod_{i=1}^{n-d} \big(\deg M(P_{d+i}) \big) \leq D^{n-d}$.

Proof: Let $J := (P_{d+1}, \dots, P_n)$. Since $<$ is in Σ, $\{P_{d+1}, \dots, P_n\}$ is a Gröbner basis of J. Take F in $k[X_1, \dots, X_n]$, and reduce it w.r.t. $\{P_{d+1}, \dots, P_n\}$. Hence we get a linear combination, with coefficients in $k[X_1, \dots, X_d]$, of the monomials $X_{d+1}^{j_{d+1}} \cdot \dots \cdot X_n^{j_n}$, $0 \leq j_{d+i} <$

$\deg M(P_{d+i})$, $i = 1, \ldots, n - d$. These monomials are linearly independent in $k[X_1, \ldots, X_d]$ (mod J). Therefore we define $\phi(F)$ to be the r-tuple (in a fixed order) of the above coefficients. Q. E. D.

Notation 2.2: We will indicate by $\mathbf{A}(P_{d+1}, \ldots, P_n)$ (or simply by $\mathbf{A}$) the set of monomials $X_{d+1}^{j_{d+1}} \cdot \ldots \cdot X_n^{j_n}$, $0 \le j_{d+i} < \deg M(P_{d+i})$.

Let $I \subseteq k[X_1, \ldots, X_n]$ be a d-dimensional ideal generated by $F_1, \ldots, F_s$ and suppose that

$$\max \deg(F_i) \le \delta.$$

We can assume that I is in *Noether normal position* (n.n.p.), i.e.:
- $I \cap k[X_1, \ldots, X_d] = (0)$;
- X_{d+i} is integral over $k[X_1, \ldots, X_d]$ (mod I), $i = 1, \ldots, n - d$.

The complexity of the algorithm which puts I in n.n.p. is bounded by $D = O(\delta^{2n})$ and parallelizable (crf. [17]), and moreover, from [4], it is known that for each $i = 1, \ldots, n - d$ there exists a polynomial $G_{d+i} \in I \cap k[X_1, \ldots, X_d, X_{d+i}]$ monic in X_{d+i} such that $\deg(G_{d+i}) \le D$. These polynomials can be computed with a truncated Gröbner bases algorithm whose complexity is simply exponential in n (crf. [17]).

Lemma 2.3. *Let $F \in k[X_1, \ldots, X_n]$ be any polynomial and let $G \in k[X_1, \ldots, X_n]$ be monic in X_n. If $\operatorname{red}(F)$ is the reduction (normal form) of F w.r.t. G, then:*

$$\deg \operatorname{red}(F) \le \deg(F) \cdot \deg(G).$$

Proof: (See also [4]) Let $Y := X_n$ and suppose that $G = Y^r - A$, $A \in k[X_1, \ldots, X_n]$, $\deg_Y(A) < r$, $\deg(A) \le \deg(G)$. If $\deg_Y(F) < r$, then $\operatorname{red}(F) = F$ and there is nothing to prove. Suppose then $r \le \deg_Y(F)$. Set $F_1 := F$. If $F_1, \ldots, F_i$ ($i \ge 1$) are defined, then define F_{i+1} as follows: suppose

$$F_i = A_{t_i} Y^{t_i} + A_{t_i-1} Y^{t_i-1} + \cdots + A_0,$$

where $A_{t_i}, A_{t_i-1}, \ldots, A_0 \in k[X_1, \ldots, X_{n-1}]$, $\deg(A_{t_i-j}) \le \deg(F_i) - t_i + j$. Let $\alpha_i, \beta_i \in \mathbf{N}$ be such that $t_i = \alpha_i r + \beta_i$, $0 \le \beta_i < r$. Then:

$$F_{i+1} := A_{t_i} Y^{\beta_i} A^{\alpha_i} + \cdots + A_0.$$

It is clear that there exists $s \ge 1$ s.t. $\deg(F_s) = \operatorname{red}(F)$. We have:

$$\deg_Y(F_{i+1}) = t_{i+1} \le t_i - \alpha_i;$$

$$\deg(F_{i+1}) \le \deg(F_i) - t_i + \beta_i + \alpha_i \deg(A) = \deg(F_i) + \alpha_i(\deg(A) - r);$$

if we add the two members of the last disequations for $j = 1, \ldots, i$, we get:

$$\deg(F_{i+1}) - \deg(F_1) \leq \sum_{j=1}^{i} \alpha_j(\deg(A) - r),$$

hence, in particular:

$$\deg(F_s) \leq \deg(F_1) + \sum_{j=1}^{s-1} \alpha_j(\deg(A) - r),$$

while from $t_{i+1} \leq t_i - \alpha_i$ in the same way we get $t_s - t_1 \leq -\sum_{j=1}^{s-1} \alpha_j$ so $\sum_{j=1}^{s-1} \alpha_j \leq t_1 - t_s \leq \deg(F_1)$. Therefore

$$\deg(F_s) \leq \deg(F_1)(\deg(A) - r + 1) \leq \deg(F) \cdot \deg(G).$$

Q. E. D.

Remark 2.4: There exist examples in which the bound given in lemma 2.3 is reached. Take for instance $F := Y^t$, $G := Y - A$, where A is a polynomial such that $\deg(A) = m(= \deg(G))$. Then $\mathrm{red}(F) = A^t$, so $\deg \mathrm{red}(F) = tm$.

Corollary 2.5. *Let* $G_{d+1}, \ldots, G_n$ *be as above, let* $<$ *be a term-order in* Σ. *If* F *is any polynomial in* $k[X_1, \ldots, X_n]$ *and* $\mathrm{red}(F)$ *is its reduction w.r.t. the Gröbner basis* $\{G_{d+1}, \ldots, G_n\}$, *then:*

$$\deg \mathrm{red}(F) \leq \deg(F) \cdot D^{n-d}.$$

In particular if ϕ *is the isomorphism given in lemma 2.1 applied to the polynomials* $P_{d+i} := G_{d+i}$, *then each component of* $\phi(F)$ *is bounded by* $\deg(F) \cdot D^{n-d}$.

Proof: Apply $n - d$ times lemma 2.3.   Q. E. D.

Let $J := (G_{d+1}, \ldots, G_n)$, then I/J is an ideal of $k[X_1, \ldots, X_n]/J$ so

$$V := \phi(I/J)$$

is a $k[X_1, \ldots, X_d]$-submodule of $k[X_1, \ldots, X_d]^r$. The set $\mathbf{A}(P_{d+1}, \ldots, P_n)$, as we showed in lemma 2.1, is a finite basis of $k[X_1, \ldots, X_n]/J$ as a $k[X_1, \ldots, X_d]$-module.

Proposition 2.6. *Let* $A_1, \ldots, A_r$ *be the elements of* $\mathbf{A}$. *Let, as usual,* $\max \deg(F_i) \leq \delta$ *and* $D = O(\delta^{2n})$ *s.t.* $\deg(G_{d+i}) \leq D$. *Then it holds:*

1) *The submodule* $V = \phi(I/J)$ *is generated, as a* $k[X_1, \ldots, X_d]$-*module, by* $\phi(A_i F_j)$, $i = 1, \ldots, r$, $j = 1, \ldots, s$;
2) *the degree of each of the* r *components of* $\phi(A_i F_j)$ *is bounded by* $(\delta + (n - d)D)D^{n-d}$.

Proof: Let $(v_1, \ldots, v_r) \in \phi(I/J)$; take $F \in I$ s.t. $\phi(F) = (v_1, \ldots, v_r)$. Let $U_1, \ldots, U_s \in k[X_1, \ldots, X_n]$ be s.t. $F = U_1 F_1 + \cdots + U_s F_s$. If

$$F' := \mathrm{red}(U_1)F_1 + \cdots + \mathrm{red}(U_s)F_s$$

(red$(\cdot)$ is, as usual, the reduction w.r.t. the Gröbner basis $\{G_{d+1}, \ldots, G_n\}$), then the vector $(v_1, \ldots, v_r) = \phi(F) = \phi(F')$ is a linear combination, with coefficients in $k[X_1, \ldots, X_d]$ of $\phi(A_i F_j)$ as required.

2) We have: $\deg(A_i) \leq (n - d)D$, $\deg(F_j) \leq \delta$, hence, from cor. 2.5 we see that each component of $\phi(A_i F_j)$ has degree bounded by $(\delta + (n - d)D)D^{n-d}$.   Q. E. D.

Proposition 2.7. *Let* $F \in k[X_1, \ldots, X_n]$ *be any polynomial. If* $F \in I$, *then there exists a representation of* F *in the form*

$$F = \sum_{i=1}^{s} Q_i F_i$$

where $\deg(Q_i)$ *is bounded by* $O((D + 1)(\deg(F) \cdot D^{2(n-d)})^{3^d})$. *In particular the bound is doubly exponential only in the dimension of the ideal.*

Proof: From lemma 2.1 applied to $P_{d+i} := G_{d+i}$ we have that $F \in I$ iff $\phi(F) \in V = \phi(I/J)$. From cor. 2.5 we have that each of the r components of $\phi(F)$ is bounded by $\deg(F)D^{n-d}$. The module V is generated by $sr \leq sD^{n-d}$ vectors, and each component of these vectors is bounded by $(\delta + (n - d)D)D^{n-d}$. Therefore the test $\phi(F) \in V$ can be converted into the solution of a linear system of $sr \leq sD^{n-d}$ variables, r equations and entries in $k[X_1, \ldots, X_d]$. If such system admits solutions, then from [9], lemma 8 (cfr. also [15]), we can find one solution z_{ij}, $i = 1, \ldots, r$, $j = 1, \ldots, s$ with $\deg(z_{ij})$ bounded by $O((\deg(F) \cdot D^{2(n-d)})^{3^d})$. From

$$\phi(F) = \sum_{ij} z_{ij} \phi(A_i F_j)$$

we get

$$H := F - \sum_{j} \left(\sum_{i} z_{ij} A_i \right) F_j \in (G_{d+1}, \ldots, G_n).$$

Since $\deg(A_i) \leq (n-d)D$, a bound for $\deg(\sum z_{ij} A_i)$ is of the same order of a bound for $\deg(z_{ij})$. Let now $w := (w_1, \ldots, w_d, w_{d+1}, \ldots, w_n)$ be a weight vector, with $w_1 = \ldots = w_d = 1$, $w_{d+1} = \ldots = w_n = D+1$ and let $\deg_w(f)$ be the total degree of f with respect to the weight w. Fix on $k[X_1, \ldots, X_n]$ a term-order $<$ w-degree compatible (i.e. such that if ϕ, ψ are monomials in $k[X_1, \ldots, X_n]$ and $\deg_w(\phi) < \deg_w(\psi)$, then $\phi < \psi$). It is immediate to verify that $\{G_{d+1}, \ldots, G_n\}$ is a Gröbner basis for the ideal it generates with respect to this term-order. Hence we can determine polynomials

$$B_{d+1}, \ldots, B_n \in k[X_1, \ldots, X_n]$$

such that:

$$H = \sum_{i=d+1}^{n} B_i G_i, \quad \deg_w(B_i G_i) \leq \deg_w(H).$$

Moreover $\deg(B_i) \leq \deg_w(B_i) \leq \deg_w(H) \leq (D+1)\deg(H)$. From [4] $G_i = \sum_{j=1}^{s} C_{ij} F_j$, where $\deg(C_{ij} F_j)$ is bounded by a function simply exponential in n. If we plug in the above representation of H these values of G_i's we get that $H = \sum_{i=1}^{s} E_i F_i$, where, as follows from the bounds computed above, $\deg(E_i) \leq O\big((D+1)(\deg(F) \cdot D^{2(n-d)})^{3^d}\big)$, and this gives the proof.Q. E. D.

Remarks 2.8: 1) The membership and representation problems have sequential complexity of order $O\big((D+1)(\deg(F) \cdot D^{2(n-d)})^{3^d}\big)$ and parallel complexity of order $O\big(n^d(\delta + \deg(F))\big)$. This last bound is easily obtained applying parallelizable linear algebra over k.

2) In case $d = 0$, V is an r-dimensional k-vector space and $\phi(F)$ is a vector in k^r. This gives a simplification in the above proof. See also the results presented in [4].

3) If $d = 1$, linear algebra computations can be done with one of the algorithms introduced for instance in [11], [12], [13] and [14] (crf. also the references given in these papers). In these cases slightly more tight bounds can be obtained.

3. The computation of the Gröbner bases

Let again $I := (F_1, \ldots, F_s) \subseteq k[X_1, \ldots, X_n]$ and $G_{d+1}, \ldots, G_n$ as in the previous section. Let $<_1$ be a term-order on $k[X_1, \ldots, X_d]$ (no request if $d = 0$), $<_2$ be a term-order on $k[X_{d+1}, \ldots, X_n]$ and $<$ the product term-order on $k[X_1, \ldots, X_n]$. As remarked in section 2, $<$ is in $\Sigma(G_{d+1}, \ldots, G_n)$. Assume that the monomials $A_1, \ldots, A_r$ (i.e. the elements of $\mathbf{A}(G_{d+1}, \ldots, G_n)$) are labelled in such a way that $A_1 < \cdots < A_r$. Finally we define a term-order $\prec$ on the free module $k[X_1, \ldots, X_d]^r$ as follows:

if ψ, ψ' are monomials in $k[X_1, \ldots, X_d]$, and if $e_1, \ldots, e_r$ is the canonical basis of the free module $k[X_1, \ldots, X_d]^r$, then

$$\psi e_i \prec \psi' e_j \text{ if } i < j \text{ or } i = j \text{ and } \psi <_1 \psi'.$$

(Clearly, if $d = 0$, we consider k^r as a vector space over k and the above condition becomes vacuous). Call H the Gröbner basis of the $k[X_1, \ldots, X_d]$-module V with respect to $\prec$.

Proposition 3.1. *The set*

$$L := \{G_{d+1}, \ldots, G_n\} \cup \phi^{-1}(H)$$

is a Gröbner basis of I with respect to the term-order $<$.

Proof: Let $F \in I$. We want to see that $F \xrightarrow{L} 0$. Let F' be the reduction of F w. r. t. $\{G_{d+1}, \ldots, G_n\}$. Hence $F' = Q_1 A_1 + \cdots + Q_s A_s$, $Q_i \in k[X_1, \ldots, X_d]$, $0 \le s \le r$, $Q_s \ne 0$. If $s = 0$ then we are done. Suppose $s > 0$. From the proof of prop. 2.6 we have that $\phi(F') \in V$, hence $\phi(F') \xrightarrow{H} 0$. As a consequence of the definition of the term-order $\prec$ we have that there exists $h \in H$, $h = (h_1, \ldots, h_s, 0, \ldots, 0)$ and there exists a monomial $\xi \in k[X_1, \ldots, X_d]$ and a constant $c \in k \setminus \{0\}$ such that

$$M_1(Q_s) = c \xi M_1(h_s)$$

($M_1(\cdot)$ is the maximal monomial w.r.t. $<_1$). But:

$$M(h_1 A_1 + \cdots + h_s A_s) = A_s M(h_s),$$

therefore: $F' \xrightarrow{L} F' - c\xi(h_1 A_1 + \cdots + h_s A_s)$

Since $M(F' - c\xi(h_1 A_1 + \cdots + h_s A_s)) < M(F')$, an iteration of this procedure gives the thesis.   Q. E. D.

From proposition 3.1 we see that an upper bound for the complexity of the Gröbner basis H will give an upper bound for the complexity of the Gröbner basis L of I. In the following two propositions we briefly sketch how to get such upper bounds. First we consider the particular cases in which $\dim I \le 1$. In these cases the computation of H can be done using specific algorithms for the linear algebra in k (and respectively in $k[X_1]$).

Proposition 3.2. *If the ideal I is of dimension $d = 0$ or $d = 1$, then there exists an upper bound for the computation of L which is simply exponential in n.*

Proof: (For the case $d = 0$ crf. also [2] and [4]) In the two cases $d = 0$ or $d = 1$ to compute H means to compute a triangulation of a matrix of dimension $D^{n-d} \times sD^{n-d}$ with entries in k or $k[X_1]$ respectively. Moreover in the case $d = 1$ we need to bound the degrees of the polynomials in such a matrix. From the result of section 2 we have that all the entries of the matrix are polynomials bounded by $(\delta + (n-1)D)D^{n-1}$. The result then follows for $d = 0$ simply as a consequence of our assumptions on the arithmetic in k, and for $d = 1$ as a consequence of the bounds for performing linear algebra given for instance in [11], [12], [13] and [14].   Q. E. D.

In the appendix we shall prove that all the polynomials in the Gröbner basis H have degree at most doubly exponential in d. As a consequence we get the following generalization of the previous proposition:

Proposition 3.3. *There exists an upper bound for the computation of the Gröbner basis L of I which is doubly exponential only in d.*

Remark 3.4: For other bounds for the computation of Gröbner bases see, for instance, [8], [19], [16].

4. On the computation of the radical of a one-dimensional ideal

In this section, for simplicity of exposition, we shall restrict ourselves to the case *char* $(k) = 0$. First of all we shall prove the following:

Lemma 4.1. *Let α be an ideal in $k[X_1, \ldots, X_n]$ such that for each $i = 1, \ldots, n$ there exists a polynomial $P_i \in \alpha \cap k[X_i]$ squarefree. Then α is a (zero-dimensional) radical ideal (crf. [20], lemma 92).*

Proof: P_i is squarefree, hence $k[X_i]/(P_i)$ is a direct sum of fields, and hence

$$\frac{k[X_1, \ldots, X_n]}{(P_1, \ldots, P_n)} \simeq \frac{k[X_1]}{(P_1)} \otimes_k \cdots \otimes_k \frac{k[X_n]}{(P_n)}$$

is a semisimple ring, finite sum of fields, so there are no nilpotents elements in it. Therefore the ring $k[X_1, \ldots, X_n]/\alpha$, which is a quotient of $k[X_1, \ldots, X_n]/(P_1, \ldots, P_n)$, is also semisimple, hence the ideal α is radical (and clearly zero-dimensional).   Q. E. D.

Assume now that the ideal $I := (F_1, \ldots, F_s)$ is zero-dimensional. As shown in the second section, it is possible to compute polynomials $G_1, \ldots, G_n \in I$ s.t. $G_i \in k[X_i]$ and $\deg(G_i) \leq D$.

Let $P_i := G_i/\gcd(G_i, G_i')$, hence P_i is squarefree, $P_i \in k[X_i]$, and it is clear that $P_i \in \sqrt{I}$ (for the computation of P_i see [1]). As a consequence of the lemma 4.1 and of section 2, we get:

Corollary 4.2. *It holds:*

$$\sqrt{I} = (F_1, \ldots, F_s, P_1, \ldots, P_n).$$

In particular there exists a bound simply exponential in n to compute a system of generators of $\sqrt{I}$. ·

Assume now that I is a one-dimensional ideal. As in the previous case, we can construct polynomials $G_2, \ldots, G_n \in I$ s.t. $G_i \in k[X_1, X_i]$, of bounded degree D and let again $P_i := G_i / \gcd(G_i, G_i')$, $i = 2, \ldots, n$. Note that, as a consequence of the Gauss lemma, the computation of the P_i's can be done in $k(X_1)[X_i]$ and they are monic in X_i. From lemma 2.1 applied to the polynomials P_i's, we see that $k[X_1, \ldots, X_n]/(P_2, \ldots, P_n)$ is a $k[X_1]$ free module of rank r (bounded by D^{n-1}). Since the module

$$\phi\left(\frac{(I, P_2, \ldots, P_n)}{(P_2, \ldots, P_n)}\right)$$

is a submodule of $k[X_1]^r$, it is free, and from prop. 2.6 applied to the ideal $(I, P_2, \ldots, P_n)$ we can determine a system of generators of it. Using the elementary divisors theorem, we can compute a new basis

$$\mathbf{b}_1, \ldots, \mathbf{b}_m, \mathbf{b}_{m+1}, \ldots, \mathbf{b}_r \in k[X_1]^r$$

and polynomials $\alpha_1, \ldots, \alpha_m \in k[X_1]$, such that

$$\alpha_1 \mid \alpha_2 \mid \ldots \mid \alpha_m$$

and $\alpha_1 \mathbf{b}_1, \ldots, \alpha_m \mathbf{b}_m$ is a free basis of $\phi\left((I, P_2, \ldots, P_n)/(P_2, \ldots, P_n)\right)$. Let

$$B_1, \ldots, B_m, B_{m+1}, \ldots, B_r \in k[X_1, \ldots, X_n]$$

be such that $\phi([B_i]) = \mathbf{b}_i$. Remark that the arithmetic cost in the computation of the elementary divisors is constant (see [5]).

Proposition 4.3. *It holds:*
1) *The ideal $I_1 := (P_2, \ldots, P_n, B_1, \ldots, B_m)$ is one-dimensional, unmixed and radical;*
2) *Let $I_0 := \sqrt{(I, P_2, \ldots, P_n, \alpha_m)}$. Then I_0 is zero-dimensional and:*

$$\sqrt{I} = I_0 \cap I_1.$$

Proof: Let's fix a primary decomposition of $\sqrt{I}$:

$$\sqrt{I} = \mathbf{M}_1 \cap \ldots \cap \mathbf{M}_p \cap \mathbf{M}_{p+1} \cap \ldots \cap \mathbf{M}_q \cap \mathbf{P}_1 \cap \ldots \cap \mathbf{P}_s$$

where $\mathbf{M}_1, \ldots, \mathbf{M}_q$ are zero-dimensional prime ideals (hence maximal), and $\mathbf{P}_1, \ldots, \mathbf{P}_s$ are prime ideals of dimension one. Suppose moreover that $\mathbf{M}_1, \ldots, \mathbf{M}_q$ are labelled in such a way that for each $i = 1, \ldots, p$ there exists $j_i \in \{1, \ldots, m\}$ s.t. $\alpha_{j_i} \in \mathbf{M}_i$ and $\alpha_j \notin \mathbf{M}_i$ for every $j = 1, \ldots, m$ and every $i = p+1, \ldots, q$ (the cases $p = 0$ or $q = 0$ are allowed). If $i \in \{1, \ldots, p\}$ then $\alpha_m = \mu \alpha_{j_i} \in \mathbf{M}_i$ (where μ is a suitable element in $k[X_1] \setminus \{0\}$), and because $P_2, \ldots, P_n \in \sqrt{I}$ we have that:

$$I \subseteq (I, P_2, \ldots, P_n, \alpha_m) \subseteq \mathbf{M}_1 \cap \ldots \cap \mathbf{M}_p.$$

Moreover we have that

$$\alpha_j B_j = \phi^{-1}(\alpha_j b_j) \in (I, P_2, \dots, P_n) \subseteq \sqrt{I},$$

so $\alpha_j B_j \in \mathbf{M}_i$, therefore $B_j \in \mathbf{M}_{p+1} \cap \dots \cap \mathbf{M}_q$ $(j = 1, \dots, m)$. From $\alpha_j B_j \in \mathbf{P}_1 \cap \dots \cap \mathbf{P}_s$, we get that

$$B_j \in \mathbf{P}_1 \cap \dots \cap \mathbf{P}_s$$

(if $B_j \notin \mathbf{P}_i$ for some i, then $\alpha_j \in P_i$, therefore $P_j \supseteq (\alpha_j, P_2, \dots, P_n)$ and this contradicts $dim P_i = 1$). Therefore:

$$I \subseteq I_1 \subseteq \mathbf{M}_{p+1} \cap \dots \cap \mathbf{M}_q \cap \mathbf{P}_1 \cap \dots \cap \mathbf{P}_s$$

hence:

$$I \subseteq (I, P_2, \dots, P_n, \alpha_m) \cap I_1 \subseteq \sqrt{I},$$

i.e.:

$$\sqrt{I} = I_0 \cap \sqrt{I_1}.$$

Now we want to show that I_1 is a radical ideal. First note that I_0 is zero-dimensional and I_1 is one-dimensional (if it were zero-dimensional then, for the equality established above, $\sqrt{I}$ would be the intersection of two zero-dimensional ideals, and this would contradict the hypothesis that $dim I = 1$). It is then clear that $k[X_1] \cap I_1 = (0)$.

Claim. $\phi(I_1)$ (i.e. $\phi(I_1/(P_2, \dots, P_n))$) is the $k[X_1]$-module generated by $\mathbf{b}_1, \dots, \mathbf{b}_m$.

Proof of the claim: $\phi(I_1)$ is generated, as a $k[X_1]$-module, by $\mu \mathbf{b}_i$, $\mu \in \mathbf{A}(P_2, \dots, P_n)$. Let:

$$(*) \qquad \mu \mathbf{b}_i = \lambda_1 \mathbf{b}_1 + \dots + \lambda_m \mathbf{b}_m + \lambda_{m+1} \mathbf{b}_{m+1} + \dots + \lambda_r \mathbf{b}_r.$$

Since $\alpha_1 \mathbf{b}_1, \dots, \alpha_m \mathbf{b}_m$ is a system of generators of the image of the ideal $(I, P_2, \dots, P_n)$, we get that $\mu \alpha_i \mathbf{b}_i \in \langle \alpha_1 \mathbf{b}_1, \dots, \alpha_m \mathbf{b}_m \rangle$, hence $\mu \alpha_i \mathbf{b}_i = f_1 \alpha_1 \mathbf{b}_1 + \dots + f_m \alpha_m \mathbf{b}_m$. If we multiply $(*)$ by α_i we get another representation of $\mu \alpha_i \mathbf{b}_i$, and as a consequence of the fact that the two representations must be equal, we get: $\lambda_{m+1} = \dots = \lambda_r = 0$, therefore $\mu \mathbf{b}_i \in \langle \mathbf{b}_1, \dots, \mathbf{b}_m \rangle$ and this shows the claim.

Now the ideal $I_1^e := I_1 k(X_1)[X_2, \dots, X_n]$ is zero-dimensional and radical (lemma 4.1). If $A \in k[X_1]$ is such that $AF \in I_1$, then $\phi(AF) = A\phi(F) \in \langle \mathbf{b}_1, \dots, \mathbf{b}_m \rangle$ and if $\phi(F) = \lambda_1 \mathbf{b}_1 + \dots + \lambda_m \mathbf{b}_m + \lambda_{m+1} \mathbf{b}_{m+1} + \dots + \lambda_r \mathbf{b}_r$, then, as before, we find that $\lambda_{m+1} = \dots = \lambda_r = 0$, i.e. $\phi(F) \in \langle \mathbf{b}_1, \dots, \mathbf{b}_m \rangle$, so $F \in I_1$. This shows that $I_1 = I_1^e \cap k[X_1, \dots, X_n]$ and therefore I_1 is the contraction in $k[X_1, \ldots, X_n]$ of an unmixed radical ideal, therefore is unmixed and radical and this completes the proof.   Q. E. D.

Proposition 4.4. *The complexity of the computation of a system of generators of the ideals*

$$I_0, \quad I_1 \quad and \quad \sqrt{I}$$

is simply exponential in n.

Proof: For I_0 the assertion comes from the previous results. Similarly for I_1, since the computation of the elementary divisors requires only linear algebra in $k[X_1]$. For the computation of a system of generators (a Gröbner basis) of $\sqrt{I} = I_0 \cap I_1$ we can procede as follows:

let's change the notations and assume that $R_1, \ldots, R_u$ is a system of generators of I_0 and $S_1, \ldots, S_v$ a system of generators of I_1. Then it is well-known that if

$$(a_{i1}, \ldots, a_{iu}, b_{i1}, \ldots, b_{iv}), \qquad i = 1, \ldots, t$$

is a system of generators of the syzygies of $(R_1, \ldots, R_u, -S_1, \ldots, -S_v)$, then $a_{i1}R_1 + \cdots + a_{iu}R_u$, $i = 1, \ldots, t$ is a system of generators of $I_0 \cap I_1$. The ideal

$$(R_1, \ldots, R_u, -S_1, \ldots, -S_v)$$

is clearly zero-dimensional, since it contains I_0, therefore, from the results of section 3 we have a simply exponential bound for the computation of a Gröbner basis of it. From this Gröbner basis we can easily compute a system of generators of the syzygies of $(R_1, \ldots, R_u, -S_1, \ldots, -S_v)$ (see [21]).   Q. E. D.

Example: Let $I := (F_1, F_2, F_3) \subseteq k[X, Y]$ be generated by:

$$F_1 := (X - Y)^2(X - 1)^3, \quad F_2 := (X - Y)^2(X - 1)Y, \quad F_3 := (X - Y)^3Y^4$$

We can take $G_2 := (X - Y)^3Y^4$, so $P_2 := Y^2 - XY$. It is clear that $\phi((I, P_2)/(P_2))$ is a $k[X]$ submodule of $k[X]^2$. $\phi(1) := (1, 0) =: e_1$, $\phi([Y]) := (0, 1) =: e_2$. We get:

$$\phi(F_1) = X(X - 1)^3[Xe_1 - e_2]$$

while

$$\phi(F_2) = \phi(F_3) = \phi(YF_1) = \phi(YF_2) = \phi(YF_3) = 0.$$

Now we apply the elementary divisors theorem and hence we change the basis of $k[X]^2$:

$$\begin{cases} b_1 = Xe_1 - e_2 \\ b_2 = e_1 \end{cases}; \quad \begin{cases} e_1 = b_2 \\ e_2 = -b_1 + X \ b_2 \end{cases}$$

In this new basis $\phi\left((I, P_2)/(P_2)\right) = \langle -X(X-1)^3 \mathbf{b}_1 \rangle$, so $\phi^{-1}(\mathbf{b}_1) = X - Y$. Therefore:

$$\sqrt{I} = \sqrt{(I, Y^2 - XY, X(X-1)^3)} \cap (Y^2 - XY, X - Y)$$

therefore, computing $\sqrt{(I, Y^2 - XY, X(X-1)^3)}$ with lemma 4.2, we get:

$$\sqrt{(I, Y^2 - XY, X(X-1)^3)} = (X, Y) \cap (X-1, Y) \cap (X-1, Y-1),$$

hence, deleting the embedded components, we have:

$$\sqrt{I} = (X-1, Y) \cap (X-Y).$$

Appendix

In this appendix we shall briefly sketch a way in which it is possible to compute bounds for the degrees of the polynomials of the Gröbner basis of a module.

Let $k[X_1, \ldots, X_n]$ be the polynomial ring in n variables and let $r \in \mathbf{N}$.

Notation: if $f \in k[X_1, \ldots, X_n]^r$, $f = (f_0, \ldots, f_{r-1})$, we set:

$$f^{[Y]} := f_0 + f_1 Y + \cdots + f_{r-1} Y^{r-1} \in k[X_1, \ldots, X_n, Y]$$

where Y is a new variable.

Let $M \subseteq k[X_1, \ldots, X_n]^r$ be a module given by a finite set of generators $v_1, \ldots, v_s$. If $\mathbf{e}_0, \ldots, \mathbf{e}_{r-1}$ is the canonical basis of $k[X_1, \ldots, X_n]^r$, we define

$$T_r := \{\mathbf{e}_i \phi \ : \ \phi \text{ is a monomial in } k[X_1, \ldots, X_n], \ i = 0, \ldots, r-1\}.$$

Any term-order $<$ on $k[X_1, \ldots, X_n, Y]$ induces a term-order $\prec$ on T_r as follows:

$$\phi \mathbf{e}_i \prec \psi \mathbf{e}_j \iff Y^i \phi < Y^j \psi.$$

In this way we get most of the term-orders of T_r (in particular we get the term-order defined in section 3).

Let $\prec$ be a term-order on T_r induced by $<$ on $k[X_1, \ldots, X_n, Y]$, as above. We can compute a Gröbner basis of M w. r. t. $\prec$ as follows:

First consider the ideal in $k[X_1, \ldots, X_n, Y]$ generated by $v_1^{[Y]}, \ldots, v_s^{[Y]}$. Then compute a Gröbner basis of this ideal w.r.t. the term-order $<$ but skip all the computations that require a multiplication by a power of Y. In this way we get a set G of polynomials which are contained in the set of all the polynomials obtained during the computation of the Gröbner basis of the ideal $(v_1^{[Y]}, \ldots, v_s^{[Y]})$ with respect to $<$. In particular the algorithm stops and when we lift the elements of G to elements of $k[X_1, \ldots, X_n]^r$, we get a Gröbner basis for the module M w.r.t. $\prec$. If Δ is any bound for the degrees of all the polynomials obtained in the computation of a Gröbner basis of the ideal $(v_1^{[Y]}, \ldots, v_s^{[Y]})$ w.r.t. $<$, then Δ is also a bound for the degrees of all the elements of G and hence for all the polynomials in the Gröbner basis of M. From the results of [19] we can get bounds Δ for all the polynomials which are involved in the computation of a Gröbner basis of an ideal. A value for Δ that we can deduce from [19] (after a rough aproximation) is doubly exponential in n.

Acknowledgments: The authors want to thank Joos Heintz, Teo Mora and Pablo Solernó for many helpful conversations on the subject.

REFERENCES

[1] B. Buchberger, G. Collins, R. Loos (eds.), "Computer Algebra: symbolic and algebraic computation," Springer Verlag, Wien; New York, 1983.

[2] L. Caniglia, A. Galligo, J. Heintz, *Some new effectivity bounds in computational geometry*, in "Proceedings of A.A.E.C.C.-6," Rome 1988, Springer L.N.C.S. **357**, 1988, pp. 131–151.

[3] A. Chistov, D. Grigor'ev, *Subexponential time solving systems of algebraic equations*, LOMI preprints E-9-83, E-10-83, Leningrad.

[4] A. Dickenstein, N. Fitchas, M. Giusti, C. Sessa, *The membership problem for unmixed polynomial ideals is solvable in single exponential time*, in "Proceedings AAECC-7," Toulouse, 1989.

[5] M. Frumkin, *Polynomial time algorithms in the theory of linear diophantine equations*, "Fundamentals of Computing Theory," M. Karpinski, ed., Springer L.N.C.S. **56**, 1977, pp. 386–392.

[6] J. Von Zur Gathen, *Parallel arithmetic computation: a survey*, in "Proc. 13ᵗʰ Conf. MFCS," 1986.

[7] P. Gianni, B. Trager, G. Zacharias, *Gröbner bases and primary decomposition of polynomial ideals*, in "Computational aspects of commutative algebra," L. Robbiano, (ed.), Acad. Press, 1989, pp. 15–33.

[8] M. Giusti, *Some effectivity problems in polynomial ideal theory*, in "EUROSAM 84," Springer L.N.C.S. **174**, 1984, pp. 159–171.

[9] J. Heintz, *Definability and fast quantifier elimination in algebraically closed fields*, Theoret. Comput. Sci. **24** (1983), 239–277.

[10] G. Hermann, *Die Frage der endlich vielen Schritte in der Theorie der Polynomideale*, Math. Ann. **95** (1926), 736–788.

[11] E. Kaltofen, M.S. Krishnamoorthy, B. Saunders, *Fast parallel computation of Hermite and Smith forms of polynomial matrices*, Siam. J. Alg. Disc. Meth. **8** (1987).

[12] E. Kaltofen, B. Saunders, *Parallel algorithms for matrix normal forms*, preprint, 1988.

[13] R. Kannan, *Solving systems of linear equations over polynomials*, Theoretical Computer Science **39** (1985), 69–88.

[14] R. Kannan, A. Bachem, *Polynomial algorithms for computing the Smith and Hermite normal forms of an integer matrix*, Siam. J. Comp **8** (1981), 499–507.

[15] D. Lazard, *Algèbre linéaire sur $K[X_1, \ldots, X_n]$ et élimination*, Bull. Soc. Math. France **105** (1977), 165–190.

[16] D. Lazard, *Gröbner bases, Gaussian elimination and resolution of systems of algebraic equations*, in "EUROCAL 83," Springer L.N.C.S. **162**, 1983, pp. 146–156.

[17] A. Logar, *A computational proof of the Noether Normalization lemma*, in "Proceedings AAECC-6," Rome, Springer L.N.C.S. **357**, 1988, pp. 259–273.

[18] E. Mayer, A. Meyer, *The complexity of the word problems for commutative semigroups and polynomials ideals*, Adv. in Math. **46** (1982), 305–329.

[19] M. Möller, T. Mora, *Upper and lower bounds for the degree of Gröbner bases*, in "EUROSAM 84," Springer L.N.C.S. **174**, 1984, pp. 172–183.

[20] A. Seidenberg, *Constructions in algebra*, Trans. Amer. Math. Soc. **197** (1974), 273–313.

[21] G. Zacharias, *Generalized Gröbner bases in commutative polynomial rings*, Bachelor Thesis, M.I.T., August 1978.

Teresa Krick
Instituto Argentino de Matemática
Consejo Nacional de Investigationes
Cientificas y Técnicas
Viamonte 1636, 1055 Buenos Aires
ARGENTINA

Alessandro Logar
Università degli Studi di Trieste
Dipartimento di Scienze Matematiche
Piazzale Europa, 1
34121 Trieste
ITALY

On the Complexity of Zero–dimensional Algebraic Systems

LAKSHMAN Y. N.*   DANIEL LAZARD[†]

Abstract. A probabilistic algorithm is given which, a zero-dimensional system of polynomials being given, computes Gröbner base for any ordering of its radical and/or all of its irreducible components in time $d^{O(n)}$, where d is the maximal degree of the polynomials and n the number of variables. With probability nearly 1, no component is lost.

This algorithm can decide zero-dimensionality with the same complexity and the same probability of success.

These complexities remain valid even if the system is not zero-dimensional at infinity.

This algorithm is a pratical one; it is probably slower than the computation of the Gröbner base for a degree ordering but faster than the same computation with a variable more.

1   Introduction.

In this paper, we consider a set of polynomials f_1, ..., f_k in n variables over a field K: $f_i \in K[y_1, \ldots, y_n]$. Let d be the maximum of the degrees of the f_i. We are concerned with the complexity of finding the set of common zeros of the f_i in an algebraic closure of K. This field K is generally the field of rationals, but may be any field on which the standard operations may be done in polynomial time, including matrix inversion and GCD of polynomials.

The story of this problem of complexity began with Lazard (1981) where it is proved that the problem has a complexity of $d^{O(n)}$ if the system is of projective dimension 0. Grigoriev and Chistov (1983) have proved that this complexity is $d^{O(n^2)}$ for any system. Both results are not complete in the sense that, for ending the resolution, one has to define the form in

---

*Work supported by N.S.F. Grants CCR-87-05363 and CDA-88-05910. This work was made while the first author was at Rensslaer Polytechnic Institute, Troy, N. Y.

[†]CNRS, LITP and GRECO de Calcul Formel

which the result is wanted and to prove that this form may be obtained
with the same complexity. Several papers in the references deal with this
aspect and will not be referred otherwise. In this paper, the output will be
a Gröbner base (for any admissible ordering) of the radical and/or all of all
the irreducible components over the field of the coefficients.

The aim of this paper is to study the complexity of this resolution when
the system is zero-dimensional (i.e. has only a finite number of solution in
an algebraic closed field) but without the restriction of being projective
zero-dimensional, i. e. the homogeneous polynomials associated with the
f_i may have infinitely many projective zeros at infinity. One of the main
tools of this study consists of a deformation which was used by Grigoriev
and Chistov (1983), Canny (1989) and probably other people.

We prove that a zero-dimensional system may be solved in time $d^{O(n)}$.
The algorithm which is described does not prove directly that the input
system is zero-dimensional. Moreover it depends on random choices which
are "good" with probability nearly 1, but we do not know how to prove
that a choice is good in time $d^{O(n)}$. However, by running the algorithm
twice, the bad choices are detected: they lead to fewer components.

The zero-dimensionality may be decided by running the algorithm with
a random linear polynomial added to the input system: In this case we get
an empty set of solutions iff the choices are bad or if the input system is
zero-dimensional.

The precise result is the following:

Theorem 1 *There exists a probabilistic algorithm which, given a zero-
dimensional system, computes the Gröbner base (for any admissible ordering) of its radical and/or all of all its associated primes over the field of
the coefficients, in time $d^{O(n)}$. With probability greater than $1 - 1/2^{d^n}$, all
irreducible components are found.*

*This algorithm can decide of zero-dimensionality at the same cost with
the same probability of success.*

*If the test of zero-dimensionality has a deterministic cost of $d^{O(n)}$, then
there exists an algorithm which outputs all solutions, the cost of which being
$d^{O(n)}$ with probability $1/2^{d^n}$; in particular, the average cost is $d^{O(n)}$.*

The same result of complexity appears already in Lakshman (1990a),
with a less practical algorithm.

This probalistic result of complexity has two interestings consequences:

The first one is that this complexity bound gives indications on the per-
formances we may hope for a practical algorithm. Moreover, the algorithm
given below is clearly a practical one. Some experiments are necessary for
comparing it with standard algorithms. We guess that it is, in general,
slower than Buchberger's one with a total degree ordering, but faster than
the same algorithm applied to problems with one variable more.

The second consequence is that the algorithm implies bounds on the size of the result, which may be needed for computing the complexity of a deterministic algorithm.

It should also be noticed that this algorithm is a set theoretical one, this means that the algebraic structure of the system around a solution is not computed. Lakshman (1990b) provides an algorithm for deducing this algebraic structure (i.e. a Gröbner base of the initial ideal) with the same complexity.

2   The algorithm.

The algorithm we present now works as a succession of steps. These steps are rather independent algorithms which are more or less described elsewhere. For this reason, we describe them in separate subsections.

2.1   Change of polynomials.

Step 1 *Let n be the number of variables and $f_1, \ldots, f_k$ be the input polynomials, sorted by decreasing degrees, with coefficients in K, which have a zero-dimensional zero-set (in an algebraic closed field).*

If $k < n$ the zero-set is empty or not zero-dimensional; the algorithm has been extended to this case, because it may decide between these properties.

If $k \leq n$, let $g_i = f_i$ for $i \leq k$ and $g_i = 0$ for $k \leq n$.

If $k > n$, let $g_i = f_i + a_{i,i+1}f_{i+1} + \cdots + a_{i,k}f_k$ for $i = 1, \ldots, n$, where the $a_{i,j}$ are random integers such that the zero-set of $(g_1, \ldots, g_n)$ is zero-dimensional.

The fact that almost all choices for the $a_{i,j}$ satisfy the wanted condition is classical; see, for example, Kaplanski (1981) or Grigoriev and Chistov (1983). If the $a_{i,j}$ are choosen one after the other, then each of them has to avoid at most d^i values where d is the maximal degree of input polynomials. In fact, if $\mathcal{P}_1, \ldots, \mathcal{P}_l$ are the minimal primes of height $i-1$ containing $g_1, \ldots, g_{i-1}$, we have to choose $a_{i,j}$ such that $f_i + a_{i,i+1}f_{i+1} + \cdots + a_{i,j}f_j$ is not contained in any of the $\mathcal{P}_\alpha$ which do not contain $f_i, \ldots, f_j$. Thus, for each $\mathcal{P}_\alpha$, the integer $a_{i,j}$ has to avoid the unique solution of a linear equation in the field of fractions of $K[X_1, \ldots, X_n]/\mathcal{P}_\alpha$. It follows that l bounds the number of bad $a_{i,j}$ and is itself bounded by d^{i-1}, by Bezout theorem.

If the $a_{i,j}$ are random integers less than $2^{d^{n+1}}$, their size is d^{n+1}; thus the computations with such integers remain polynomial in d^n (with a bigger degree). On the other hand, the probability of a good choice is greater than

$$(1 - \frac{d^n}{2^{d^{n+1}}})^{n^2} > 1 - \frac{n^2 d^n}{2^{d^{n+1}}} > 1 - \frac{1}{2^{d^n}}$$

2.2   Deformation and homogeneization.

Step 2 *Let $h'_i = g_i + sx_i^{d_i}$ where s is a new indeterminate and d_i is the degree of g_i. Let h_i be the homogeneous polynomial associated with h'_i by adding a new variable x_0.*

This is Canny's (1989) idea: the h_i, when viewed as polynomials with coefficients rational in s have no common zeros at infinity and only a finite number of common zeros. Moreover, the common zeros of the g_i are the limit (when $s \rightarrow 0$) of the common zeros of the h_i which have a finite limit.

Thus, in the next steps, we will compute the zeros of the h_i in a convenient form, take the limit, and remove the zeros which are not zeros of all f_i.

2.3   U-resultant.

For this step, we need Lazard's (1981) algorithm. We give as an appendix a simpler version of it.

When applied to the h_i, it computes square, polynomial in s, matrices $M_1, \ldots, M_n$ such that, if $U_0, \ldots, U_n$ are new indeterminates, the determinant R of $U_0 M_0 + \cdots + U_n M_n$ is Macaulay's U-resultant of the h_i, up to the product by an element of $K(s)$.

Canny (1988) has proved that the coefficient of the part of R of least degree in s is a product of linear (in U) factors over an algebraic closure of $K(s)$. The factors $\alpha_0 U_0 + \cdots + \alpha_n U_n$ with $\alpha_0 \neq 0$ correspond to the zeros of $(g_1, \ldots, g_n)$ by the map

$$(\alpha_0, \ldots, \alpha_n) \rightarrow (\alpha_1/\alpha_0, \ldots, \alpha_n/\alpha_0)$$

The direct computation of R is too costly. Thus, we will compute only specializations of it, in the subsequent steps. The present step, which uses Gröbner base computations and linear algebra on big matrices over $K(s)$ is probably the most expensive one of the algorithm presented here.

Step 3 *Apply to the h_i the algorithm in the appendix, in order to compute the matrices $M_0, \ldots, M_n$.*

2.4   Change of variable.

Step 4 *Choose random integers $b_1, \ldots, b_n, c_2, \ldots, c_n$ and set $N_0 := M_0$, $N_1 := M_1 + b_2 M_2 + \cdots + b_n M_n + b_1 M_0$ and $N_i := M_i + c_i M_0$, for $i \geq 2$.*

The matrix M_i is the one of multiplication by x_i in some space. Thus, above change of matrices correspond to following change of variables in the initial set of polynomials: $y_1 = x_1 + b_2 x_2 + \cdots + b_n x_n + b_1$ and $y_i = x_i + c_i$ for $i \geq 2$.

Let $(\alpha_{1,j}, \ldots, \alpha_{n,j})$, for $j = 1, \ldots, k_1$ be the common zeros of the g_i on these new coordinates, and let $(\alpha_{1,j}, \ldots, \alpha_{n,j})$ for $j = k_1 + 1, \ldots, k_2$ be the limit directions of those of the common zeros of the h_i which tends to infinity. By Bezout theorem, we have $k_2 \le d^m$.

The above change of variable has to satisfy, for each $i \ge 2$, the following condition:

For $j \le k_2$ and $j_1, j_2 \le k_1$, we have $\alpha_{1,j} \ne 0$ and the relation $\alpha_{i,j_2}/\alpha_{1,j_1} = \alpha_{i,j}/\alpha_{1,j}$ implies $j = j_1 = j_2$.

Clearly, this condition implies that the $\alpha_{1,j}$ are all different for $j \le k_1$: it suffices to take $j_2 = j$.

To realize the condition $\alpha_{1,j} \ne 0$, it suffices to choose $b_1, b_2, \ldots b_n$ outside of a set of k_2 hyperplanes, whose equations have the $\alpha_{i,j}$ as coefficients. When the b_i are choosen, each c_i has to avoid less than k_2 values for satisfying the last condition.

Remark 1 Unlike step 1, this step may be made deterministic without changing the nature of the complexity.

In fact, the validity of the change of variables is automatically tested in next steps, first for the choice of the b_i, then, *independently* for each c_i. Using this independence and the fact that k points in a space of dimension n may define more than k/n distinct hyperplanes, it is easy to insure that a "good" change of variables may be found in less than $k_2^2 + nk_2^3 \le d^{4n}$ trials (if $d > 1$).

2.5 Specializing U-resultant.

Step 5 *For $i = 1$ to n, let $s^{k_i} p_i(y_i)$ be the part of least degree in s of the determinant of $y_i N_0 - N_i$ (we choose k_i in order that p_i is independant of s).*

For $i = 2$ to n, let $s^{l_i} r_i(y_1, y_i)$ be the part of least degree in s of the determinant of $y_i N_1 - y_1 N_i$.

Tests: If the k_i are not all equal, then all the isolated common zeros of the g_i are at infinity on some hyperplane $x_i = 0$.

If $k_1 < k_i$ or $l_i \ne k_1$ or y_1 divides r_i for some i, then the choice of the b_i was a bad one (condition $\alpha_{1,j} \ne 0$ not satisfyed).

If p_1 is a (not 0) constant, then the initial system has an empty zeroset or (unprobably) does not have any zero-dimensional component. In this case, the algorithm stops here.

These specialisations of the U-resultant are choosen in order that the roots of p_i are the $\alpha_{i,j}$ and the roots of $r_i(1, y_i)$ are the $\alpha_{i,j}/\alpha_{1,j}$. The meaning of the tests follows easily from this interpretation and Canny's theorem.

Remark 2 The p_i and the r_i are deduced from the characteristic polynomials of the matrices $N_i N_0^{-1}$ and $N_i N_1^{-1}$ with almost no computations.

Moreover, the minimal polynomials of these matrices may be used instead of the characteristic polynomials.

This may be useful for improving the efficiency of an implementation. However, most of the above tests lose their significance.

2.6   Gröbner base of the radical of the g_i.

Step 6 *For* $i = 1$ *to* n, *let* p'_i *be the square-free kernel of* p_i, *i.e. the quotient of* p_i *by its* GCD *with its derivative with respect to* y_i.

Let $q_1 := p'_1$ *(this is a polynomial in* y_1*).*

For $i = 2$ *to* n, *let* $y_i - q_i(y_1)$ *be the monic GCD of* p'_i *and* r_i *viewed as polynomials in* y_i *with coefficients defined modulo* q_1. *This may be computed by reducing by* q_1 *the polynomial subresultant of degree 1 of* p'_i *and* r_i *viewed as polynomials in* y_i, *and inverting modulo* q_1 *the leading coefficient of the result.*

$(q_1, y_2 - q_2, \ldots, y_n - q_n)$ *is a Gröbner base of the radical of the ideal generated by the polynomials obtained from the* g_i *by above change of variables.*

Test: If the subresultant reduces to 0 *or if its leading term has a not trivial* GCD *with* q_1, *then the choice of* c_i *was a bad one.*

The linear factors of p'_i are all distincts (by square-free computation) and of the form $y_i - \alpha_{i,j}$. Those of r_i are of the form $\alpha_{i,j}y_1 - \alpha_{1,j}y_i$. Thus the GCD over $K(\alpha_{1,j})$ contains $y_i - \alpha_{i,j}$. The fact that it does not have another factor is exactly the condition of preceding section. If it would have another factor, the leading coefficient of the subresultant would vanish modulo some factor of q_1, and this is exactly what is tested.

The fact that we obtain a Gröbner base of the radical follows from the well known fact that such a base has the form we have obtained, under some conditions of genericity which are implied by our conditions on the change of variables.

2.7   Coming back to the input polynomials.

Step 7 *If the number* k *of initial polynomials is greater than the number* n *of variables, then for* $j = 1, \ldots, k$, *let* f'_j *be the polynomial obtained from* f_j *by substituting* $q_i - c_i$ *to* x_i *for* $i > 1$ *and* $y_1 - b_2 q_2 - \cdots - b_n q_n - b_1 - c_2 - \cdots - c_n$ *(this is the change of variables of step 4 followed by the substitution of* q_i *to* y_i *for* $i > 1$*).*

For $j = 1, \ldots, k$ *replace* q_1 *by its* GCD *with* f'_j *and* q_i, *for* $i > 1$, *by its remainder by the new* q_1.

If q_1 *is a constant, then the initial system has no solution. Stop here.*

If ones wants the irreducible primes of the radical, then factorize q_1 *and, for each factor* $q_{1,j}$ *produce the base* $(q_{1,j}, y_2 - q_{2,j}, \ldots, y_n - q_{n,j})$ *where the* $q_{i,j}$ *are the remainders of the* q_i *by the* $q_{1,j}$.

Apply a variant of FGLM algorithm (Faugère et al., (1989) for finding Gröbner bases of the radical of initial ideal and its primes for any admissible ordering.

The beginning of this step consists of only removing the zeros of the g_j which are not zeros of all f_j.

The variant of FGLM algorithm which is necessary differs from the original one only by the fact that the monomials in the x_i are evaluated in the y_i before being reduced by the old base. See Lakshman (1990a) for details.

3 Appendix

In this appendix, we give a short description of Lazard (1981) algorithm in which we omit the recursive row reduction which allows a deterministic decision of the zero-dimensionality: this reduction is rather complicated and is not needed here where we know the zero-dimensionality from the deformation.

Procedure 1 Input: $f_1, \ldots, f_k$: *homogeneous polynomials in $x_0, \ldots, x_n$ over a field K, sorted by decreasing degree.*

Output: *"not zero dimensional" or "no solution" or $(M_0, \ldots, M_n)$ a list of $n+1$ squares matrices such that, if $U_0, \ldots, U_n$ are new indeterminates, then the determinant of $U_0 M_0 + \cdots + U_n M_n$ is a product of linear factors over the algebraic closure of K, which correspond bijectively with the projective zero-set of the f_i by the map $\alpha_0 U_0 + \cdots \alpha_n U_n \to (\alpha_0, \ldots, \alpha_n)$, which is compatible with multiplicities.*

Begin

> *If $k < n$ then return "not zero-dimensional" ;*
>
> *for $i = 1$ to n do $d_i := \text{degree}(f_i)$;*
>
> *if $k = n - 1$ then $d_n := 1$;*
>
> *$D := d_0 + \cdots + d_n - 1$;*
>
> *compute the elements of degree atmost D of a reduced Gröbner base G of $(f_0, \ldots, f_k)$ (for any degree ordering) ;*
>
> *let $m_1, \ldots, m_C$ (resp. $n_1, \ldots, n_R$) be the monomials of degree $D - 1$ (resp. D) which are irreducible by G ;*
>
> *if $r = 0$ then return "no solution" ;*
>
> *for $i = 0$ to n, let M_i be the $R \times C$ matrix in which the entry in the r-th row and the c-th column is the coefficient of n_r in the normalform of $x_i m_c$ with respect to G ;*

$i_0 := 0$;

for $i = 1$ to n do

> *by operations on columns $c_{i-1} + 1, \ldots, C$ on all M_i together, transform these matrices in order that the submatrix of columns $c_{i-1} + 1, \ldots, c_i$ of M_i has rank $c_i - c_{i-1}$ and the columns $c_i + 1, \ldots, C$ of M_i contain only 0 ;*

if $c_n \neq r$ then return "not zero-dimensional" ;

remove the $C - c_n$ last columns of $M_0, \ldots, M_n$;

if $a_0 M_0 + \cdots + a_n M_n$ has rank R for some (random) integers, then zero-dimensionality is proved ;

if a deterministic decision on zero-dimensionality is needed, then use the row reduction of Lazard (1981) ;

return $(M_0, \ldots, M_n)$;

end.

Let A be the graduate ring $K[X_0, \ldots, X_n]/(f_0, \ldots, f_k)$ and A_d be its homogeneous part of degree d. Let $Y = a_0 X_0 + \cdots + a_n X_n$ be a generic element (i.e. sufficiently general) of degree 1 in A. Let us denote by x_i (resp. y) the multiplication by the image of X_i in A.

The $R \times C$ matrices M_i are the matrices of $x_i : A_{D-1} \rightarrow A_D$ on the bases of irreducibles monomials. The column reduction consists in a change of base in A_{D-1} in order that the last elements of the new base generate the intersection of the kernels in A_{D-1} of all x_i. Let us denote this intersection by N.

It is proved in appendices 1 and 2 of Lazard (1981) that, if y is surjective on some A_d then $(f_0, \ldots, f_k)$ is projective zero-dimensional and, conversely, that if this ideal is zero-dimensional then $y : A_{D-1} \rightarrow A_D$ is surjective, $y : A_D \rightarrow A_{D+1}$ is bijective and the dimension of A_D is the number of common zeros of the f_i in the projective space over an algebraic closure of K.

It follows that, if the ideal is zero-dimensional, N is the kernel in A_{D-1} of y, because, if $y(v) = 0$ for v in A_{D-1}, then $x_i(y(v)) = y(x_i(v)) = 0$ and $x_i(v) = 0$, by the bijectivity of y. Thus the rank of $y : A_{D-1} \rightarrow A_D$ is R and N is of dimension $C - R$.

We refer to Lazard (1981) for the proof of the property of the factorization of determinant $(U_0 M_0 + \cdots + U_n M_n)$ asserted in the algorithm. The fact that, when $k = n - 1$, this determinant is Macaulay U-resultant (modulo the product by a constant in K) follows from its description as a product of linear factors. It may also be deduced from the interpretation of the Gröbner base computation and of the the the definition of the matrix $U_0 M_0 + \cdots + U_n M_n$ as a reduction of the generalized Sylvester matrix from which the U-resultant is computed (see Lazard, 1983). In fact, if the linear polynomial depending on the U_i is the last one, the Macaulay resultant is

REFERENCES

[1] Canny, J., *Generalized characteristic polynomials*, in "Symbolic and algebraic computation, proc. ISSAC'88," Lect. Notes in Comp. Sci. **358**, Springer Verlag, 1989, pp. 293–299.

[2] Faugère, J.C., Gianni, P., Lazard, D., Mora, T., *Efficient Computation of Zero-dimensional Gröbner Bases by Change of Ordering*, (submitted), J. Symb. Comp. (1989). Technical Report LITP 89-52

[3] Gianni, P. and Mora T., *Algebraic Solution of Systems of Polynomial Equations using Gröbner Bases*, Preprint, 1988.

[4] Grigoriev, D., Chistov,, *Subexponential-time Solving Systems of Algebraic Equations*, Preprints LOMI E-9-83 and E-10-83, Leningrad 1983.

[5] Ierardi, D., *The complexity of quantifier elimination in the theory of algebraically closed fields*, Ph.D. Thesis, Cornell University, 1989.

[6] Kaplanski, I., "Commutative rings," Allyn and Bacon, Boston, 1981.

[7] Kobayashi, H., Fujise, T., Furukawa, A., *Solving Systems of Algebraic Equations by General Elimination Method*, J. Symb. Comp. **5** (1988), 303–320.

[8] Kobayashi, H., Moritsugu, S., Hogan, R.W., *On Solving Systems of Algebraic Equations*, in "Proc. ISSAC 88," 1988.

[9] Lakshman, Y. N., *On the complexity of computing a Gröbner basis for the radical of a zero dimensional ideal*, in "Proc. STOCS'90," 1990.

[10] Lakshman, Y. N., *A single exponential bound on the complexity of computing Gröbner bases of zero dimensional ideals*, (in these proceedings).

[11] Lazard, D., *Résolution des Systèmes d'Equations Algébriques*, Theor. Comp. Sci. **15** (1981), 77–110.

[12] Lazard, D., *Gröbner Bases, Gaussian Elimination and Resolution of Systems of Algebraic Equations*, in "Proc. EUROCAL 83," Lect. Notes in Comp. Sci. **162**, Springer, 1983, pp. 146–157.

[13] Lazard, D., *Solving zero-dimensional algebraic systems*, (submitted), J. Symb. Comp. (1989). Technical Report LITP 89-48.

[14] Macaulay, F.S., "The Algebraic Theory of Modular Systems," Cambridge Tracts in Math. **19**, Cambridge Univ. Press, 1916. Stechert-Hafner (1964)

Lakshman Y. N.
Deptartment of Computer and Information Sciences
University of Delaware, 103, Smith Hall
Newakk, DE 19716, U. S. A.
lakshman@math.udel.edu

Daniel Lazard
Informatique, Université Paris VI
4, place Jussieu,
75252 Paris Cedex 05, France
dl@frunip11.bitnet or lazard@litp.ibp.fr

A Single Exponential Bound
on the Complexity of Computing Gröbner Bases
of Zero Dimensional Ideals

LAKSHMAN Y. N.

Introduction

Let $\mathcal{R} = \mathbf{Q}[x_1, x_2, \ldots, x_n]$ denote the ring of polynomials in n variables over the rational numbers $\mathbf{Q}$. Let $f_1, f_2, \ldots, f_r \in \mathcal{R}$, $r \geq n$ with $\deg(f_i) = d_i$ and let $d = \max(d_i)$. Let $\mathcal{I} = (f_1, f_2, \ldots, f_r)$ be the ideal generated by f_i. $\mathcal{I}$ is assumed to be zero–dimensional. Let $\mathcal{Q}_i$, $i = 1, \ldots, k$ be primary ideals with associated primes $\mathcal{P}_i$ such that $\mathcal{I} = \bigcap_{i=1}^{k} \mathcal{Q}_i$.

We describe an algorithm to compute a reduced Gröbner basis for $\mathcal{I}$ under any admissible term ordering. We assume that we are given reduced Gröbner bases for each of the $\mathcal{P}_i$ under some admissible term ordering. We establish an upper bound on the number of arithmetic operations performed by the algorithm that is polynomially bounded by the dimension of the residue class ring $\mathcal{R}/\mathcal{I}$ seen as a rational vector space. This, combined with the upper bounds established in [Lak, LL] on the complexity of constructing reduced Gröbner bases for $\mathcal{P}_i$, implies a bound that is polynomial in d^n on the complexity of constructing reduced Gröbner bases for zero-dimensional polynomial ideals. Previously known bounds are not as strong ($r^3 d^{O(n^3)}$, see [CGH]) and are obtained using entirely different methods. Recent related work can be found in [KL] and [DS].

Our algorithm for constructing a Gröbner basis for $\mathcal{I}$ works in two stages. In the first stage, the algorithm constructs Gröbner bases for each $\mathcal{Q}_i$ separately from the given basis for $\mathcal{I}$ and a Gröbner basis for $\mathcal{P}_i$. In the second stage, the Gröbner bases of $\mathcal{Q}_i$ are put together to obtain a Gröbner basis for $\mathcal{I}$.

This material is based on work supported by the National Science Foundation under Grant No. CCR-87-05363 and under Grant No. CDA-8805910 at Rensslaer Polytechnic Institute, Troy, N. Y. as part of the author's doctoral dissertation.

From Primes to Primaries

Our construction is based on the following well known theorem (see [vdW]).

Theorem 1. *If the ideal $\mathcal{I}$ has an isolated primary component Q whose associated prime $\mathcal{P}$ is maximal, then, if ρ is the index of Q, then, for any $\sigma \geq \rho$,*

$$Q = (\mathcal{I}, \mathcal{P}^\sigma).$$

An algorithm for primary decomposition based on this theorem has been reported by Kredel [Kre]. However, no analysis of the algorithm is provided there.

Let Q be a zero–dimensional primary ideal in *general position* with associated prime $\mathcal{P}$. Let $\{q_1, q_2, \ldots, q_s\}$ be a basis for Q with $\varrho = \max\{\deg(q_i)\}$. Since Q is in *general position*, the reduced Gröbner basis for $\mathcal{P}$ under the purely lexicographic ordering $>_l$ with $x_1 < x_2 < \ldots < x_n$ looks like

$$x_n - g_n(x_1), \ldots, x_2 - g_2(x_1), g_1(x_1)$$

[GTZ]. Let $\deg(g_1(x_1)) = \delta$. We now state the algorithm for computing a reuced Gröbner basis for Q.

Algorithm "Primary Component":

Input: A basis for $\mathcal{I}$ and a reduced Gröbner basis for $\mathcal{P}$, a prime ideal containing $\mathcal{I}$.

Output: A reduced Gröbner basis for Q, the primary component of $\mathcal{I}$ whose associated prime is $\mathcal{P}$.

$$\mathcal{C} := \mathcal{P};$$
$$\mathcal{B} := \mathcal{GB}(\mathcal{I}, \mathcal{P}^2);$$
$$\texttt{while}\ \ \mathcal{C} \neq \mathcal{B}\ \ \texttt{do}\ \ \ \mathcal{C} := \mathcal{B};\ \ \ \mathcal{B} := \mathcal{GB}(\mathcal{I}, \mathcal{P} \cdot \mathcal{B})\ \ \ \texttt{od}$$
$$\texttt{return}\ (\mathcal{B});$$

The correctness of the algorithm follows from theorem 1 and the observation that

$$(\mathcal{I}, \mathcal{P}^{j+1}) = (\mathcal{I}, \mathcal{P} \cdot \mathcal{I}, \mathcal{P}^{j+1}) = (\mathcal{I}, \mathcal{P} \cdot (\mathcal{I}, \mathcal{P}^j)).$$

This observation is crucial for the following analysis.

Procedure $\mathcal{GB}(\mathcal{A}_1, \mathcal{A}_2)$ is very similar to Buchberger's algorithm for finding the reduced Gröbner basis for $(\mathcal{A}_1, \mathcal{A}_2)$ except that the polynomials of the basis $\mathcal{A}_1 \cup \mathcal{A}_2$ are first mutually reduced as much as possible (bounded Gröbner basis reduction). Following lemmas lead to upper bounds on the time complexity of *Primary component*.

Lemma 1. *The sequence* $(\mathcal{I}, \mathcal{P}^i), i = 0, 1, \ldots$ *is strictly increasing until* $i = \rho$, *the index of the primary component of* $\mathcal{I}$ *whose associated prime is* $\mathcal{P}$. *Q. E. D.*

Lemma 2. *If* $\widehat{Q}$ *is any primary ideal whose associated prime is* $\mathcal{P}$, *then all the lead terms in the reduced Gröbner basis of* $\widehat{Q}$ *under* $>_l$ *are of the form*

$$z^{i_1} x_2^{i_2} \ldots x_n^{i_n}, \quad i_j \geq 0$$

where $z = x_1^\delta$.

Proof: Let f be the polynomial with the smallest lead term among the polynomials in the reduced Gröbner basis of $\widehat{Q}$ under $>_l$ whose lead terms are not of the form $z^{i_1} x_2^{i_2} \ldots x_n^{i_n}$, $i_j \geq 0$. Let

$$f = x_1^\gamma m + m f' + f''$$

where m is a monomial not depending on x_1, γ mod $\delta \neq 0$, f' depends on only x_1, f'' is not divisible by m and only has terms smaller than $x_1^\gamma m$. The univariate polynomial in x_1 in the Gröbner basis of $\widehat{Q}$ (call it g) has to be a pure power of g_1. Let $g = x_1^\Gamma + f_1$ where Γ is some multiple of δ and f_1 has lower terms. Consider the s-polynomial of f and g i.e.,

$$x_1^{\Gamma - \gamma} f - mg = x_1^{\Gamma - \gamma}(mf' + f'') - m f_1$$
$$= m(x_1^{\Gamma - \gamma} f' - f_1) + x_1^{\Gamma - \gamma} f''.$$

Because of the choice of f, the only polynomial that can be used to reduce $m(x_1^{\Gamma - \gamma} f' - f_1)$ is f (if it is not zero already). Complete reduction of the terms divisible by m leads us to the equation

$$mp(x^\gamma + f') = mg$$

where p is a univariate polynomial in x_1. Follows that $x^\gamma + f'$ divides g. But g is a power of the irreducible polynomial g_1. Therefore $x^\gamma + f'$ is itself some power of g_1 or a constant which implies that $\gamma = 0$ mod (δ) contradicting the original assumption. *Q. E. D.*

Lemma 3. *The dimension of the vector space* $\mathcal{R}/\widehat{Q}$ *(which is the same as the number of reduced monomials with respect to any Gröbner basis for* $\widehat{Q}$*) is a multiple of* δ.

Proof: Consider any reduced monomial of the form $m = z^{i_1} x_2^{i_2} \ldots x_n^{i_n}$, $i_j \geq 0$. The monomials $x_1^l m, l = 0, 1, \ldots, \delta - 1$ are also reduced by the lemma above. We can identify all these monomials with a class labelled by m. Similarly, every monomial can be identified with a unique class. Each class has exactly δ monomials. The lemma follows. *Q. E. D.*

Analysis of *Primary Component.*

By lemma 3, the number of reduced monomials with respect to any Gröbner basis for Q is $\rho\delta, \rho \geq 1$. By lemma 1, the number of reduced monomials with respect to C is increasing after each iteration of the *while-loop* except the last in algorithm *Primary Component.* Again, by lemma 3, this increase in each step is by a multiple of δ. Therefore, ρ is an upper bound on the number of iterations of the *while-loop* in *Primary Component.* This is a tight bound as the following example demonstrates. Let

$$Q = (q_1^{\delta_1}, q_2^{\delta_2} - q_1, \ldots, q_n^{\delta_n} - q_{n-1})$$

where $q_1(x_1)$ is an irreducible polynomial in x_1 of degree δ and $q_i(x_i) = x_i - g_i(x_1), i > 1$. Q is primary with associated prime $\mathcal{P} = (q_1, q_2, \ldots, q_n)$. The above basis for Q is a reduced Gröbner basis under $>_l$ and the number of reduced monomials with respect to the basis is $\delta\delta_1\delta_2\ldots\delta_n$. For any $\Delta < \delta_1\delta_2\ldots\delta_n$, $\mathcal{P}^\Delta \not\subset Q$ (clear from looking at the normal form of q_n^Δ with respect to Q) while $\mathcal{P}^{\delta_1\delta_2\ldots\delta_n} \subset Q$. Now, we can bound the main step in the loop, namely, $\mathcal{GB}(\mathcal{I}, \mathcal{P} \cdot \mathcal{B})$, which is essentially this:

Algorithm $\mathcal{GB}(\mathcal{I}, \mathcal{P} \cdot \mathcal{B})$:
1) Mutually reduce all polynomials in $\mathcal{I} \cup \mathcal{P} \cdot \mathcal{B}$ as much as possible. Let I' be the resulting set of normal forms (they need not be unique.)
2) Perform Buchberger's algorithm on I'.

At the end of each iteration of the *while-loop* in *Primary Component*, $\mathcal{B}$ is a reduced Gröbner basis for some primary ideal $\widehat{Q}$ that divides Q. Therefore, the dimension of the vector space $\mathcal{R}/\widehat{Q} \leq \rho\delta$.

Bounding the number of s-polynomial reductions.

Lemma 4. *The number of reduced monomials and the number of lead terms with respect to the basis $\mathcal{P} \cdot \mathcal{B}$ (which is not a Gröbner basis) are bounded by $(n+1)\rho\delta$ and $n^2\rho$ respectively.*

Proof: If the number of reduced monomials with respect to $\mathcal{B}$ is $k\delta$, for some $k \leq \rho$, they can be put in k classes as before and since the lead monomials can only be *simple multiples* of the class labels (i.e., every lead monomial is of the form zm or x_jm for some $j > 1$ where m is a class label, a consequence of lemma 2), the maximum number of lead terms in $\mathcal{B}$ is nk. In the basis $\mathcal{P} \cdot \mathcal{B}$, the old lead terms become reduced and the number of reduced monomials is therefore bounded by

$$\delta\rho + nk\delta \leq \delta\rho + n\rho\delta = (n+1)\rho\delta \qquad (= D, say).$$

Since $\mathcal{B}$ has atmost nr lead terms and $\mathcal{P}$ has n elements, the basis $\mathcal{P} \cdot \mathcal{B}$ has atmost $n^2k \leq n^2\rho$ lead terms. Q. E. D.

Lemma 5. *The number of s-polynomial reductions in step 2 of $\mathcal{GB}(\mathcal{I}, \mathcal{P} \cdot \mathcal{B})$ is $O((nD + r)^2)$.*

Proof: By the previous lemma, the number of lead terms in $(\mathcal{I}, \mathcal{P} \cdot \mathcal{B})$ before starting Buchberger's algorithm is bounded by $n^2\rho + r = T(\text{say})$. As usual, consider all possible pairs of lead terms – $\binom{T}{2}$ of them. Create 2 sets, *Tried* and *Untried*. Initially all the pairs are in *Untried*. The algorithm picks pairs from *Untried* and after performing the corresponding s-polynomial reduction, puts that pair in *Tried*. A "Step" is counted as all the action between two s-polynomial reductions that add a new polynomial to the basis. Each s-polynomial reduction, if not a zero reduction, decreases the number of reduced monomials by at least one. The number of reduced monomials is bounded from below by δ. Hence the number of irredundant s-polynomial reductions i.e., the number of "Steps" is no more than $D - \delta$. At the end of a "Step", we have a new polynomial g; augment the basis with it and also add (g, f) to *Untried* for every f in the current basis. We add atmost $T + D - \delta$ new pairs. The algorithm halts when *Untried* is empty. Now,

$$\max(|\textit{Untried} \cup \textit{Tried}|)$$

$$\leq \binom{T}{2} + \sum_{1}^{D-\delta}(T + i)$$

$$\leq (nD + r)^2 + (nD + r)D + D^2$$

which is a bound on the number of s-polynomial reductions performed.
Q. E. D.

Bounding the cost of a single s-polynomial reduction.

Lemma 6. *A single s-polynomial reduction can be performed using $O(D^4 + \varrho tr D^2)$ field operations.*

Proof: A monomial of degree θ can be completely reduced with respect to $\mathcal{P} \cdot \mathcal{B}$ in $O(\theta D^2)$ steps, regarding each normal form as a vector of length D. If $\mathcal{I}$ has r polynomials each bounded in degree by ϱ and having at most t terms, then the first step in $\mathcal{GB}(\mathcal{I}, \mathcal{P} \cdot \mathcal{B})$ takes $O(\varrho tr D^2)$ steps.

Each s-polynomial reduction in the second step can be regarded as a normal form reduction of a polynomial of degree less than $2D$, and having at most $2D$ terms. A complete reduction takes $O(D^4)$ steps by similar analysis as before.
Q. E. D.

Corollary. *The the number of field operations performed by the algorithm $\mathcal{GB}$ is $O((D^4 + \varrho tr D^2)(nD + r)^2)$.*

Proof: Immediate from lemmas 5 and 6.
Q. E. D.

We can compute reduced Gröbner bases for all the primary components $Q_1, \ldots, Q_k$ of $\mathcal{I}$ in this manner. By the above analysis, the cost of computing a reduced Gröbner basis for Q_i is polynomially bounded by $\dim(\mathcal{R}/Q_i)$. Noting that $\dim(\mathcal{R}/\mathcal{I}) = \sum_i \dim(\mathcal{R}/Q_i)$, we have:

Theorem 2. *Given a basis for a zero–dimensional ideal $\mathcal{I}$ and reduced Gröbner bases for all the prime ideals $\mathcal{P}_i$ containing $\mathcal{I}$, reduced Gröbner bases for all the primary ideals containing $\mathcal{I}$ can be constructed in time polynomial in the dimension of the vector space $\mathcal{R}/\mathcal{I}$.*

The next step is to put together the reduced Gröbner basis for $\mathcal{I}$ under the desired term ordering.

From Primaries to the Full Ideal

At this stage, we have computed reduced Gröbner bases for the primary ideals $Q_1, Q_2, \ldots, Q_k$ where $\mathcal{I} = Q_1 \cap Q_2 \cap \ldots \cap Q_k$. Let G_i be the Gröbner basis of Q_i. This step is based on the simple observation that *a polynomial f belongs to $\mathcal{I}$ iff it belongs to each Q_i* and using this observation in conjunction with the change of basis algorithm of Faugère et al [FGLM]. The algorithm proceeds as in the change of basis algorithm, computing normal forms of monomials in the required order with respect to each G_i separately. However, while looking for a linear relation among the normal forms, we look for one that holds simultaneously in each G_i. Precise description follows.

Let $\mathcal{N}_i(m)$ denote the normal form of monomial m with respect to G_i.
NewBasis: Gröbner basis being built.
ReducedMons: Set of monomials that are known to be reduced with respect to *NewBasis*; Initialized to $\{1\}$.
NextMonom: Function that returns the smallest monomial (under the desired admissible term ordering) that is neither in *ReducedMons* nor is a multiple of some lead term in *NewBasis*. Returns false if no such monomial exists.

```
ReducedMons := {1};
NewBasis := { };
while (m := NextMonom()) do
    If there exist m₁,...,mₛ in ReducedMons, and λⱼ ∈ Q
    such that 𝒩ᵢ(m) + ∑ⱼ₌₁ˢ λⱼ𝒩ᵢ(mⱼ) = 0,   i = 1,...,k then,
        NewBasis := NewBasis ∪{m + ∑ⱼ₌₁ˢ λⱼmⱼ}
    else
        ReducedMons := ReducedMons ∪{m};
        Save 𝒩ᵢ(m),   i = 1,...,k.
    fi
od end;
```

The total cost is bounded by the cost of finding linear relations among the normal forms. $\mathcal{N}_i(m)$ is a vector of length $\dim(\mathcal{R}/\mathcal{Q}_i)$. Therefore, the time for solving the linear systems required to find elements of *NewBasis* is polynomially bounded by $\dim(\mathcal{R}/\mathcal{I})$. Suppose d is the maximum of the degrees of the polynomials in the initial basis, then $\dim(\mathcal{R}/\mathcal{I}) \leq d^n$. Combining the results of [Lak], [LL] and the results presented in this paper so far, we have the following theorem.

Theorem 3. *Given a zero dimensional ideal*

$$\mathcal{I} = (f_1, f_2, \ldots, f_r) \subseteq \mathbb{Q}[x_1, x_2, \ldots, x_n]$$

the reduced Gröbner basis of $\mathcal{I}$ under any admissible term ordering can be computed in time polynomial in nd^n where $d = \max(\deg(f_i))$.

Remark 1: It was helpful for the analysis to assume that the ideal $\mathcal{I}$ is in general position. This however, is not required for the algorithm.

Remark 2: We assumed that we have a complete decomposition of the radical of $\mathcal{I}$ available. This is not necessary, since, if $\mathcal{Q}_1, \mathcal{Q}_2$ are primary components of $\mathcal{I}$ with distinct associated primes $\mathcal{P}_1, \mathcal{P}_2$, then

$$\mathcal{Q}_1 \mathcal{Q}_2 = (\mathcal{I}, (\mathcal{P}_1 \mathcal{P}_2)^\sigma)$$

where σ is at least as large as the maximum of the indices of $\mathcal{Q}_1$ and $\mathcal{Q}_2$. This might be helpful if one wants to trade factorization for longer lifting.

Remark 3: We have not taken into account the sizes of the rational numbers that arise during the algorithm in our analysis. For theorem 3 to hold while counting the number of bit operations instead of field operations, it is necessary to prove a polynomial bound (in d^n and the sizes of coefficients in the initial basis) on the sizes of numbers arising in the computations.

Conclusion

We have presented a single exponential time algorithm to construct a reduced Gröbner basis for a zero–dimensional ideal. At this time, the bottle neck is actually getting the radical. The methods proposed in [Lak], [LL] make use of the u-resultant and a change of coordinates which makes the problem *dense*. If these can be avoided, one might have a sparse, more practical algorithm for computing a reduced Gröbner basis for zero–dimensional ideals. We view the efforts described in this paper as a first step towards that goal.

Acknowledgements: I wish to thank my advisor Prof. Erich Kaltofen for several useful discussions and Prof. Daniel Lazard for his encouragement and suggestions.

REFERENCES

[CGH]   Caniglia L., Galligo A., Heintz J., *Some New Effectivity Bounds in Computational Geometry*, in "Proc. 6th AAECC," LNCS 357, Springer-Verlag, pp. 131-152..

[DS]   Dickenstein A.M., Sessa C., *Duality Methods for Membership Problem*, these proceedings.

[FGLM] Faugère J.C., Gianni P., Lazard D., Mora T., *Efficient Computation of Zero-dimensional Gröbner Bases by Change of Ordering*, Tech. Report., LITP, Universite Paris, July 1989.

[GTZ]   Gianni P., Trager B., Zacharias G., *Gröbner bases and Primary Decomposition of Polynomial Ideals*, Jour. Symb. Comp. 6 (1988), 149–167.

[Kre]   Kredel H., *Primary Ideal Decomposition*, in "Proc. EUROCAL '87 Leipzig," LNCS vol.378, Springer-Verlag, 1987, pp. 270–281.

[KL]   Krick T., Logar A., *Membership Problem, Representation Problem and the Computation of the Radical for One-dimensional Ideals*, these proceedings.

[Lak]   Lakshman Y.N., *On the Complexity of Computing a Gröbner Basis for the Radical of a Zero Dimensional Ideal*, in "Proc. of 22nd ACM Symposium on Theory of Computing (STOC)," May 1990 (to appear).

[LL]   Lakshman Y.N., Lazard D., *On the Complexity of Zero-dimensional Algebraic Systems*, these proceedings.

[vdW]   van der Waerden B. L.,, "Algebra," vol.2, Frederick Ungar Pub. Co., 1970.

Lakshman Y. N.
Deptartment of Computer and Information Sciences
University of Delaware
103, Smith Hall
Newakk, DE 19716
U. S. A.
lakshman@math.udel.edu

Algorithms for a Multiple Algebraic Extension

Abstract. We give fast algorithms for computing product and inverse in a multiple algebraic extension of the rational numbers. The algorithms are almost linear in terms of the output length, *i.e.* they work in time $O(d^{1+\delta})$, for all $\delta > 0$, where d is an a priori bound on the length of the output. Since we require time $\Omega(d)$ just to write down the output the algorithms are close to optimal. The algorithm for inverse uses a technique referred to as dynamic evaluation for computing in algebraic extensions defined by reducible polynomials.

1 Introduction

We say that an algorithm is *almost optimal* if it has computing time $O(d^{1+\delta})$ binary operations, for all $\delta > 0$, where d is the best known a priori bound on the length of the output. We use the bitwise computation model (Aho *et al.* [5]). Previously we have given almost optimal algorithms for product, inverse and univariate polynomial product for a single algebraic extension of the rational numbers (Langemyr [15]). These results were in turn used to obtain an almost optimal probabilistic algorithm for polynomial greatest common divisor (GCD) over a single algebraic extension of the rational numbers [15], a refinement of an algorithm (Langemyr and McCallum [16]) which did not fully take advantage of dynamic evaluation (Duval [12]).

In this paper we generalize the results on multiplication and inverse to multiple algebraic extensions. Our work will result in almost optimal algorithms for computing product and inverse in a multiple algebraic extension. Similarly to the case of a single extension we obtain these results by using modular techniques originally applied to polynomial factorization over algebraic number fields (Weinberger and Rothschild [19]). The main contributions of this paper are the size bound for product and inverse in

*Supported by ESPRIT BRA 3012 CompuLog and Fakultetsnämnden KTH. *Formerly with:* The Royal Institute of Technology, 100 44 Stockholm, Sweden.

a multiple algebraic extension and that we obtain almost optimal algorithms. For our inverse algorithm the application of dynamic evaluation to the inverse algorithm seems essential.

Our motivation for seeking the algorithms of the present paper stems from our interest in the Cylindrical algebraic decomposition (CAD) algorithm (Collins [11], and Arnon *et al.* [6]). The CAD algorithm decomposes real Euclidean space into regions where a set of polynomials have constant sign, and furthermore computes a point in each such region. Such a point may be represented as an element of a multiple algebraic extension. Algorithms similar to the ones in this paper have been used in an implementation of the CAD algorithm which uses multiple algebraic extensions instead of single extensions combined with primitive element computations as originally suggested by Collins. The multiple algebraic extension package also contains sign determination for elements of a multiple algebraic extension number field, and univariate polynomial GCD, factoring and root isolation. Experiments have shown that the improvement to the CAD algorithm is significant. An extended version of the present paper describing the full capabilities of the multiple extension package is in preparation together with a paper describing the CAD implementation.

Results on an implementation of product and inverse for a multiple algebraic extension have been previously published (Abbott *et al.* [3]). The paper contained timings of different versions of the algorithms. A technique similar to ours for bounding sizes of elements in a multiple algebraic extension is described by Abbott [2].

2 Prerequisites and Notation

We consider a multiple algebraic extension of the rational numbers. We have $K_i = K_{i-1}(\alpha_i)$, for $i = 1, \ldots, n$, where $K_0 = \mathbf{Q}$, and where α_i is a root of the monic polynomial $r_i(y_i) \in \mathbf{Z}[\alpha_1, \ldots, \alpha_{i-1}][y_i]$ of degree m_i in y_i irreducible over K_{i-1}. Thus we insist that the algebraic numbers used to build the algebraic extension are algebraic integers. We thus have a corresponding tower of ring extensions $A_i = A_{i-1}[\alpha_i]$, for $i = 1, \ldots, n$, where $A_0 = \mathbf{Z}$. These rings are integral domains.

We now discuss how an element of this multiple algebraic extension is represented in the computer. First we note that any element of K_n can be represented by a pair of a rational number and a representative of an element in the ring A_n. We therefore restrict our attention to how to represent an element of A_n. We can view the polynomials $r_n(y_n), \ldots, r_1(y_1)$ as an ideal $I_n = (r_n(y_n), \ldots, r_1(y_1))$ in the polynomial ring $\mathbf{Q}[y_1, \ldots, y_n]$, and then apply the theory of Gröbner bases (Buchberger [8]). Since I_n is a Gröbner basis in the lexicographical variable ordering $y_n > y_{n-1} > \cdots > y_2 > y_1$ we see that an element of $\mathbf{Q}[y_1, \ldots, y_n]$ has a canonical form modulo the ideal I_n, which can be obtained by reducing modulo the minimal polynomials until no further reductions are possible. We can use polynomial division for performing the reductions. Since the leading coefficients of the polynomials r_i are 1 we do not introduce rational numbers. We thus obtain a canonical representation of an element in A_i as an element of the quotient ring $R_i = \mathbf{Z}[y_1, \ldots, y_i]/I_i$. We denote $R_0 = \mathbf{Z}$. Our representation coincides with the one used by Abbott [1].

Throughout the paper we assume that the polynomials used for the representation of the algebraic numbers are dense. We deem this reasonable because of the fact that the polynomials which we use as representatives for the algebraic numbers quickly become dense when we start computing with them.

3 Product

Consider $a, b \in R_n$. It is not difficult to device an algorithm for computing $c = ab \in R_n$ which uses classical techniques, but there is a non-trivial issue which one has to consider—in particular in what order polynomial multiplication and divisions are made. We have two polynomials a and b in $\mathbf{Z}[y_1, \ldots, y_n]$ both fully reduced modulo I_n. Our task is to compute $c \in R_n$. This can be done by an algorithm which recursively multiplies the coefficients for $i = 1, \ldots, n$ as elements of R_i, by first performing polynomial multiplication in y_i over R_{i-1}, and then performing polynomial division in y_i over R_{i-1}. This will be our first algorithm. Our second algorithm first multiplies a and b in $\mathbf{Z}[y_1, \ldots, y_n]$, and then recursively reduces the result by performing polynomial division in y_i over $\mathbf{Z}[y_1, \ldots, y_{i-1}]$ for $i = n, n-1, \ldots, 1$, recursing over coefficients. We investigate which one of these approaches is the fastest one. The analysis of this question was left open by Abbott et al. [3]. We may note that in the first algorithm the degrees of the intermediate results were kept at minimal level, thereby incurring some coefficient growth. In the second approach the size of the integers in the intermediate result are kept to a minimum by letting the degree increase. The analysis of the computing times for both algorithms and the size analysis for the second algorithm is made in the full version of this paper [14]. The analysis of computing times shows that the first approach is the superior one and, we therefore list the first algorithm here. In fact, the second method renders exponential degree blowup, which was unforseen by us before our analysis was made.

Algorithm 1 *Multiple algebraic extension product, first method.*

- *Inputs:* $a, b \in R_i$.

- *Output:* $c = ab \in R_i$.

(1) [Multiply.] View a and b as polynomials over R_{i-1} in y_i of degree at most $m_i - 1$. Perform polynomial multiplication over R_{i-1}, using

Algorithm 1 recursively for coefficient multiplication in R_{i-1}. Obtain $c'(y_i) \in R_{i-1}[y_i]$ of degree at most $2m_i - 2$ in y_i.

(2) [Reduce.] Divide the polynomials $c'(y_i)$ and $r_i(y_i)$ as polynomials in y_i over R_{i-1} again using Algorithm 1 recursively for coefficient multiplication in R_{i-1}. We obtain c □

We now apply explicit tracing of this algorithm for obtaining a size bound on the integers in the product. Given an element $a \in R_i$ we use the notation $|a|$ to denote the maximum absolute value of the integers in the representation of a. Abusing notation we also use this notation for denoting the maximum absolute value of the integers in the representation of a polynomial over R_{i-1} in y_i.

Lemma 2 *Let* $a, b \in R_n$, $n \geq 0$. *Let* $|r_i| \leq E_i$, $1 \leq i \leq n$. *Then for* $c = ab \in R_n$ *we have* $|c| \leq |a| |b| S_n$, *where* $S_0 = 1$, *and*

$$
\begin{aligned}
S_1 &= m_1 S_0 (1 + E_1)^{m_1 - 1}, \\
S_2 &= m_2 S_1 (1 + S_1 E_2)^{m_2 - 1}, \\
&\vdots \\
S_i &= m_i S_{i-1} (1 + S_{i-1} E_i)^{m_i - 1}, \\
&\vdots \\
S_n &= m_n S_{n-1} (1 + S_{n-1} E_n)^{m_n - 1}.
\end{aligned}
$$

(1)

Further, $|a| |b| S_i$ *bounds the integers at the recursive level* $n - i + 1$ *in Algorithm 1.*

Proof: Consider the recursive level $n - j + 1$, $0 \leq j \leq i$, in Algorithm 1. We compute the product of polynomials in y_j over R_{j-1} of degree $m_j - 1$. An absolute value bound on the integers in the coefficients after step 1 is given by S_{j-1}. Then at most m_j integers, each one bounded by S_{j-1}, are added, and then reduction modulo r_j takes place. Each division step increases the absolute value with a factor $(1 + S_{j-1} E_j)$ and this occurs $m_j - 1$ times.

Q. E. D.

It now remains to solve (1). After the simplification $Q_0 = 1$ and $Q_i = m_i Q_{i-1}^{m_i} (2E_i)^{m_i}$, for $i = 1, \ldots, n$ having $S_i \leq Q_i$, we obtain the solution:

Theorem 3 *Consider* $c = ab \in R_n$. *Let* $|r_i| \leq E_i$ *for* $i = 1, \ldots, n$. *We then have* $|c| \leq |a| |b| Q_n$, *where*

$$
Q_n = \prod_{i=1}^{n} m_i^{\prod_{j=i+1}^{n} m_j} (2E_i)^{\prod_{j=i}^{n} m_j}.
$$

(2)

Thus $\log(Q_n)$ is $O(q_n)$, where $q_n = \sum_{i=1}^{n} e_i \prod_{j=i}^{n} m_j$, $e_i = \log(E_i)$, $1 \leq i \leq n$. We will use this notation throughout this paper.

We have also used different techniques for obtaining bounds on the length of the integers in the product. Using Hadamard's bound on the determinant formula for the pseudo-remainder (Loos [17]) renders a formula which is not as good as the one obtained with the above explicit technique, e.g. with $n = 1$ we obtain the bound $(2m_1 - 1)D_1 D_2((m_1 + 1)E_1)^{m_1 - 1}$, which is $O(d_1 + d_2 + m_1(e_1 + \log(m_1)))$ bits. We suspect that the technique of Landau [13] for bounding the representation of an algebraic number given is absolute value interpreted as a complex number similarly would yield inferior bounds.

Having obtained a bound on the size of the product, we have the possibility of giving a modular algorithm, which is asymptotically fast. This means that we will compute the modular image of the product in the ring $\tilde{R}_n = R_n/(p)$, for a sufficient number of rational odd primes p, and then restore the result to R_n using the Chinese remainder theorem. By Cantor and Kaltofen [9] we can apply fast Fourier techniques for multiplication in $\tilde{R}_n$ since $1 + 1$ is a unit in this ring for all rational odd primes. We present the algorithm.

Algorithm 4 *Multiple Algebraic Extension Product. Modular algorithm.*

- *Inputs: a, $b \in R_n$, $r_i \in R_{i-1}(y_i)$, for $i = 1, \ldots, n$.*

- *Output: $c = ab \in R_n$.*

(1) [Compute bounds.] Compute $K = |a| \, |b| \, Q_n$ according to Theorem 3.

(2) [Compute prime list.] Compute an array of rational odd primes $\{p_i\}_{i=1}^{s}$, for which we have $\prod_{i=1}^{s} p_i > 2K$

(3) [Compute modular images.] For $i \leftarrow 1, 2, \ldots, s$ compute $\tilde{a}_i \leftarrow a \bmod p_i$ and $\tilde{b}_i \leftarrow b \bmod p_i$ and for for $j \leftarrow 1, 2, \ldots, n$ compute $\tilde{r}_i(y_i) \leftarrow r_i(y_i) \bmod p_i$.

(4) [Compute modular products.] For $i \leftarrow 1, 2, \ldots, s$ compute $c_i \leftarrow a_i b_i$.

(5) [Apply the Chinese remainder algorithm.] Solve the system of congruences
$$(3) \qquad\qquad c \equiv c_i \pmod{p_i} \qquad 1 \leq i \leq s$$
using the Chinese remainder algorithm. In the above system we regard the c_i as a polynomial in $\mathbf{Z}[y_1, \ldots, y_n]$. Thus the congruence "$c \equiv c_i \bmod p_i$" is really a system of integer congruences, one for each integer coefficient. When performing modular arithmetic in the ring $\mathbf{Z}_m$ of integers modulo m we take as the computational model of $\mathbf{Z}_m$ the set $\{n \in \mathbf{Z} | -m/2 < n < m/2\}$.

We analyze the computing time of Algorithm 4. We have to know the number of rational primes used, and also the length of the primes. We apply some results due to Rosser and Schoenfeld [18]. Let x be a natural number. Let $\pi(x)$ be the number of primes less than or equal to x, and $\vartheta(x)$ be the logarithm of the product of those primes. We then have that $\pi(x)$ is $\theta(x/\log(x))$ and $\vartheta(x)$ is $\theta(x)$. Thus if we use the smallest possible primes forming a product exceeding K it is sufficient to use $O(k/\log(k))$ primes, where $k = \log(K)$, having at most $O(\log(k))$ bits. In order to shorten the formulae we use the notation $\mathcal{L}(x) = \log x \log\log x$ and $\mathcal{L}_2(x) = \log^2 x \log\log x$. For Step 2 we use the sieve of Eratosthenes. By Proposition 2.44 in Langemyr [15] this can be done using $O(k\,\mathcal{L}(k))$ binary operations. For Step 3 we compute the modular images of $\prod_{i=1}^{n} m_i$ integers in a and b respectively. Further, we compute the modular images of $\sum_{i=1}^{n} \prod_{j=1}^{i} m_j$ integers in all $r_i(y_i)$. We obtain $O(k\,\mathcal{L}_2(k) \sum_{i=1}^{n} \prod_{j=1}^{i} m_j)$ binary operations by applying Theorem 8.9 in Aho et al. [5]. In Step 4 we apply Cantor and Kaltofen [9]. Since $1+1$ is a unit in $\tilde{R}_n$ for all odd primes p_i we can apply their results. We thus have that multiplication of polynomials of degree at most m_j over the rings $\tilde{R}_{j-1} = R_{j-1}/(p_i)$ can be done in $C'm_j\,\mathcal{L}(m_j)$ multiplications in $\tilde{R}_{j-1}$, for some constant C'. Further, by applying Algorithm 8.3 in Aho et al., we can do the reduction modulo r_j in $C''m_j\,\mathcal{L}(m_j)$ multiplications in $\tilde{R}_{j-1}$. Summing the cost over the primes (forming a k bit product) we obtain $O(k\,\mathcal{L}(\log(k))C^n \prod_{i=1}^{n} m_i\,\mathcal{L}(m_i))$ binary operations for this step, where C is $C' + C''$. In Step 5 we apply Theorem 8.21 also in Aho et al. We can then perform the lifting in $O(k\,\mathcal{L}_2(k) \prod_{i=1}^{n} m_i)$ binary operations. We have proved the following theorem.

Theorem 5 *Let $a, b \in R_n$, and let $k = \log(|a||b|Q_n)$ as in Theorem 3. We can then compute $c = ab \in R_n$ in*

$$(4) \qquad O(k\,\mathcal{L}_2(k)(\sum_{i=1}^{n} \prod_{j=1}^{i} m_j) + k\,\mathcal{L}(\log(k))C^n \prod_{i=1}^{n} m_i\,\mathcal{L}(m_i)).$$

binary operations, where C is the constant from the fast multiplication algorithm.

We note that Algorithm 4 is almost optimal since the bound on the size of c is $O(k \prod_{i=1}^{n} m_i)$ and $k\,\mathcal{L}(\log(k))C^n$ is $O(k^{1+\delta})$ and $k\,\mathcal{L}_2(k)n$ is also $O(k^{1+\delta})$, for all $\delta > 0$. We should note that if we let n vary, we must enforce $C^n = O(k^{\delta'})$, $\delta' < \delta$, e.g. by letting $m_i > C^{1/\delta'}$, in order to obtain the result above.

4 Inverse

Consider $a \in A_n$ represented as an element in R_n. In general, of course, inverses in A_n do not exist, but using the embedding in the quotient field

K_n we have $a^{-1} \in \frac{1}{c}A_n$, for some $c \in \mathbf{Z}$. We present an analysis of the size of the integers in the inverse. We use the results on the size of the integers in the product from the previous section for determining the size of the integers in the inverse. Using this bound, we present an asymptotically fast algorithm for computing the inverse. We view our representative of a in R_n as a polynomial in y_n over R_{n-1} of degree less than m_n, $a = a_n = \sum_{i=0}^{m_n-1} a_{n,i} y_n^i$. We can obtain the inverse by iterating the equation

(5) $$b_i(y_i) a_i(y_i) \equiv a_{i-1} \pmod{r_i(y_i)}$$

for $i = n, n-1, \ldots, 1$, over R_{i-1}, where $a_{i-1} = \text{res}(a_i(y_i), r_i(y_i))$, where $\text{res}(\cdot, \cdot)$ denotes the resultant. This can be written as a system of linear equations over R_{i-1}.

(6)
$$
\begin{bmatrix}
a_{i,m_i-1} & & & a_{i,0} & & \\
& \ddots & & & \ddots & \\
a_{i,m_i-1} & & & a_{i,0} & \\
r_{i,m_i} & & & r_{i,0} & \\
& \ddots & & & \ddots & \\
& & r_{i,m_i} & & & r_{i,0}
\end{bmatrix}
\begin{bmatrix}
b_{i,m_i-1} \\
\vdots \\
b_{i,0} \\
r'_{i,m_i-2} \\
\vdots \\
r'_{i,0}
\end{bmatrix}
=
\begin{bmatrix}
0 \\
\vdots \\
0 \\
0 \\
\vdots \\
a_{i-1}
\end{bmatrix},
$$

where $b_i(y_i) = \sum_{j=0}^{m_i-1} b_{i,j} y_i^j$. Since r_i is assumed to be irreducible the system is always non-singular. We obtain the following lemma.

Lemma 6 *Consider $a \in R_n$. The inverse $a^{-1} \in \frac{1}{c}R_n$ is*

(7) $$\frac{1}{a_0} b_1(y_1) b_2(y_2) \cdots b_{n-1}(y_{n-1}) b_n(y_n),$$

where $b_i(y_i)$, $i = 1, \ldots, n$, and a_0 is obtained from the above system of equations (6), and thus $c = a_0$, and $b = \prod_{i=1}^{n} b_i(y_i) \in R_n$.

We can now give the size of the integers in a_{i-1} and b_i by applying our multiplication size bound Q_i from Theorem 3 in Cramer's rule for solving a system of linear equations.

Lemma 7 *Let $a \in R_n$. Assume $|a| \leq D$, and further that $|r_i| \leq E_i$. Then for b_i we have the bound $|b_i| \leq I_{n,i}$ according to the equations*

(8)
$$
\begin{aligned}
I_{n,n} &= (2m_n - 1)! \, D^{m_n} E_n^{m_n-1} Q_{n-1}^{2m_n-2}, \\
I_{n,n-1} &= (2m_{n-1} - 1)! \, I_{n,n}^{m_{n-1}} E_{n-1}^{m_{n-1}-1} Q_{n-2}^{2m_{n-1}-2}, \\
&\vdots \\
I_{n,i} &= (2m_i - 1)! \, I_{n,i+1}^{m_i} E_i^{m_i-1} Q_{i-1}^{2m_i-2}, \\
&\vdots \\
I_{n,1} &= (2m_1 - 1)! \, I_{n,2}^{m_1} E_1^{m_1-1}.
\end{aligned}
$$

For c an absolute value bound is $I_{n,1}$ also. For $b \in R_n$ a bound is given by

(9) $$I_n = I_{n,1} I_{n,2} Q_1 \cdots Q_{n-3} I_{n,n-1} Q_{n-2} I_{n,n} Q_{n-1}$$

so that $|b| \leq I_n$.

Solving the equations in the above lemma we obtain the result which we have aimed for.

Theorem 8 *Let $a \in R_n$. Assume $|a| \leq D$, and further that $|r_i| \leq E_i$. We then have the absolute value bound*

(10)
$$I_n = D^{\sum_{k=1}^{n} \prod_{i=k}^{n} m_i} \prod_{i=1}^{n-1} Q_i \prod_{i=1}^{n} [(2m_i - 1)! \, E_i^{m_i-1} Q_{i-1}^{2m_i-2}]^{\sum_{k=1}^{i} \prod_{j=k}^{i-1} m_j}$$

for the integers in b, and

(11) $$I_{n,i} = D^{\prod_{j=i}^{n} m_j} \prod_{j=i}^{n} [(2m_j - 1)! \, E_j^{m_j-1} Q_{j-1}^{2m_j-2}]^{\prod_{k=i}^{j-1} m_k},$$

for the integers in b_i and a_i (and thus $c = a_0$). By $I_i(x)$ we denote the value of I_i if x would have replaced D.

We can now insert the result from Theorem 3 obtaining that $\log(I_n)$ is $O(l)$, where

(12) $$l = d \sum_{i=1}^{n} \prod_{j=i}^{n} m_j + \sum_{i=1}^{n} e_i \sum_{j=i}^{n} m_j (\sum_{k=1}^{j} \prod_{l=k}^{j-1} m_l)(\prod_{l=i}^{j} m_l),$$

where $d = \log(D)$ and $e_i = \log(E_i)$, $1 \leq n \leq n$.

Also for the case of inverse computations in R_n we will use modular techniques for obtaining an asymptotically fast algorithm. Again this means that we will compute in the ring $\tilde{R}_n = R_n/(p)$, for a sufficient number of rational odd primes p. For the case of multiplication in R_n this caused no problems since only multiplications in $\tilde{R}_n$ were needed. For inverse computations we need to solve the linear system (6) modulo p. If $\tilde{R}_{i-1} = R_{i-1}/(p)$ had been a field we could do this by applying the Half-extended version of the algorithm of Figure 8.8 in Aho *et al.* [5]. For our case this would imply having to compute the inverse of certain elements in $\tilde{R}_{i-1}$ which we cannot be sure exist. We describe below how to handle this situation efficiently.

Let r_1 have the irreducible factor $\bar{r}_1$ mod p. Then $\bar{R}_1 = \mathbf{Z}_p[y_1]/(\bar{r}_1(y_1))$ is a field. We can go on by letting $\bar{r}_2$ be an irreducible factor over $\bar{R}_1$ of r_2 mod $(p, \bar{r}_1)$. Then $\bar{R}_2 = \bar{R}_1[y_2]/(\bar{r}_2(y_2))$ is also a field. We obtain a tower of field extensions $\bar{R}_i$, $i = 1, \ldots, n$, where

(13) $$\bar{R}_i = \bar{R}_{i-1}[y_i]/(\bar{r}_i(y_i)),$$

and $\bar{r}_i$ is an irreducible factor over $\bar{R}_{i-1}$ of $r_i \bmod (p, \bar{r}_{i-1}, \ldots, \bar{r}_1)$. We call such a tower of fields an irreducible tower of $r_n, \ldots, r_1$. We can also define a partially factored tower of rings, $\hat{R}_i = \hat{R}_{i-1}[y_i]/(\hat{r}_i(y_i))$, for $i = 1, \ldots, n$, where $\hat{r}_i$ is a factor of $r_i \bmod (p, \hat{r}_{i-1}, \ldots, \hat{r}_1)$. Since all $\hat{r}_i$ are monic the division test can be performed without using division, and thus we will have that $\hat{r}_i$ divides r_i modulo $(p, \bar{r}_{i-1}, \ldots, \bar{r}_1)$ for all irreducible towers of $\hat{r}_{i-1}, \ldots, \hat{r}_1$.

Having computed all the irreducible towers of $r_n, \ldots, r_1$ by e.g., Berlekamp [7], we could do the following for all such towers: Let $\bar{a} = a \bmod (p, \bar{r}_n, \ldots, \bar{r}_1)$ and provided that $\bar{a}$ is non-zero compute the inverse in $\bar{R}_n$. Then lift the result back into $\tilde{R}_n$ by using using the Chinese remainder theorem for polynomials from Aho et al. [5]Theorem 8.13. (In order to be able to apply the Chinese we need a restriction on p which we give later.) For the case of polynomial GCD computations over $\mathbf{Z}[\alpha]$ it was shown in Langemyr [15] that neither applying the best known deterministic nor probabilistic algorithm for factoring the minimal polynomial of α modulo the required number of primes made us able to give an almost optimal algorithm for computing the GCD. We claim that this would also be the case for the inverse computation if we would construct all the irreducible towers by using polynomial factorization. We will therefore use a technique referred to as dynamic evaluation (Duval [12]) for computing in partially factored towers. Using this technique finding a factor of $\hat{r}_i$ modulo $(p, \hat{r}_{i-1}, \ldots, \hat{r}_1)$ splits the current partially factored tower into two partially factored towers.

We have to consider the possibility that

$$(14) \qquad\qquad a \equiv 0 \bmod (p, \bar{r}_n, \ldots, \bar{r}_1)$$

for an irreducible tower. This means that a and r_n have a common factor modulo $(p, \bar{r}_{n-1}, \ldots, \bar{r}_1)$. Primes for which this happens cannot be used for computing the inverse over $\tilde{R}_n$. A sufficient condition for this not to happen for any choice of factors is

$$(15) \qquad\qquad p \nmid \operatorname{res}(\ldots \operatorname{res}(\operatorname{res}(a, r_n), r_{n-1}) \ldots, r_1).$$

This can be seen by noting that the system (6) becomes singular for i iff

$$(16) \qquad\qquad \operatorname{res}_{y_i}(r(y_i), a_i(y_i)) \equiv 0 \bmod (p, \bar{r}_{i-1}, \ldots, \bar{r}_1),$$

for some irreducible tower. Using the notation of Abbott and Davenport [4] the norm is exactly the resultant from (15). Using their notation we can write the condition as $p \nmid N_Q^{K^*}(a)$.

Having computed the image of b_n, modulo $(p, \bar{r}_{n-1}, \ldots, \bar{r}_1)$, for all irreducible towers of $r_{n-1}, \ldots, r_1$, we could recover b_n by recursively using the Chinese remainder theorem for polynomials from Aho et al. [5]Theorem 8.13. In order to be able to apply the theorem we have to assure that all r_i

has to be square-free modulo $(p, \bar{r}_{i-1}, \ldots, \bar{r}_1)$ for all irreducible towers of $r_{i-1}, \ldots, r_1$. A sufficient condition for this not to happen is that

$$(17) \qquad p \nmid \operatorname{res}(\ldots \operatorname{res}(\operatorname{res}(r_i(y_i), r_i'(y_i)), r_{i-1}) \ldots, r_1),$$

for $i = 1, \ldots, n$. This can be seen by replacing a_n by r_n' in the argument for (15) above, and similarly for $i = n - 1, \ldots 1$. Again using the notation of Abbott and Davenport we see that the condition above can be expressed as $p \nmid N_Q^{K_i}(r_i'(y_i))$, for $i = 1, \ldots, n$. Furthermore following the definitions of Abbott and Davenport this is equivalent to p not dividing the discriminant of the basis of the number field K_n.

In order to obtain an algorithm which computes $a^{-1} \bmod (p, r_n, \ldots, r_1)$ iff $a^{-1} \bmod (p, \bar{r}_n, \ldots, \bar{r}_1)$ exists for all possible irreducible towers, we apply the principle of dynamic evaluation (Duval [12]) to our particular modular setting. We apply the principle to the Half-extended version of the algorithm of Figure 8.8 in Aho et al. [5]. We have to adjust the result similarly to Collins [10] so that we obtain $\operatorname{res}(a_i, r_i)$ and b_i according to

$$(18) \qquad b_i a_i \equiv \operatorname{res}(a_i, r_i) \bmod (p, \tilde{r}_i, \ldots, \tilde{r}_1)$$

for all factorizations automatically generated by the converted algorithm. If we obtain two different factors $\tilde{r}'_i$ and $\tilde{r}''_i$ of r_i modulo some factorization $(p, \tilde{r}_{i-1}, \ldots, \tilde{r}_1)$, which are not relatively prime this means that the condition (17) is not satisfied for p. In a full version of this paper [14] we present an implementation of the dynamic evaluation version of the algorithm of Figure 8.8 in Aho et al. [5].

From above we know that we may not be able to use primes which divide $N_Q^{K_i}(r_i'(y_i))$, for $i = 1, \ldots, n$, and that primes which divide $N_Q^{K_n}(a)$ cannot be used for computing the correct modular image of the inverse. We also know that the integers in the inverse are bounded by the number I_n in absolute value. Since $I_i(m_i E_i)$ bounds $N_Q^{K_i}(r_i'(y_i))$ and $I_{n,n}$ bounds $N_Q^{K_i}(a)$ (by Theorem 8) in absolute value, we have that the product of primes used by our modular inverse algorithm must be

$$(19) \qquad L = I_n I_{n,n} \prod_{i=1}^{n} I_{n,i}(m_i E_i),$$

if we want it to have a sufficient number of primes for all cases. Applying Theorem 8 and (12) we see that the expression is also $O(l)$ bits. We are now ready to present the algorithm.

Algorithm 9 *Multiple Algebraic Extension Inverse. Modular algorithm.*

- *Inputs:* $a \in R_n$, $r_i \in R_{i-1}(y_i)$, for $i = 1, \ldots, n$.

- *Output:* $b \in R_n$, $c \in \mathbf{Z}$, such that $a^{-1} = \frac{1}{c}b$.

(1) [Compute bounds.] Compute the bound $L = I_n I_{n,n} \prod_{i=1}^n I_{n,i}(m_i E_i)$.

(2) [Compute prime list.] Compute an array of odd primes $\{p_i\}_{i=1}^s$, for which we have $\prod_{i=1}^s p_i > 2L$. Initialize an array $t_i \leftarrow 0$, for $i = 1$, $\ldots$, s.

(3) [Compute modular images.] For $i \leftarrow 1, 2, \ldots, s$ compute $\tilde{a}_i \leftarrow a \bmod p_i$, and for $j \leftarrow 1, 2, \ldots, n$ compute $\tilde{r}_{ij}(y_i) \leftarrow r_i(y_i) \bmod p_j$.

(4) [Compute a_i and b_i] For $i \leftarrow 1, 2, \ldots, s$ unless $t_i = 1$ do all what follows: Let $\tilde{a}_{i,n} \leftarrow \tilde{a}_i$. For $j = n, n-1, \ldots, 1$, apply the dynamic evaluation principle to the Half extended version of the algorithm of Figure 8.8 in Aho *et al.* [5] with inputs $\tilde{a}_{i,j}$ and $\tilde{r}_j$ modulo $(\tilde{r}_{j-1}, \ldots, \tilde{r}_1)$. Having obtained a set of $\tilde{b}_{i,j,k}$, $a_{i,j-1,k}$ for each k corresponding to a partially factored tower of $r_j, \ldots, r_1$. Restore the results $\tilde{b}_{i,j}$, $\tilde{a}_{i,j-1}$ by recursively applying the polynomial version of the Chinese remainder algorithm (Aho *et al.* [5]Theorem 8.13). If this is not possible because some r_i was not square-free modulo p or because $\tilde{a}_{i,j-1} \equiv 0$ for some tower set $t_i \leftarrow 1$ and continue at the next i. Otherwise let $\tilde{c}_i \leftarrow \tilde{a}_{i,0}$.

(5) [Compute product] For $i \leftarrow 1, 2, \ldots, s$ unless $t_i = 1$ compute $\tilde{b}_i = \prod_{j=1}^n \tilde{b}_{ij}$.

(6) [Apply the Chinese remainder algorithm.] Solve the system of congruences

(20) $\qquad b \equiv \tilde{b}_i \qquad c \equiv \tilde{c}_i \pmod{p_i} \qquad 1 \le i \le s$

using the Chinese remainder algorithm, for $i = 1, \ldots, s$ unless $t_i = 1$.

We must prove the algorithm correct.

Lemma 10 *Algorithm 9 correctly computes the inverse of a.*

Proof: In Step 1 we compute the bound $L = I_n I_{n,n} \prod_{i=1} I_{n,i}(m_i E_i)$ which clearly bounds the product of $N_Q^{K_i}(r_i'(y_i))$ for $i = 1, \ldots, n$, and and $N_Q^{K_n}(a)$ by Theorem 8. Thus the product of primes suffices to restore the coefficients in the inverse, discard the primes which divide $N_Q^{K_n}(a)$ and further, to exhaust the primes which may cause the Chinese remainder algorithm in Step 4 to fail. $\qquad$ Q. E. D.

For the correctness in applying the dynamic evaluation principle we refer to Duval [12]. In the full version of this paper we will give a listing of the resulting algorithm together with a correctness proof.

We now analyze the computing time of Algorithm 9. For Step 2 we use the sieve of Eratosthenes. By Proposition 2.44 in Langemyr [15] this can be done using $O(l\,\mathcal{L}(l))$ binary operations. For Step 3 we compute the modular images of $\prod_{i=1}^{n} m_i$ integers in a. Further, we compute the modular images of $\sum_{i=1}^{n} \prod_{j=1}^{i} m_j$ integers in all $r_i(y_i)$. We obtain $O(l\,\mathcal{L}_2(l) \sum_{i=1}^{n} \prod_{j=1}^{i} m_j)$ binary operations by applying Theorem 8.9 in Aho et $al.$ [5]. In Step 4 we apply the principle of dynamic evaluation. Following Duval [12] we know that the cost will be bounded by the cost of applying the algorithm of Figure 8.8 in Aho et $al.$, had $\tilde{r}_n, \ldots, \tilde{r}_1$ been an irreducible tower. Summing over the primes (forming an l bit product) we obtain $O(l\,\mathcal{L}_2(\log(l))\bar{C}^n \prod_{i=1}^{n} m_i\,\mathcal{L}_2(m_i))$ binary operations including the cost of the Chinese remainder step (also including the cost of preconditioning). $\bar{C}$ is a constant. Finally, by applying Theorem 8.21 from Aho et $al.$ we can perform the lifting in $O(l\,\mathcal{L}_2(l) \prod_{i=1}^{n} m_i)$ binary operations. We have proved the following theorem.

Theorem 11 *Let $a \in R_n$. Assume $|a| \leq D$, and further that $|r_i| \leq E_i$. We can then compute $b \in R_n$ and $c \in \mathbf{Z}$ such that $a^{-1} = \frac{1}{c}b$ in*

$$(21) \qquad O(l\,\mathcal{L}_2(l) \sum_{i=1}^{n} \prod_{j=1}^{i} m_j + l\,\mathcal{L}_2(\log(l))\bar{C}^n \sum_{i=1}^{n} \prod_{j=1}^{i} m_j\,\mathcal{L}_2(m_j))$$

binary operations, where l is given by (12).

Similarly to the case of multiplication we note that the algorithm is almost optimal.

5 Conclusion

We have given almost optimal algorithms for computing product and inverse in a multiple algebraic extension. We have implemented algorithms similar to the ones described above in SAC-2. In Langemyr and McCallum [16] it was observed that factors of the minimal polynomial (only a single extension was treated) was almost never discovered if the prime numbers are of the size of a normal computer word (32 bits). It turns out that this is also true in our case and we can thus in practice skip implementing the dynamic evaluation, and just ignore the particular prime if a factor is discovered. The principle of dynamic evaluation is theoretically important however since it seems difficult to obtain almost optimal algorithms if it is not utilized (Langemyr [15]).

Further algorithms have been implemented in SAC-2 and a full paper describing the present results, univariate GCD, and further an analysis of the second product algorithm is available [14]. Further papers on the factoring and root isolation algorithms are planned.

Acknowledgement: The author is grateful to Rüdiger Loos for providing a good working environment in Tübingen. Thanks to Johan Håstad for suggestions.

REFERENCES

[1] J. A. Abbott, *Factorization of Polynomials Over Algebraic Nubmer Fields*, PhD thesis, University of Bath, Bath, 1989.

[2] J. A. Abbott, *Recovery of algebraic numbers from their p-adic approximations*, in "Proc. ISSAC '89," ACM, 1989, pp. 112–120.

[3] J. A. Abbott, R. J. Bradford, and J. H. Davenport, *The Bath algebraic number package*, in "Proc. SYMSAC '86," ACM, 1986, pp. 250–253.

[4] J. A. Abbott and J. H. Davenport, *Polynomial factorization: an exploration of Lenstra's algorithm*, in "Proc. EUROCAL '87," J. H. Davenport, editor; Lecture Notes in Computer Science 378, Springer-Verlag, Berlin-Heidelberg-New York, 1989, pp. 391–402.

[5] A. Aho, J. E. Hopcroft, and J. D. Ullman, "The Design and Analysis of Computer Algorithms," Addison-Wesley, Reading, Mass., 1974.

[6] D. S. Arnon, G. E. Collins, and S. McCallum, *Cylindrical algebraic decomposition*, SIAM Journal on Computing 13 (1984), 865–877, 878–889.

[7] E. R. Berlekamp, *Factoring polynomials over large finite fields*, Math. Comp. 24 (1970), 713–735.

[8] B. Buchberger, *Basic features and development of the critical pair completion procedure*, in "Rewriting Techniques and Applications," J. P. Jouannaud, editor; Lecture Notes in Computer Science 202, Springer-Verlag, Berlin-Heidelberg-New York, 1986, pp. 1–45.

[9] D. G. Cantor and E. Kaltofen, *Fast multiplication over arbitrary rings*, Manuscript, 1986.

[10] G. E. Collins, *The calculation of multivariate polynomial resultants*, Journal of the ACM 18 (1971), 515–532.

[11] G. E. Collins, *Quantifier elimination for real closed fields by cylindrical algebraic decomposition*, in "Second GI Conf. Automata Theory and Formal Languages," Lecture Notes in Computer Science 33, Springer-Verlag, 1975, pp. 134–183.

[12] D. Duval, *Diverse Questions relatives au Calcul Formel avec des Nombres Algébriques*, PhD thesis, L'université scientifique, technologique, et médicale de Grenoble, Grenoble, 1987.

[13] S. Landau, *Factoring polynomials over algebraic number fields*, SIAM Journal on Computing 14 (1985), 184–195.

[14] L. Langemyr, *Algorithms for a Multiple Algebraic Extension*, Technical Report, Wilhelm-Schickard-Institut für Informatik, D-7400 Tübingen, 1990.

[15] L. Langemyr, *Computing the GCD of two Polynomials Over an Algebraic Number Field*, PhD thesis, Royal Institute of Technology, Stockholm, 1988.

[16] L. Langemyr and S. McCallum, *The computation of polynomial greatest common divisors over an algebraic number field*, J. Symbolic Comp. 8 (1989), 429–448.

[17] R. G. K. Loos, *Generalized polynomial remainder sequences.*, in "Computer Algebra, Symbolic and Algebraic Computation," B. Buchberger, G. E. Collins, and R. G. K. Loos, editors, Springer-Verlag, Wien-New York, 1982, pp. 115–137.

[18] J. B. Rosser and L. Schoenfeld, *Approximate formulas for some functions of prime numbers*, Illinois J. Math. 6 (1962), 64–94.

[19] P. J. Weinberger and L. P. Rothschild, *Factoring polynomials over algebraic number fields*, ACM Transactions on Mathematical Software 2 (1976), 335–350.

Lars Langemyr
Wilhelm-Schickard-Institut für Informatik
Universität Tübingen
Auf dem Sand 13
D–7400 Tübingen, FRG

Elementary constructive theory
of ordered fields

HENRI LOMBARDI MARIE-FRANÇOISE ROY

1. Introduction

The classical theory of ordered fields (Artin-Schreier theory) makes intensive use of non-constructive methods, in particular of the axiom of choice. However since Tarski (and even since Sturm and Sylvester) one knows how to compute in the real closure of an ordered field $\mathbf{K}$ solely by computations in $\mathbf{K}$. This apparent contradiction is solved in this paper.

We give here a constructive proof of the first results of the theory of ordered fields, including the existence of the real closure.

The proofs can be interpreted in the particular philosophy of each reader. In a classical point of view for example, the effective procedures in the definitions may be interpreted as given by oracles. Hence one gets the existence of the real closure of an arbitrary ordered field without the axiom of choice. In a constructive framework "à la Bishop" one gets the existence of the real closure of a discrete ordered field. The reference for discrete fields is [MRR]. From the point of view of classical recursive theory the proofs give uniformly primitive recursive algorithms for Turing machines with oracles (cf [Kl]).

The essential tools needed are the following: a constructive version of the mean value theorem in an ordered field, the notions of prime cone (see [BCR]) and of ordered d-closed field.

The use of algorithm IF from [CR] gives a concrete representation for elements of the real closure, with no need of primitive elements.

Through the paper "A real root calculus" of Zassenhauss ([Za]), we discovered recently Holkott 's thesis [Ho]. Holkott's method and ideas are, sometimes surprisingly, very similar to ours. Our paper can be considered as a modern and, we hope, clearer presentation of Holkott's results. Thanks to L. Gonzalez for communicating the reference [Za] and to T. Sander for translating to us decisive parts of [Ho].

Tomas Sander also studied recently and independently the existence of the real closure without the axiom of choice ([Sa]).

2. Preliminaries

Ordered fields.

Definition 1: A set is *discrete* when the equality of two elements is decidable. A *discrete field* is a field $\mathbf{K}$, discrete as a set and in which the laws of addition, multiplication, opposite and inverse are computable. A discrete ordered field $\mathbf{K}$ is a discrete field where the sign of an element is decidable. From now on, all fields and ordered fields considered are assumed to be discrete.

Remark 1: An ordered field with an oracle giving results of arithmetic operations and sign of elements is a discrete ordered field. A codable ordered field where elements are represented by a finite data structure and where arithmetic operations and sign determinations are given by algorithms is a discrete ordered field.

Let $\mathbf{K}$ be an ordered field. An open interval is by definition a set

$$]a,b[= \{x \in \mathbf{K} \mid a < x < b\}$$

where a and b are in $\mathbf{K}$ or equal to $+\infty$ or $-\infty$.

Theorem 1 (constructive mean value theorem). *Let $\mathbf{K}$ be an ordered field, a and b two elements of $\mathbf{K}$ with $a < b$.*
There exist two families $(\lambda_{n,i})_{n\in\mathbb{N},i=1,2,\ldots,n}$ and $(r_{n,i})_{n\in\mathbb{N},i=1,2,\ldots,n}$ of rational numbers in $]0,1[$ such that, for every polynomial P of $\mathbf{K}[X]$ of degree $\leq n$, the following equality holds:

$$P(a) - P(b) = (a-b) \sum_{i=1,\ldots,n} r_{n,i}.P'(a + \lambda_{n,i}(b-a)).$$

In particular
1) if P' is positive on an interval, P is increasing on this interval,
2) on every bounded interval the function defined by P is Lipschitz.

Proof: The theorem is an immediate consequence of the following lemma:

Lemma. *There exist two families $(\lambda_{n,i})_{i=1,2,\ldots,n}$ and $(r_{n,i})_{i=1,2,\ldots,n}$ of rational numbers in $]0,1[$ such that, for every polynomial P in $\mathbb{Q}[X]$ of degree $\leq n$, the following equality holds:*

$$P(a) - P(b) = (a-b) \sum_{i=1,\ldots,n} r_{n,i}.P'(a + \lambda_{n,i}(b-a)).$$

The lemma gives algebraic identities about variables a, b, and the coefficients of the polynomial which are valid in any commutative ring which is a $\mathbb{Q}$-algebra, and in particular in fields of characteristic zero.

Let us prove the lemma. Using an affine change of coordinates one may suppose $a = -1$ and $b = 1$. Let the degree n be fixed. The function sending P to $P(1) - P(-1)$ is a linear form where the constant coefficient plays no role. Such linear forms constitue a vector space of dimension n. For every choice of n different rational numbers $(\lambda_{n,i})_{i=1,\ldots,n}$, the linear forms sending P to $P'(\lambda_{n,i})$ are independent in this space. So to this choice corresponds rational numbers $r_{n,i}$ making the formula true. The only difficult point is to choose $\lambda_{n,i}$ in $]0,1[$ such that the corresponding $r_{n,i}$ are still in $]0,1[$. Gauss formulas (where one has to consider zeroes of Legendre polynomials, cf. [L]) correspond to such a choice, but with real numbers and not rational numbers. A choice of $\lambda_{n,i}$ rational numbers close enough to the $\lambda_{n,i}$ of Gauss ensures that the corresponding $r_{n,i}$ are still positive.

Remark 2: Explicit upper and lower bounds for P' are easy to compute on a bounded interval, hence a Lipschitz modulus for P.

Definition 2: A *sign condition* is a member of $\{> 0, = 0, < 0\}$. A *generalized sign condition* is a member of $\{< 0, \leq 0, = 0, > 0, \geq 0\}$. When a sign condition < 0 or > 0 is replaced by the corresponding generalized sign condition ≤ 0 or ≥ 0, the sign condition is said to have been *relaxed*.

A subset of an ordered field is *open* if it is a union of open intervals. A function from **K** to **K** is *continuous* if the inverse image of an open set is open.

Lemma. *A polynomial function from an ordered field into itself is continuous.*

2. Prime cones

Definitions 3 (see [BCR]:
  a) A *prime cone* of a ring **A** is a subset α such that
       1) $\forall x \in A, x^2 \in \alpha$,
       2) $\alpha + \alpha \subset \alpha$,
       3) $\alpha.\alpha \subset \alpha$,
       4) $\forall x \in \mathbf{A}, \forall y \in \mathbf{A}, xy \in \alpha \Rightarrow x \in \alpha$ or $-y \in \alpha$.
  b) The *support* of α, $\mathrm{Supp}(\alpha) = \alpha \cap -\alpha$, is a prime ideal whose residue field $k(\mathrm{Supp}(\alpha))$ is ordered: positive or zero elements of $k(\mathrm{Supp}(\alpha))$ are images of elements of α.
  c) Let **K** be an ordered field and **A** a **K**-algebra. The prime cone α is *compatible with the order of* **K** if moreover
       5) $\alpha \cap \mathbf{K} = \{x \in \mathbf{K} \mid x \geq 0\}$.
  The field $k(\mathrm{Supp}(\alpha))$ is then an ordered extension of **K**. Let **L** be an ordered extension of **K** and f a ring homomorphism of **A** in **L**. **L** is an ordered extension of $k(\mathrm{Supp}(\alpha))$ if and only if $\{x \in \mathbf{A} \mid f(x) \geq 0 \text{ in } \mathbf{L}\} = \alpha$.

d) When $k(\text{Supp}(\alpha))$ is an algebraic extension of $\mathbf{K}$, α is *algebraic over* $\mathbf{K}$.

e) Let us denote by α_0, α_+ and α_- the subsets of $\mathbf{A}$ of elements whose images in $k(\text{Supp}(\alpha))$ are 0, $+1$ and -1. Then $\alpha_0 = \text{Supp}(\alpha)$ and $\alpha = \alpha_0 \cup \alpha_+$.

Axioms 1), 2), 3) and 4) can be rewritten as

1') $\mathbf{A}$ is the disjoint union of α_0, α_+ and α_-, and $\alpha_- = -\alpha_+$,

2'a) $\alpha_0 + \alpha \subset \alpha$,

2'b) $\alpha_+ + \alpha_+ \subset \alpha_+$,

3'a) $\alpha_0.\alpha \subset \alpha_0$,

3'b) $\alpha_+.\alpha_+ \subset \alpha_+$.

f) When $\mathbf{A} = \mathbf{K}[X]$ one writes X_α for the image of X in $k(\text{Supp}(\alpha))$. When moreover α is algebraic over $\mathbf{K}$ one writes $\mathbf{K}[X_\alpha]$ for the ordered field $k(\text{Supp}(\alpha))$.

3. d-closed ordered fields

3.1. Definitions.

Definition 4: A field is *real* if -1 is not a sum of squares.

An ordered field is *d-closed* (where $d \geq 1$) if every polynomial P of degree $\leq d$ such that $P(a)P(b) < 0$ has a root on the interval $]a, b[$.

In the classical situation, this definition is equivalent to the definition of d-real closed field in [B].

Remark 3: Every ordered field is real and 1-closed. Every real field is of characteristic zero.

Comment: In the classical theory, using Zorn's lemma it is possible to prove that any real field can be ordered. This is no longer true from a constructive point of view. More concretely it is impossible to prove constructively that in a real field it is possible to add a real square root to a or to $-a$ and get a real extension: it would be necessary to assert that a or $-a$ is not a sum of squares. This would clearly imply the "lesser limited principle of omniscience" (LLPO) which is not constructively valid (cf [MRR], Chapter 1). An example of recursive real field not recursively orderable appears in [MN].

3.2. Construction of the 2-closure of an ordered field.

Definition 5: An ordered extension $\mathbf{R}$ of an ordered field $\mathbf{K}$ is an *ordered 2-closure* of $\mathbf{K}$ if it is a 2-closed ordered field and if every element of $\mathbf{R}$ is obtained starting from elements of $\mathbf{K}$ by repetition of arithmetic operations and extraction of the real square root of a positive element.

The next proposition is training for the proof of the existence of the real closure that will be proved later along the same lines.

Proposition 1. *Every ordered field* **K** *has an ordered 2-closure, unique up to (unique)* **K**-*isomorphism of ordered fields.*

Proof: If a is a positive element of **K**, it is easy to see that there exists an ordered extension **K** obtained by adding a positive real square root of a: without taking into consideration the fact that **K** might or might not have had such a positive real root, one may give without ambiguity a sign to each expression $x + y\sqrt{a}$, where x and y are in **K**, hence also to every expression $Q(\sqrt{a})$ where $Q \in \mathbf{K}[X]$, by considering the remainder of the division of $Q(X)$ by $X^2 - a$; this defines a prime cone of $\mathbf{K}[X]$, the corresponding residue field is denoted by $\mathbf{K}[\sqrt{a}]$.

If **L** is an ordered extension of **K** in which a has a positive square root a', there exists a unique **K**-isomorphism of ordered fields from $\mathbf{K}[\sqrt{a}]$ to $\mathbf{K}[a']$ (the subfield of **L** generated by **K** and a').

This implies the following lemma:

Lemma. *Let a and b be two positive elements of an ordered field* **K**. *The ordered fields* $\mathbf{K}[\sqrt{a}][\sqrt{b}]$ *and* $\mathbf{K}[\sqrt{b}][\sqrt{a}]$ *are isomorphic as ordered extensions of* **K**.

Let us consider now the union of all $\mathbf{K}[\sqrt{a_1}][\sqrt{a_2}]\cdots[\sqrt{a_i}]$ with a_j, $(j = 1, \ldots, i)$ positive in $\mathbf{K}[\sqrt{a_1}][\sqrt{a_2}]\cdots[\sqrt{a_{j-1}}]$. The ordered 2-closure we look for, will be the quotient of this union by an equivalence relation.

Let us define this equivalence relation. Let

$$\mathbf{K}_1 = \mathbf{K}[\sqrt{a_1}][\sqrt{a_2}]\cdots[\sqrt{a_i}]$$

with $a_j(j = 1, \ldots, i)$ positive in $\mathbf{K}[\sqrt{a_1}][\sqrt{a_2}]\cdots[\sqrt{a_{j-1}}]$ and

$$\mathbf{K}_2 = \mathbf{K}[\sqrt{b_1}][\sqrt{b_2}]\cdots[\sqrt{b_{i'}}]$$

with $b_j(j = 1, \ldots, i')$ positive in $\mathbf{K}[\sqrt{b_1}][\sqrt{b_2}]\cdots[\sqrt{b_{j-1}}]$. Let us define

$$\mathbf{K}' = \mathbf{K}_1[\sqrt{b_1}][\sqrt{b_2}]\cdots[\sqrt{b_{i'}}].$$

Using several times the lemma one has a unique **K**-isomorphism from $\mathbf{K}'$ to

$$\mathbf{K}'' = \mathbf{K}_2[\sqrt{a_1}][\sqrt{a_2}]\cdots[\sqrt{a_i}].$$

By definition, elements of $\mathbf{K}_1$ and $\mathbf{K}_2$ are equivalent if their images in $\mathbf{K}'$ and $\mathbf{K}''$ coincide up to the isomorphism. This defines an equivalence relation compatible with the ordered field structure: reflexivity and symmetry are immediate. Transitivity involves three extensions. The ordered 2-closure is then the quotient of the union of $\mathbf{K}[\sqrt{a_1}][\sqrt{a_2}]\cdots[\sqrt{a_i}]$ (with a_j, $(j = 1, \ldots, i)$ positive in $\mathbf{K}[\sqrt{a_1}][\sqrt{a_2}]\cdots[\sqrt{a_{j-1}}]$) by this equivalence relation.

3.3. Sign conditions.

Definition 6: Let $L = [P_1, P_2, \ldots, P_k]$ be a list of polynomials of $\mathbf{K}[X]$ of degrees less than or equal to d, where $\mathbf{K}$ is a subfield of a d-closed field $\mathbf{R}$. The complete list of signs of the list L is known when the roots of P_i in $\mathbf{R}$ have been computed, they are in increasing order, and the sign of each of the polynomials in each of these roots and on each interval between these roots is computed.

Theorem 2. *Let $\mathbf{K}$ be an ordered field, subfield of a d-closed ordered field $\mathbf{R}$. Let*
$$L = [P_1, P_2, \ldots, P_k]$$
be a list of polynomials of $\mathbf{K}[X]$ of degrees less than or equal to d. It is possible to compute the complete list of signs of L.

Proof: Because of theorem 1 and of the intermediate value theorem for polynomials of degree less than or equal to d, we have all the tools needed to apply Hörmander's method to L (cf. [BCR] Chapter 1).

Comment: For a constructivist this theorem has the following provoking corollary: *in a d-closed ordered field, the roots of a polynomial of degree $\leq d$ form a finite set.*

Theorem 3 (Thom's lemma). *Let $\mathbf{K}$ be an ordered field contained in a d-closed ordered field $\mathbf{R}$, P be a polynomial of $\mathbf{K}[X]$, of degree $n \leq d$, and $[\sigma_0, \sigma_1, \ldots, \sigma_n]$ be a list of sign conditions other than $= 0$. The set*

$$A_\sigma = \{x \in \mathbf{R} \mid P(x)\sigma_0, P'(x)\sigma_1, \ldots, P^{(i)}(x)\sigma_i, \ldots, P^{(n-1)}(x)\sigma_{n-1}\}$$

is either empty, or an open interval with endpoints $+\infty$, $-\infty$, or a root of one of the polynomials $P, P', P'', \ldots$. If the sign conditions are relaxed, and if the open A_σ were a non empty interval, one gets the corresponding closed interval. If now the first condition is $= 0$, the set has zero or one point.

Proof: One can perform the usual proof by induction on the degree of P (cf. [BCR]).

3.4. Sturm's algorithm.

Definition 6: Let $\mathbf{K}$ be an ordered field. Let P and Q be two polynomials with coefficients in $\mathbf{K}$ and R be the remainder of the euclidean division of $P'Q$ by P. Let a and b be two elements of $\mathbf{K}$ with $a < b$ (or possibly $a = -\infty$, $b = +\infty$), a and b not being roots of P.

The Sturm sequence of P and Q is defined by

$\mathrm{Stu}_0(P,Q) = P$

$\mathrm{Stu}_1(P,Q) = R$

$\mathrm{Stu}_{i+1}(P,Q) = -\,\mathrm{Remainder}(\mathrm{Stu}_i(P,Q),\mathrm{Stu}_{i-1}(P,Q))$

The Sturm sequence of P is obtained when $Q = 1$.

One denotes by $v_{\mathrm{St}}(P,Q,a,b)$ the difference between the number of sign variations in the Sturm sequence at a and at b.

Theorem 4 (Sturm-Sylvester in degree $\leq d$ in a d-closed ordered field). *Let* **K** *be an ordered field, subfield of a d-closed ordered field* **R**. *Let P and Q be two polynomials with coefficients in* **K** *with P of degree less than or equal to d. Using the preceding notations, the number $v_{\mathrm{St}}(P,Q,a,b)$ is equal to the difference between the number of roots of P between a and b with $Q > 0$ and the number of roots of P between a and b with $Q < 0$.*

Proof: The classical proof (see for example [**GLRR**]) works because of theorem 2.

Remark 4: There are examples of ordered fields with polynomials P of constant sign on an interval, but with the number of roots predicted by Sturm non zero: add to **Q** an infinitely small positive element ϵ, and consider the polynomial $P = (X^2 - \epsilon^3).(X^3 - \epsilon^4)$ and the interval $[\epsilon^2, \epsilon]$.

Proposition 2 (polynomial of degree $d+1$ in an ordered d-closed field). *Let P be a polynomial of degree $d + 1$ in an ordered d-closed field* **K** *and let* $]a,b[$ $(a < b)$ *be an interval of the field* **K** *such that P is not 0 at a and at b. If P is square free $v_{\mathrm{St}}(P,1,a,b)$ gives the number of sign changes of P on* $]a,b[$. *In particular $v_{\mathrm{St}}(P,1,a,b)$ is always positive or zero and the number of roots of P in* **K** *over* $]a,b[$ *is less than or equal to $v_{\mathrm{St}}(P,1,a,b)$. If P is reducible in* **K**$[X]$ *(in particular if it is not square-free) $v_{\mathrm{St}}(P,1,a,b)$ is equal to the number of roots of P in* **K** *over* $]a,b[$.

Proof: When P is square-free, consider the roots of all polynomials in the Sturm-sequence except P in **K** and repeat the usual proof. When P is reducible in **K**$[X]$ repeat the usual proof (see for example [**GLRR**]).

3.5. Algorithm IF.

Algorithm IF ("inégalités formelles") proposed in [**CR**] (on the basis of [**BKR**]) in order to determine, by computations in **K** (only arithmetic operations and sign determinations) the signs of a list of polynomials at the roots of a polynomial of degree less than or equal to d may be applied in any ordered field **K** with d-closed ordered extension **R** because of preceding theorems.

Algorithm IF, applied to P (of degree less than or equal to d) and its derivatives, is called RAN (Real Algebraic Number) and works in any ordered field **K** with d-closed ordered extension **R**: that is to say that, to every sign condition on the derivatives predicted by RAN, there corresponds effectively a root of P in **R** satisfying these sign conditions.

One may also use systems of equations.

A triangular system of equations (of degrees less than or equal to d) over the field **K** is given by a list of polynomials

$$P = [P_1, P_2, \ldots, P_k]$$

with

$$P_1 \in \mathbf{K}[X_1], P_2 \in \mathbf{K}[X_1, X_2], \ldots, P_k \in \mathbf{K}[X_1, X_2, \ldots, X_k]$$

each P_j being monic of degree d_j as polynomial in X_j with $d_j \geq 2$ for every j and $d_{X_h}(P_j) < d_h$ for every $h < j$. A real solution of the system defined by the list P is a k-tuple $x = [x_1, x_2, \ldots, x_k]$ in an ordered extension of **K**, with:

$$P_1(x_1) = 0, P_2(x_1, x_2) = 0, \ldots, P_k(x_1, x_2, \ldots, x_k) = 0.$$

If **K** has a d-closed ordered extension **R**, and if all the d_i are less than or equal to d, a root in **R** of the triangular system may be characterized à la Thom, by the list of signs of the derivatives of the $P_i(x_1, x_2, \ldots, x_{i-1}, X)$ at $X = x_i$, by computations only in **K**.

The computation goes as follows: the case of one variable corresponds to algorithm RAN above. In the case of a triangular system one applies the preceeding algorithm IF in an iterative way (with respect to the number of variables) and determines, by computations in **K**, all the codings à la Thom of the solutions $(x_1, x_2, \ldots, x_k)$ in $\mathbf{R}^k$ of the system.

Theorem 6. *Let* **K** *be an ordered field contained in a d-closed ordered extension* **R**. *It is possible, by computations in* **K**, *to characterize à la Thom the roots in* **R** *of a triangular system of equations with coefficients in* **K** *(of degrees less than or equal to d) and to decide the sign of every polynomial* $\mathbf{K}[X_1, \ldots, X_k]$ *at these roots.*

4. Real closure

4.1. Real closed field.

Definition 7: A field **K** is *real closed* if it is ordered, if every positive element is a square, and if every polynomial of odd degree has a root.

Theorem 7. *Let* **K** *be a field. The following properties are equivalent*
  a) **K** *is real closed,*
  b) **K** *is ordered, and d-closed for every integer d,*
  c) **K** *is real and* $\mathbf{K}[\sqrt{-1}]$ *is algebraically closed,*

d) **K** *is real and every polynomial is decomposable in factors of degree one or two,*

e) **K** *is ordered and the number of roots on an interval* $]a, b[$ $(a < b)$ *coincides with the number given by applying the Sturm's Theorem.*

Proof:

a) $\Rightarrow$ b) is clear (cf. [**BCR**] page 9).

b) $\Rightarrow$ a) is immediate.

a) $\Rightarrow$ c) as in [**BCR**] page 9.

c) $\Rightarrow$ d) group the conjugate roots.

d) $\Rightarrow$ a) one starts by proving that for every a, a or $-a$ is a square: it is sufficient to decompose the polynomial $T^4 - a$ as a product of two monic polynomials of degree 2 and to equate coefficients; hence **K** is ordered and 2-closed; one constructs easily the sign table of any polynomial, and it is then clear that it has a root on every interval where its sign changes (irreducible factors of degree 2 have no influence on the sign table).

a) $\Rightarrow$ e) after theorem 4

e) $\Rightarrow$ b) Sturm's algorithm prescribes two roots to a polynomial $X^2 - c$ with $c > 0$ hence **K** is 2-closed. Then one proves by induction on d that **K** is d-closed using Proposition 2.

4.2. How to add one root.

Proposition 3. *Let* **K** *be a d-closed ordered field, P be a polynomial of degree $d + 1$, a and b, $a < b$, be two elements of* **K**. *Let us suppose that $P(a).P(b) < 0$ and that P' is of constant sign over $]a, b[$. There exists a unique prime cone α of* **K**$[X]$ *algebraic over* **K** *such that X_α satisfies $P(X_\alpha) = 0$ and $a < X_\alpha < b$. Moreover in any ordered extension* **L** *of* **K**, *with a root c of P in $]a, b[$, there exists a unique* **K**-*isomorphism of ordered fields from* **K**$[X_\alpha]$ *to the subfield* **K**$[c]$ *of* **L**.

Proof: Let suppose for example that P' is positive over the interval. Let Q be a polynomial of **K**$[X]$ and let us decide whether it belongs to α. Let Q_1 be the remainder of the division of Q by P. If Q_1 is zero (case 1) one has $Q \in \alpha$. Else, let us compute the subdivision defined by the roots of Q_1 over the interval $]a, b[$, and so the ordered list $[a = u_0, u_1, \ldots, u_n = b]$. The successive values of P are in strictly increasing order (by theorem 1). If $P(u_i) = 0$ for some i, (case 2), one has to take $Q \in \alpha$. Else P passes from sign $-$ to sign $+$ over one of the subintervals $[u_i, u_{i+1}]$, and Q_1 is of known constant sign σ over the interval $]u_i, u_{i+1}[$ (case 3). One has to take then $Q \in \alpha$ if σ is > 0.

Let us verify that we have defined a prime cone. Let us make two preliminary remarks. First, in the case when P has a root c in **K** on $]a, b[$, Q belongs to α_0 (resp. α_+, α_-) if and only if $Q(c)$ is 0 (resp. > 0, < 0) and it is clear that we have a prime cone. So we never have to consider case 2.

For the same reason we never have to consider in the proof cases where P is 0 at the root of a polynomial of degree $\leq d$. Second, if there exists an ordered extension $\mathbf{L}$ of $\mathbf{K}$ in which P has a root c on $]a, b, [$, P belongs to α_0 (resp. α_+, α_-) if and only if $P(c)$ is 0 (resp. > 0, < 0). This implies that α is a prime cone, as well as the existence of a unique $\mathbf{K}$-isomorphism from $\mathbf{K}[X_\alpha]$ to the subfield $\mathbf{K}[c]$ of $\mathbf{L}$.

Conditions 1') and 5) of definition 3 are trivially verified. Let us look at conditions 2'a) 2'b), 3'a), 3'b).

2'a) and 3'a): Let us suppose that Q is in α_0 (case 1). Then $Q + S$ and S have the same remainder modulo P, this implies $\alpha_0 + \alpha \subset \alpha$. Also QS is 0 modulo P hence $\alpha_0 . \alpha \subset \alpha_0$.

2'b) $\alpha_+ + \alpha_+ \subset \alpha_+$: Let Q and S be in α_+ (case 3), Q_1 and S_1 be the remainders of their euclidean division by P. Let us denote by $[u_0, u_1, \ldots, u_n]$ and $[v_0, v_1, \ldots, v_m]$ the subdivisions introduced by the roots of Q_1 and S_1 respectively. Let us join them in one subdivision, $[w_0, w_1, \ldots, w_l]$. The two polynomials Q_1 and S_1 are positive over the open interval of this subdivision where P changes sign. Hence $Q_1 + S_1$ is also positive on this interval and the interval is a subinterval of those considered for the assignment of a sign to $Q + S$ via $Q_1 + S_1$.

3'b) $\alpha_+ . \alpha_+ \subset \alpha_+$: The case of the product is slightly more complicated. It is necessary to introduce R, the remainder of the division of QS by P, which is also the remainder of the division of $Q_1 S_1$ by P. One can consider the subdivision $[t_0, t_1, \ldots, t_s]$ associated to R and join the subdivisions u, v and t in one subdivision l. Let us define A as the quotient of $Q_1 S_1$ by P, that is by the equality $Q_1 S_1 = AP + R$. One has $\deg(A) < d$. Over the minimal open interval of the subdivision l where P changes sign, one knows that Q_1 and S_1 are > 0, hence if A is zero R is > 0 which means that QS is in α_+. Else it is necessary to consider also the subdivision associated to A and join it with l in a subdivision m. Over the interval of the subdivision $]c, d[$ where P changes sign A has a sign σ and we chose the endpoint of the interval where P has sign $-\sigma$. Since P is continuous, there exists a point c' of the interval where P has again sign $-\sigma$. The sign of R over the interval, which is constant, is then the same as the sign of $R(c') = (Q_1 S_1 - AP)(c')$, hence > 0.

Comment: We have not supposed P irreducible and we do not use factorization. It is well known that the existence of a factorization is not in general guaranteed from the constructive or computational point of view [Se].

4.3. Construction of the real closure.

Definition 8: A *real closure of an ordered field* $\mathbf{K}$ is an algebraic ordered extension of $\mathbf{K}$ which is a real closed field. An extension $\mathbf{R}$ of an ordered field $\mathbf{K}$ is an *ordered d-closure* of $\mathbf{K}$ if it is a d-closed ordered field and if every element of R can be obtained from elements of K by repetition of arithmetic operations and addition of a root of a polynomial of degree $\leq d$.

Theorem 7. *It is possible to construct a real closure for every ordered field K. The real closure is unique up to unique K-isomorphism of ordered fields.*

Proof: The proof is by induction on d, in order to show that:

(H_d) for every ordered field L we can construct a d-closure $L^{(d)}$, unique up to unique L-isomorphism of ordered fields. Moreover if M is a d-closed ordered extension of L there exists a unique increasing L-morphism from $L^{(d)}$ to M.

For $d = 1$, there is nothing to prove.

Let us suppose the hypothesis (H_d) true for d. If K is an ordered field, if P is a monic polynomial of degree $d + 1$ in $K^{(d)}[X]$, and if a and b are two consecutive roots of P' (or at infinity) satisfying $P(a).P(b) < 0$, we shall denote by $K^{(d)}[X_\alpha]^{(d)}$ the d-closure of the field $K^{(d)}[X_\alpha]$ with X_α root of P in $]a, b[$.

This ordered extension of K is unique up to (unique) K-isomorphism of ordered fields as d-closed ordered extension of $K^{(d)}$ containing a root of P over $]a, b[$. More precisely hypothesis (H_d) and proposition 3 show the following lemma.

Lemma. *If M is a d-closed extension of K there exists an algebraic ordered extension $M[X_\alpha]$ of M such that there exists a (unique) increasing K-morphism from $K^{(d)}[X_\alpha]$ into $M[X_\alpha]$.*

Let us use the following obvious notation when iterating the construction:

$$K^{(d)}[X_{\alpha_1}]^{(d)}[X_{\alpha_2}]^{(d)} \cdots [X_{\alpha_i}]^{(d)}.$$

To obtain $K^{(d+1)}$ one has to glue together all these extensions: which means introducing a good equivalence relation over their disjoint union. If

$$K_1 = K^{(d)}[X_{\alpha_1}]^{(d)}[X_{\alpha_2}]^{(d)} \cdots [X_{\alpha_i}]^{(d)}$$

and

$$K_2 = K^{(d)}[X_{\beta_1}]^{(d)}[X_{\beta_2}]^{(d)} \cdots [X_{\beta_j}]^{(d)}$$

are two extensions as before, there exists a unique K-isomorphism of ordered fields of the composite extension

$$K' = K_1[X_{\beta_1}]^{(d)}[X_{\beta_2}]^{(d)} \cdots [X_{\beta_j}]^{(d)}$$

to

$$K'' = K_2[X_{\alpha_1}]^{(d)}[X_{\alpha_2}]^{(d)} \cdots [X_{\alpha_i}]^{(d)}.$$

An element of K_1 will be considered as equal (in $K^{(d+1)}$) to an element of K_2, if and only if their images in K' and K'' coincide up to the isomorphism. This defines an equivalence relation compatible with the ordered field structure: reflexivity and symmetry are immediate. Transitivity involves three extensions. It is clear that one gets in this way a $(d+1)$-closed extension and that it is unique up to unique K-isomorphism of ordered fields.

It would be interesting to have a more direct proof of the following corollary.

Corollary. *In every ordered field, the Sturm algorithm prescribes a number of roots positive or zero.*

4.4. Data structure for the real closure.

The preceding theorem does not give immediately a finite data structure for the elements of the real closure since it is necessary to construct a lot of d-closures. Thinking a little about the proof one sees that the whole d-closure is not needed and that it would suffice to add a finite number of roots of polynomials of degree $\leq d$ (essentially the polynomials needed in Hörmander's method (cf [BCR], Chapter 1)). This point of view would lead to a much more technical proof of the existence of the real closure.

Since we proved that every ordered field may be embedded in a real closed field, it will be possible now to give a more concrete description of the real closure.

We have the following result:

Proposition 4. *The subfield of the real closure* R *of* K *consisting of the roots in* R *of triangular systems with coefficients in* K *is a real closed field equal to* R.

Proof: The ring structure is clear. The existence of an inverse is shown by induction on the number k of equations of a triangular system. Finally it is clear that by adding one variable one can represent the square root of a positive number and the roots of polynomials of odd degree with coefficients in K.

If one deals with a codable field it is thus possible to represent an element of the real closure as a polynomial expression of a real root coded à la Thom of a triangular system. One has to note that a given element of the real closure admits several representations and that it is possible to test by algorithm IF (with computations only in K) whether two representations correspond or not to the same element. The computer algebra system SCRATCHPAD where one may use ordered fields as parameters will be necessary to implement our point of view.

5. Constructive theory of real closed fields

Theorem 10 (Tarski-Seidenberg principle). *Let K be an ordered field, subfield of a real closed field R and Φ be a formula of the language of ordered fields in $n+1$ variables with coefficients in K, and without quantifiers. There exists a formula Ψ of the language of ordered fields with coefficients in K in n variables without quantifiers such that*

$$\{y \in R^n \mid \exists x \in R \ \Phi(x,y)\} = \{y \in R^n \mid \Psi(y)\}$$

Proof: As in [BCR] by using Hörmander's method since all the tools needed are available.

It is not difficult to mimic the previous proofs in the framework of the formal intuitionistic theory of discrete real closed fields with parameters in K. The general excluded-middle principle is not used, but one has a restricted excluded-middle of the form:

$$\forall x \ x > 0 \text{ or } x = 0 \text{ or } x < 0$$

which is a formal translation of the discrete character of the order considered. It is not possible to put immediately every formula under prenex form. Nevertheless the Tarski-Seidenberg principle above implies the possibility of eliminating one quantifier $\exists$ (before a quantifier free formula), hence to eliminate quantifiers even in formulas not in prenex form. So that the theory is also complete. The existence of a model (the real closure of K) gives a constructive proof of the consistency of this formal theory. In short, as far as first order statements are concerned, one can use either classical logic or intuitionistic logic in a real closed field. Let us note also that a direct proof of the consistency and of the completeness of the formal intuitionistic theory considered would not give a method for constructing the real closure of K, as we can see in the example of the theory of discrete algebraically closed fields (the "completeness theorem" is not valid constructively; on the contrary the consistency of the theory is assured as soon as any denumerable field has an algebraic closure).

Theorem 11. *Let K be an ordered field and $T_1(K)$ be the formal intuitionistic theory of real closed discrete fields with parameters in K. Then $T_1(K)$ is decidable, complete and non contradictory. In particular, for every formula F, "F or not F" is a theorem.*

REFERENCES

[BCR]   J. Bochnak, M. Coste, M.-F. Roy, "Géométrie algébrique réelle," Springer Verlag, 1987.

[BKR]   M. Ben-Or, D. Kozen, J. Reif, *The complexity of elementary algebra and geometry*, J. of Computation and Systems Sciences 32 (1986), 251–264.

[B]   S. Boughattas, "L'arithmétique ouverte et ses modèles non-standards," Thèse, Université Paris VI, 1987.

[CR]   M. Coste, M.-F. Roy, *Thom's lemma, the coding of real algebraic numbers and the computation of the topology of semi-algebraic sets*, J. Symbolic Computation 5 (1988), 121–129.

[GLRR] L. Gonzalez, H. Lombardi, T. Recio, M.-F. Roy, *Spécialisation de la suite de Sturm et sous-résultants*, To appear in RAIRO Informatique théorique. Detailed version, in CALSYF Journées du GRECO de Calcul Formel 1989.

[Ho]   A. Hollkott, "Finite Konstruktion geordneter algebraischer Erweiterungen von geordneten Grundkörpen," Dissertation. Hamburg, 1941.

[Kl]   S. C. Kleene, "Introduction to Metamathematics.," Van Nostrand, 1952.

[L]   J. Legras, "Méthodes numériques," Dunod, 1963.

[MN]   G. Metakides, A. Nerode, *Effective content of field theory*, Annals of Math. Logic 17 (1979), 289–320.

[MRR]   R. Mines, F. Richman, W. Ruitenburg, "A Course in Constructive Algebra," Universitext, Springer-Verlag, 1988.

[Sa]   T. Sander, *Existence and uniqueness of the real closure of an ordered field*, Journal of Pure and Applied Algebra (to appear).

[Se]   A. Seidenberg, *Constructions in algebra*, Transactions of AMS 197 (1974), 273–313.

[Za]   H. Zassenhauss, *A real root calculus*, in "Computational aspects in abstract algebra," Proceedings of a Conference held at Oxford, Pergamon Press, 1967, pp. 383–392.

Henri Lombardi
Mathématiques UFR des Sciences et Techniques
Université de Franche-Comté
25 030 Besançon cédex
France

Marie-Françoise Roy
I R M A R Université de Rennes 1
Campus de Beaulieu
35 042 Rennes cedex
France

Effective real Nullstellensatz and variants

HENRI LOMBARDI

Abstract. We give a constructive proof of the real Nullstellensatz. So we obtain, for every ordered field **K**, a uniformly primitive recursive algorithm that computes, for the input "a system of generalized signs conditions (gsc) on polynomials of $\mathbf{K}[X_1, X_2, \ldots, X_n]$ impossible to satisfy in the real closure of **K**, an algebraic identity that makes this impossibility evident. The main idea is to give an "algebraic identity version" of universal and existential axioms of the theory of real closed fields, and of the simplest deduction rules of this theory (as Modus Ponens). We apply this idea to the Hörmander algorithm, which is the conceptually simplest test for the impossibility of a gsc system in the real closure of an ordered field.

1) Introduction

This paper is the direct successor of [**LR**], where we develop the constructive elementary theory of ordered fields, in particular the constructive proof of the existence of the real closure of an ordered field **K** when one has a test for the sign of an element of **K**.

Here, we give a constructive proof of the real Nullstellensatz and its variants.

The fundamental theorem from which one can deduce the real Nullstellensatz and its variants is the following (cf [**BCR**] theorem 4.4.2): let **R** be the real closure of an ordered field **K**, $\mathbf{K}[\mathbf{X}]$ the ring $\mathbf{K}[X_1, X_2, \ldots, X_n]$, I a finitely generated ideal of $\mathbf{K}[\mathbf{X}]$, $\mathcal{P}$ a finitely generated cone of $\mathbf{K}[\mathbf{X}]$ (containing the positive elements of **K**), $\mathcal{M}$ a finitely generated multiplicative monoid in $\mathbf{K}[\mathbf{X}]$; let us consider the semialgebraic subset S of $\mathbf{R}^n$ defined by:

$$S = \{x \in \mathbf{R}^n : f(x) = 0 \text{ for } f \in I, \ g(x) \geq 0 \text{ for } g \in \mathcal{P},$$
$$h(x) \neq 0 \text{ for } h \in \mathcal{M}\};$$

then S is empty if and only if there exist $f \in I$, $g \in \mathcal{P}$, $h \in \mathcal{M}$ with $f + g + h^2 = 0$.

The general idea of our constructive proof is the following one. For an ordered field **K** there is an algorithm, conceptually very simple, for testing if a system of *gsc* (generalized sign conditions) on polynomials in many variables is possible or impossible in the real closure of **K**. It is the Hörmander algorithm (cf. the proof of the Tarski-Seidenberg principle in [**BCR**] chap. 1), applied iteratively to diminish one at time the number of variables in the *gsc*. If one inspects the arguments on which the impossibility proof is based (in case of impossibility), one sees that there are essentially algebraic identities (corresponding to euclidean divisions), the mean value theorem and the existence of a root for a polynomial on an interval where it changes of sign. So the effective real ... -stellensatz will be obtained if one succeeds to "algebraicize" the basic arguments of the proof and the methods of deduction.

An important step has already been made with the algebraic version of the mean value theorem for polynomials (cf. [**LR**]). One can also verify that the purely universal axioms in the theory of ordered fields can be expressed in the form of *strong implications* (i.e. in an "algebraic identity" form, i.e. also in a "stellensatzised" form).

Another step consists in translating into a form of *constructions of strong implications* certain elementary methods of deduction (as: if $A \implies B$ and $B \implies C$ then $A \implies C$). It is necessary moreover to find an "algebraic identity" version of existential axioms in the theory of real closed fields. This is made through the notion of *potential existence*.

Let us point out also that an important simplification in the construction of the real Nullstellensatz is obtained through an "algebraic identity" version of the Thom's lemma, given by what we call the mixed Taylor formulas.

Although we adopt a priori a constructive framework "à la Bishop", as developed in [**MRR**] concerning the theory of discrete fields, since we don't define the meaning of the words "effective" and "decidable", all the proofs can be read through glasses adapted to the philosophy or to the working framework of any particular mathematician.

If one accepts the "classical" point of view for example, the effective procedures in the initializing definitions can be considered as given by oracles. So, the proofs given provide a proof in the classical framework, and *without the axiom of choice*, of the real Nullstellensatz (and variants) in an arbitrary ordered field.

If one accepts the classical "recursive" theory point of view, the proofs given provide uniformly primitive recursive algorithms, "uniformly" understood in relation to an oracle giving the structure of the field of coefficients of the *gsc* system.

In Bishop's framework, we obtain the real Nullstellensatz and its variants for an arbitrary discrete ordered field.

In this paper we give the theorems, and some relevant proofs. Detailed proofs, and more constructive comments, can be found in [**Lom**].

2) Strong incompatibilities, evidence and implications

Strong incompatibilities (definitions).
We consider an ordered field $\mathbf{K}$, and $\boldsymbol{X}$ denotes a list of variables $X_1, X_2 \ldots, X_n$. We then denote by $\mathbf{K}[\boldsymbol{X}]$ the ring $\mathbf{K}[X_1, X_2, \ldots, X_n]$. If F is a finite subset of $\mathbf{K}[\boldsymbol{X}]$, we let F^{*2} be the set of squares of elements in F, $\mathcal{M}(F)$ be the *multiplicative monoid generated by* $F \cup \{1\}$, and we shall let $\mathcal{M}_2(F) := \mathcal{M}(F^{*2})$ and $\mathcal{M}_1(F)$ be the part of $\mathcal{M}(F)$ formed by products where each element of appears at most once.

$Cp(F)$ is the *positive cone generated by* F (the additive monoid generated by elements of type $p \cdot P \cdot Q^2$ where p is positive in $\mathbf{K}$, P is in $\mathcal{M}(F)$, Q is in $\mathbf{K}[\boldsymbol{X}]$).[1] We note that we may assume that P is in $\mathcal{M}_1(F)$.

Finally, let $I(F)$ be the ideal generated by F.

Definition: Consider 4 finite subsets of $\mathbf{K}[\boldsymbol{X}]$: $F_>, F_\geq, F_=, F_\neq$, containing polynomials for which we want respectively the sign conditions $> 0, \geq 0$, $= 0, \neq 0$: we say that $\mathbf{F} = [F_>; F_\geq; F_=; F_\neq]$ is *strongly incompatible* in $\mathbf{K}$ if we have in $\mathbf{K}[\boldsymbol{X}]$ an equality of the following type:

(1) $S + P + Z = 0$ with $S \in \mathcal{M}(F_> \cup F_\neq^{*2})$,
$$P \in Cp(F_\geq \cup F_>), \ Z \in I(F_=)$$

Every strong incompatibility written in the form (1) can be rewritten as a strong incompatibility in the following form (2):

(2) $S + P + Z = 0$ with $S \in \mathcal{M}(F_>^{*2} \cup F_\neq^{*2})$,
$$P \in Cp(F_\geq \cup F_>), Z \in I(F_=)$$

We can indeed multiply the first equality by a suitable element of $\mathcal{M}_1(F_>)$ to obtain each polynomial (in the first term S) with an even exponent.

It is clear that a strong incompatibility is a very strong form of incompatibility. In particular, it implies it is impossible to give the indicated signs to the polynomials, in any extension of $\mathbf{K}$. If one considers the real closure $\mathbf{R}$ of $\mathbf{K}$, the previous impossibility is testable by Hörmander's algorithm, for example. Moreover it is then constructively equivalent to its formulation in form of various implications: for example "$P = 0 \implies Q > 0$" is equivalent to "$P = 0, -Q \geq 0$ is impossible". We shall speak thus of *strong incompatibility, strong implication*, or *strong evidence*, meaning always implicitly a strong incompatibility.

---

[1] It would be more correct to denote $Cp(F, \mathbf{K}^+; \mathbf{K}[\boldsymbol{X}])$ in order to state: a) the positive elements of $\mathbf{K}$ are in the positive cone, and b) the positive cone is the one generated in $\mathbf{K}[\boldsymbol{X}]$.

Notation: We use the following notation for a strong implication:

$$*([S_1 > 0, \ldots, S_i > 0, P_1 \geq 0, \ldots, P_j \geq 0, Z_1 = 0, \ldots, Z_k = 0,$$
$$N_1 \neq 0, \ldots, N_h \neq 0] \implies Q \, \tau \, 0)^*$$

Note that if one takes $1 = 0$ in the right-hand side in the above strong implication, and applies the definitions, one obtains exactly the strong incompatibility for the left-hand side of the implication. Thus we can formulate any strong incompatibility in form of a strong implication.

Notation: Let us denote by **H** the left-hand side in the preceding notation. Let us denote by **H**′ a system of *gsc* : $Q_1 \, \tau_1 \, 0, \ldots, Q_k \, \tau_k \, 0$. We then write: *(**H** $\implies$ **H**′)* to mean

$$*(\mathbf{H} \implies Q_1 \, \tau_1 \, 0)^* \text{ and } \ldots \text{ and } *(\mathbf{H} \implies Q_k \, \tau_k \, 0)^*$$

Remark: We could have an algebraic identity version for any quantifier free formula in the language of ordered rings with constants in **K**.

The Nullstellensatz and its variants.

The different variants of the Nullstellensatz in the real case are consequences of the following general theorem:

Theorem. *Let* **K** *be an ordered field and* **R** *a real closed extension of* **K**. *The three following facts, concerning a gsc system on polynomials of* **K**[**X**], *are equivalent:*

> *strong incompatibility in* **K**
> *impossibility in* **R**
> *impossibility in all the ordered extensions of* **K**

This Nullstellensatz was first proved in 1974 ([Ste]). Less general variants were given by Krivine ([Kri]), Dubois ([Du]), Risler ([Ris]), Efroymson ([Efr]). All the proofs up to now used the axiom of choice. The first formulations were geometric: affirmation of the existence of an algebraic identity insuring that a given polynomial satisfied a given *gsc* on an algebraic or semialgebraic given set.

One speaks of *Nullstellensatz* when one considers the condition for a polynomial to belong to the ideal of a given algebraic variety (i.e. an implication: "equalities to zero imply an equality to zero"); of weak *Nullstellensatz* when one considers the condition for a given algebraic variety to be empty (i.e. "equalities to zero are incompatible"), of *Positivestellensatz* when one considers the condition for a polynomial to be strictly positive on a given semi-algebraic variety (i.e. the general form of the incompatibility between *gsc*, seen as an implication, the conclusion of which is a strictly positive sign), of *Nichtnegativestellensatz* when one considers the condition for a polynomial to be nonnegative on a given semi-algebraic variety (i.e. the general form of the incompatibility between *gsc* seen as an implication the conclusion of which is a nonnegative sign). Let us for example give the general Positivestellensatz.

Theorem (Positivestellensatz). *Let* **K** *be an ordered field and* **R** *a real closed extension of* **K**. *Let* A *be the semi algebraic set in* $\mathbf{R}^n$ *defined as:*

$$A = \{x \in \mathbf{R}^n : S_1(x) > 0, \ldots, S_i(x) > 0, P_1(x) \geq 0, \ldots, P_j(x) \geq 0,$$
$$Z_1(x) = 0, \ldots, Z_k(x) = 0, N_1(x) \neq 0, \ldots, N_h(x) \neq 0\}$$

Let Q *be a polynomial in* **K**$[X]$. *Then* Q *is positive at each point of* A *if and only if there is algebraic identity:* $Q \cdot P = S \cdot N^2 + R + Z$ *where:*

P *and* R *are in the positive cone of* **K**$[X]$: $Cp(S_1, \ldots, S_i, P_1, \ldots, P_j)$;
Z *is in the ideal of* **K**$[X]$: $I(Z_1, \ldots, Z_k)$;
S *is in the monoid* $\mathcal{M}(S_1, \ldots, S_i)$ *and*
N *is in the monoid* $\mathcal{M}(N_1, \ldots, N_h)$

Some trivial strong implications.

Proposition 2. *We have the following strong implications.*

$$^*([U > 0, V > 0] \implies [U + V > 0, U \cdot V > 0])^*$$
$$^*([U + V \geq 0, U \cdot V > 0] \implies [U > 0, V > 0])^*$$
$$^*([U > 0, V \geq 0] \implies U + V > 0)^*$$
$$^*([U \geq 0, U \cdot V > 0] \implies V > 0)^*$$
$$^*(U \neq 0 \implies U^2 > 0)^*$$
$$^*(U^2 > 0 \implies U \neq 0)^*$$
$$^*(U = 0 \implies U \cdot V = 0)^*$$
$$^*(U = V \implies P(X, U) = P(X, V))^*$$
$$^*([U = V, V \, \tau \, 0] \implies U \, \tau \, 0)^* \quad (\cdot \tau 0 \text{ is a } gsc)$$
$$^*([W = 0, U = V + W \cdot Z] \implies U = V)^*$$
$$^*([W = 0, U = V + W \cdot Z, V \, \tau \, 0] \implies U \, \tau \, 0)^*$$
$$^*([\ ] \implies [1 + U^2 > U, 1 + U^2 > -U])^*$$

One proof: Let us prove the last but one strong implication when τ is $>$. We have to give a strong incompatibility between the following *gsc*:

$$W = 0, V + W \cdot Z - U = 0, V > 0, -U \geq 0$$

we can take:

$$V^2 + ((-U) \cdot V) + ((Z \cdot V) \cdot W + (-V) \cdot (V + W \cdot Z - U)) = 0$$

with

$$V^2 \in \mathcal{M}(F >^{*2} \cup F_{\neq}^{*2}), (-U) \cdot V \in Cp(F_\geq \cup F_>),$$
$$(Z \cdot V) \cdot W + (-V) \cdot (V + W \cdot Z - U) \in I(F_=)$$

Q. E. D.

Proposition 3 (substitution principle). *If, in a strong implication, one replaces each occurrence of one variable by a fixed polynomial, one obtains again a strong implication.*

The proof is trivial. So, the strong implications of proposition 2, stated for variables U and V, are also valid for polynomials $U(X)$ and $V(X)$.

Constructions of strong implications.

Definition 4: We speak of construction of a strong implication from other strong implications when we have an algorithm that constructs the first from the others. So it is a logical implication in the constructive meaning. We denote it by the symbol: $\vdash_{cons}$ For example we give explicitly (theorem 8) the construction which proves:

$$[{}^*(H \implies H')^* \text{ and } {}^*(H' \implies H'')^*] \vdash_{cons} {}^*(H \implies H'')^*$$

As another example, we can state the principle of substitution in the form:

$${}^*(H(X,W) \implies H'(X,W))^* \vdash_{cons} {}^*(H(X,P(X)) \implies H'(X,P(X)))^*$$

Lemma 5. *Let H be a gsc system on polynomials of $K[X]$, Q an element of $K[X]$. Then each strong implication of the form ${}^*(H \implies Q\,\tau\,0)^*$ (where τ is $=, <$ or $>$) can be interpreted as any "weaker" strong implication ${}^*(H \implies Q\,\tau'\,0)^*$. For example, one has*

$${}^*(H \implies Q > 0)^* \vdash_{cons} {}^*(H \implies Q \geq 0)^*$$

Proposition 6. *Let H be a gsc system on polynomials of $K[X], Q$ be an element of $K[X]$, then:*

$$[{}^*(H \implies Q \leq 0)^* \text{ and } {}^*(H \implies Q \geq 0)^*] \vdash_{cons} {}^*(H \implies Q = 0)^*.$$

Likewise:

$$[{}^*(H \implies Q \leq 0)^* \text{ and } {}^*(H \implies Q \neq 0)^*] \vdash_{cons} {}^*(H \implies Q < 0)^*$$

and

$$[{}^*(H \implies Q = 0)^* \text{ and } {}^*(H \implies Q \neq 0)^*] \vdash_{cons} {}^*(H \implies 1 = 0)^*.$$

Proof: Let us give the first construction. Call $F_>$, $F_\geq$, $F_=$, $F_\neq$ the 4 finite subsets of $K[X]$ containing polynomials for which we have respectively the sign conditions $> 0, \geq 0, = 0, \neq 0$ in the hypothesis H.

The hypothesis ${}^*(H \implies Q \leq 0)^*$ can be rewritten in the form ${}^*([H, Q > 0] \implies 1 = 0)^*$ and means that we have an equality:

$$S + P + Z = 0$$

with

$$S \in \mathcal{M}(F_>^{*2} \cup F_\neq^{*2} \cup \{Q^2\}), \quad P \in \mathcal{C}p(F_\geq \cup F_> \cup \{Q\}), \quad Z \in I(F_=)$$

i.e. also:

$$Q^{2n} \cdot S_1 + Q \cdot P_1 + R_1 + Z_1 = 0$$

with

$$S_1 \in \mathcal{M}(F_>^{*2} \cup F_{\neq}^{*2}), \ P_1, R_1 \in Cp(F_\geq \cup F_>), \ Z_1 \in I(F_=)$$

Likewise the hypothesis $^*(H \implies Q \geq 0)^*$ means we have an equality:

$$Q^{2m} \cdot S_2 - Q \cdot P_2 + R_2 + Z_2 = 0$$

with

$$S_2 \in \mathcal{M}(F_>^{*2} \cup F_{\neq}^{*2}), \ P_2, R_2 \in Cp(F_\geq \cup F_>), \ Z_2 \in I(F_=)$$

We rewrite the two equalities:

$$-Q \cdot P_1 = Q^{2n} \cdot S_1 + R_1 + Z1, \ Q \cdot P_2 = Q^{2m} \cdot S_2 + R_2 + Z_2$$

and we multiply: so

$$-Q^2 \cdot P_1 \cdot P_2 = Q^{2n+2m} \cdot S_1 \cdot S_2 + [Q^{2n} \cdot S_1 \cdot R_2 + Q^{2m} \cdot S_2 \cdot R_1 + R_1 \cdot R_2] + W$$

with $W \in I(F_=)$, so $Q^{2n+2m} \cdot S_1 \cdot S_2 + V + W = 0$ with:

$$S_1 \cdot S_2 \in \mathcal{M}(F_>^{*2} \cup F_{\neq}^{*2}), \ V \in Cp(F_\geq \cup F_>), \ W \in I(F_=)$$

and this is precisely a strong implication $^*(H \implies Q = 0)^*$.   Q. E. D.

The following theorem is a corollary of proposition 6.

Theorem 7 (proof by cases, according to the sign of a polynomial). *To show that* **H** *is strongly incompatible, it is sufficient to construct a strong incompatibility for each of the 3 cases: $Q > 0$, $Q < 0$, $Q = 0$.*

Theorem 8 (transitivity of strong implications). *Let* **H**, **H'**, **H''** *be three gsc systems on polynomials of* $K[X]$.
Then: $[^*(H \implies H')^*$ *and* $^*([H, H'] \implies H'')^*] \underset{\text{cons}}{\vdash} {}^*(H \implies H'')^*$

Proof: It is sufficient to remove one after the other the hypothesis of **H'** in:

$$^*([H, H'] \implies H'')^*.$$

Thus one may assume that **H'** contains a unique hypothesis $Q \tau 0$.

It is thus sufficient to show that if one has two strong implications $^*(H \implies Q\tau 0)^*$, and $^*([H, Q\tau 0, A] \implies 1 = 0)^*$, then one can construct the strong implication: $^*([H, A] \implies 1 = 0)^*$ (where A is a *gsc* on a polynomial). But this can be done by cases according to the sign of Q. Q. E. D.

Combining the transitivity of the strong implications and trivial strong implications, one obtains many corollaries, for example:

Corollary (example).

$$^*(H \implies [P \cdot Q > 0, Q \geq 0])^* \underset{\text{cons}}{\vdash} {}^*(H \implies P > 0)^*$$

Mixed Taylor formulas (strong evidence of Thom's lemma)

Let us at first recall Thom's lemma and the coding "à la Thom":

Thom's lemma. *Let* $\mathbf{R}$ *be a real closed field,* $P \in \mathbf{R}[T]$, *of degree* d, $\sigma_1, \sigma_2, \ldots, \sigma_d$ *a list of* $>$ *or* $<$. *The set*

$$\{t \in \mathbf{R} : P'(t)\,\sigma_1\,0, \ldots, P^{(i)}(t)\,\sigma_i\,0, \ldots, P^{(d)}(t)\,\sigma_d\,0\}$$

is either empty or an open interval. In the latter case, its closure is obtained by weakening the signs σ_i. *In the same way, the set:* $\{\tau \in \mathbf{R} : P(t) = 0, P'(t)\,\sigma_1\,0, \ldots, P^{(i)}(t)\,\sigma_i\,0, \ldots, P^{(d)}(U)\,\sigma_d\,0\}$ *is either empty or reduces to one point.*

Definition of the "coding à la Thom".

Let $\mathbf{K}$ be an ordered field, $\mathbf{R}$ its real closure. An element ζ of $\mathbf{R}$ is said to be *coded à la Thom* (in $\mathbf{K}$) if it is given as a root of a polynomial P of $\mathbf{K}[X]$, specifying the strict[2] signs of $P'(\zeta), P''(\zeta), etc\ldots$

An open unbounded interval of $\mathbf{R}$ is said to be *coded à la Thom* (in $\mathbf{K}$) if it is given as the set of elements ξ which give specified strict signs to a list of polynomials $[P, P', P'', etc\ldots]$, the finite bound a of the interval being obtained for $P(a) = 0$.

A bounded open interval of $\mathbf{R}$ is said to be *coded à la Thom* (in $\mathbf{K}$) if it is given as the set of elements ξ which give specified strict signs to two lists of polynomials $[P, P', P'', etc\ldots]$ and $[Q, Q', Q'', etc\ldots]$, the bounds α and β of the interval being obtained for $P(\alpha) = 0$ and $Q(\beta) = 0$.

NB: Each point of $\mathbf{R}$, but only few open intervals of $\mathbf{R}$, can be coded à la Thom in $\mathbf{K}$. The important fact is that the minimal open intervals of Hörmander tableaux (cf. §4) are naturally coded à la Thom.

One considers now two variables U and V and one lets $\Delta : U - V$. One considers a polynomial P with coefficients in an ordered field $\mathbf{K}$ or more generally in a commutative ring A which is a $\mathbf{Q}$-algebra.

If $\deg(P) \leq 3$, one has the following 4 mixed Taylor formulas:

$$P(U) - P(V) = \Delta \cdot P'(V) + (1/2) \cdot \Delta^2 \cdot P''(V) + (1/6) \cdot \Delta^3 \cdot P^{(3)}$$
$$P(U) - P(V) = \Delta \cdot P'(V) + (1/2) \cdot \Delta^2 \cdot P''(U) - (1/3) \cdot \Delta^3 \cdot P^{(3)}$$
$$P(U) - P(V) = \Delta \cdot P'(U) - (1/2) \cdot \Delta^2 \cdot P''(V) - (1/3) \cdot \Delta^3 \cdot P^{(3)}$$
$$P(U) - P(V) = \Delta \cdot P'(U) - (1/2) \cdot \Delta^2 \cdot P''(U) + (1/6) \cdot \Delta^3 \cdot P^{(3)}$$

Assume now that U and V give the same strict sign σ to P', and the same strict sign σ'' to P''. Then, if we give a sign to Δ and $P^{(3)}$, one of the 4

---

[2]We say a sign to be strict when it is $+1$ or -1.

mixed Taylor formula is strong evidence showing that $P(U) - P(V)$ and $\Delta \cdot P'(U)$ have the same sign. For example, if $\sigma = +1, \sigma'' = -1$ and if $\Delta > 0, P^{(3)} < 0$, the third mixed Taylor formula can be reread:

$$P(U) - P(V) = \Delta \cdot (P'(U) - (1/3) \cdot \Delta^2 \cdot P^{(3)}) - (1/2) \cdot \Delta^2 \cdot P''(V)$$

Conversely these mixed Taylor formulas provide strong evidence in order to obtain the sign of Δ from the sign of $P(U) - P(V)$. In particular, they provide strong evidence that two roots of P coded à la Thom by the same sign sequence are equal. If $\deg(P) \leq 4$, one has the following 8 mixed Taylor formulas:

$P(U) - P(V) =$
$$\Delta \cdot P'(V) + (1/2) \cdot \Delta^2 \cdot P''(V) + (1/6) \cdot \Delta^3 \cdot P^{(3)}(V) + (1/24) \cdot \Delta^4 \cdot P^{(4)}$$
$P(U) - P(V) =$
$$\Delta \cdot P'(V) + (1/2) \cdot \Delta^2 \cdot P''(V) + (1/6) \cdot \Delta^3 \cdot P^{(3)}(U) - (1/8) \cdot \Delta^4 \cdot P^{(4)}$$
$P(U) - P(V) =$
$$\Delta \cdot P'(V) + (1/2) \cdot \Delta^2 \cdot P''(U) - (1/3) \cdot \Delta^3 \cdot P^{(3)}(V) - (5/24) \cdot \Delta^4 \cdot P^{(4)}$$
$P(U) - P(V) =$
$$\Delta \cdot P'(V) + (1/2) \cdot \Delta^2 \cdot P''(U) - (1/3) \cdot \Delta^3 \cdot P^{(3)}(U) + (1/8) \cdot \Delta^4 \cdot P^{(4)}$$
$P(U) - P(V) =$
$$\Delta \cdot P'(U) - (1/2) \cdot \Delta^2 \cdot P''(V) - (1/3) \cdot \Delta^3 \cdot P^{(3)}(V) - (1/8) \cdot \Delta^4 \cdot P^{(4)}$$
$P(U) - P(V) =$
$$\Delta \cdot P'(U) - (1/2) \cdot \Delta^2 \cdot P''(V) - (1/3) \cdot \Delta^3 \cdot P^{(3)}(U) + (5/24) \cdot \Delta^4 \cdot P^{(4)}$$
$P(U) - P(V) =$
$$\Delta \cdot P'(U) - (1/2) \cdot \Delta^2 \cdot P''(U) + (1/6) \cdot \Delta^3 \cdot P^{(3)}(V) + (1/8) \cdot \Delta^4 \cdot P^{(4)}$$
$P(U) - P(V) =$
$$\Delta \cdot P'(U) - (1/2) \cdot \Delta^2 \cdot P''(U) + (1/6) \cdot \Delta^3 \cdot P^{(3)}(U) - (1/24) \cdot \Delta^4 \cdot P^{(4)}$$

As all the possible sign combinations appear, one obtains: if U and V give the same sign sequence to the successive derivatives of a polynomial P of degree ≤ 4, then one has strong evidence that $P(U) - P(V)$ and $(U - V) \cdot P'(U)$ have the same sign. Likewise, if U and V don't give the same sign sequence to P and its successive derivatives (P of degree ≤ 4), one of the mixed Taylor formulas for $P, P', P'',$ or $P^{(3)}$ provides strong evidence for the sign of Δ from the signs of $P^{(i)}(V)$ and $P^{(i)}(U)(i = 0, \ldots, 4)$.

Theorem 9 (mixed Taylor formula). *For each degree d, there are 2^{d-1} mixed Taylor formulas and all the possible sign combinations do appear.*

Proof: Linear algebra shows there is a mixed Taylor formula for each choice ($i \mapsto P^{(i)}(U)$ or $P^{(i)}(V), i = 1, \ldots, d - 1$). The difficult point is showing that all the possible sign combinations do appear. From the algebraic mean value theorem for polynomials, we can obtain the following result: if we choose $P^{(i)}(V)$ the signs of the coefficients of $P^{(i)}$ and $P^{(i+1)}$ are equal, and if we choose $P^{(i)}(U)$ the signs of the coefficients of $P^{(i)}$ and $P^{(i+1)}$ are opposite.   Q. E. D.

Theorem 10 (strong evidence of Thom's lemma). *Let $P \in \mathbf{K}[\mathbf{X}][T]$, of degree d in T, $\sigma_1, \sigma_2, \ldots, \sigma_d$ a list of $>$ or $<$. Denote by $\mathsf{H}(\mathbf{X}, U)$ the gsc system: $P'(\mathbf{X}, U)\,\sigma_1\,0, \ldots, P^{(i)}(\mathbf{X}, U)\,\sigma_i\,0, \ldots, P^{(d)}(\mathbf{X}, U)\,\sigma_d\,0$, (derivatives with respect to T). Write $\mathsf{H}(U)$ for $\mathsf{H}(\mathbf{X}, U)$, $P(U)$ for $P(\mathbf{X}, U)$ and so on: One has then the following strong evidence:*

(1) $^*([\mathsf{H}(U), \mathsf{H}(V), P(U) = P(V)] \implies U = V)^*$

(2) $^*([\mathsf{H}(U), \mathsf{H}(V)] \implies \mathrm{sign}((U - V) \cdot P'(U)) = \mathrm{sign}(P(U) - P(V)))^*$

(3) $^*([\mathsf{H}(U), \mathsf{H}(V), (W - U) \cdot (W - V) \leq 0] \implies \mathsf{H}(W))^*$

(4) $^*([\mathsf{H}(U), \mathsf{H}(V), P^{(i)}(W) \not\sigma_i 0] \implies (W - U) \cdot (W - V) > 0)^*$
$$(i = 1, \ldots, d)$$

Let H' be the gsc system obtained from H by weakening all the sign conditions except those referring to $P^{(d)}$. One has then the following strong evidence:

(5) $^*([\mathsf{H}'(U), \mathsf{H}'(V), U < W < V] \implies \mathsf{H}(W))^*$

Note that (1) is one of the six strong implications written in the abbreviated form (2), and that the d strong incompatibilities in (4) are the same as the ones in (3).

Proof: (1) and (2) are obtained from mixed Taylor formulas as has been explained for degree 3. The d strong implications (3):

$$^*([\mathsf{H}(U), \mathsf{H}(V), (W - U) \cdot (W - V) > 0] \implies P^{(i)}(W)\,\sigma_i\,0)^*(i = 1, \ldots, d)$$

are proved by steps, for i decreasing from d to 1, using at step $P^{(i)}$ a mixed Taylor formula for $P^{(i)}$, and applying transitivity for strong implications. (5) is proved in the same way.   Q. E. D.

Note that theorem 10 doesn't capture entirely Thom's lemma in form of strong evidence: statements concerning the bounds of the interval are missing. This gap is fulfilled in the section on the Hörmander tableaux, and requires the notion of potential existence.

3) Potential existence

Notations and definitions.

A strong implication $^*(\mathsf{H} \implies \mathsf{H}')^*$ is a strong form (in an algebraic identity form) for the corresponding *universal* implication: $\forall \boldsymbol{X}\ (\mathsf{H} \implies \mathsf{H}')$. But the theory of real closed fields has axioms which are not purely universal. So, we require a "stellensatzised" form for statements of the following type:

$$\forall \boldsymbol{X}\ \exists T\ \mathsf{H}(\boldsymbol{X}, T).$$

We should like to speak of potential existence when a *gsc* system is not strongly incompatible.

In fact, we want a little more. The non impossibility of the equation $P(\boldsymbol{X}) = T^2$ taken in isolation is not the same thing as the non impossibility of the equation $P(\boldsymbol{X})^2 = T^4$. Indeed, in the second case, contrarily to the first, whatever hypothesis is made on $\boldsymbol{X}$, adding the equation cannot introduce a contradiction. This distinction is translated in logic by an alternation of quantifiers:

$$\forall \boldsymbol{X}\ \exists T\ P(\boldsymbol{X})^2 = T^4.$$

A "direct translation" of this alternation in terms of strong implication would seem to be: for each not strongly incompatible specification à la Thom of the X_i, the system $\mathsf{H}(\boldsymbol{X}, T)$ is not strongly incompatible. But, in a general proof, a specification of the X_i may depend on values of parameters Y_j. So we are led to the following definition.

Definition 11: Let $\mathsf{H}_1(\boldsymbol{X})$ be a *gsc* system on polynomials of $\mathsf{K}[\boldsymbol{X}]$ and $\mathsf{H}_2(\boldsymbol{X}, \boldsymbol{T})$ a *gsc* system on polynomials of $\mathsf{K}[\boldsymbol{X}, T_1, T_2, \ldots, T_m] = \mathsf{K}[\boldsymbol{X}, \boldsymbol{T}]$. We shall say that *the hypothesis* $\mathsf{H}_1(\boldsymbol{X})$ *allow the existence of* T_i *satisfying* $\mathsf{H}_2(\boldsymbol{X}, \boldsymbol{T})$ when, for all *gsc* systems $\mathsf{H}(\boldsymbol{X}, \boldsymbol{Y})$ on polynomials of $\mathsf{K}[\boldsymbol{X}, \boldsymbol{Y}]$, one has the construction of the strong implication:

$$^*([\mathsf{H}_2(\boldsymbol{X}, \boldsymbol{T}), \mathsf{H}(\boldsymbol{X}, \boldsymbol{Y})] \implies 1 = 0)^* \underset{\text{cons}}{\vdash} {}^*([\mathsf{H}_1(\boldsymbol{X}), \mathsf{H}(\boldsymbol{X}, \boldsymbol{Y})] \implies 1 = 0)^*.$$

We shall speak also of the potential existence of T_i satisfying H_2 under the hypothesis H_1.

NB: The condition on H is that no variable $T_1, T_2, \ldots, T_m$ is in it; but this is possible for other variables, distinct from $X_1, X_2, \ldots, X_n$, hence the $\mathsf{K}[\boldsymbol{X}, \boldsymbol{Y}]$.

Notation: We shall denote this potential existence by:

$$^*(\mathsf{H}_1 \implies \exists \boldsymbol{T}\ \mathsf{H}_2)^*.$$

We can specify the variables in the gsc systems, we then write:

$$^*(H_1(X) \implies \exists T\, H_2(X, T))^*.$$

When the system H_1 is empty, we shall use the notation: $^*(\exists T\, H_2)^*$.
  For example, we shall show:

$$^*(P(X, U) \cdot P(X, V) < 0 \implies \exists W P(X, W) = 0)^*$$

Note that the substitution principle stated in the preceding paragraph can be rewritten in the form:

$$^*(H(X, P(X)) \implies \exists W\, H(X, W))^*$$

Remarks:
  1) At first, we insist on the constructive reading of the above definition: the construction of the strong implication is to be provided by a uniform algorithmic process.
  2) The notation is to be read as a unit (contrarily to the notation concerning constructions of strong implications).
  3) If **L** is a given ordered extension of **K** there is not any obvious a priori relation between a statement $^*(H_1(X) \implies \exists T\, H_2(X, T))^*$ read in **K** and the same statement read in **L**. In fact, after the Nullstellensatz' proof, it is clear that the two statements are equivalent to the statement: $\forall x(H_1(x) \implies \exists t\, H_2(x, t)\,)$ read in the real closure of **K**.

Some rules of manipulation for potential existence statements.
  Among the following rules, only the substitution rule is not immediate.

Lemma 12. *Any potential existence* $^*(H_1(X) \implies \exists T\, H_2(X, T))^*$ *remains true:*
  a) *if one weakens the conclusion,*
  b) *if one strengthens the hypothesis, or*
  c) *if one suppresses behind $\exists$ some variables that are not in* $H_2(X, T)$.

Proposition 13. (simultaneous reinforcement of the hypothesis and the conclusion).
  If
$$^*(H_1(X) \implies \exists T\, H_2(X, T))^*$$
then
$$^*([H_1(X), H_3(X)] \implies \exists T\, [H_2(X, T), H_3(X)])^*$$
(recall of the hypothesis in the conclusion).
  If
$$^*(H_1(X) \implies \exists T\, H_2(X, T))^*$$

then

$$^*(H_1(X) \implies \exists T \, [H_2(X, T), H_1(X)])^*$$

Proposition 14 (potential existence as a generalization of strong implication). *Assume that the gsc systems H_1 and H_2 act only on the variables X. Then $^*(H_1(X) \implies \exists T \, H_2(X))^*$ if and only if $^*(H_1(X) \implies H_2(X))^*$.*

Proposition 15 (rule of proof by cases). *Let Q be a polynomial of $K[X]$. In order to settle a potential existence: $^*(H_1(X) \implies \exists T \, H_2(X, T))^*$ it is sufficient to prove the potential existence*

$$^*([H_1(X), Q \, \sigma \, 0] \implies \exists T \, H_2(X, T))^*$$

for the three σ possible.

Proposition 16 (existence implies potential existence). *Let*

$$P_1, P_2, \ldots, P_m \in K[X]$$

and let us denote $P(X)$ for $P_1(X), \ldots, P_m(X)$.
   If $^*(H_1(X) \implies H_2(X, P(X)))^*$ then $^*(H_1(X) \implies \exists T \, H_2(X, T))^*$

Theorem 17 (transitivity of potential existence). *One considers variables $X_1, X_2, \ldots, X_n, T_1, T_2, \ldots, T_m, U_1, U_2, \ldots, U_k$ and gsc systems $H_1(X)$, $H_2(X, T)$ and $H_3(X, T, U)$. If one has*

$$^*(H_1(X) \implies \exists T \, H_2(X, T))^*$$

and

$$^*([H_1(X), H_2(X, T)] \implies \exists U \, H_3(X, T, U))^*$$

then one has also

$$^*(H_1(X) \implies \exists T, U \, [H_1(X), \, H_2(X, T), \, H_3(X, T, U)])^*$$

Remark 4: Combining the preceding theorem and proposition 14, one obtains some variants. A strong implication followed by a potential existence gives a potential existence. A potential existence followed by a strong implication gives a potential existence. Note also that we may see proposition 14 as a particular case of lemma 12 c).

Theorem 18 (substitution principle in potential existence). *One considers variables $X_1, X_2, \ldots, X_n, Z_1, Z_2, \ldots, Z_k, T_1, T_2, \ldots, T_m$, and polynomials $P_1, P_2, \ldots, P_n$ of $K[Z]$. Let us write $P(Z)$ for $P_1(Z), \ldots, P_n(Z)$.*
   If one has

(a) $$^*(H_1(X) \implies \exists T \, H_2(X, T))^*$$

then one has also

(b) $^*(\mathsf{H}_1(P(Z)) \implies \exists \; \boldsymbol{TH_2}(P(Z), \boldsymbol{T}))^*$

Proof: Assume

(1) $^*([\mathsf{H}_2(P(Z), T), \mathsf{H}(Z, Y)] \implies 1 = 0)^*$

We want to construct

(2) $^*([\mathsf{H}_1(P(Z)), \mathsf{H}(Z, Y)] \implies 1 = 0)^*$

But:

(3) $^*([\mathsf{H}_2(X, T), \mathsf{H}(Z, Y), X = P(Z)] \implies [\mathsf{H}_2(P(Z), T), \mathsf{H}(Z, Y)])^*$

Transitivity on (1) and (3) gives:

(4) $^*([\mathsf{H}_2(X, T), \mathsf{H}(Z, Y), X = P(Z)] \implies 1 = 0)^*$

The definition of potential existence gives:

(5) $^*([\mathsf{H}_1(X), \mathsf{H}(Z, Y), X = P(Z)] \implies 1 = 0)^*$

We have also (trivial strong implications):
(6)
$^*([\mathsf{H}_1(P(Z)), \mathsf{H}(Z, Y), X = P(Z)] \implies [\mathsf{H}_1(X), \mathsf{H}(Z, Y), X = P(Z)])^*$

Transitivity on (5) and (6) gives:

(7) $^*([\mathsf{H}_1(P(Z)), \mathsf{H}(Z, Y), X = P(Z)] \implies 1 = 0)^*$

If we substitute $P(Z)$ for X in (7), we obtain (2)   Q. E. D.

Remark 5: The proofs of potential existence can generally be given directly in the form (b). Theorem 18 merely allows one to see more clearly the structure of theorems stating potential existence.

Remark 6: If one applies theorem 18 once again, one can substitute some X_j for some Z_i. One sees thus that the hypothesis that the X_j and the Z_i are distinct is in fact useless.

Fundamental potential existence

Theorem 19 (authorization to add the square root of a positive element).

$$^*(U \geq 0 \implies \exists T \ U = T^2)^*$$

Theorem 20. *(authorization to add the inverse of a nonzero element):*

$$^*(U \neq 0 \implies \exists T \ 1 = U \cdot T)^*$$

Proof: Assume without loss of generality that U is the variable X_n. Consider a *gsc* system $\mathbf{H}(\boldsymbol{X})$. Notation is as in the proof of proposition 6.

Let us write Cp', I' when we consider the positive cone or the ideal generated in the polynomial ring with the extra variable T: $\mathbf{K}[\boldsymbol{X}, T] = \mathbf{K}[X_1, X_2, \ldots, X_n, T]$.

We want to give the construction:

$$^*([\mathbf{H}, 1 - U \cdot T = 0] \implies 1 = 0)^* \underset{\text{cons}}{\vdash} {}^*([\mathbf{H}, U \neq 0] \implies 1 = 0))^*.$$

The hypothesis is an algebraic identity:

$$S_1(\boldsymbol{X}) + P_1(\boldsymbol{X}, T) + (1 - U \cdot T) \cdot Y_1(\boldsymbol{X}, T) + Z_1(\boldsymbol{X}, T) = 0$$

where

$$S_1 \in \mathcal{M}(F_>^{*2} \cup F_{\neq}^{*2}), P_1 \in Cp'(F_{\geq} \cup F_>), Z_1 \in I'(F_=).$$

More precisely:

$$S_1(\boldsymbol{X}) + \sum_{i=1}^{h} Q_i(\boldsymbol{X}) \cdot V_i^2(\boldsymbol{X}, T)) + (1 - U \cdot T) \cdot Y_1(\boldsymbol{X}, T)$$

$$+ \sum_{j=1}^{r} N_j(\boldsymbol{X}) \cdot W_j(\boldsymbol{X}, T)) = 0$$

where $Q_i(\boldsymbol{X}) \in Cp(F_{\geq} \cup F_>)$ and $N_j(\boldsymbol{X}) \in F_=$. Informally: let us work modulo $(1 - U \cdot T)$. Replace everywhere in V_i and W_j, T by $1/U$ so that T disappears, then multiply by a suitable U^{2m} in order to suppress the denominator. More precisely: the same final result should be obtained if we first multiply by U^{2m} ($m \geq \deg_T(V_i)$ and $2m \geq \deg_T(W_j)$) and then replace each $U^k \cdot T^k$ in V_i and W_j by 1 modulo $(1 - U \cdot T)$.

One obtains thus an algebraic indentity:

$$S_1(\boldsymbol{X}) \cdot U^{2m} + \sum_{i=1}^{h} Q_i(\boldsymbol{X}) \cdot A_i^2(\boldsymbol{X}) + (1 - U \cdot T) \cdot Y_2(\boldsymbol{X}, T) +$$

$$\sum_{j=1}^{r} N_j(\boldsymbol{X}) \cdot C_j(\boldsymbol{X}) = 0$$

One now has $Y_2(X, T) = 0$ (otherwise consider in Y_2 the monomial of maximum degree in T). The remaining algebraic identity is a strong implication $*([\mathbf{H}, U \neq 0] \implies 1 = 0))^*$:

$$S_1(X) \cdot U^{2m} + \sum_{i=1}^{h} Q_i(X) \cdot A_i^2(X) + \sum_{j=1}^{r} N_j(X) \cdot C_j(X) = 0$$

Q. E. D.

Corollary 1 (authorization to add the inverse of the square root of a strictly positive element).

$$*(U > 0 \implies \exists T\, 1 = U \cdot T^2)^*$$

Corollary 2 (the weak real Nullstellensatz implies the other real ... stellensatz). *Assume that for each natural number n and all systems of equalities to 0 on polynomials of $\mathbf{K}[X]$, the impossibility in $\mathbf{R}$ (real closure of $\mathbf{K}$) implies the strong incompatibility in $\mathbf{K}$. Then, for all gsc systems on polynomials of $\mathbf{K}[X]$, the impossibility in $\mathbf{R}$ implies the strong incompatibility in $\mathbf{K}$.*

Remarks 7: Theorems 19 and 20 "give the authorization" to add the root(s) of an equation of degree 1 or 2. Theorem 19 is also a consequence of theorem 21. Corollary 1 can be proved as in theorems 19 and 20. Corollary 2 is thus "directly" provable without the general theory of potential existence, as in the algebraically closed case.

Theorem 21 (authorization to add a root on an interval where a polynomial changes sign). *Denote by $P(U)$ a polynomial $P(X, U)$. Then we have:*

$$*([P(U) \cdot P(V) < 0, U < V] \implies$$
$$\exists W[P(W) = 0, P(U) \cdot P(V) < 0, U < W < V])^*$$

Proof: Notation is as in the proof of proposition 6.

FIRST PART: We prove the potential existence

$$*(P(U) \cdot P(V) \leq 0 \implies \exists W\, P(W) = 0)^*$$

We give a proof by induction[3] on the degree in T of $P(X, T)$. When $\deg(P) = 0$ or -1, the result is easy.

---

[3]This proof "recopies" the classical proof of "if we have an ordered field and if $P(u) \cdot P(v) < 0$ with P irreducible, then the field $\mathbf{K}[W]/P(W)$ is real".

One may assume the variables U and V to be two variables X_i.[4] For any gsc system H without the variable W we have to give a construction:

$$^*([H, P(X,W) = 0] \implies 1 = 0)^* \underset{\text{cons}}{\vdash}$$

$$^*([H, P(X,U) \cdot P(X,V) \leq 0] \implies 1 = 0)^*$$

which we may reread:

$$^*(H \implies P(X,W) \neq 0)^* \underset{\text{cons}}{\vdash} {}^*(H \implies P(X,U) \cdot P(X,V) > 0)^*$$

Assume at first P is monic. The strong implication $^*(H \implies P(X,W) \neq 0)^*$ is written as an algebraic identity:

$$S_1(X) + \sum_{i=1}^{h} Q_i(X) \cdot B_i^2(X,W) - P(X,W) \cdot G(X,W) +$$

$$\sum_{j=1}^{r} N_j(X) \cdot C_j(X,W) = 0$$

with $Q_i(X) \in Cp(F_\geq \cup F_>)$ and $N_j(X) \in F_=$. The polynomials $B_i(X,W)$ and $C_j(X,W)$ may be taken modulo P in W (because P is monic), and so $\deg_W(G) \leq \deg_W(P) - 2$. The same equality may be reinterpreted as various strong implications:

(1) $\qquad\qquad\qquad {}^*(H \implies G(X,W) \neq 0)^*$

(2) $\qquad\qquad\qquad {}^*(H \implies P(X,W) \cdot G(X,W) > 0)^*$

Then, by substitution in (2),

$$^*(H \implies P(X,U) \cdot G(X,U) > 0)^*, \quad {}^*(H \implies P(X,V) \cdot G(X,V) > 0)^*$$

Hence,

$$^*(H \implies P(X,U) \cdot G(X,U) \cdot P(X,V) \cdot G(X,V) > 0)^*$$

By the induction hypothesis, (1) implies that we can construct a strong implication:

$$^*(H \implies G(X,U) \cdot G(X,V) > 0)^*.$$

But by trivial strong implications:

$$^*([P(X,U) \cdot G(X,U) \cdot P(X,V) \cdot G(X,V) > 0,\ G(X,U) \cdot G(X,V) > 0] \implies$$
$$P(X,U) \cdot P(X,V) > 0)^*$$

We conclude the proof by transitivity of strong implications.

---

[4]According to the substitution principle for potential existence, we may actually assume we are in the generic case where U, V and the coefficients of P are all independent variables X_i.

Assume now P is not monic.

Let $C(X) \cdot W^n$ be the leading monomial of $P(X, W)$. Let $R(X, W) = P(X, W) - C(X) \cdot W^n$. (so $\deg_W(R) < \deg_W(P)$). Consider a new variable T, and consider the polynomial $P_1(X, T, W) = T \cdot R(X, W) + W^n$. We give a proof of the potential existence by cases, according to the sign of $C(X)$.

1st case: $C(X) = 0$. We have

$$ {}^*([P(X, U) \cdot P(X, V) \le 0, C(X) = 0] \implies R(X, U) \cdot R(X, V) \le 0)^* $$

and by the induction hypothesis

$$ {}^*(R(X, U) \cdot R(X, V) \le 0 \implies \exists W\ R(X, W) = 0)^* $$

As

$$ {}^*([R(X, W) = 0, C(X) = 0] \implies P(X, W) = 0)^* $$

we conclude the proof by transitivity.

2nd case: $C(X) \ne 0$. We have

$$ {}^*(C(X) \ne 0 \implies \exists T\ 1 = C(X) \cdot T)^*, $$
$$ {}^*(1 = C(X) \cdot T \implies T \cdot P(X, W) = P_1(X, T, W))^* $$

and

$$ {}^*(1 = C(X) \cdot T \implies P(X, W) = C(X) \cdot P_1(X, T, W))^* $$

so

$$ {}^*([P(X, U) \cdot P(X, V) \le 0, C(X) \ne 0] \implies $$
$$ \exists T\ [1 = C(X) \cdot T, P_1(X, T, U) \cdot P_1(X, T, V) \le 0])^* $$

As P_1 is monic,

$$ {}^*(P_1(X, T, U) \cdot P_1(X, T, V) \le 0 \implies \exists W\ P_1(X, T, W) = 0)^* $$

By transitivity,

$$ {}^*([P(X, U) \cdot P(X, V) \le 0, C(X) \ne 0] \implies $$
$$ \exists T, W\ [1 = C(X) \cdot T, P_1(X, T, W) = 0])^* $$

Hence,

$$ {}^*([P(X, U) \cdot P(X, V) \le 0, C(X) \ne 0] \implies \exists T, W\ P(X, W) = 0)^*, $$

where we may remove T.

SECOND PART: Proof of the potential existence stated in the theorem.

We don't give the detailed proof. One may mimick the classical proof: if a root w of P is not between u and v, then we consider the polynomial $P(Z)/(Z - w)$, which also changes sign between u and v. So we may conclude by induction on $\deg(P)$.   Q. E. D.

4) Strong evidence of the facts
readable from a Hörmander tableau

Recall at first Hörmander's algorithm.

Proposition 22 (Hörmander tableau). *Let* **K** *be an ordered field, subfield of a real closed field* **R**.

Let $L = [P_1, P_2, \ldots, P_k]$ *a list of polynomials of* **K**$[\mathbf{X}]$.

Let $\mathcal{P}$ *be the polynomial family generated by the elements of* L *and by the operations* $P \mapsto P'$, *and* $(P, Q) \mapsto Rem(P, Q)$. *Then:*

1) $\mathcal{P}$ *is finite.*

2) One can set up the complete sign tableau for $\mathcal{P}$ *using only the following information:*

   a) the degree of each polynomial in the family;

   b) the diagrams of the operations $P \mapsto P'$, *and* $(P, Q) \mapsto Rem(P, Q)$
      (where $\deg(P) \geq \deg(Q)$*) in* $\mathcal{P}$ *; and*

   c) the signs of the constants of $\mathcal{P}$.[5]

Proof: 1) is easy.

2) We number the polynomials in $\mathcal{P}$ in order of nondecreasing degree. Let $\mathcal{P}_n$ be the subfamily of $\mathcal{P}$ made of polynomials numbered 1 to n. Let us denote by T_n the Hörmander tableau corresponding to the family $\mathcal{P}_n$: i.e. the tableau where all the real roots of the polynomials of $\mathcal{P}_n$ are defined via a coding à la Thom, listed in increasing order, and where all the signs of the polynomials of $\mathcal{P}_n$ are indicated, at each root, and on each interval between two consecutive roots (or between $-\infty$ and the first root, or between the last root and $+\infty$).

Then by induction on n it is easy to prove that one can construct the tableau T_n from the allowed information.   Q. E. D.

We are going to give a sufficiently faithful sketch for the proof of:

Theorem 23 (real Nullstellensatz in one variable). *Let* **K** *be an ordered field and* **R** *its real closure.*

Let $\mathcal{P}$ *be a family of polynomials of* **K**$[X]$ *and* **H**(X) *be a gsc system on elements of* $\mathcal{P}$.

   Then:

   either **H**(x) *is impossible in* **R** *and then* *(**H** $\implies$ $1 = 0$)* *in* **K**, *and thus* **H**(x) *is impossible in any ordered extension of* **K**.

   or **H**(x) *is possible in* **R** *and then* *($\exists X$ **H**(X)))* *in* **K** *and in any ordered extension of* **K**.

---

[5]Note that the constants in $\mathcal{P}$ are essentially the leading coefficients of polynomials in $\mathcal{P}$, and the values $P(\xi)$ where P is a polynomial in $\mathcal{P}$ and ξ a root of a degree one polynomial in $\mathcal{P}$.

One may assume that the family $\mathcal{P}$ is closed under the operations "remainder" and "derivation". The impossibility of $H(x)$ in $\mathbf{R}$ or the existence of x in $\mathbf{R}$ verifying $H(x)$ is directly readable from the Hörmander tableau of the family, and can be tested solely by computations in $\mathbf{K}$. We are going to show how the construction of the Hörmander tableau can be transformed, step by step, into strong evidence and potential existence which translate all the facts readable from the Hörmander tableau. If one now considers a given extension L of $\mathbf{K}$, one may apply to H, L and its real closure, the results obtained for H, $\mathbf{K}$ and $\mathbf{R}$: as the test is made solely by computations in $\mathbf{K}$ the possibility or the impossibility will be equivalent in the two cases.

When the field K is real closed.

Thus one has $\mathbf{R} = \mathbf{K}$. Let $\nu_1, \nu_2, \ldots, \nu_k$ be the finite list of points in the Hörmander tableau of the family $\mathcal{P}$. One can compute ν_0 and ν_{k+1} in $\mathbf{R}$ such that the strong evidence for the signs of all the $P \in \mathcal{P}$ is easy to state for $x \leq \nu_0$ and for $x \geq \nu_{k+1}$.

The possibility or otherwise in $\mathbf{R}$ for a given gsc system is immediately readable. Possibility occurs either for a ν_i, or for an $x = (\nu_i + \nu_{i+1})/2$ and this implies the potential existence. The incompatibility in $\mathbf{R}$ for a gsc system H is also readable from the Hörmander tableau, but the strong incompatibility requires a new argument. One argues by cases, and it is sufficient to state the strong incompatibility for at least one gsc in H: at each point ν_i on the one hand, on each open interval $]\nu_i, \nu_{i+1}[$ on the other hand, and finally for $X < \nu_0$ and for $X > \nu_{k+1}$. At a point ν_i the sign of each $P(\nu_i)$ is strongly evident in $\mathbf{R}$ (since $\nu_i \in \mathbf{R}$). On an interval $]\nu_i, \nu_{i+1}[$, the signs, constant and non zero, of the $P \in \mathcal{P}$ are all strongly evident from the signs at the end-points modulo a suitable mixed Taylor formula (cf. theorem 10(5)).

In the coefficients field.

We want to state, for all the facts readable from the Hörmander tableau, strong incompatibility and potential existence in $\mathbf{K}$. We have now to follow the Hörmander algorithm step by step, i.e. introducing the points in the Hörmander tableau one after the other. We begin by computing a and b in $\mathbf{K}$, such that for $x \leq a$ and for $x \geq b$, the signs of the polynomials in $\mathcal{P}$ are strongly evident. These 2 elements of $\mathbf{K}$ will replace $-\infty$ and $+\infty$ in the Hörmander tableau.

One first proves the lemma:

Lemma 24 (strong evidence and potential existence for the elementary facts readable from a Hörmander tableau). *Let* $\mathbf{K}$ *be an ordered field and* $\mathbf{R}$ *its real closure. Let* $\mathcal{P}$ *be a family of polynomials of* $\mathbf{K}[X]$ *closed under the operations "remainder" and "derivation", and let* $\mathcal{T}$ *be its Hörmander tableau.*

1) *the points of the Hörmander tableau, defined à la Thom by their construction, satisfy the potential existence for their coding à la Thom.*[6]

2) *the comparison of 2 points of the tableau is strongly evident from their coding à la Thom.*

3) *at each point α in the tableau, the signs of all the polynomials in the family are strongly evident from the coding à la Thom for α.*

4) *at each point of a minimal open interval in the tableau, the signs of all the polynomials previously introduced are strongly evident either from the coding à la Thom for the interval bounds (if the interval is unbounded, only the finite bound is to be considered) and from the fact that the point is between the bounds, or also from the coding à la Thom for the interval.*

Proof of the lemma: We prove the lemma for the family $\mathcal{P}_n$ and the tableau $\mathcal{T}_n$, by induction on n. The lemma is evident when all the polynomials are constants.

Let us go from n to $n + 1$. If λ is a point of $\mathcal{T}_n$, we shall denote by $Q_\lambda(X)$ the first polynomial of which λ is a root, and $H_\lambda(X)$ the *gsc* system which is its coding à la Thom (λ is the only point of $\mathbf{R}$ verifying $H_\lambda(\lambda)$). Let now P be the polynomial numbered $n + 1$, of degree $d \geq 1$.

In the following proof we examine only bounded open intervals. The modifications for the other case are easy.

For each point λ of $\mathcal{T}$, we introduce a new variable X_λ. In order to have a more readable proof, we shall write λ for X_λ.[7]

point 1): The only problematic points are roots of P. The sign of P at a point λ of $\mathcal{T}_n$ is strongly evident from the sign of $\text{Rem}(P, Q_\lambda)(\lambda)$ and from the fact that $Q_\lambda(\lambda) = 0$; thus also, by the induction hypothesis (3), from $H_\lambda(\lambda)$. Let ζ be a root of P on the minimal open interval $]\alpha, \beta[$ of $\mathcal{T}_n$. We have thus

$$\text{*}(H_\alpha(\alpha) \implies P(\alpha) > 0)\text{*} \quad \text{and} \quad \text{*}(H_\beta(\beta) \implies P(\beta) < 0)\text{*} \quad \text{or vice-versa.}$$

By the induction hypothesis (2) we have

$$\text{*}([H_\alpha(\alpha), H_\beta(\beta)] \implies \alpha < \beta)\text{*}$$

Theorem 21 and transitivity of potential existence give us

$$\text{*}([H_\alpha(\alpha), H_\beta(\beta)] \implies \exists X[\alpha < X < \beta, P(X) = 0])\text{*}$$

---

[6] A single point could be coded à la Thom via distinct polynomials. The coding we consider here is the first that appears in the tableau construction.
[7] The λ that we must read as X_λ are clear from the context.

Again by the induction hypothesis (3), there are $\tau_i \in \{<, >\}(i = 1, \ldots, d)$ such that, if we call τ_i' the sign $\leq$ or $\geq$ associated to τ_i, we have:[8]

$$^*(\mathsf{H}_\alpha(\alpha) \implies [P^{(i)}(\alpha)\,\tau_i'\,0(i = 1, \ldots, d-1), P^{(d)}(\alpha)\,\tau_d\,0])^*$$
$$^*(\mathsf{H}_\beta(\beta) \implies [P^{(i)}(\beta)\,\tau_i'\,0(i = 1, \ldots, d-1), P^{(d)}(\beta)\,\tau_d\,0])^*$$

Let us apply mixed Taylor formulas (theorem 10(5)) and transitivity:

$$^*([\mathsf{H}_\alpha(\alpha), \mathsf{H}_\beta(\beta)] \implies$$
$$\exists X[\alpha < X < \beta, P(X) = 0, P^{(i)}(X)\,\tau_i\,0(i = 1, \ldots, d)])^*$$

We have previously

$$^*(\exists \alpha, \beta[\mathsf{H}_\alpha(\alpha), \mathsf{H}_\beta(\beta)])^*$$

By transitivity,

$$^*(\exists X[P(X) = 0, P^{(i)}(X)\,\tau_i\,0(i = 1, \ldots, d)])^*$$

We rewrite this potential existence

$$^*(\exists \zeta\ \mathsf{H}_\zeta(\zeta))^*$$

point 2): We have already the strong implications

$$^*(\mathsf{H}_\alpha(\alpha) \implies P^{(i)}(\alpha)\,\tau_i'\,0)^*, ^*(\mathsf{H}_\alpha(\alpha) \implies P(\alpha) > 0)^*$$

(or < 0) and

$$^*(\mathsf{H}_\alpha(\alpha) \implies P^{(d)}(\alpha)\,\tau_d\,0)^*,$$

So the sign of $\alpha - \zeta$ is strongly evident from the codings à la Thom of α and ζ via theorem 10 (2), (idem for $\beta - \zeta$):

$$^*([\mathsf{H}_\alpha(\alpha), \mathsf{H}_\zeta(\zeta)] \implies \alpha < \zeta)^*$$

Point 2) for $\mathcal{T}_{n+1}$ can then be deduced from point 2) for $\mathcal{T}_n$: if for example $\lambda \in \mathcal{T}_n$ with $\lambda < a$ the induction hypothesis shows:

$$^*([\mathsf{H}_\alpha(\alpha), \mathsf{H}_\lambda(\lambda)] \implies \lambda < \alpha)^*$$

Thus

$$^*([\mathsf{H}_\alpha(\alpha), \mathsf{H}_\lambda(\lambda), \mathsf{H}_\zeta(\zeta)] \implies \lambda < \alpha < \zeta)^*$$

But $(\exists \alpha\ \mathsf{H}_\alpha(\alpha))^*$, so

$$^*([\mathsf{H}_\lambda(\lambda), \mathsf{H}_\zeta(\zeta)] \implies \lambda < \zeta)^*$$

point 3): The sign of P at each point λ of $\mathcal{T}_n$ is already strongly evident from the coding à la Thom of λ. It remains to see that the sign of $Q \in \mathcal{P}_n$ at a new point (as ζ in 1)) is strongly evident from its coding à la Thom. From 2) we have:

$$^*([\mathsf{H}_\alpha(\alpha), \mathsf{H}_\beta(\beta), \mathsf{H}_\zeta(\zeta)] \implies \alpha < \zeta < \beta)^*$$

The induction hypothesis (3) shows that the sign of Q in α and β is strongly evident from $\mathsf{H}_\alpha(\alpha)$ and $\mathsf{H}_\beta(\beta)$.

---

[8]The statements concerning $P^{(d)}(.)$ are trivial since $P^{(d)}$ is a constant, we give them here essentially for the rereading of this proof when the coefficients of P will depend on parameters.

Again by theorem 10 (5) and transitivity,

$$^*([\mathsf{H}_\alpha(\alpha), \mathsf{H}_\beta(\beta), \mathsf{H}_\zeta(\zeta)] \implies Q(\zeta)\,\tau\,0)^* \quad \text{with } \tau \in \{<, >\}$$

But

$$^*(\exists \alpha, \beta \, [\mathsf{H}_\alpha(\alpha), \, \mathsf{H}_\beta(\beta)])^*$$

so we obtain

$$^*(\mathsf{H}_\zeta(\zeta) \implies Q(\zeta)\,\tau\,0)^* \quad \text{with } \tau \in \{<, >\}$$

point 4): Let us denote by $\mathsf{H}_{\lambda,\mu}(X)$ the coding à la Thom for a minimal open interval of T_{n+1}. It is obtained from $\mathsf{H}_\lambda(X)$, $\mathsf{H}_\mu(X)$ by replacing sign conditions $Q_\lambda(X) = 0$ and $Q_\mu(X) = 0$ by the suitable strict sign conditions. Applying theorem 10 (2) we obtain:

$$^*([\mathsf{H}_\mu(\mu), \mathsf{H}_\lambda(\lambda), \mathsf{H}_{\lambda,\mu}(X)] \implies \lambda < X < \mu)^*$$

If now Q is an arbitrary polynomial in $\mathcal{P}_{n+1}$ we argue as in 3) for the sign of Q in ζ and we obtain the strong evidence for the sign of $Q(X)$ under the hypothesis $\mathsf{H}_{\lambda,\mu}(X)$ Q. E. D.

To finish the proof of theorem 23, we can recopy (using lemma 24), with the usual cautions, all that we have done in the case of a real closed field. The disjunction of cases will be sound because of (2). The sign evaluation for a polynomial at a point of the tableau will be replaced by the strong evidence of the sign for this polynomial etc ...   Q. E. D.

5) Effective real Nullstellensanz and variants

When one has shown the "strong implication" version of the axioms and of the deduction rules in the formal theory of real closed fields with elements of **K** as constants, it is natural to wish to translate in form of strong implication every statement provable in this formal theory.

So to speak, the hardest part has been done with the validation of "proof by cases", the transitivity of strong implications and the authorization to add a root to a polynomial on an interval where it changes sign. In fact, as we have no "strong implication" version for statements with too many quantifier alternations, this is not completely straightforward.

The proof of the Nullstellensatz consists therefore of verifying that the algorithm for deciding a purely universal statement in the formal theory of real closed fields doesn't make use of logical arguments using statements with too many quantifier alternations.

Proposition 25 (parametrized Hörmander tableau). *Let* **K** *be an ordered field, subfield of a real closed field* **R**. *Let* $L = [Q_1, Q_2, \ldots, Q_k]$ *a list of polynomials of* $\mathbf{K}[U_1, U_2, \ldots, U_n][X]$. *One can construct a finite family* $\mathcal{F}$ *of polynomials in* $\mathbf{K}[U_1, U_2, \ldots, U_n]$ *such that, for all* $u_1, u_2, \ldots, u_n$ *in* **K**, *if we set* $P_i(X) = Q_i(u_1, u_2, \ldots, u_n; X)$, *the complete sign tableau for* $L = [P_1, P_2, \ldots, P_k]$ *is computable from the signs of the* $S(u_1, u_2, \ldots, u_n)$ *for* $S \in \mathcal{F}$.

Proof: The constants in the Hörmander algorithm (cf. proposition 22) are all obtained as rational fractions in the coefficients of polynomials of L. Otherwise, the computation of the family $\mathcal{P}$ is "uniform" except that a remainder computation, e.g. of $Rem(P, Q)$, depends on the degree of Q. As the Q coefficients are rational fractions in the coefficients of polynomials of L, the degree of Q, for a given specialization $u_1, u_2, \ldots, u_n$ of $U_1, U_2, \ldots, U_n$, depends on the vanishing of some polynomials in the coefficients of polynomials of L. Thus we include in the family $\mathcal{F}$ all the polynomials which appear in the numerator or the denominator of a coefficient of any polynomial of a family $\mathcal{P}$, for all the possible families $\mathcal{P}$.   Q. E. D.

Theorem 26 (parametrized Hörmander tableau, strong implications and potential existence). *Let* $\mathbf{K}$ *be an ordered field, subfield of a real closed field* $\mathbf{R}$. *Let* $L = [Q_1, Q_2, \ldots, Q_k]$ *a list of polynomials of* $\mathbf{K}[U_1, U_2, \ldots, U_n][X]$. *One constructs the finite family* $\mathcal{F}$ *of polynomials in* $\mathbf{K}[U_1, U_2, \ldots, U_n]$ *as in proposition 25. Let* $\mathbf{H}(U_1, U_2, \ldots, U_n, X)$ *be a gsc system on polynomials in the list* L. *Let* $\sum = (\sigma_S)_{S \in \mathcal{F}}$ *in* $\{-1, 0, +1\}^{\mathcal{F}}$.
One denotes by $\mathbf{H}_\Sigma(U_1, U_2, \ldots, U_n)$ *the gsc system*

$$[S(U_1, U_2, \ldots, U_n) \equiv \sigma_S;\ S \in \mathcal{F}].$$

Assume that there exist $u_1, u_2, \ldots, u_n \in \mathbf{R}$ *satisfying* $\mathbf{H}_\Sigma(u_1, u_2, \ldots, u_n)$. *Then:*
either

$$\forall u_1, u_2, \ldots, u_n \in \mathbf{R}\ (\mathbf{H}_\Sigma(u_1, u_2, \ldots, u_n) \implies \exists x \in \mathbf{R}\ \mathbf{H}(u_1, u_2, \ldots, u_n, x))$$

and then

$$^*(\mathbf{H}_\Sigma(U_1, U_2, \ldots, U_n) \implies \exists X\ \mathbf{H}(U_1, U_2, \ldots, U_n, X))^*\ (read\ in\ \mathbf{K})$$

or

$$\forall u_1, u_2, \ldots, u_n, x \in \mathbf{R}(\mathbf{H}_\Sigma(u_1, u_2, \ldots, u_n)\ and\ \mathbf{H}(u_1, u_2, \ldots, u_n, x))$$
$$\implies 1 = 0$$

and then

$$^*([\mathbf{H}_\Sigma(U_1, U_2, \ldots, U_n), \mathbf{H}(U_1, U_2, \ldots, U_n, X)] \implies 1 = 0)^*\ (in\ \mathbf{K}).$$

Proof: The sign conditions $\mathbf{H}_\Sigma$ prescribe the degrees of the polynomials in the family (closed under remainder and derivation) generated by L, and also prescribe the Hörmander tableau of the family. We can then repeat with the usual cautions the reasonings in the proof of theorem 23, and we obtain theorem 23 "with parameters", i.e. theorem 26.   Q. E. D.

The real effective Nullstellensatz is now easy.

Theorem 27 (Effective real Nullstellensatz, Positivestellensatz and Nicht-negativestellensatz). *Let* **K** *be an ordered field, subfield of a real closed field* **R**. *Let* $H(U_1, U_2, \ldots, U_n)$ *be a gsc system for a finite family of polynomials in* $K[U_1, U_2, \ldots, U_n]$. *This system is impossible in* **R** *if and only if it is strongly incompatible in* **K**.

In more formal terms:
If $\forall u_1, u_2, \ldots, u_n \in R \; H(u_1, u_2, \ldots, u_n)$ *is absurd, then:*

$$^*(H(U_1, U_2, \ldots, U_n) \implies 1 = 0)^* \quad (in \; K \;).$$

If

$$^*(H(U_1, U_2, \ldots, U_n) \implies 1 = 0)^* \quad (in \; K \;),$$

then the gsc $H(u_1, u_2, \ldots, u_n)$ *are impossible to realize in any ordered extension of* **K**.

Proof: The "converse" part is evident. For the "forward" part, one argues by induction on the number of variables. For $n = 1$, this is theorem 23. Let us go from n to $n + 1$. Let us call X the $(n + 1)^{\text{st}}$ variable. In order to construct the strong implication, one argues case by case, according to the signs of the polynomials in the family $\mathcal{F}$ and one uses theorem 26.Q. E. D.

One has also immediately:

Theorem 28 (uniformly primitive recursive real Nullstellensatz and variants). *Let* **K** *be an ordered field, subfield of a real closed field* **R**. *Let* $H(U_1, U_2, \ldots, U_n)$ *be a gsc system for a finite family of polynomials in* $K[U_1, U_2, \ldots, U_n]$. *Let* $(c_i)_{i \in I}$ *be the finite family of coefficients of polynomials in* **H**. *Suppose that the structure of ordered field of* $Q((c_i)_{i \in I}$ *is given by an oracle that answers to the question: " what is the sign of* $P((c_i)_{i \in I})$?", *where the input is the polynomial* $P \in Z[(C_i)_{i \in I}]$. *There exists a uniformly primitive recursive algorithm that says wether* **H** *is impossible in* **R** *and constructs, in the case of a positive answer, a strong implication* $^*(H \implies 1 = 0)^*$ *(in* **K***)*.

Remark 8: It would be easy to prove, by induction on the number of variables, an improvement of theorem 27, that should state: existence in **R** implies potential existence read in **K**, and vice versa. In fact, the Nullstellensatz having been proved, one can deduce immediately the following interpretation for potential existence under conditions: Let H_1 be a *gsc* system on polynomials of $K[X] = K[X_1, X_2, \ldots, X_n]$, and H_2 a *gsc* system on polynomials of $K[X, T_1, T_2, \ldots, T_m] = K[X, T]$. Then one has

$$^*(H_1(X) \implies \exists T \; H_2(X, T))^* \quad (\text{read in } K)$$

if and only if

$$\forall x \in R^n(H_1(x) \implies \exists t \in R^m H_2(x, t))$$

Remark 9: The same methods, simplified, could be applied in field theory (the only sign conditions are $= 0$ and $\neq 0$). One can thus obtain a direct constructive proof for the Hilbert Nullstellensatz, with a uniformly primitive recursive algorithm, (for the discrete case), without having to develop the constructive noetherian theory.

Acknowledgements: I thank Marie-Françoise Roy for her many comments and helpful suggestions.

REFERENCES

[BCR]   Bochnak J., Coste M., Roy M.-F., "Géométrie Algébrique réelle," A series of Modern Surveys in Mathematics 11, Springer-Verlag, 1987.

[Du]   Dubois, D. W., *A nullstellensatz for ordered fields*, Arkiv for Mat. **8** (1969), 111–114, Stockholm.

[Efr]   Efroymson, G., *Local reality on algebraic varieties*, J. of Algebra **29** (1974), 113–142.

[Kri]   Krivine, J. L., *Anneaux préordonnés*, Journal d'analyse mathématique **12** (1964), 307–326.

[LR]   Lombardi H., Roy M.-F., *Théorie constructive élémentaire des corps ordonnés*. English version in these proceedings

[Lom]   Lombardi H., *Théorème des zéros réel effectif et variantes*, Publications Mathématiques de Besançon 88-89. Théorie des nombres. Fascicule 1.

[MRR]   Mines R., Richman F., Ruitenburg W., "A Course in Constructive Algebra," Universitext, Springer-Verlag, 1988.

[Ris]   Risler, J.-J., *Une caractérisation des idéaux des variétés algébriques réelles*, C.R.A.S. Paris, série A **271** (1970), 1171-1173.

[Ste]   Stengle, G., *A Nullstellensatz and a Positivestellensatz in semialgebraic geometry*, Math. Ann. **207** (1974), 87–97.

Henri Lombardi
Laboratoire de Mathématiques.
UFR des Sciences et Techniques.
Université de Franche-Comté.
25 030 Besançon cédex
France

Algorithms for the Solution of Systems of Linear Equations in Commutative Rings

GENNADI I. MALASHONOK

Abstract. Solution methods for linear equation systems in a commutative ring are discussed. Four methods are compared, in the setting of several different rings: Dodgson's method [1], Bareiss's method [2] and two methods of the author — method by forward and back-up procedures [3] and a one-pass method [4].

We show that for the number of coefficient operations, or for the number of operations in the finite rings, or for modular computation in the polynomial rings the one-pass method [4] is the best. The method of forward and back-up procedures [3] is the best for the polynomial rings when we make use of classical algorithms for polynomial operations.

Introduction

Among the set of known algorithms for the solution of systems of linear equations, there is a subset, which allows us to carry out computations within the commutative ring generated by the coefficients of the system. Recently, interest in these algorithms grew due to computer algebra computations. These algorithms may be used (a) to find exact solutions of systems with numerical coefficients, (b) to solve systems over the rings of polynomials with one or many variables over the integers or over the reals, (c) to solve systems over finite fields.

Let

$$(1) \qquad\qquad Ax = a$$

be the given system of linear equations over a commutative ring R, $A \in R^{n \times (m-1)}$, $a \in R^n$, $x \in R^{m-1}$, $n < m$, with extended coefficient matrix $A^* = (a_{i,j})$, $i = 1, \ldots, n$, $j = 1, \ldots, m$.

The solution of such a system may be written according to Cramer's rule $x_i = (\delta_{im}^n - \sum_{j=n+1}^{m-1} x_j \delta_{ij}^n)/\delta^n$, $i = 1, \ldots, n$, where x_j, $j = n+1, \ldots, m$ are free variables and $\delta^n \neq 0$. $\delta^k = |a_{ij}|$, $i = 1, \ldots, k$, $j = 1, \ldots, k$, $k = 1, \ldots, n$, denote corner minors of the matrix A of order k, δ_{ij}^k denotes minors obtained after a substitution in the minors δ^k of the column i by the column j of the matrix A^*, $k = 1, \ldots, n$, $i = 1, \ldots, k$, $j = 1, \ldots, m$.

We examine four algorithms [1]–[4] for solving system (1) over the fraction field of the ring R, assuming that R does not have zero divisors and all corner minors δ^k, $k = 1, \ldots, n$, of the matrix A are different from zero. Each of the algorithms is in fact a method for computing the minors δ^n and δ^n_{ij} in the ring R. For each of the algorithms we evaluate:

1. The general time for the solution taking into consideration only arithmetic operations and assuming moreover, the execution time for multiplication, division and addition/subtraction of two operands, the first of which is a minor of order i and the second one, a minor of order j, will be M_{ij}, D_{ij}, A_{ij} correspondingly.

2. The exact number of operations of multiplications, division and addition/substraction over the coefficients of the system.

3. The number of operations of multiplication/division ($M^{\mathbf{R}}$), when $R = \mathbf{R}[x_1, \ldots, x_r]$ is a ring of polynomials with r variables with real coefficients and only one computer word is required for storing any one of the coefficients.

4. The number of operations of multiplication/division ($M^{\mathbf{Z}}$), when $R = \mathbf{Z}[x_1, \ldots, x_n]$ is a ring of polynomials with r variables with integer coefficients and these coefficients are stored in as many computer words as are needed.

5. The number of operations of multiplication/division ($M^{\mathbf{Z}M}$), when $R = \mathbf{Z}[x_1, \ldots, x_n]$ is a ring of polynomials with r variables with integer coefficients, but for the solution the modular method is applied, which is based on the remainder theorem.

1. Dodgson's Algorithm

Dodgson's algorithm [1], the first algorithm to be examined, appeared more than 100 years before the others.

The first part of the algorithm consists of $n - 1$ steps. In the first step, all minors of order 2 are computed

$$\hat{a}^2_{ij} = a_{i-1,j-1}a_{ij} - a_{i-1,j}a_{i,j-1}, \qquad i = 2, \ldots, n, j = 2, \ldots, n$$

formed from four neighboring elements, located at the intersection of lines $i - 1$ and i and of columns $j - 1$ and j. In the k-th step, $k = 2, \ldots, n - 1$, according to formula

(2)
$$\hat{a}^{k+1}_{ij} = (\hat{a}^k_{i-1,j-1}\hat{a}^k_{ij} - \hat{a}^k_{i-1,j}\hat{a}^k_{i,j-1})/\hat{a}^{k-1}_{i-1,j-1},$$
$$i = k+1, \ldots, n, \ j = k+1, \ldots, m,$$

all minors $\hat{a}^{k+1}_{ij}$ of order $k + 1$ are computed, these minors are formed by the elements located at the intersection of the lines $i - k, \ldots, i$ and columns $j - k, \ldots, j$ of the matrix A^*.

In this way the minors $\hat{a}_{nj}^n = \delta_{nj}^n$, $j = n, \ldots, m$, will be computed.

Remark that it is essential that all minors $\hat{a}_{ij}^k$, $k = 1, \ldots, n-1$, $i = k, \ldots, n-1$, $j = k, \ldots, m-1$, appearing in the denominator of expression (2), be different from 0. This, of course, narrows down the set of solvable problems.

The second part of Dodgson's algorithm is feasible only under the assumption that $m = n + 1$ and all the searched for unknown variables x belong to R. Then the value $x_n = \delta_{nm}^n/\delta_{nn}^n$ may be computed and substituted in the initial system. After eliminating the last equation and recalculating the column of free members, a system of order $n - 1$ is obtained, to which the algorithm described above may be applied again to find x_{n-1}. During this process, it suffices to recalculate only those minors in which the column of free members appears. This process continues until all solutions are obtained.

Let us evaluate the computing time of the algorithm. The first part of the algorithm is executed in time

$$(3) \quad T_1^1 = (n-1)(m-1)(2M_{1,1} + A_{2,2})$$
$$+ \sum_{i=2}^{n-1}(n-i)(m-i)(2M_{i,i} + A_{2i,2i} + D_{2i,i-1}).$$

The second part of the algorithm, when $m = n + 1$, is executed in time

$$T_1^2 = \sum_{i=1}^{n-1} i(M_{1,1} + A_{2,1}) + \sum_{j=1}^{n-2} j(2M_{1,2} + A_{3,3})$$
$$+ \sum_{i=2}^{n-2}\sum_{j=1}^{i}(2M_{i,i+1} + A_{2i+1,2i+1} + D_{2i+1,i-1})$$

In this we suppose that the time needed for the recalculation of one free member is $M_{1,1} + A_{2,1}$.

2. Bareiss's Algorithm

The forward procedure of Bareiss's algorithm [2] differs from Dodgson's algorithm only in the selection of the leading (pivoting) minors. In the first step all minors of second order are computed

$$(4) \qquad a_{ij}^2 = a_{11}a_{ij} - a_{1j}a_{i1}, \quad i = 2, \ldots, n, \quad j = 2, \ldots, m,$$

which surround the corner element a_{11}. At the k-th step, $k = 2, \ldots, n-1$, and according to the formula

$$(5) \qquad a_{ij}^{k+1} = (a_{kk}^k a_{ij}^k - a_{ik}^k a_{kj}^k)/a_{k-1,k-1}^{k-1},$$
$$i = k+1, \ldots, n, \quad j = k+1, \ldots, m,$$

all the minors a_{ij}^{k+1} of order $k+1$ are computed, which are formed by surrounding the corner minor δ^k by row i and column j, that is, minors which are formed by the elements, located at the intersection of row $1, \ldots, k, i$ and of columns $1, \ldots, k, j$. Obviously, $a_{nj}^n = \delta_{nj}^n$, $j = n, \ldots, m$, holds.

Comparing the above procedure with Dodgson's algorithm, it is seen that here, it suffices that only the corner minors $\delta^k = a_{kk}^k$, $k = 1, \ldots, n-1$, be different from zero and zero divisors. In order to do so, they can be controlled by choice of the pivot row or column.

The back-up procedure consists of $n-1$ steps, where at the k-th step, $k = 1, \ldots, n-1$, all the minors

$$(6) \qquad \delta_{ij}^{k+1} = (a_{k+1,k+1}^{k+1}\delta_{ij}^k - a_{k+i,j}^{k+1}\delta_{ij}^k)/a_{k,k}^k,$$

$$i = k+1, \ldots, n, \quad j = k+1, \ldots, m,$$

are computed.

The computing time of the forward procedure is the same as that of the first part of Dodgson's algorithm (3), $T_2^1 = T_1^1$.

The computing time of the back-up procedure is

$$T_2^2 = \sum_{i=1}^{n-1} i(m-i-1)(2M_{i,i+1} + A_{2i+1,2i+1} + D_{2i+1,i}).$$

3. Algorithm of Forward and Back-up Procedures

The forward procedure in this algorithm [3] is the same as the forward procedure in Bareiss's algorithm, and is based on formulae (4) and (5). The back-up procedure is more economical, and consists of immediate computation of the values δ_{ij}^n according to the formulæ

$$\delta_{ij}^n = (a_{nn}^n a_{ij}^i - \sum_{k=i+1}^n a_{ik}^i \delta_{kj}^n)/a_{ii}^i, \qquad i = n-1, \ldots, 1, j = n+1, \ldots, m.$$

The computing time of the forward procedure is the same as that of the first part of Dodgson's algorithm (3), $T_3^1 = T_1^1$.

The computing time of the back-up procedure is

$$T_3^2 = (m-n)\sum_{i=1}^{n-1}((n+1-i)M_{n,i} + (n-i)A_{i+n,i+n} + D_{i+n,i}).$$

Let us note that when $m = n+1$, T_3^2 is a quantity of order n^2, and T_2^2 is a quantity of order n^3.

4. One-pass Algorithm

Algorithms [2] and [3] consist of two parts (two passes). They make zero elements under the main diagonal of the coefficient matrix during the first pass. And during the second pass, they make zero elements up to the main diagonal.

Algorithm [4] requires one pass, consisting of $n - 1$ steps. In this algorithm, we make diagonalisation of the coefficient matrix minor-by-minor and step-by-step.

In the first step the minors of second order are computed

$$\delta_{2j}^2 = a_{11}a_{2j} - a_{21}a_{1j}, \quad j = 2, \ldots, m,$$
$$\delta_{1j}^2 = a_{1j}a_{22} - a_{2j}a_{12}, \quad j = 3, \ldots, m.$$

In the k-th step, $k = 2, \ldots, n - 1$, the minors of order $k + 1$ are computed according to the formulae (and see the Appendix)

(8) $$\delta_{k+1,j}^{k+1} = a_{k+1,k+1}\delta_{kk}^k - \sum_{p=1}^{k} a_{k+1,p}\delta_{pj}^k, \quad j = k + 1, \ldots, m,$$

(9) $$\delta_{ij}^{k+1} = (\delta_{k+1,k+1}^{k+1}\delta_{ij}^k - \delta_{k+1,j}^{k+1}\delta_{i,k+1}^k)/\delta_{k,k}^k,$$
$$i = 1, \ldots, k, \ j = k + 2, \ldots, m.$$

In this way, at the k-th step the coefficients of the first $k + 1$ equations of the system take part.

The general computing time of the solution is

$$T_4 = (2m - 3)(2M_{1,1} + A_{2,2}) + \sum_{k=2}^{n-1}(m - k)((k + 1)M_{k,1} + kA_{k+1,k+1}) +$$

$$+ \sum_{k=1}^{n-1} k(m - k - 1)(2M_{k,k+1} + A_{2k+1,2k+1} + D_{2k+1,k}).$$

5. Evaluation of the Quantity of Operations
over the System Coefficients

We begin the comparison of the algorithms considering the general number of multiplication NM^m, divisions NM^d and additions/subtractions NM^a, which are necessary for the solution of the system of linear equations (1) of order $n \times m$. Moreover, we will not make any assumptions regarding the computational complexity of these operations; that is we will consider that during the execution of the whole computational process, all multiplications of the coefficients are the same, as are the same all divisions and all additions/subtractions. The quantity of operations, necessary for Bareiss's algorithm will be

$$NM_2^m = 2n^2m - n^3 - 2nm + n,$$
$$NM_2^d = (2n^2m - n^3 - 4nm + 2m + 3n - 2)/2,$$
$$NM_2^a = (2n^2m - n^3 - 2nm + n)/2.$$

The quantity of operations, necessary for the algorithm of forward and back-up procedure, is

$$NM_3^m = (9n^2m - 5n^3 - 3nm - 3n^2 - 6m + 8n)/6,$$
$$NM_3^d = (3n^2m - n^3 - 3nm - 6n^2 + 13n - 6)/6,$$
$$NM_3^a = (6n^2m - 4n^3 - 6nm + 3n^2 + n)/6.$$

The quantity of operations, necessary for the one-pass algorithm, is

$$NM_4^m = (9n^2m - 6n^3 - 3nm - 6m + 6n)/6,$$
$$NM_4^d = (3n^2m - 2n^3 - 3nm - 6m + 2n + 12)/6$$
$$NM_4^a = (6n^2m - 4n^3 - 6nm + 3n^2 + n)/6.$$

In the case when the number of equations and unknowns in the system is the same and equal to n, we can compare all four algorithms.

al.	quantity of operations		
#	multiplication	division	add./substr.
1	$(2n^3 - n^2 - n)/2$	$(n^3 - 4n^2 - 5n - 2)/2$	$(n^3 - n)/2$
2	$n^3 - n$	$(n^3 - 2n^2 + n)/2$	$(n^3 - n)/2$
3	$(4n^3 + 3n^2 - n - 6)/6$	$(2n^3 - 6n^2 + 10n - 6)/6$	$(2n^3 + 3n^2 - 5n)/6$
4	$(n^3 + 2n^2 - n - 2)/2$	$(n^3 - 7n + 6)/6$	$(2n^3 + 3n^2 - 5n)/6$

In this way, according to this evaluation, the fourth algorithm (one-pass) is to be preferred. Bareiss's and Dodgson's algorithms are approximately equal regarding the quantity of operations, and each one of them requires three times more divisions and two times more multiplications, as does the one-pass algorithm. The third algorithm lies somewhere in between. If we evaluate according to the general quantity of multiplication and division operations, considering only the third power, then we obtain the evaluation $3n^3/2 : 3n^3/2 : n^3 : 2n^3/3$.

6. Evaluation of the Algorithms
in the Ring $R[x_1, x_2, \ldots, x_r]$

Let R be the ring of polynomials of r variables over an integral domain and let us suppose that every coefficient a_{ij} of the system (1) is a polynomial of degree p in each variable

$$
(10) \qquad a_{ij} = \sum_{u=0}^{p} \sum_{v=0}^{p} \cdots \sum_{w=0}^{p} a_{u,v,\ldots,w}^{ij} X_1^u X_2^v \cdots X_r^w .
$$

Then it is possible to define, how much time is required for the execution of the arithmetic operations over polynomials which are minors of order i and j of the matrix A^*

$$
\begin{aligned}
A_{ij} &= (jp+1)^r a_{ij}, \\
M_{ij} &= (ip+1)^r (jp+1)^r (m_{ij} + a_{i+j,i+j}), \\
D_{ij} &= (ip - jp + 1)^r (d_{ij} + (jp+1)^r (m_{i-j,j} + a_{ii})).
\end{aligned}
$$

Here we assume, that the classical algorithms for polynomial multiplication and division are used. And also, we consider that the time necessary for execution of the arithmetic operations of the coefficients of the polynomials, — when the first operand is coefficient of the polynomial, which is a minor of order i, and the second — of order j, — is m_{ij}, d_{ij}, a_{ij}, for the operations of multiplication, division and addition/subtraction, respectively.

Let us evaluate the computing time for each of the four algorithms, considering that the coefficients of the polynomials are real numbers and each one is stored in one computer word. We will assume that $a_{ij} = 0$, $m_{ij} = d_{ij} = 1$, $A_{ij} = 0$, $M_{ij} = i^r j^r p^{2r}$, $D_{ij} = (i-j)^r j^r p^{2r}$, and we will consider only the leading terms in m and n:

$$
M_2^R = \rho n^r \left(\frac{3m}{2r+1} - \frac{3n}{2r+2} \right),
$$

$$
M_3^R = \rho n^r \left(\frac{3m}{2r+1} - \frac{3m+3n}{2r+2} + \frac{3n}{2r+3} + \frac{m-n}{r+1} - \frac{m-n}{r+2} \right),
$$

$$
M_4^R = \rho n^r \left(\frac{3m}{2r+2} - \frac{3n}{2r+3} \right) + \rho \left(\frac{m}{r+2} - \frac{n}{r+3} \right),
$$

where $\rho = n^{r+2} p^{2r}$.

For $m = n + 1$ it is possible to compare all four algorithms: $N_1^R = 3\sigma$, $N_2^R = (2r+3)\sigma$, $N_3^R = 2\sigma$, $N_4^R = (2r+1)\sigma + \rho n/(r+2)(r+3)$, where $\sigma = 3n^{2r+3} p^{2r}/(2r+1)(2r+2)(2r+3)$. For $r = 0$ we obtain the same evaluation as in the previous section. For $r \neq 0$ we obtain $3 : (2r+3) : 2 : (2r+1)$.

7. Evaluation of the Algorithms
in the Ring $\mathbf{Z}[x_1, x_2, \ldots, x_r]$, Classical Case

As before we suppose that every coefficient of the system is a polynomial of the form (10). However, the coefficients of these polynomials are now integers and each one of these coefficients $a_{uv,\ldots,w}^{ij}$ is stored in l computer words. Then, the coefficients of the polynomial, which is a minor of order i, are integers of length il of computing words.

Under the assumption that classical algorithms are used for the arithmetic operations on these long integers, we obtain: $a_{ij} = 2jla$, $m_{ij} = ijl^2(m + 2a)$, $d_{ij} = (il - jl + 1)(d + jl(m + 2a))$, where a, m, d are the execution time of the single-precision operations of addition/subtraction, multiplication, and division.

Assuming that $a = 0$, $m = d = 1$, we obtain the following evaluation of the execution times of polynomial operations: $M_{ij} = ijl^2(ijp^2)^r$, $D_{ij} = (i-j)^{r+1}j^{r+1}l^2p^{2r}$, $A_{ij} = 0$.

In this way, the evaluation of the time for solution will be the same as that for the ring $\mathbf{R} = \mathbf{R}[x_1, x_2, \ldots, x_r]$ (section 6), if we replace everywhere r by $r + 1$ and p^r by lp^r.

Therefore, for $m = n + 1$ we obtain $N_1^Z = 3\psi$, $N_2^Z = (2r + 5)\psi$, $N_3^Z = 2\psi$, $N_4^Z = (2r+3)\psi$, where $\psi = 3n^{2r+5}p^{2r}l^2/(2r+3)(2r+4)(2r+5)$, $l \geq 1$, $r \geq 0$.

8. Evaluation of the Algorithms
in the Ring $\mathbf{Z}[x_1, x_2, \ldots, x_r]$, Modular Case

Let us evaluate the time for the solution of the same problem, for the ring of polynomials with r variables with integer coefficients $\mathbf{R} = \mathbf{Z}[x_1, \ldots, x_r]$, when the modular method is applied, based on the remainder theorem. In this case we will not take into consideration the operations for transforming the problem in the modular form and back again.

It suffices to define the number of moduli, since the exact quantity of operations on the system coefficients for the case of a finite field has already been obtained in section 4.

We will consider that every prime modulus m_i is stored in exactly one computer word, so that, in order to be able to recapture the polynomial coefficients, which are minors of order n, $n(l + \log(np^3)/2\log m_i)$ moduli are needed. It is easy to see due to Hadamard's inequality.

Further, we need up moduli for each unknown x_j, which appears with maximal degree np. There are r such unknowns, and therefore, in all, $\mu = prn^2(l + \log(np^3)/2\log m_i)$ moduli are needed.

If we now make use of the table in section 5, denote the time for modular multiplication by m and the time for modular division by d, then not considering addition/subtraction and considering only leading terms in

n, we obtain for $m = n + 1$: $N_1^{ZM} = (6m + 3d)\nu$, $N_2^{ZM} = (6m + 3d)\nu$, $N_3^{ZM} = (4m + 2d)\nu$, $N_4^{ZM} = (3m + d)\nu$, where $\nu = \mu n^3/3$.

Conclusions

We see, that modular methods are better then non-modular ones, as usual. And the one-pass method [4] is the best for modular computation.

The method of forward and back-up procedures [3] are better for non-modular computation in polynomial rings. And Dodgson's method [1] stands nearly it for such polynomial computations.

Appendix
Foundation of the One-pass Algorithm

Identity (8) is an expansion of the minor $\delta_{k+1,j}^{k+1}$ according to line $k + 1$, and therefore it suffices to prove only identity (9).

In order to do so, let us consider the following determinant identity

$$\begin{vmatrix} A_{si} & A_{00} \\ N_{s0} & A_{tj} \end{vmatrix} = \begin{vmatrix} A_{0i} & N_{-t,-j} \\ N_{s0} & A_{tj} \end{vmatrix}$$

where s, i, t, j are column numbers of the matrix A^*, A_{uv} is a submatrix of the matrix A^*, this submatrix of order k will be at the left upper corner of a matrix A^*, if we replace columns s and i by columns u and v respectively. Here $u = 0$ denotes a column consisting of zeros, and $u = -t$ denotes a column obtained by changing the signs of all the elements of column t. Matrix N_{uv} is obtained from matrix A_{uv} if, in addition, all remaining $k - 2$ columns are replaced by zero ones.

In order to obtain the determinant on the right side, it is necessary in the determinant on the left side to subtract from the first (block) line the second one.

If we expand each one of the determinants of order $2k$ by the first k lines according to Laplace's rule, then we obtain the columns-substitution identity

(11) $$\delta^k \delta_{st;ij}^k = \delta_{st}^k \delta_{ij}^k - \delta_{sj}^k \delta_{it}^k,$$

where $\delta_{st;ij}^k$ is a minor formed from the minor δ^k after substituting columns s by t and i by j in the matrix A^*.

To prove identity (9) it remains to expand the existing minors of order $k + 1$ by row $k + 1$

$$\delta_{ij}^{k+1} = a_{k+1,k+1}\delta_{ij}^k - a_{k+1,j}\delta_{i,k+1}^k - \sum_{s=1,s\neq i}^{k} a_{k+1,s}\delta_{s,k+1;i,j}^k$$

$$\delta_{k+1,j}^{k+1} = a_{k+1,j}\delta^k - \sum_{s=1}^{k} a_{k+1,s}\delta_{s,j}^k,$$

and make use of the columns-substitution identity (11).

REFERENCES

[1] Dodgson C. L., *Condensation of determinants, being a new and brief method for computing their arithmetic values*, Proc. Royal Soc. Lond. A. **15** (1866), 150–155.

[2] Bareiss E. N., *Sylvester's identity and multistep integer-preserving Gaussian elimination*, Math. Comput. **22** (1968), 565–578.

[3] Malashonok G. I., *On the solution of a linear equation system over commutative ring*, Math. Notes of the Acad. Sci. USSR **42**, **N4** (1987), 543–548.

[4] Malashonok G. I., *A new solution method for linear equation systems over the commutative ring*, in "Int. Algebraic Conf., Novosibirsk," Aug. 21–26, 1989, Theses on the ring theory, algebras and modules, 1989, p. 82.

Gennadi I. Malashonok
Institute for Applied Problems of Mechanics and Mathematics
Academy of Sciences
Vul. Naukova 3-b, 290053 Lvov, USSR

Une conjecture sur les anneaux de Chow
$A(G, \mathbf{Z})$
renforcée par un calcul formel

ROGER MARLIN

I. Introduction

Dans son article *Torsion in cohomology of compact Lie groups* Victor Kac écrivait: "Remark: ... $A(G, \mathbf{Z})$ for $G = Spin_n$, SO_n, G_2 and F_4 was computed by R. Marlin. His thesis was very helpful for finding the general pattern."

Cette remarque m'a évidemment encouragé à pousser plus avant divers calculs et à développer des programmes informatiques permettant d'alléger cette fastidieuse besogne. Il y a dans ce domaine, comme en bien d'autres, matière à étayer ou à infirmer de nombreuses théories par le calcul. L'environnement s'y prête particulièrement bien puisque la classification des algèbres de Lie semi-simples est exhaustive et particulièrement claire.

II. Position du problème

II.1. Problème général

Soit X un espace projectif, G un groupe algébrique affine opérant transitivement sur X; alors il existe G' groupe algébrique semi-simple et P sous-groupe parabolique de G' tels que X soit isomorphe à l'espace homogène G'/P. En effet le radical R de G est normal et résoluble, donc R a un point fixe dans X (théorème de Borel) et $G' = G/R$ (qui est un groupe algébrique semi-simple) agit transitivement sur X. Soit B un sous-groupe de Borel de G'; il a, à son tour, un point fixe dans X. Le stabilisateur de ce point fixe est un sous-groupe parabolique P et X est isomorphe à G'/P. De la suite exacte $0 \longrightarrow P/B \longrightarrow G'/B \longrightarrow G'/P \longrightarrow 0$, on déduit que G'/P est le quotient de G'/B par P/B. Soit L le sous-groupe de Levi correspondant à P: on a l'isomorphe $P/B \sim L/(B \cap L)$. Ce dernier espace est lui-même de la forme quotient d'un semi-simple par un Borel. On voit donc toute l'importance de l'étude de la topologie des espaces homogènes, et plus particulièrement du cas quotient d'un semi-simple par un Borel.

II.2. *Exemple*: Soit X l'espace de drapeaux de l'espace vectoriel E de dimension 13 formé par un 5-espace contenu dans un 9-espace contenu dans un 10-espace contenu dans un 12-espace. C'est un espace projectif sur lequel $GL(13)$ opère transitivement. Le radical R de $GL(13)$ est constitué des matrices scalaires, les drapeaux sont tous invariants par homothétie. D'où action bien naturelle du quotient $G' = Sl(13)$ des matrices de déterminant 1. Fixons $(e_1, e_2, e_3, \ldots, e_{13})$ une base de E. Les matrices triangulaires supérieures dans cette base constituent un Borel. Le drapeau constitué par le 5-espace engendré par $(e_1, e_2, \ldots, e_5)$, le 9-espace engendré par $(e_1, e_2, \ldots, e_9)$, le 10-espace engendré par $(e_1, e_2, \ldots, e_{10})$ et le 12-espace engendré par $(e_1, e_2, \ldots, e_{12})$, est invariant par le Borel B. Son stabilisateur est le parabolique P décrit ci-dessous:

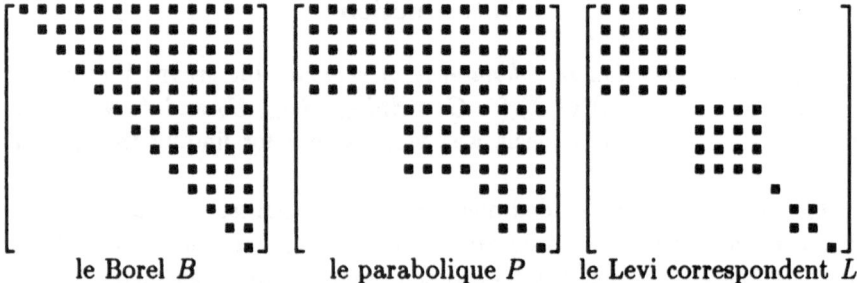

le Borel B le parabolique P le Levi correspondent L

Les éléments non marqués d'un ■ sont nuls. Le Levi L est le produit de $Sl(5)$ par $Sl(4)$ par $Sl(2)$ par un tore de dimension 2. L'espace G'/B peut alors être présenté comme l'espace des drapeaux complets (droite vectorielle contenue dans un plan vectoriel ... contenu dans un hyperplan vectoriel). L'espace X est isomorphe à G'/P.

II.3. Approche

L'approche, développée ici, est de nature algébrique. Elle s'appuie sur la classification des groupes semi-simples et sur les systèmes de racines correspondants.

III. Résultats classiques

III.1. Notations

- K un groupe de Lie compact,
    - G son complexifié (groupe algébrique semi-simple),
    - T un tore maximal de K,
    - T son complexifié,
    - $W = Norm(T)/T$ le groupe de Weyl,
    - B un Borel de G contenant T (on a alors $G/B = K/T$),

- $A(G/B)$ anneau de Chow de l'espace homogène,
- c l'homomorphisme qui, à un caractère de T, associe sa classe caractéristique dans $H^2(K/T, \mathbf{Z})$ $(= A^1(G/B))$,
- $X(T)$ le groupe des caractères de T, pour $\lambda : T \longrightarrow U(1)$ caractère de T, on a $K \longrightarrow K/\mathrm{Ker}\lambda \longrightarrow K/T$ fibration de groupe $U(1)$ d'où classe caractéristique associée à λ dans $H^2(K/T, \mathbf{Z})$

$$c : X(T) \longrightarrow H^2(K/T, \mathbf{Z}) = A^1(G/B)$$

III.2. Racines, racines positives, racines simples

L'algèbre de Lie $\mathfrak{g}$ de G est un T-module via la représentation adjointe. Soit R l'ensemble des poids non nuls de cette représentation. Ce sont les racines du couple (G, T). L'algèbre de Lie $\mathfrak{b}$ de B est aussi un T-module. Ces poids sont des racines du couple (G, T): ce sont les racines négatives du triplet (G, T, B). Les autres racines sont les racines positives R_+ du triplet (G, T, B). De R_+ on peut extraire une partie S qui constitue une base du $\mathbf{Z}$-module libre engendré par les racines et telle que toute racine positive soit combinaison à coefficient positifs d'éléments de S. Ce sont les racines simples du triplet (G, T, B). On peut aussi les caractériser en terme d'algèbre de Lie, elles correspondent aux espaces propres de $\mathfrak{g}$ qui ajoutés à $\mathfrak{b}$ forment une sous algèbre de Lie de $\mathfrak{g}$.

III.3. Groupe de Weyl

A chaque racine α est associé un élément s_α de W. Considéré comme opérant sur $X(T)$ c'est une symétrie. Le groupe de Weyl $W = Norm(T)/T$ est un groupe de réflexions (groupe de Coxeter). Il est engendré par les symétries par rapport aux racines simples associées à un Borel. Etant donné un système de racines simples, chaque élément w de W est produit de symétries par rapport à des racines simples. Le nombre minimum de symétries par rapport à des racines simples nécessaire pour obtenir un élément w est la longueur de cet élément, notée $l(w)$.

III.4. Classification

La classification répartit les groupes semi-simples en quatre familles infinies et cinq cas exceptionnels (l'indice est le rang du groupe $(\dim T)$):

$$A_n \ (n \geq 1), \ B_n \ (n \geq 2), \ C_n \ (n \geq 3), \ D_n \ (n \geq 4), \ E_6, \ E_7, \ E_8, \ F_4, \ G_2.$$

Pour chaque cas, il y a un groupe simplement connexe et quelque groupes qui ne le sont pas (2 ou 4). Ces groupes différent par le réseau $X(T)$.

III.5. Base du Groupe de Chow

Les adhérences des cellules de Bruhat constituent une base du $\mathbf{Z}$-module libre $A(G/B)$. Les adhérences des cellules de Bruhat sont indexées par le éléments du groupe de Weyl. Plus précisément, si w_0 est l'élément de plus grande longueur de W, Bw_0wB/B est une cellule de Bruhat de codimension la longueur de w.

III.6. Structure multiplicative

Le groupe de Chow (groupe des cycles à équivalence rationnelle près) n'a pas toujours de structure multiplicative. Dans le cas présent, le lemme de déplacement nécessaire à la définition d'un produit se montre en utilisant la décomposition de G/B en cellules de Bruhat (G/B est réunion disjointe d'espaces affines dont les adhérences s'obtiennent par réunion avec des cellules plus petites, voir [M3] par exemple). Le produit est donc obtenu par intersection de cycles après les avoir mis en position "transverse". Pour décrire la structure multiplicative de $A(G/B)$, on dispose de deux formules.

III.6.1. Cycles de dimensions complémentaires

Soit w et w' des éléments de W tel que $l(w) + l(w') = \dim(G/B)$. On a alors [D2]: $Z_w \cdot Z_{w'} = B/B$ si $l(ww') = \dim(G/B)$ où B/B est la classe du point $= 0$ sinon

III.6.2. Produit par un cycle de codimension 1 (formule de Chevalley)

Soit λ un caractère de T et w un élément du groupe de Weyl. Si Z_w est l'adhérence de la cellule de Bruhat correspondant à w et si $c(\lambda)$ est le cycle de codimension 1 correspondant à λ, on a:

III.6.3. Produit de trois cycles de dimensions complémentaires

Une troisième formule serait très appréciée: soit w, w' et w'' des éléments de W tel que $l(w) + l(w') + l(w'') = \dim(G/B)$. On a alors: $Z_w \cdot Z_{w'} \cdot Z_{w''} = \phi(w, w', w'') \cdot B/B$, avec $\phi(w, w', w'') \in \mathbf{Z}$. Cette formule permettrait en effet, grâce au produit de cycles de dimensions complémentaires donné précédemment, d'avoir le produit de deux cycle de manière tout à fait générale. En effet si w, w' et w'' sont des éléments de W avec $l(w'') = l(w) + l(w')$, le coefficient $Z_{w''}$ de dans $Z_w \cdot Z_{w'}$ serait $\phi(w, w', w_0 w''^{-1})$ où w_0 désigne l'élément de plus grande longueur de W. Aucune formule explicite de ce genre n'existe actuellement. Martine Smolders a écrit un ensemble de fonction LeLisp qui permet de calculer cette formule.

IV. Structuration des objets

IV.1. Racines et poids

Ce sont des éléments de $\mathbf{Z}$-modules libres; on les représentera donc par la liste de leurs coordonnées dans une base ou comme élément d'un $\mathbf{Z}^n$, où n est le rang (éventuellement plus un).

IV.2. Groupe de Weyl

Selon le but poursuivi, on peut utiliser différentes structurations, dont en particulier:

IV.2.1. Produit de symétries simples

La représentation d'un élément du groupe de Weyl par un produit de symétries simples peut paraître naturelle. Elle est en fait peu pratique, lourde (gestion d'une liste), et surtout difficile d'utilisation puisqu'un même élément a de multiples écritures, pas toutes minimales. Par exemple, dans A^2, on a $s_1 s_2 s_1 s_2 = s_2 s_1$ et $s_1 s_2 s_1 = s_2 s_1 s_2$. La longueur maximale de cette liste est la dimension de G/B.

IV.2.2. Matrice

Pour résoudre le problème de représentation multiple, on peut associer à un élément du groupe de Weyl la matrice de son action sur $X(T)$. Dans le cas A_n, le groupe de Weyl est le groupe des permutation des $n+1$ lettres et la représentation est donc de taille $n+1$. Dans les autres cas, le groupe de Weyl est plus compliqué et demande donc une représentation de plus grande taille, sans toutefois excéder le carré du rang (c'est un endomorphisme de $X(T)$).

IV.2.3. Image d'un poids

Les chambres de Weyl étant indexées par les éléments du groupe de Weyl, l'image par w d'un élément invariant par aucune symétrie est caractéristique de w. La demi somme ρ des racines positives (= somme des poids fondamentaux) est un tel élément. On peut donc représenter l'élément w par la valeur de $w(\rho)$.

IV.2.4. Comment choisir

La structuration 1 n'a d'intérêt que pour communiquer à l'utilisateur la valeur d'un élément du groupe de Weyl, mais elle ne peut être utilisée efficacement que couplée à une représentation du type 2 ou 3. La représentation 2 sera préférée quand on aura réellement besoin du produit de deux éléments quelconques du groupe de Weyl. Mais le calcul de la longueur du résultat pose problème et, pour être efficace, la structuration doit tenir compte du cas considéré (il serait dommage de structurer une matrice de permutation en tableau carré!). La représentation 3 est compacte, indépendante du cas considéré mais ne permet que la multiplication par des symétries simples (sauf si elle est associé à la 1, mais alors, elle n'est plus compacte!). La structuration à retenir sera donc fonction de ce que l'on a à faire. Dans la majeure partie des cas, la représentation 3 m'a donné pleine satisfaction.

IV.2.5. Exemple de la formule de Chevalley

Dans la formule de Chevalley,

$$c(\lambda) \cdot Z_w = \Sigma_{\alpha \in R^+, \ell(w) = \ell(w s_\alpha) + 1} \langle \alpha^\vee \mid \lambda \rangle Z_{w s_\alpha}$$

on voit que l'important est de savoir quels sont les éléments de longueur un de plus que w. La représentation des éléments du groupe de Weyl par l'image d'un poids est ici parfaitement adaptée. Plus précisément la gestion simultanée de la liste des éléments de longueur l et de longueur $l + 1$ est commode puisque le produit d'un élément de longueur $l+1$ par une symétrie simple est de longueur l ou $l + 2$. On peut donc, de proche en proche, gérer ce couple de listes.

IV.3. Anneau de Chow

L'élément $\sum x_i Z_{w_i}$ de $A(G/B)$ sera représenté par la liste des couples (x_i, w_i).

V. Boîte à outils

Il s'agit de se donner les moyens de mettre en oeuvre les résultats présentés précédemment.

V.1. Donnée d'un cas.

Se donner un cas, c'est fixer un triplet (G, T, B). Algébriquement cela revient à se donner:
— Un système de racines simples
— Un système de poids
   Par exemple $SO(7)$ est donné par les deux listes

$$\text{racines\_simples} = ((2 \ -2 \ 0) \ (0 \ 2 \ -2) \ (0 \ 0 \ 2))$$
$$\text{poids} = ((2 \ -2 \ 0) \ (0 \ 2 \ -2) \ (0 \ 0 \ 2))$$

le groupe simplement connexe correspondant serait donné par

$$\text{poids} = ((1 \ -1 \ -1) \ (1 \ 1 \ -1) \ (1 \ 1 \ 1))$$

V.2. Racines positives

Etant donné un système de racines simples, on obtient les autres racines en appliquant des symétries. Soit α une racine simple et w un élément du groupe de Weyl; $w(\alpha)$ est une racine positive si $l(ws_\alpha) = l(w) + 1$, négative sinon. On peut associer aux racines positives un niveau, c'est le nombre minimal de symétries simples qu'il faut appliquer à une racine simple pour les obtenir. Les racines simples sont de niveau 0.

V.2.1. Procédure Calcul-Racines-Positives

V.2.1.1. Extérieurs

Paramètre d'entrée: Racines-Simples: liste-de-racines

Paramètre de sortie: Racines-Positives: liste-de-racines

Antécédent: Racines-Simples est une liste de racines correspondant à un système existant

Rôle de la procédure: calcule dans Racines-Positives la liste des racines positives correspondant à Racines-Simples.

V.2.1.2. Algorithme

variable: liste-0, liste-1, liste-2: liste-de-racines

liste-0 = liste des racines positives de niveau k

liste-1 = liste des racines positives de niveau $k - 1$

liste-2 = liste des racines positives de niveau $k - 2$

Racines-Positives = liste des racines positives de niveau au plus $k - 3$

    initialiser-à-vide Racines-Positives

    copier Racines-Simples dans liste-2

    calculer dans liste-1 la liste des 1-racines

    répéter

        calculer liste-0 à partir de liste-1 et de liste-2

        ajouter liste-2 à Racines-Positives

        pointer liste-2 sur liste-1

        pointer liste-1 sur liste-0

    jusqu'à liste-2 vide

V.2.2. Calcul des racines de niveau 1

On obtient les racines de niveau 1 en considérant l'ensemble des $s_\alpha(\beta)$ pour α et β racines simples différentes.

V.2.3. Procédure Calcul-k-Racines

Cette procédure permet de calculer les racines de niveau k connaissant les racines de niveau $k - 1$ et $k - 2$. Supposons que l'image d'une racine de niveau $k - 1$ par une symétrie simple s_α soit une racine β de niveau h. Soit $\beta = w(\gamma)$ avec γ racine simple et $l(w) = h$. On a alors $\alpha = (s_\alpha w)(\gamma)$ donc $l(s_\alpha w) \geq k - 1$ et donc $l(w) \geq k - 2$. L'image d'une racine de niveau $k - 1$ par une symétrie simple ne peut donc être qu'une racine de niveau $k - 2$ ou k.

V.2.3.1. Extérieurs

    Paramètres d'entrée: liste-1, liste-2: liste-de-racines

    Paramètre de sortie: liste-0: liste-de-racines

    Antécédent: liste-1 contient la liste des racines de niveau $k - 1$

    liste-2 contient la liste des racines de niveau $k - 2$

Rôle de la procédure: calcule dans liste-0 les images des éléments de liste-1 par des symétries simples quand ces images ne sont pas dans liste-2, c'est à dire la liste des racines de niveau k

V.2.3.2. Algorithme

    initialiser-à-vide liste-0

    pour tout α dans Racines-Simples faire

        pour tout β dans liste-1 faire

            si l'image de β par $s\alpha$ n'est ni dans liste-2 ni dans liste-0

            ajouter la à liste-0

V.3. Groupe de Weyl

V.3.1. Procédure Calcul-Weyl

Par cette procédure, il s'agit de gérer la liste des éléments du groupe de Weyl de longueur donnée. Plus précisément de gérer le couple des listes des éléments de deux longueurs consécutives.

V.3.1.1. Extérieurs

    Paramètres d'entrée: liste-1, liste-2: liste-d'él-de-Weyl

    Paramètres de sortie: liste-1, liste-2: liste-d'él-de-Weyl

    Antécédent:

    liste-1 contient les éléments du groupe de Weyl de longueur $k - 2$

    liste-2 contient les éléments du groupe de Weyl de longueur $k - 1$

    Rôle de la procédure:

    liste-1 contient les éléments du groupe de Weyl de longueur $k - 1$

    liste-2 contient les éléments du groupe de Weyl de longueur k

V.3.1.2. Algorithme

variable: liste: liste-d'él-de-Weyl

initialiser-à-vide liste

pour tout α dans Racines-Simples faire

    pour tout w dans liste-2 faire

        si $s\alpha w$ n'est ni dans liste-1 ni dans liste ajouter le à liste

pointer liste-1 sur liste-2

pointer liste-2 sur liste

V.4. Formule de Chevalley

V.4.1. Procédure Calcul-Chevalley

V.4.1.1. Extérieurs

Paramètres d'entrée: λ: poids; w: Weyl; liste: liste-d'él-de-Weyl

Paramètres de sortie: a: élément-de-Chow

Antécédent:

liste contient la liste des éléments du Groupe de Weyl de longueur $l(w)+1$

Rôle de la procédure: a contient le produit $c(\lambda) \cdot Zw$

V.4.1.2. Algorithme

initialiser-à-vide a

pour tout α dans Racines-Positives faire

    si $s_\alpha(w)$ est dans liste ajouter $(\langle \breve{\alpha} \mid \lambda \rangle, s_\alpha(w))$ à a

V.5. Diagonalisation des morphismes de Z-modules libres

V.5.1. Procédure Calcul-Diagonale

Théorème de Smith: Soit A et B deux Z-modules libres de type fini et f un morphisme de Z-modules de A dans B. Il existe une base de A et une base de B telles que la matrice de f soit diagonale dans ces bases. On peut donc sur la matrice dans une telle base lire les dimensions des noyaux et images, et voir l'éventuelle torsion du conoyau.

V.5.1.1. Extérieurs

Paramètres d'entrée: m:matrice $[1 \ldots N, 1 \ldots M]$ d'entiers

Paramètres de sortie: l:liste d'entiers

Antécédent:

Rôle de la procédure: la liste l contient les valeurs diagonales de m

V.5.1.2. Algorithme

pour chaque i de 1 à N faire

    si le bloc $[i \ldots N, i \ldots M]$ ne contient que des zéros

        sortir de l'itération

    par permutation de lignes et de colonnes amener en position $[i, i]$

        un élément non nul du bloc $[i \ldots N, i \ldots M]$

    répéter

        pour chaque j de $i+1$ à N tel que $m[j, i] \neq 0$ faire

            calculer l'identité de Bezout pour $m[i, i]$ et $m[j, i]$

                $\mathrm{pgcd}(m[i, i] \, \text{et} \, m[j, i]) = d = a \cdot m[i, i] + b \cdot m[j, i]$

                $m[i, i] = x \cdot d \; m[j, i] = y \cdot d \; \{x \cdot a + y \cdot b = 1\}$

            remplacer la colonne i par $a \cdot \text{colonne}\, i + b \cdot \text{colonne}\, j$

            remplacer la colonne j par $y \cdot \text{colonne}\, i - x \cdot \text{colonne}\, j$

            $\{m$ est équivalente à la matrice de départ $\}$

            $\{m[i, co] = 0$ pour $i \neq co$ et $(co \leq j)\}$

        pour chaque j de $i+1$ à N tel que $m[i, j] \neq 0$ faire

            calculer l'identité de Bezout pour $m[i, i]$ et $m[i, j]$

                $\mathrm{pgcd}(m[i, i] \, \text{et} \, m[i, j]) = d = a \cdot m[i, i] + b \cdot m[i, j]$

                $m[i, i] = x \cdot d \; m[i, j] = y \cdot d \; \{x \cdot a + y \cdot b = 1\}$

            remplacer la ligne i par $a \cdot \text{ligne}\, i + b \cdot \text{ligne}\, j$

            remplacer la ligne j par $y \cdot \text{ligne}\, i - x \cdot \text{ligne}\, j$

            m est équivalente à la matrice de départ

            $m[li, i] = 0$ pour $i \neq li$ et $(li \leq j)$

    jusqu'à ce que $m[i, j] = 0$, pour tout j tel que $i+1 \leq j \leq M$

        $\{m$ est équivalente à la matrice de départ $\}$

        $\{m[li, co] = 0$ pour $li \neq co$ et $(li \leq i$ ou $co \leq i)\}$

La sortie de la boucle "répéter" est garantie par le fait que $m[i, i]$ diminue strictement en valeur absolue à chaque fois que la colonne i est modifiée dans la première boucle "pour" ou que la ligne i l'est dans la seconde. Un entier naturel ne pouva nt indéfiniment décroître il stationne, mais alors la deuxième boucle pour met la ligne i à zéro sans perturber la ligne i que la première boucle "pour" avait mis à zéro.

VI. Utilisation

VI.1. Calcul de $A(G)$

Une remarque classique de Grothendieck [G] permet d'obtenir l'anneau de Chow $A(G)$ comme quotient de $A(G/B)$ par l'idéal engendré par l'image de l'homomorphisme caractéristique

$$c : X(T) \to H^2(K/T, \mathbf{Z}) = A^1(G/B) \text{ avec } c(\lambda) = \sum_{\alpha \in S} \langle \alpha^{\vee} \mid \lambda \rangle Z_{s_\alpha}$$

$A(G)$ est donc le conoyau d'un morphisme de $\mathbf{Z}$-modules libres de $X(T) \otimes A(G/B)$ dans $A(G/B)$. La formule de Chevalley permet d'expliciter ce morphisme, [M2].

VII. Conjecture

Soit sur $A(G/B) \otimes \wedge X(T)$ la graduation définie en posant degré($A^i(G/B) \otimes \wedge^j X(T)$) = $2i + j$. Soit d l'antidérivation de degré 1 de $A^i(G/B) \otimes \wedge^j X(T)$ dans $A^{i+1}(G/B) \otimes \wedge^{j-1} X(T)$ obtenue en posant $d(w \otimes 1) = 0$ pour tout w dans $A(G/B)$ et $d(1 \otimes r) = c(r) \otimes 1$ pour tout r dans $X(T)$; elle fait de $A(G/B) \otimes \wedge X(T)$ un complexe.

On a alors: l'homologie de $A(G/B) \otimes \wedge X(T) \otimes F$ est isomorphe (module) à $H^*(K, F)$ ($F = \mathbf{R}, \mathbf{Z}, \mathbf{Z}/p\mathbf{Z}$) et à $H^*(K)$ par dualité.

Le calcul montre que jusqu'en rang 4 (G_2, F_4, $Spin_n$ et SO_n pour $n < 9$) la conjecture est vraie. Ensuite, des problèmes de calculs intermédiaires faisant intervenir des entiers trop grands ne m'ont pas permis d'aboutir. Reste à prouver la conjecture! Le cas critique est $F = \mathbf{Z}$ et G un groupe avec torsion (B_n, D_n, G_2, F_4, E_6, E_7, E_8). Les autres cas découlant d'interprétation géométrique de la fibration $T \longrightarrow K \longrightarrow K/T$, [M5].

VII.1. Un exemple: G_2

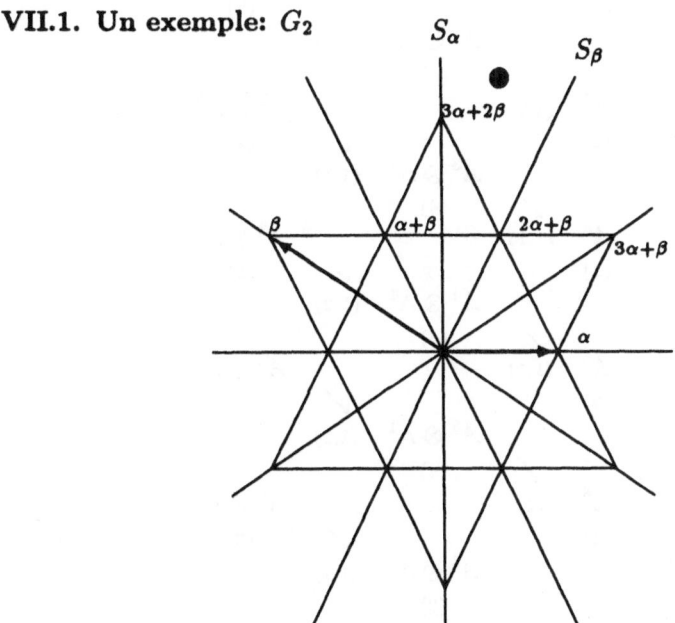

Le système G_2 admet pour racines simples (α, β) les autres racines positives étant ($\alpha + \beta$, $2\alpha + \beta$, $3\alpha + \beta$, $3\alpha + 2\beta$). Le groupe de Weyl est engendré par S_α et $S\beta$ avec la relation: $S_\alpha S_\beta S_\alpha S_\beta S_\alpha S_\beta = S_\beta S_\alpha S_\beta S_\alpha S_\beta S_\alpha$. L'anneau de Chow a donc les dimensions suivantes: dim $A^0(G/B) = 1$,

$\dim A^i(G/B) = 2$ pour $1 \le i \le 5$, $\dim A^6(G/B) = 1$. On a $\langle \breve{\alpha} \mid \alpha \rangle = 2$; $\langle \breve{\beta} \mid \beta \rangle = 2$; $\langle \breve{\alpha} \mid \beta \rangle = -3$; $\langle \breve{\beta} \mid \alpha \rangle = -1$; Donc si $\lambda = x * \alpha + y * \beta$, on a: $c(\lambda) = (2x - 3y)Z_{S_\alpha} + (-x + 2y)Z_{S_\beta}$. En particulier: $c(2 * \alpha + \beta) = Z_{S_\alpha}$ et $c(3 * \alpha + 2 * \beta) = Z_{S_\beta}$.

Le cas G_2 est particulièrement intéressant. C'est un petit cas $\dim(G/B) = 6$, $\dim(A(G/B)) = 12$, mais il a de la torsion. On le voit par exemple en appliquant le théorème de Smith à la flèche de $A^2(G/B) \otimes \wedge^1 X(T)$ dans $A^3(G/B)$. On obtient les valeurs 1 et 2: il y a donc de la 2-torsion. Le diagramme suivant montre comment le complexe $A(G/B) \otimes \wedge X(T)$ se décompose.

0	A^0			
	(1)			
1		$A^0 \otimes \wedge^1$		
		(2)		
2	A^1 (1,1)		$A^0 \otimes \wedge^2$	
	(2)		(1)	
3		$A^1 \otimes \wedge^1$ (1)		
		(4)		
4	A^2 (1,1)		$A^1 \otimes \wedge^2$	
	(2)		(2)	
5		$A^2 \otimes \wedge^1$ (1,1)		
		(4)		
6	A^3 (1,2)		$A^2 \otimes \wedge^2$	
	(2)		(2)	
7		$A^3 \otimes \wedge^1$ (1,1)		
		(4)		
8	A^4 (1,1)		$A^3 \otimes \wedge^2$	
	(2)		(2)	
9		$A^4 \otimes \wedge^1$ (1,2)		
		(4)		
10	A^5 (1,1)		$A^4 \otimes \wedge^2$	
	(2)		(2)	
11		$A^5 \otimes \wedge^1$ (1,1)		
		(4)		
12	A^6 (1)		$A^5 \otimes \wedge^2$	
	(1)		(2)	
13		$A^6 \otimes \wedge^1$ (1,1)		
		(2)		
14			$A^6 \otimes \wedge^2$	
			(1)	

Le module $A^i(G/B) \otimes \wedge^j X(T)$ est représenté par $A^i \otimes \wedge^j$. Les lignes correspondent aux degrés de la graduation; les valeurs entre parenthèses sous $A^i(G/B) \otimes \wedge^j X(T)$ indiquent les dimensions de ces modules. En fin les valeurs entre parenthéses sous les flèches donne le résultat du théorème de Smith. On voit donc que dans le cas G_2 l'homologie du complexe $A(G/B) \otimes \wedge X(T)$ est composé d'un facteur $\mathbb{Z}$ en degré 0, 3, 11 et 14 ainsi que d'un facteur $\mathbb{Z}/2\mathbb{Z}$ en degré 6 et 9. Ceci correspond bien à la cohomologie de G_2 à valeur dans $\mathbb{Z}$. Toutefois, on voit aussi que la structure de produit de la cohomologie n'a aucune chance de se retrouver sur ce diagramme puisque aprés tensorisation par $\mathbb{Z}/2\mathbb{Z}$, on a bien un élément en 3 dans $A^1(G/B) \otimes \wedge^1 X(T)$, mais l'élément de degré 6 qui devrait être son carré est dans $A^3(G/B)$! On l'aurait préféré dans $A^2(G/B) \otimes \wedge^2 X(T)$....

REFERENCES

[B]   Bourbaki, N., "Groupes et algèbres de Lie," chap 4 à 8, Hermann, Paris, 1968 et 1975.

[D1] Demazure, M., *Invariants symétriques entiers des groupes de Weyl et torsion.*, Inventiones Mathematicae 21 (1973), 273–301.

[D2] Demazure, M., *Désingularisation des variétés de Schubert générali-sées*, Ann. scient. Ec. Norm. Sup. 6 (1974), 53–88.

[G]   Grothendieck, A., "Théorie des intersections et théorème de Rie-mann-Roch," SGA 6, Exposé 0, App:RRR, p. 20-57 et exposé 14, p. 667–689 (Lecture Notes in Math., 225), Springer-Verlag, Berlin, 1971.

[K]   Kac, V. G., *Torsion in cohomology of Lie groups and Chow rings of reductive algebraic groups*, Inventiones Mathematicae 80 (1985), 69–79.

[M1] Marlin, R., *Anneaux de Chow des groupes algébriques...*, C.R.Acad. Sci. Paris, 279 (1974).

[M2] Marlin, R., *Anneaux de Chow des groupes algébriques...*, U.E.R. de Math. 1974, II et 35p. multigr. (Université Paris XI, n. 95-7419).

[M3] Marlin, R., *Comparaison de l'anneau de Chow et de l'anneau de Gro-thendieck*, Astérisque 36–37 (1975).

[M4] Marlin, R., *Cohomologie des espaces homogènes*, III + 70 pages mul-tigraphiées Nice 1979.

[M5] Marlin, R., *Cohomologie des groupes de Lie compact et anneau de Chow: une conjecture*, (en préparation).

Roger Marlin
Université de Nice
Parc Valrose
F-06034 NICE cedex
rm@cerisi.cerisi.fr

Construction de courbes de genre 2 à partir de leurs modules

Soient $\mathcal{A}$ la variété des modules des courbes de genre 2, $\mathcal{R}$ la surface de $\mathcal{A}$ correspondant aux courbes ayant une involution autre que l'involution hyperelliptique, et P un point de $\mathcal{A} - \mathcal{R}$ défini sur un corps k de caractéristique 0.

Nous donnons dans cet article un algorithme permettant d'obtenir, à partir de P, une équation explicite d'une courbe de genre 2 lui correspondant. En général, une telle courbe n'est pas définissable sur k, mais seulement sur certaines extensions quadratiques convenables de k.

Nous construisons, à partir de P, une conique définie sur k qui est k-isomorphe à la droite projective si et seulement s'il existe une courbe de genre 2 définie sur k correspondant à P.

De telles constructions utilisent de façon essentielle la théorie des invariants et des covariants des formes binaires, que nous rappelons brièvement dans le premier paragraphe.

Le second paragraphe est consacré à l'étude des courbes de genre 2, notamment à la construction explicite de la conique dont il est fait mention plus haut, et à une étude succincte de divers types de réduction $\mod p$ des courbes de genre 2, donnés en fonction du comportement p-adique de leurs invariants.

1 Rappels sur quelques résultats de Clebsch

1.1 Invariants et covariants simultanés de plusieurs formes binaires

Si

$$
\begin{aligned}
f_1 &= \quad \sum a_0^{(1)} x^{n_1} + a_1^{(1)} x^{n_1-1} y + \ldots + a_{n_1}^{(1)} y^{n_1}, \\
&\vdots \\
f_m &= \quad \sum a_0^{(m)} x^{n_m} + a_1^{(m)} x^{n_m-1} y + \ldots + a_{n_m}^{(m)} y^{n_m}
\end{aligned}
$$

sont m formes binaires de degré respectif $n_1, \ldots, n_m$, à coefficients dans un corps K de caractéristique 0, un *covariant* de ces formes (*cf.* [3], par exemple) est un polynôme $C((a)^{(1)}, \ldots, (a)^{(m)}, x, y)$ qui, lorsqu'on transforme $f_1, \ldots, f_m$ par des transformations $M = \begin{pmatrix} \alpha & \beta \\ \gamma & \delta \end{pmatrix}$ de $\mathrm{GL}_2(\overline{K})$

$$\begin{cases} x &= \alpha x' + \beta y' \\ y &= \gamma x' + \delta y' \end{cases}$$

est telle que

$$C((a')^{(1)}, \ldots, (a')^{(m)}, x', y') = (\det M)^{-k} C((a)^{(1)}, \ldots, (a)^{(m)}, x, y).$$

(Par exemple, les formes f_i elles-mêmes sont des covariants).

Une telle expression C est alors homogène en x, y. Disons ici que le degré total de C en x, y est l'*ordre* du covariant C; d'autre part, C est homogène en $(a)^1$ (resp. $(a)^2$, etc.). Le *degré* de C relativement à f_1 (resp. f_2, etc.) est le degré total de C relativement à $(a)^1$ (resp. $(a)^2$, etc.). Enfin, k est l'*indice* de C.

On a évidemment des relations entre ordre, degrés et indice. Si par exemple $m = 1$, et si l (resp. r) est l'ordre (resp. le degré) de C, on a $k = \dfrac{n_1 r - l}{2}$.

De plus, on peut montrer qu'un covariants de covariants de formes binaires est un covariant des formes initiales.

Un *invariant* simultané de $f_1, \ldots, f_m$ est un covariant d'ordre 0. (Par exemple, le discriminant d'un forme binaire est un invariant de cette forme, le résultant de deux formes est un invariant simultané de ces 2 formes, etc.)

Un *invariant absolu* de f est un quotient de deux invariants de même degré.

Soit I un invariant de degré r d'une forme binaire $f(x, y) = \sum a_0 x^n + \ldots + a_n y^n$ d'ordre n. Si l'on affecte les coefficients a_i du poids i, I est un polynôme en les a_i isobare de poids égal à son indice k. On a $k = nr/2$.

L'algèbre engendrée par les covariants est de type fini. Un système fini de générateurs homogènes de cette algèbre est parfois appelé un système de *covariants fondamentaux*.

Exemple: Dans le cas des formes quintiques, il existe un système de 23 covariants fondamentaux, dont voici l'ordre et le degré:

ordre	degré
0	$4, 8, 12, 18$
1	$5, 7, 11, 13$
2	$2, 6, 8$
3	$3, 5, 9$
4	$4, 6$
5	$1, 3, 7$
6	$2, 4$
7	5
9	3

On a donc 4 invariants fondamentaux (dont celui dû à Hermite, de degré 18). Le covariant d'ordre 5 et de degré 1 est évidemment la forme quintique elle-même.

Automorphismes d'une forme binaire

Soit G le sous-groupe de $GL_2(K)$ qui transforme une forme binaire f en uf, $u \in K^*$; G contient les homothéties. Par définition, le groupe des K-automorphismes de f est l'image de G dans $PGL_2(K)$.

En général, ce groupe est trivial.

1.2 Calcul symbolique

Soient f et g deux formes binaires de degrés respectifs n et m. Clebsch définit une opération (qu'il appelle *Ueberschiebung*, et que Faa di Bruno, dans son **Traité des formes binaires**, traduit par *translation*, et souvent désignée aujour'hui sous le nom de *transvectant*) définie par

$$(fg)_k = \frac{(m-k)!(n-k)!}{m!n!}\left(\frac{\partial f}{\partial x}\frac{\partial g}{\partial y} - \frac{\partial f}{\partial y}\frac{\partial g}{\partial x}\right)^k.$$

Il s'agit de notations symboliques, en ce sens que, dans le développement binômial de l'expression ci-dessus, on remplace $\left(\frac{\partial f}{\partial x}\right)^l\left(\frac{\partial f}{\partial y}\right)^m$ par $\frac{\partial^{l+m} f}{\partial x^l \partial y^m}$.

On définit de manière analogue des produits $(fg)_k(fh)_l$, etc. Par exemple,

$$(fg)_1(fh)_1 = f''_{x^2}g'_y h'_y + f''_{y^2}g'_x h'_x - f''_{xy}(g'_y h'_x + g'_x h'_y).$$

Lorsque $f = g$, on emploie le symbole $(ff')_k$ de préférence à $(ff)_k$: on développe le binôme conformément à la règle ci-dessus, sans identifier f et f'. Ensuite seulement, on remplace f' par f.

Calculons par exemple $(ff')_2$, pour une forme quadratique $f = ax^2 + bxy + cy^2$. Alors

$$
\begin{aligned}
4(ff')_2 &= \left(\frac{\partial f}{\partial x}\frac{\partial f'}{\partial y} - \frac{\partial f}{\partial y}\frac{\partial f'}{\partial x}\right)^2 \\
&= \frac{\partial^2 f}{\partial x^2}\frac{\partial^2 f'}{\partial y^2} + \frac{\partial^2 f}{\partial y^2}\frac{\partial^2 f'}{\partial x^2} - 2\frac{\partial^2 f}{\partial x \partial y}\frac{\partial^2 f'}{\partial x \partial y} \\
&= 2\frac{\partial^2 f}{\partial x^2}\frac{\partial^2 f}{\partial y^2} - 2\left(\frac{\partial^2 f}{\partial x \partial y}\right)^2 \\
&= 2(4ac - b^2).
\end{aligned}
$$

L'"*Ueberschiebung*" a les propriétés suivantes:

a) $(fg)_k(fh)_l \ldots$ est un covariant simultané des formes f, g, ... Par exemple, $(fg)_k$ est un covariant de f et g d'ordre $m + n - 2k$, où m (resp. n) est le degré de f, (resp. g).

En particulier, si g et h sont eux-mêmes des covariants d'ordre (resp. de degré) m, n (resp. r, s) d'une forme f, $(gh)_k$ est un covariant de f d'ordre $m + n - k$ et de degré $r + s$.

b) Tous les covariants d'une forme f peuvent être obtenus de cette manière, par des "*Ueberschiebung*" successifs.

Exemple 1:   Soit f une forme quartique. Alors $C_{4,2} = (ff')_2$ est un covariant d'ordre 4 et de degré 2, $C_{6,3} = (C_{4,2}f)_1$ est d'ordre 6 et de degré 3, $I_2 = (ff')_4$ et $I_3 = (fC_{4,2})_4$ sont les invariants de degré 2 et 3 bien connus dans la théorie des courbes elliptiques.

On peut montrer que f, $C_{4,2}$, $C_{6,3}$, I_2 et I_3 engendrent tous les covariants de f. Ils sont reliés par la relation de dépendance algébrique

$$12C_{6,3}^2 = -6C_{4,2}^3 + 3I_2C_{4,2}f^2 - 2I_3f^3,$$

d'où un revêtement explicite de degré 2 de la courbe de genre 1 $u^2 = f(x, 1)$ sur sa jacobienne E, d'équation $12v^2 = -6X^3 + 3I_2X - 2I_3$, donné par $(x, u) \mapsto (X, v) = (C_{4,2}u^{-2}, C_{6,3}u^{-2})$.

Exemple 2:   Soient $f = ax^2 + bxy + cy^2$ et $g = a'x^2 + b'xy + c'y^2$ deux formes quadratiques binaires.

Alors $i = 2(fg)_1 = (ab' - ba')x^2 + 2(ac' - a'c)xy + (bc' - b'c)y^2$ est un covariant quadratique de degré 2. Le résultant de f et g est égal à $\frac{1}{2}(ii')_2$.

Clebsch (*loc. cit.*, p. 197 − 201) démontre de jolies formules reliant f, g, i, $D = (ff')_2$, $D'' = (gg')_2$ et $D' = (fg)_2$, par exemple

$$
\begin{aligned}
-2i^2 &= \quad Df^2 - 2D'fg + D''g^2 \\
2(if)_1 &= \quad Dg - D'f \\
2(ii')_2 &= \quad DD'' - D'^2.
\end{aligned}
$$

Exemple 3 :   Soient x_1, x_2 et x_3 trois formes quadratique binaires.

Les trois formes quadratiques

$$x_1^* = (x_2x_3)_1, \quad x_2^* = (x_3x_1)_1, \quad x_3^* = (x_1x_2)_1$$

vérifient les formules:

$$x_1x_1^* + x_2x_2^* + x_3x_3^* = 0, \quad \sum_{i,j} A_{ij}x_i^*x_j^* = 0,$$

où $A_{ij} = (x_ix_j)_2$ $(1 \leq i, j \leq 3)$.

Posons $R = -(x_1x_2)_1(x_2x_3)_1(x_3x_1)_1$. Les formes x_i sont linéairement indépendantes si et seulement si $R \neq 0$; R est en effet le déterminant de x_1, x_2 et x_3 dans la base x^2, xy, y^2. D'autre part, le déterminant de la matrice A_{ij} est égal à $2R^2$.

Si f est une forme quadratique quelconque, on a

$$Rf = (fx_1)_2x_1^* + (fx_2)_2x_2^* + (fx_3)_2x_3^*.$$

Plus généralement, si f est une forme binaire de degré pair $2n$, on a

$$R^nf = ((fx_1)_2x_1^* + (fx_2)_2x_2^* + (fx_3)_2x_3^*)^{(n)},$$

l'élévation à la puissance n étant symbolique, en ce sens que les termes $(fx_i)(fx_j)\ldots$ y sont des *"Ueberschiebung"*.

Si l'on pose $a_{i_1\ldots i_n} = (fx_{i_1})\ldots(fx_{i_n})$ $(i_1,\ldots,i_n \in \{1,2,3\})$, la formule ci-dessus devient

$$\sum_{i_1,\ldots,i_n} a_{i_1\ldots i_n} x_{i_1}^* \ldots x_{i_n}^* = R^n f.$$

Supposons désormais x_1, x_2 et x_3 linéairement indépendants; R est alors non nul, et l'application

$$\varphi: \ \mathbf{P}_1 \longrightarrow C$$

qui au point (x,y) de P_1 associe le point (x_1^*, x_2^*, x_3^*) de la conique C d'équation homogène $\sum A_{ij}x_i x_j = 0$ est un isomorphisme.

Si T est la courbe d'équation homogène

$$\sum_{i_1,\ldots,i_n} a_{i_1\ldots i_n} x_{i_1}^* \ldots x_{i_n}^* = 0,$$

l'image par φ du diviseur somme des zéros de la forme binaire f est le diviseur intersection de T et C.

Si de plus les x_i sont des **covariants** de f, les A_{ij} et $a_{i_1\ldots i_n}$ sont des **invariants** de f. La conique C et la courbe T ont donc des équations à coefficients des invariants de f.

1.3 Le cas des formes sextiques

Soit f une forme sextique. On obtient alors, par *"Ueberschiebung"* répétés, les covariants suivants:

covariants		ordre	degré
i	$= (ff')_4$	4	2
Δ	$= (ii')_2$	4	4
y_1	$= (fi)_4$	2	3
y_2	$= (iy_1)_2$	2	5
y_3	$= (iy_2)_2$	2	7
A	$= (ff')_6$	0	2
B	$= (ii')_4$	0	4
C	$= (i\Delta)_4$	0	6
D	$= (y_3 y_1)_2$	0	10
x_1	$= (y_2 y_3)_1$	2	12
x_2	$= (y_3 y_1)_1$	2	10
x_3	$= (y_1 y_2)_1$	2	8
A_{ij}	$= (y_i y_j)_2 \qquad (1 \le i,j \le 3)$	0	$2(i+j+1)$
a_{ijk}	$= (fy_i)_2(fy_j)_2(fy_k)_2 \quad (1 \le i,j,k \le 3)$	0	$2(i+j+k+2)$
R	$= -(y_1 y_2)(y_2 y_3)(y_3 y_1)$	0	15

Les quatre invariants A, B, C et D de degrés respectifs 2, 4, 6 et 10 forment une base des invariants de degré pair.

En particulier, les A_{ij} définis ci-dessus sont des polynômes en A, B, C, D. Clebsch (*loc. cit.*, p.299) donne les formules:

$$A_{11} = 2C + \frac{1}{3}AB \qquad A_{23} = \frac{1}{3}B(B^2 + AC) + \frac{1}{3}C(2C + \frac{1}{3}AB)$$
$$A_{22} = D \qquad A_{31} = D$$
$$A_{33} = \frac{1}{2}BD + \frac{2}{9}C(B^2 + AC) \qquad A_{12} = \frac{2}{3}(B^2 + AC).$$

De même, les a_{ijk} sont des polynômes en les A, B, C, D. Après quelques calculs, on trouve

$$36a_{111} = 8(A^2C - 6BC + 9D)$$
$$36a_{112} = 4(2B^3 + 4ABC + 12C^2 + 3AD)$$
$$36a_{113} = 36a_{122} = 4(AB^3 + \frac{4}{3}A^2BC + 4B^2C + 6AC^2 + 3BD)$$
$$36a_{123} = 2(2B^4 + 4AB^2C + \frac{4}{3}A^2C^2 + 4BC^2 + 3ABD + 12CD)$$
$$36a_{133} = 2(AB^4 + \frac{4}{3}A^2B^2C + \frac{16}{3}B^3C$$
$$\qquad\qquad + \frac{26}{3}ABC^2 + 8C^3 + 3B^2D + 2ACD)$$
$$36a_{222} = 4(3B^4 + 6AB^2C + \frac{8}{3}A^2C^2 + 2BC^2 - 3CD)$$
$$36a_{223} = 2(-\frac{2}{3}B^3C - \frac{4}{3}ABC^2 - 4C^3 + 9B^2D + 8ACD)$$
$$36a_{233} = 2(B^5 + 2AB^3C + \frac{8}{9}A^2BC^2 + \frac{2}{3}B^2C^2 - BCD + 9D^2)$$
$$36a_{333} = -2B^4C - 4AB^2C^2 - \frac{16}{9}A^2C^3$$
$$\qquad\qquad - \frac{4}{3}BC^3 + 9B^3D + 12ABCD + 20C^2D.$$

L'algèbre des invariants est engendrée par A, B, C, D et R. D'après l'exemple 3 du paragraphe précédent, la relation liant R^2 (de degré 30) à A, B, C, D est donnée par

$$R^2 = \frac{1}{2}\begin{vmatrix} A_{11} & A_{12} & A_{31} \\ A_{12} & A_{22} & A_{23} \\ A_{31} & A_{23} & A_{33} \end{vmatrix}.$$

Remarque 1 : **Automorphismes des formes sextiques.**

Bolza [2] a décrit les divers groupes d'automorphismes des formes sextiques. Il nous suffit ici de savoir que, si une forme sextique a un groupe G d'automorphismes non trivial, on est dans l'un des deux cas suivants:

- G contient une involution. Ceci est équivalent à l'équation $R = 0$. La forme sextique est alors isomorphe à $f(x, y) = x^6 + ux^4y^2 + vx^2y^4 + y^6$.

- G contient un élément d'ordre 5. On a alors $A = B = C = 0$, $D \neq 0$. La forme sextique est $\overline{K}$-isomorphe à $f(x, y) = x(x^5 + y^5)$.

Remarque 2 : Igusa ([6], p. 620) définit des invariants, qu'il note A, B, C, D, et que nous notons ici A', B', C', D', d'une forme sextique non homogène $f = u_0 x^6 + \ldots + u_6$, définis en fonction des racines de f par

$$A' = u_0^2 \sum (12)^2 (34)^2 (56)^2$$
$$B' = u_0^4 \sum (12)^2 (23)^2 (31)^2 (45)^2 (56)^2 (64)^2$$
$$C' = u_0^6 \sum (12)^2 (23)^2 (31)^2 (45)^2 (56)^2 (64)^2 (14)^2 (25)^2 (36)^2$$
$$D' = u_0^{10} \prod (ij)^2.$$

D' est donc le discriminant de la sextique. Une phrase ambigüe, p. 623, l. 25, donne à penser qu'il s'agit des mêmes invariants que ceux de Clebsch. Il n'en est rien. Les formules de passage des invariants de Clebsch à ceux d'Igusa sont

$$A' = -120A$$
$$B' = -720A^2 + 6750B$$
$$C' = 8640A^3 - 108000AB + 202500C$$
$$D' = -62208A^5 + 972000A^3B + 1620000A^2C - 3037500AB^2$$
$$-6075000BC - 4556250D.$$

L'algèbre des covariants est engendrée par les 5 invariants A, B, C, D, R et par 21 covariants, dont il n'est pas besoin ici de donner la liste (cf. Clebsch, *loc. cit.*, p.296).

D'après l'exemple 3 du paragraphe précédent, on a

$$A_{11}x_1^2 + A_{22}x_2^2 + A_{33}x_3^2 + 2A_{13}x_1x_3 + 2A_{12}x_1x_2 + 2A_{23}x_1x_3 = 0$$
(1)
$$\sum a_{ijk} x_i x_j x_k = R^3 f.$$

Les covariants quadratiques x_1, x_2, x_3 (resp. y_1, y_2, y_3) sont linéairement indépendants si et seulement si $R \neq 0$. Dans ce cas, cela permet à Clebsch de montrer que deux formes f d'invariants A, B, C, D et f_1 d'invariants A_1, B_1, C_1, D_1 sont équivalentes sous GL_2 si et seulement s'il existe $r \neq 0$ tel que

$$(2) \qquad A_1 = r^2 A, B_1 = r^4 B, C_1 = r^6 C, D_1 = r^{10} D,$$

où $R \neq 0$. Bolza a étendu ce résultat au cas où $R = 0$.

En revanche, ni Clebsch ni Bolza ne démontrent qu'à tout système de A, B, C, D correspond une sextique les admettant comme invariants. Il a fallu attendre Igusa (*loc. cit.*) pour le démontrer, par des méthodes de nature différente.

1.4 Détermination des racines multiples d'une forme sextique à partir de ses invariants

Soit $g(x)$ un polynôme de degré 6, et $f(x, y) = y^6 g(x/y)$ la forme sextique associée. Les invariants de f fournissent les renseignements suivants sur la multiplicité des racines de g:

Cas d'un corps de caractéristique ≥ 5.

- Type $6 \Longleftrightarrow D' \neq 0$.

- Type $1^2 4 \Longleftrightarrow D' = 0, E \neq 0$,

- Type $1^2 1^2 2 \Longleftrightarrow D' = E = 0, 16B' - A'^2 \neq 0, B' \neq 0$,

- Type $1^2 1^2 1^2 \Longleftrightarrow 64C_1 - A'^3 = 16B' - A'^2 = 0, A' \neq 0$,

- Une racine exactement triple $\Longleftrightarrow B' = C' = D' = 0$, $A' \neq 0$,

- Une racine d'ordre $\geq 4 \Longleftrightarrow A' = B' = C' = D' = 0$,

où l'on a posé $C_1 = 3C' - A'B'$, $E = B'^3 - C_1^2$.

Cas d'un anneau de valuation discrète

Soit R un anneau de valuation discrète de caractéristique résiduelle ≥ 3, v sa valuation, t une uniformisante.

Soit $g \in R[X]$ un polynôme unitaire de degré 6, d'invariants A', B', C', D'. Posons encore $C_1 = 3C' - A'B'$, et $E = B'^3 - C_1^2$. Supposons D' non nul, et étudions $g \mod t$:

- Type 6 (i.e. 6 racines distinctes) $\Longleftrightarrow v(D') = 0$.

- Type $1^2.4 \Longleftrightarrow v(D') = m > 0, v(E) = 0$.

  La racine double est d'épaisseur m (i.e. g est de la forme $((x - a)^2 + t^m) P_4(x)$).

- Type $1^2 1^2 2$: $v(D') = l > 0, v(E) = m > 0, v(B') = v(16B' - A'^2) = 0$.

  Les racines doubles sont d'épaisseur respective

  $$r_1 = \inf(\frac{l}{2}, m) \text{ et } r_2 = l - r_1.$$

- Type $1^2 1^2 1^2 \Longleftrightarrow v(D') = l > 0, v(16B' - A'^2) = m > 0, v(64C_1 - A'^3) > 0, v(A') = 0$.

  Les épaisseurs respectives des 3 racines doubles $\mod t$ sont

  $$r_1 = \inf(\frac{l}{3}, \frac{n}{2}, m), \ r_2 = \inf(n - r_1, \frac{l - r_1}{2}), \ r_3 = l - r_1 - r_2,$$

  où $n = v(E)$.

1.5   La conique L et la cubique M

Pour notre propos, il est souhaitable de modifier comme suit les équations de Clebsch (1).

Nous supposons ici que la forme f n'a pas d'automorphisme non trivial, et que D', son discriminant, est non nul. Donc $R \neq 0$, et A, B, C ne sont pas tous trois nuls.

Il existe alors un invariant U non nul de degré 12: si A (resp. B, resp. C) est non nul, on peut prendre $U = A^6$ (resp. B^3, resp. C^2). Si l'on pose

$$\begin{aligned}
x_1 &= U D'^4 X_1 \\
x_2 &= D'^5 X_2 \\
x_3 &= U^4 X_3,
\end{aligned}$$

la relation $\sum A_{ij} x_i x_j = 0$ devient $\sum B_{ij} X_i X_j = 0$, les B_{ij} étant des invariants de degré 110. Comme $D' \neq 0$, on peut diviser par D'^{11}, d'où la relation

(3) $$\sum C_{ij} X_i X_j = 0,$$

où les $C_{ij} = B_{ij} D'^{-11}$ sont des invariants absolus, et les X_i, $i = 1, 2, 3$, des covariants quadratiques de f.

Si L est la conique de $\mathbf{P}^2$ d'équation homogène (3), on a ainsi un isomorphisme

$$\varphi : (x, y) \mapsto (X_1, X_2, X_3)$$

de $\mathbf{P}^1$ sur L.

Cette conique est non dégénérée, de discriminant $2R^2 U^{10} D'^{-15}$.

L'équation $R^3 f = \sum a_{ijk} x_i x_j x_k$ devient $R^3 f = \sum b_{ijk} X_i X_j X_k$, les b_{ijk} étant des invariants de degré 166. On définit M comme étant la cubique de $\mathbf{P}^2$ d'équation homogène

$$\sum c_{ijk} X_i X_j X_k = 0,$$

où les $c_{ijk} = b_{ijk} U^{-18} D'^{-1} U^{-13}$ sont des invariants absolus.

Il est clair que les points d'intersection de L et M sont les images par φ des solutions de $f(x, y) = 0$.

Remarque : A partir des résultats de Clebsch, il ne doit pas être difficile de prouver des résultats analogues pour toute forme binaire f de degré pair $2m$, à savoir qu'il existe trois covariants quadratiques x_1, x_2, x_3, une forme quadratique $Q(X, Y, Z)$ et une forme $T(X, Y, Z)$ de degré m, dont les coefficients sont des invariants de f, telles que

$$\begin{aligned}
Q(x_1, x_2, x_3) &= 0 \\
T(x_1, x_2, x_3) &= R^m f,
\end{aligned}$$

où R est un invariant qui s'annule pour les formes admettant une involution.

Par les mêmes manipulations que ci-dessus, au moins dans le cas où f n'a pas d'automorphisme non trivial, ceci permettrait de construire une conique non dégénérée L et une courbe M de degré m, dont les coefficients sont des invariants absolus de f, et un isomorphisme

$$\varphi : \mathbf{P}^1 \to L,$$

tel que les points d'intersections de L et M sont les images des solutions de $f(x, y) = 0$.

2   Courbes de genre 2

2.1   Rappels sur les courbes hyperelliptiques

Soit C une courbe hyperelliptique de genre $g \geq 2$, définie sur un corps k, w son involution hyperelliptique (définie sur k), $\omega_1, \ldots, \omega_g$ une base des différentielles de première espèce de C, définies sur k, et

$$h : C \to \mathbf{P}^{g-1}$$

le morphisme associé (défini sur k).

Si ρ est l'application canonique de C sur C/w, il existe un morphisme $\tilde{h} : C/w \to \mathbf{P}^{g-1}$, défini sur k, tel que $h = \tilde{h}\rho$.

L'image du diviseur canonique de C par ρ est donc de la forme $2D$, où D est un diviseur de degré $g - 1$, rationnel sur k.

Supposons à présent, et dans toute la suite du paragraphe, g pair.

Le diviseur $D + \dfrac{g-2}{2}K$, où K est un diviseur canonique, défini sur k, de C/w, est de degré 1. Par Riemann-Roch, on en déduit l'existence d'un diviseur effectif de degré 1 sur C/w, c'est-à-dire d'un point de C/w rationnel sur k; C/w est donc k-isomorphe à $\mathbf{P}^1$. On en déduit facilement les assertions suivantes:

- Si car $k \neq 2$, les courbes de genre g définie sur k sont en bijection avec les classes de diviseurs effectifs sans multiplicité de degré $2g + 2$ de $\mathbf{P}^1$, rationnels sur k, à $\overline{k}$-automorphisme projectif près.

Il existe un modèle plan de C, défini sur k, d'équation homogène

$$z^2 y^{2g} = f(x, y),$$

où f est un polynôme homogène de degré $2g + 2$. L'unique point singulier de ce modèle est le point de coordonnées $(x, y, z) = (0, 0, 1)$. L'involution hyperelliptique est donnée par $(x, y, z) \mapsto (x, y, -z)$. Les $2g + 2$ racines de f sont les abscisses des points de Weierstrass de la courbe hyperelliptique C. Deux équations non homogènes $y^2 = f(x)$ et $y'^2 = f'(x)$ correspondent à deux courbes k-isomorphes si et seulement si on passe de l'une à l'autre par une transformation

$$(x', y') = (\frac{ax+b}{cx+d}, \frac{uy}{(cx+d)^{g+1}}),$$

$a, b, c, d, u \in k$, $\Delta = ad - bc$ et u non nuls.

Pour tout invariant I_s de degré s des formes binaires de degré $2g+2$, on a alors

$$I_s(f) = M^s I_s(f),$$

avec $M = \Delta^{g+1} u^{-2}$.

Réciproquement, s'il existe $r \in \overline{k}^*$ tel que, pour tout invariant I_s de degré pair s des formes binaires de degré $2g+2$, $I(f) = r^s I(f')$, les courbes d'équation $y^2 = f(x)$ et $y^2 = f'(x)$ sont $\overline{k}$-isomorphes.

Si $y^2 = f(x)$ est un modèle d'une courbe hyperelliptique C de genre g, soit $(\omega_1, \ldots, \omega_g)$ la base de $\Omega_1(C)$ définie par $\omega_i = x^{i-1} \frac{dx}{y}$, $0 \leq i \leq g-1$. Posons $\omega = \omega_1 \wedge \ldots \wedge \omega_g$.

Si $y^2 = f'(x)$ est un autre modèle de C, obtenu par le changement de base ci-dessus, et ω_i' la base de $\Omega_1(C)$ correspondante, on a

$$\omega_i' = \Delta u^{-1}(ax+b)^i (cx+d)^{g-1-i}\frac{dx}{y}.$$

Après un petit calcul, on trouve

$$\omega' = \Delta^{g(g+1)/2} u^{-g}\omega = M^{g/2}\omega.$$

(Rappelons qu'on a supposé g pair.)

Dans ce qui suit, nous appelons *invariant* d'un modèle $y^2 = f(x)$ d'une courbe de genre g définie sur k, car $k \neq 2$, un invariant de la forme sextique homogène $y^{2g+2} f(x/y)$ associée.

Pour $g = 2$, (et seulement dans ce cas!) le calcul précédent montre qu'un invariant de degré m peut être vu comme une loi qui, à toute courbe de genre 2, associe une section de $\omega_C^{\otimes m}$, (où $\omega_C = \Lambda^2 \Omega^1$.)

Supposons à présent k parfait. Soit L une courbe de genre 0, définie sur k, et $D = \sum P_i$ un diviseur effectif de L de degré $2g+2$, sans multiplicité, rationnel sur k; D détermine donc une courbe hyperelliptique C de genre g, dont les invariants absolus sont définis sur k. Supposons que C n'ait pas d'autre automorphisme que son involution hyperelliptique w. Ceci équivaut à dire que le seul $\overline{k}$-automorphisme de L qui préserve D est l'identité.

Lemme 1 *Avec les notations et hypothèses précédentes, il existe un modèle de C défini sur k si et seulement si $L(k) \neq \emptyset$.*

Si $L(k) \neq \emptyset$, L est k-isomorphe à $\mathbf{P}^1$, et l'assertion est évidente. Réciproquement, supposons C définie sur k; C/w est alors k-isomorphe à $\mathbf{P}^1$. Soit D_1 le diviseur de C/w image des points de Weierstrass de C; D_1 est rationnel sur k. Choisissons un $\overline{k}$-isomorphisme $\varphi : C/w \rightarrow L$ qui envoie D_1 sur D. Si $\sigma \in \mathrm{Gal}(\overline{k}/k)$, définissons comme à l'ordinaire $\sigma(\varphi) = \sigma \circ \varphi \circ \sigma^{-1}$. Alors $\sigma(\varphi)$ est un $\overline{k}$-isomorphisme de C/w sur L, qui envoie encore D_1 sur D, et et $\sigma(\varphi)^{-1} \circ \varphi$ est un automorphisme de C/w, qui préserve D_1. Donc c'est l'identité, et $\sigma(\varphi) = \varphi$. Comme on a supposé k parfait, φ est défini sur k. Donc L est k-isomorphe à C/w, elle-même k-isomorphe à $\mathbf{P}^1$, et $L(k) \neq \emptyset$.

• Si car $k = 2$, k parfait, on obtient un modèle d'Artin-Schreier de C de la forme $y^2 + y = V(x)$, où $V \in k(x)$. On ne peut plus définir d'invariants de C à l'aide des formes binaires. Néanmoins, dans le cas où $g = 2$, Igusa (*loc. cit.*, p. 622) définit des invariants $\tilde{J}_2$, $\tilde{J}_4$, $\tilde{J}_6$, $\tilde{J}_8$ et $\tilde{J}_{10}$ d'une telle courbe, qui classifient tout aussi bien qu'en caractéristique $\neq 2$ les courbes de genre 2 sur k.

• En toute caractéristique, Igusa montre que, pour k algébriquement clos, si C est une courbe de genre 2, à tout couple (P, Q) de points de C, où P (resp. Q) est (resp. n'est pas) un point de Weierstrass, on peut associer un modèle de C d'équation

$$XY^2 + (1 + aX + bX^2)Y + X^2(c + dX + X^2) = 0,$$

où $Q = (0,0)$ et P est le point à l'infini. Une telle normalisation (P et Q fixés) est unique modulo les transformations $(a, b, c, d) \mapsto (\zeta a, \zeta^2 b, \zeta^3 c, \zeta^4 c)$, avec $\zeta^5 = 1$.

2.2   Le schéma de modules des courbes de genre 2

En caractéristique 2, les invariants A', B', C', D' des formes binaires de degré 6 ne sont pas adéquats pour classifier les courbes de genre 2, et Igusa introduit des invariants modifiés

$$
\begin{aligned}
J_2 &= 2^{-3}A' & J_4 &= 2^{-5}3^{-1}(4J_2^2 - B') \\
J_6 &= 2^{-6}3^{-2}(8J_2^3 - 160J_2J_4 - C') & J_8 &= 2^{-2}(J_2J_6 - J_4^2) \\
J_{10} &= 2^{-12}D'
\end{aligned}
$$

qui conviennent en toute caractéristique (*loc. cit.*).

(En particulier, les invariants se "réduisent bien" $\mod 2$. On obtient ainsi les $\tilde{J}_i$ ci-dessus.) Ceci permet à Igusa de prouver que le schéma de modules des courbes de genre 2 est le schéma affine $\mathcal{A} = \operatorname{Spec} N$, où N est l'anneau des éléments invariants de $\mathbf{Z}[X_1, X_2, X_3, X_4]$ par la transformation $X_i \mapsto \zeta^i X_i$, avec ζ est une racine cinquième de l'unité, X_1, X_2, X_3 des variables indépendantes et $4X_4 = X_1 X_3 - X_2^2$.

Les éléments de N sont appelés les *invariants absolus* des courbes de genre 2. En caractéristique 0, ils coïncident avec les invariants absolus définis plus haut à l'aide des formes sextiques.

D'une façon générale, à tout invariant absolu $J_2^{e_1} J_4^{e_2} J_6^{e_3} J_8^{e_4} J_{10}^{-e_5}$, avec $e_1 + 2e_2 + 3e_3 + 4e_4 = 5e_5$, $e_i \geq 0$, est associé l'élément $X_1^{e_1} X_2^{e_2} X_3^{e_3} X_4^{e_4}$ de N.

N est un anneau intègre, noethérien, intégralement clos, à 10 générateurs sur $\mathbf{Z}$. On peut par exemple choisir
$$J_2^5 J_{10}^{-1},\ J_2^3 J_4 J_{10}^{-1},\ J_2^2 J_6 J_{10}^{-1},\ J_2 J_8 J_{10}^{-1},\ J_4 J_6 J_{10}^{-1},\ J_4 J_8^2 J_{10}^{-2},\ J_6^2 J_8 J_{10}^{-2},$$
$$J_6^5 J_{10}^{-3},\ J_6 J_8^3 J_{10}^{-3},\ J_8^5 J_{10}^{-4}.$$

2.3 Réduction des courbes de genre 2

Il s'agit dans ce paragraphe de donner des résultats très fragmentaires sur les divers types de réduction d'une courbe de genre 2 définie sur un anneau de valuation discrète.

Nous adoptons ici les définitions de [4]:

- Soient g un entier ≥ 2, et S un schéma. Une *courbe stable de genre g sur S* est un morphisme propre et plat $C \to S$ dont les fibres géométriques sont des schémas C_s réduits, connexes, de dimension 1, tels que:

  i) C_s n'a que des points doubles ordinaires.

  ii) toute composante rationnelle non singulière de C_s rencontre les autres composantes de C_s en au moins 2 points.

  iii) $\dim H^1(\mathcal{O}_{C_s}) = g$.

- Soient R un anneau de valuation discrète, K son corps de fractions, k son corps résiduel, η (resp. s) le point générique (resp. fermé) de $\operatorname{Spec} R$. Si C est une courbe lisse, géométriquement irréductible et propre sur K, de genre $g \geq 2$, on note J sa jacobienne, $\mathcal{J}$ le modèle de Néron de J sur R, $\mathcal{C}$ le modèle minimal de C sur R.

  Par définition,

  - *J a réduction stable sur R si $\mathcal{J}_s$ n'a pas de radical unipotent.*
  - C a réduction stable sur R si l'une des 2 conditions équivalentes suivantes est réalisée:

    a) $\mathcal{C}_s$ est réduit et n'a que des points doubles ordinaires.

    b) Il existe une courbe $\mathcal{C}'$ stable sur R, de fibre générique C.

On a alors:

Théorème. *(Deligne-Mumford, loc. cit., p. 89).- J a réduction stable si et seulement si C a réduction stable.*

Modèle minimal de C.

Ogg [8] énumère les divers types de fibres possibles dans un pinceau de courbes de genre 2.

Namikawa et Ueno [7] complètent la classification de Ogg (d'un point de vue analytique, mais la situation est certainement identique dans le cas arithmétique, au moins si car $k \geq 7$); en fait, ils classifient non seulement les types de fibres, mais les pinceaux de courbes de genre 2

$$\pi : X \to D,$$

où D est le disque unité, π lisse sur $D - \{0\}$.

Ils trouvent 120 types de pinceaux possibles, qui parfois ont le même type de fibre spéciale, et qui sont décrits par 3 invariants: un point de la variété de modules des courbes stables de genre 2, compactifiée comme dans Deligne-Mumford (*loc. cit.*), une matrice de monodromie ($\in \mathrm{Sp}(2, \mathbf{Z})$), définie à conjugaison près, et un entier, le "degré" de la famille. Dans le cas des pinceaux de courbes elliptiques, les deux premiers invariants suffisent, notamment parce qu'un pinceau de courbes elliptiques à fibre singulière a toujours une monodromie non triviale. Ce n'est plus le cas ici: par exemple, la courbe $y^2 = (x^3 + \alpha x + 1)(x^3 + \beta t^8 x + t^{12})$ a un modèle minimal dont la fibre spéciale est singulière (réunion de deux courbes elliptiques se coupant transversalement en un point), et pourtant sa monodromie est triviale.

Remarque: Ogg, dans son article, donne une formule conjecturale, dans le cas où car $k \geq 7$, reliant le discriminant "minimal" Δ d'une courbe C définie sur K (i.e. le discriminant d'une équation $y^2 = f(x)$ de C à coefficients dans R de discriminant minimal) et le nombre de composantes n_p (sans compter la multiplicité) de C_s: $v_p(\Delta) = n_p + \varepsilon_p - 1$, où $0 \leq \varepsilon_p \leq 4$ est un entier mesurant, grosso modo, la mauvaise réduction des formes différentielles de C. D'après Namikawa et Ueno, cette formule n'est pas vraie telle quelle (*cf.* néanmoins [1]).

D'un point de vue pratique, si on se donne un modèle $y^2 = f(x)$ de C sur K, car $k > 5$, on le ramène facilement à l'un des modèles "standard" que Namikawa et Ueno associent à chaque type de pinceau trouvé, et on lit dans leur table le type de réduction du modèle minimal de C. Dans le cas où car $k \leq 5$, il faut soi-même faire le travail de désingularisation.

Dans le cas des courbes elliptiques, les invariants $c_4(E, \omega)$ et $c_6(E, \omega)$ déterminent la courbe E à K-isomorphisme près. Il serait intéressant d'obtenir, en fonction des invariants d'Igusa J_2, J_4, J_6, J_8 et J_{10}, une classification arithmétique analogue des divers types de fibres spéciales possibles.

Il y a néanmoins une difficulté, qui n'a pas lieu dans le cas des courbes elliptiques: soit C une courbe de genre 2 définie sur un corps L, sans automorphisme non trivial, de modèle $y^2 = f(x)$; si $u \in L^*$, $u \notin L^2$, soient C_1 la courbe d'équation $y^2 = f(ux) = g(x)$ (C_1 est donc k-isomorphe à C), et C_2 la courbe d'équation $y^2 = u^3 f(x)$, qui n'est pas L-isomorphe à C. Donc C_1 et C_2 ne sont pas L-isomorphes, mais ont les mêmes invariants.

Voici quelques cas très simples (mais qui nous suffisent dans la suite) où la connaissance de l'équation de la courbe, ou simplement de ses invariants, permet de conclure.

Soit $f(x, y) = 0$ une équation plane d'une courbe C de genre 2, de la forme $P_1(x)y^2 + P_2(x)y + P_3(x) = 0$, avec $P_i \in R[X]$, deg $P_i \le 3$. Soit J_{2i}, $1 \le i \le 5$, les invariants de la forme binaire $P_2^2 - 4P_1 P_3$.

Si $v(J_{10}) = 0$, la courbe C a bonne réduction. Nous supposons donc ci-dessous que $v(J_{10}) > 0$.

- S'il existe i tel que $v(J_{2i}) < 2i$, l'équation est minimale.

- La courbe C a bonne réduction potentielle si et seulement si $5v(J_{2i}) \ge iv(J_{10})$ pour $1 \le i \le 4$.

Supposons maintenant car $k \ne 2$. On emploie alors les invariants A', B', C', D'. On peut supposer l'équation de C de la forme $y^2 = f(x)$.

- f a une racine d'ordre ≥ 4 équivaut à $v(A'), v(B'), v(C') > 0$.

- f a une racine α d'ordre exactement 3 équivaut à $v(A') = 0$, $v(B'), v(C') > 0$.

  C n'a donc pas bonne réduction potentielle.

  Par contre, dans ce cas, la jacobienne de C peut avoir bonne réduction: supposons par exemple que de plus $v(D') = 2$. Alors C n'a pas réduction stable. Si on suppose que $K = K_{nr}$ (où K_{nr} désigne l'extension maximale non ramifiée de K), l'extension minimale L de K sur laquelle C acquiert une réduction stable est cyclique de degré 6. La jacobienne de C a bonne réduction sur L. La fibre spéciale d'un modèle minimal de C sur L est la réunion de 2 courbes elliptiques se coupant transversalement en un point.

- $v(B')$ ou $v(C') = 0$. La courbe C a réduction stable.

  Dans ce cas, l'étude du type de réduction d'un polynôme de degré 6 en fonction de ses invariants faite précédemment nous permet, au moins en caractéristique ≥ 5, d'obtenir le tableau suivant (où $C_1 = 3C' - A'B'$ et $E = B'^3 - C_1^2$):

  $- v(D') = m > 0, v(E) = 0$.

Le nombre de composantes connexes de $\mathcal{J}_s$ est $\Phi = m$. Si $m = 1$, la fibre spéciale de $\mathcal{C}$ est une courbe elliptique avec un point double ordinaire:

Si $m > 1$, c'est la réunion d'une courbe elliptique et de $(m - 1)$ droites de multiplicité 1:

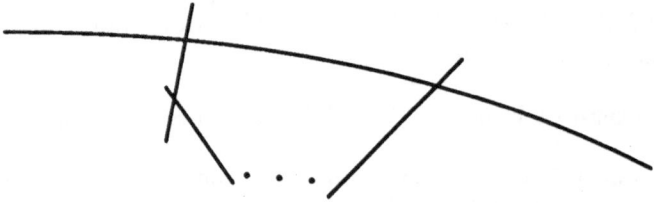

$- \ v(D') = l > 0, v(E) = m > 0, v(B') = v(J_4) = 0.$

Posons

$$r_1 = \inf(\frac{l}{2}, m) \text{ et } r_2 = l - r_1.$$

Alors $\Phi = r_1 r_2$.

Si $r_2 = 1$, $\mathcal{C}_s$ est une courbe rationnelle avec deux points doubles ordinaires:

Si $r_1 = 1, r_2 > 1$, $\mathcal{C}_s$ est de la forme

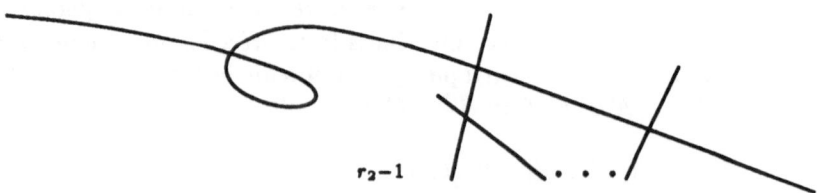

Si $r_1 > 1$, C_s est de la forme

$- v(D') = l > 0, v(J_4) = m > 0, v(J_6) > 0, v(A') = 0.$

Posons

$$r_1 = \inf(\frac{l}{3}, \frac{n}{2}, m), \ r_2 = \inf(n - r_1, \frac{l - r_1}{2}), \ r_3 = l - r_1 - r_2,$$

où $n = v(E)$.

Alors $\Phi = r_1 r_2 + r_2 r_3 + r_3 r_1.$

Si $r_3 = 1$, C_s est de la forme

Si $r_2 = 1, r_3 > 1$, C_s est de la forme

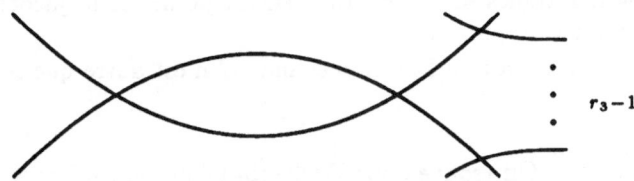

Si $r_1 = 1, r_2 > 1$, C_s est de la forme

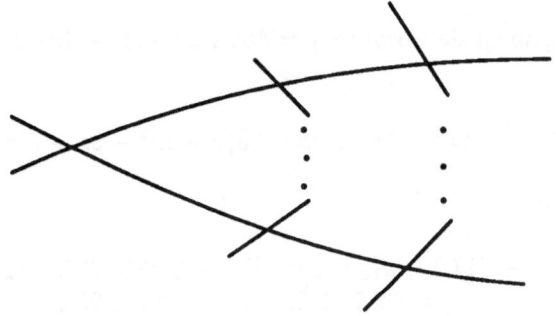

Si $r_1 > 1$, C_s est de la forme

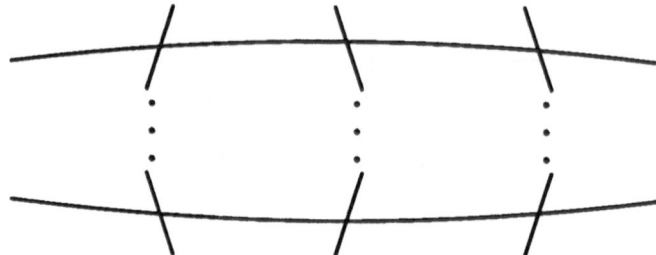

Le calcul de Φ a été fait en calculant le déterminant d'une matrice d'intersections, par la méthode de Raynaud [9].

Exemple 1:   Soit C la courbe d'équation

$$y^2 + y(x^3 - x^2 - 1) + x(1 - x) = 0.$$

On calcule

$$A' = 8,\ B' = 2^2 13.61,\ C' = 2^3 13.17^2,\ D' = -2^{12} 13^2,\ J_{10} = -13^2.$$

La courbe a donc bonne réduction en dehors de 13. D'après ce qui précède, sa réduction n'est pas stable en 13, mais le devient sur $K = \mathbf{Q}(\cos(2\pi/13))$; la fibre singulière de $C_K = C \otimes K$ est formée de deux courbes elliptiques se rencontrant en un point, et la jacobienne de C_K a bonne réduction partout.

Tout ceci était en fait bien connu: C n'est autre que $X_1(13)$.

Exemple 2:   On trouve dans Fricke [5] l'équation suivante de $X_0(23)$:

$$y^2 = x^6 - 14x^5 + 57x^4 - 106x^3 + 90x^2 - 16x - 19.$$

Le changement de variables $y = 46z + x^3 - 7x^2 + 4x - 25$ donne comme l'équation

$$23z^2 + (x^3 - 7x^2 + 4x - 25)z + 3x^2 - 2x + 7 = 0.$$

On calcule

$$A' = -2^3 11.13,\ J_{10} = 23^6,\ E = -2^{12} 3^4 5.23^3 373.27947,$$
$$J_4 = -2.7.19.23,\ J_6 = -2^5 .3.5.23^2.$$

On en déduit que la courbe a bonne réduction en 2. La situation sur $\mathbb{Z}_{23}$ se lit dans le tableau ci-dessus: $\mathbf{C}$ est stable, $\mathcal{C}_s$ est de la forme

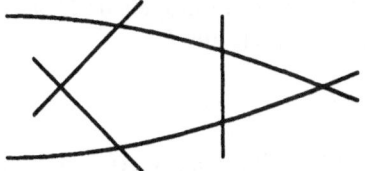

avec $r_1 = \inf(2, 3/2, 1) = 1$, $r2 = \inf(2, 5/2) = 2$, $r_3 = 6 - 3 = 3$. Le nombre de composantes connexes de $\mathcal{J}_s$ est

$$\Phi = r_1 r_2 + r_2 r_3 + r_3 r_1 = 11.$$

On retrouve le fait bien connu que le nombre de composantes connexes de la fibre en p de $J_0(p)$ est le numérateur de $(p-1)/12$. Le résultat $r_1 = 1$, $r_2 = 2$, $r_3 = 3$ reflète le fait qu'il y a trois courbes elliptiques supersingulières $\bmod 23$, d'invariant modulaire égal à 19 (resp. 1728, resp. 0), qui ont respectivement deux, quatre et six automorphismes.

2.4   Corps de définition des courbes de genre 2

Scholie :   *Soit k un corps parfait de caractéristique $\neq 2$. Les points de $\mathcal{A}(k)$ sont en bijection avec les classes de $\overline{k}$-isomorphismes de couples (L, D), où L est une courbe de genre 0 définie sur k, et D un diviseur effectif de degré 6, sans multiplicité, rationnel sur k.*

Soit $C \in \mathcal{A}(k)$, et (L, D) un couple lui correspondant. Supposons que (L, D) n'a pas d'automorphisme. Alors il existe un modèle de C défini sur k si et seulement si $L(k) \neq \emptyset$.

*Par ailleurs, au moins si car $k = 0$, l'application qui à un point de $\mathcal{A}(k)$ (sans automorphisme) associe (L, D) est **effectivement** calculable.*

Le problème que nous voulons résoudre est le suivant: si k est un corps, k_1 une extension de k, et C un point de $\mathcal{A}(k)$, comment calculer l'obstruction à obtenir un modèle de C sur k_1, et, si cette obstruction est nulle, comment construire un tel modèle?

Dans ce qui suit, on considère un corps k de caractéristique nulle (bien qu'il ne semble pas difficile de traiter le cas de caractéristique $\neq 2$.)

Soit donc un point C de $\mathcal{A}(k)$, et choisissons un modèle de C d'équation

$$z^2 = f(x).$$

Notons w l'involution hyperelliptique de C, et supposons que C n'a pas d'autre automorphisme que w.

Nous avons construit dans 1.4 une conique non dégénérée L et une cubique M, définies sur k, et un isomorphisme $\varphi : C/w \to L$, défini sur $\overline{k}$, tel que les images par φ des 6 points de Weierstrass de C sont les points d'intersection de L et M.

D'après le lemme 1, l'existence d'un modèle de C sur k_1 est équivalente à l'existence de points de L rationnels sur k_1.

Calcul de la conique L

1. Courbes admettant des invariants de degré 2 non nuls.

   (Il s'agit du cas général: les courbes qui ne sont pas dans ce cas forment une surface de $\mathcal{A}$; cf. 2. ci-dessous.)

   Le corps des fractions de N est $k(B/A^2, C/A^3, D/A^5)$.

   Posons

   $$x = \frac{B}{3A^2}, \quad y = \frac{C}{3A^3} \quad z = \frac{D}{2A^5},$$

   et soit $\mathcal{C}$ un point de $\mathcal{A}(k)$ tel que $x(\mathcal{C})$, $y(\mathcal{C})$ et $z(\mathcal{C})$ sont finis (c'est-à-dire que, sur un (donc tout) modèle de C, l'invariant A est non nul.)

   On peut alors prendre comme conique L la conique d'équation

   $$\sum C_{ij} x_i x_j = 0,$$

   où les C_{ij} sont les coefficients de la matrice

   $$\begin{pmatrix} x + 6y & 6x^2 + 2y & 2z \\ 6x^2 + 2y & 2z & 9x^3 + 4xy + 6y^2 \\ 2z & 9x^3 + 4xy + 6y^2 & 6x^2y + 2y^2 + 3xz \end{pmatrix}$$

   Si T est le discriminant de L, on a

   $$T = -81x^7 - 72x^5y - 702x^6y - 16x^3y^2 - 756x^4y^2 - 216x^2y^3$$
   $$-648x^3y^3 - 8y^4 - 324xy^4 - 216y^5 + 108x^5z + 108x^3yz + 24xy^2z$$
   $$+216x^2y^2z + 72y^3z + 6x^2z^2 + 36xyz^2 - 8z^3.$$

   Il reste à voir si $L(k)$ est vide. Supposons par exemple que k est un corps local, ou un corps de nombres; on est alors ramené à calculer un invariant de Hilbert.

   - Si $z = 0$, $L(k)$ contient le point $(x_1, x_2, x_3) = (4(3x^2 + y), -x - 6y, 0)$: C est définissable sur k.
   - si $d_2 = 9(C_{12}^2 - C_{11}C_{22}) = (6x^2 + 2y)^2 - 2z(x + 6y)$ est nul, $L(k)$ contient le point $(-2z, 3x^2 + y, 0)$: C est définissable sur k.

- si $zd_2 \neq 0$, on peut prendre comme invariant de Hilbert de la conique L

$$u = (d_2, -2zT).$$

La courbe $\mathcal{C}$ est définissable sur une extension k_1 de k si et seulement si $u = 0$ sur k_1. (En particulier, $\mathcal{C}$ est définissable sur $k(\sqrt{d_2})$.)

Le symbole u n'est pas identiquement nul: si $x = z = 1$, $y = 2$, il n'est pas trivial sur $\mathbf{Q}_2$.

Remarque :   Je n'ai pas tenu compte ici du fait que la courbe $\mathcal{C}$ peut avoir un discriminant D' nul.

2. **Courbes dont le degré minimal des invariants non nuls est 4.**

L'ensemble de ces courbes est l'ouvert $x \neq 0$ de la surface de $\mathcal{A}$ paramétrée par

$$x = 2\frac{B^5}{D^2}, \quad y = \frac{BC}{D}.$$

Des calculs identiques aux précédents montrent qu'on peut prendre comme conique L la conique d'équation

$$\sum C_{ij} x_i x_j = 0,$$

où les C_{ij} sont les coefficients de la matrice

$$\begin{pmatrix} y & x & 3 \\ x & 9x & 3(x + 4y^2) \\ 3 & 3(x + 4y^2) & 4y + 9 \end{pmatrix}$$

dont le déterminant T est égal à

$$T = -81x + 9x^2 + 81xy - 13x^2 y + 108xy^2 - 72xy^3 - 144y^5.$$

Si $x = 9y$, le point $(x, -y, 0)$ appartient à $L(k)$. Sinon, comme $x \neq 0$, le symbole de Hilbert de L peut être écrit sous la forme

$$(x(x - 9y), -xT).$$

Ce symbole n'est pas toujours trivial (par exemple, si $x = y = 1$, il est non trivial sur $\mathbf{Q}_7$).

Remarque: Si D' est le discriminant des courbes $\mathcal{C}$ ci-dessus, on a

$$\frac{D'}{D} = -2.3^5 5^5 (3 + 4y).$$

Donc, pour $y = -3/4$, on n'obtient pas une courbe de genre 2.

3. Courbes dont les invariants de degré ≤ 4 sont tous nuls.

Soit $\mathcal{C}$ telle que $A = B = 0$. La courbe de $\mathcal{A}$ qui classifie de telles courbes est paramétrée par $x = \dfrac{C^5}{3D^3}$. En procédant comme dans 1), on voit qu'on peut prendre comme conique L la conique d'équation

$$2X(X + Z) + xY(3Y + 2Z) = 0.$$

Le point $(1, 0, -1)$ appartient à $L(k)$, donc $\mathcal{C}$ est définissable sur k.

REFERENCES

[1] S. Bloch, *De Rham cohomology and conductors of curves*, Duke Math. J. **54** (1987), 295–308.

[2] O. Bolza, *On binary sextics with linear transformations into themselves*, Amer. J. Math. **10** (1888), 47–70.

[3] A. Clebsch, "Theorie der Binären Algebraischen Formen," Verlag von B.G. Teubner, Leipzig, 1872.

[4] P. Deligne, D. Mumford, *The irreducibility of the space of curves of given genus*, Publ. Math. IHES **36** (1969), 75–110.

[5] R. Fricke, "Lehrbuch der Algebra," Volume III, F. Vieweg & Sohn, Braunschweig, 1928.

[6] J. Igusa, *Arithmetic variety of moduli for genus two*, Ann. of Math. **72** (1960), 612–649.

[7] Y. Namikawa, K. Ueno, *The complete classification of fibres in pencils of curves of genus two*, Manuscripta Math. **9** (1973), 143–186.

[8] A.P. Ogg, *On pencils of curves of genus two*, Topology **5** (1966), 355–362.

[9] M. Raynaud, *Spécialisation du foncteur de Picard*, Publ. Math. IHES **38** (1970), 27–76.

Jean-François Mestre
Université de Paris VII
2 place Jussieu, 75005 Paris
e-mail: MESTRE at FRMAP711.BITNET
et
Département de Mathématiques et d'Informatique
École normale supérieure
45 rue d'Ulm, 75230 Paris Cedex 05
e-mail: MESTRE at FRULM11.BITNET

Computing Syzygies à la Gauß-Jordan

H. MICHAEL MÖLLER

Abstract. Let U be a submodule of $\mathcal{P}^s$, $\mathcal{P}$ a ring of multivariate polynomials, let $G_1, \ldots, G_r$ generate U, and let $F_1 := \sum b_{1j} G_j, \ldots, F_m := \sum b_{mj} G_j$ with $b_{ij} \in \mathcal{P}$. If a set of generators for the module of syzygies w. r. t. $(G_1, \ldots, G_r)$ is given, then one problem is to compute a set of generators for the module of syzygies w. r. t. $(F_1, \ldots, F_m)$ and to decide whether $F_1, \ldots, F_m$ also generate U. For this purpose an algorithm is presented which is similar to the Gauß-Jordan algorithm in linear algebra. It uses Gröbner bases techniques for modules and allows to solve some constructive problems in connection with modules.

1   Introduction.

Consider the set of all solutions $(v_1, \ldots, v_m) \in \mathcal{P}^m$ of the system of linear equations

$$f_{11} v_1 + \cdots + f_{1m} v_m = f_{10},$$

(1)

$$\cdots$$

$$f_{s1} v_1 + \cdots + f_{sm} v_m = f_{s0},$$

where all f_{ij} are in $\mathcal{P} := K[x_1, \ldots, x_n]$, K a field. This set can be computed by a procedure proposed by Furukawa, Sasaki, and Kobayashi (1986). It amounts in finding all representations $F_0 = \sum_{i=1}^m v_i F_i$, with $F_i := (f_{1i}, \ldots, f_{si}) \in \mathcal{P}^s$. If $F_1, \ldots, F_m$ is a Gröbner basis, then a particular solution $F_0 = \sum_{i=1}^m w_i F_i$ can be found by reducing F_0 modulo $\{F_1, \ldots, F_m\}$ to 0. Then the general solution is $F_0 = \sum_{i=1}^m (w_i + h_i) F_i$ where $(h_1, \ldots, h_m)$ is an arbitrary element of the module of syzygies

$$Syz(F_1, \ldots, F_m) := \{(u_1, \ldots, u_m) \in \mathcal{P}^m \mid \sum u_i F_i = 0\}.$$

A basis of $Syz(F_1, \ldots, F_m)$ is easily obtained by Gröbner reductions, see section 2. The general case requires first the transformation of $F_1, \ldots, F_m$ to a Gröbner basis $G_1, \ldots, G_r$. The solutions $F_0 = \sum v_i F_i$ are then obtained by expressing the particular solution $F_0 = \sum w_i G_i$ in terms of $F_1, \ldots, F_m$ which requires the knowledge of the transformation matrix $A := (a_{ij})$

with $G_i = \sum a_{ij} F_j$ and by calculating a basis for $Syz(F_1, \ldots, F_m)$ from $Syz(G_1, \ldots, G_r)$. (For this calculation the knowledge of $B := (b_{ij})$ with $F_i = \sum b_{ij} G_j$ is helpful, which can be obtained again easily by reducing every F_i modulo the Gröbner basis to 0.)

This problem can be formulated in a slightly more general way without requiring explicitly that one of the bases is a Gröbner basis.

Problem 1. Let $\{G_1, \ldots, G_r\}$ generate the module $U \subseteq \mathcal{P}^s$. Assume $B := (b_{ij})$ with $F_i := \sum b_{ij} G_j$ and $b_{ij} \in \mathcal{P}$ and a basis for $Syz(G_1, \ldots, G_r)$ are given. Compute a basis for $Syz(F_1, \ldots, F_m)$ and compute, if $F_1, \ldots, F_m$ also generates U, a matrix $A := (a_{ij})$ with $G_i = \sum a_{ij} F_j$.

The case that a basis for the module of syzygies of a proper submodule can be computed seems to be not considered before. For the case that $G_1, \ldots, G_r$ and $F_1, \ldots, F_m$ generate the same module, Furukawa, Sasaki, and Kobayashi (1986) proposed Buchberger's algorithm with tracing of the dependence of all s-tuples on the input tuples $F_1, \ldots, F_m$ in order to compute A. Then they used a result of Zacharias (1978) who described a basis of $Syz(F_1, \ldots, F_m)$ using A, B, and a basis of $Syz(G_1, \ldots, G_r)$.

Unfortunately it is sometimes difficult or even impossible due to insufficient technical support to compute a Gröbner basis for $Mod(F_1, \ldots, F_m)$, the module generated by $F_1, \ldots, F_m$. Then there is no hope to find the matrix A by tracing. If a Gröbner basis $G_1, \ldots, G_r$ is for instance obtained by multimodular calculations and reconstruction, see for instance Traverso (1988) or Melenk, Möller, Neun (1988), then even $Mod(F_1, \ldots, F_m) = Mod(G_1, \ldots, G_r)$ is not known a priori. Since the tests of $F_i \in Mod(G_1, \ldots, G_r)$ are easy and give (if it exists) the matrix B and since a basis for $Syz(G_1, \ldots, G_r)$ is as well easily computed, see section 2, it remains to find an efficient constructive solution to the following.

Problem 2. Let $Mod(F_1, \ldots, F_m) \subseteq Mod(G_1, \ldots, G_r)$ and let a matrix $B := (b_{ij})$ with $F_i = \sum b_{ij} G_j$ and polynomials $b_{ij} \in \mathcal{P}$ and a basis of $Syz(G_1, \ldots, G_r)$ be given. Then decide $Mod(F_1, \ldots, F_m) = Mod(G_1, \ldots, G_r)$.

Again, $\{G_1, \ldots, G_r\}$ is not necessarily a Gröbner basis in this problem.

This paper deals with the solution of these two problems. I will present an algorithm which needs as input an arbitrary finite set $\{G_1, \ldots, G_r\} \subset \mathcal{P}^s$, a basis for $Syz(G_1, \ldots, G_r)$ and a matrix $B = (b_{ij})$ with $b_{ij} \in \mathcal{P}$. Then it decides whether $Mod(G_1, \ldots, G_r) = Mod(F_1, \ldots, F_m)$ for $F_i := \sum b_{ij} G_j$ holds true and if so, it computes A. Independently from equality or inequality of these modules it also computes a basis for $Syz(F_1, \ldots, F_m)$. Neither $\{G_1, \ldots, G_r\}$ nor $\{F_1, \ldots, F_m\}$ need to be a Gröbner basis. But if $\{G_1, \ldots, G_r\}$ is Gröbner, then the calculations in the algorithm are simplified, as it will shown in section 3.

The algorithm is similar to the Gauß-Jordan algorithm in linear algebra, see for instance Maeß (1984), which transforms a given relation $Px = Iy$ into $Ix = P^{-1}y$ by successively exchanging a component x_i

by a component y_j. Here P is an invertible matrix and I a unit matrix. The presented new algorithm is initialized with the rows of B and with a basis $d_1, \ldots, d_K$ of $Syz(G_1, \ldots, G_r)$ as input, such that initially $\binom{B}{D}G = \binom{I}{0}F$ holds, where D consists of the M rows $d_1, \ldots, d_K$, I the $m \times m$ unit matrix, $G := (G_1, \ldots, G_r)^T$, $F := (F_1, \ldots, F_m)^T$. If and only if $Mod(G_1, \ldots, G_r) = Mod(F_1, \ldots, F_m)$, the algorithm terminates with rows $a_1, \ldots, a_r$ and $c_1, \ldots, c_L$, such that $\binom{I}{0}G = \binom{A}{C}F$ holds, where A consists of the r rows a_i and C of the L rows c_i, and such that $c_1, \ldots, c_L$ generate $Syz(F_1, \ldots, F_m)$

The main part of the algorithm is an algorithm for computing Gröbner bases for modules. At the moment there is no efficient variant of Buchberger's algorithm for modules installed in REDUCE. Therefore I conclude the paper by giving just one example, in which the computational details are explained.

2   Gröbner bases for modules.

The Gröbner basis concept for modules was systematically investigated by Möller and Mora (1986). For consistency of this paper I resume briefly the notions and elementary results needed in the forthcoming sections without quoting this reference everytimes.

Let $\mathcal{P}$ denote the polynomial ring $K[x_1, \ldots, x_n]$, K a field. As usual for Gröbner basis techniques the set of terms

$$T := \{x_1^{i_1} \cdots x_n^{i_n} \mid i_1, \ldots, i_n \geq 0\}$$

is equipped with a so called *admissible* ordering, i. e. an ordering $<_T$ with $1 \leq_T t$, and $t_1 <_T t_2 \Rightarrow t \cdot t_1 <_T t \cdot t_2$ for all $t, t_1, t_2 \in T$.

The finite rank free module $\mathcal{P}^s$ contains the set of module terms

$$T^{(s)} := \{t e_i^{(s)} \mid i \in \{1, \ldots, s\}, \ t \in T\},$$

where $e_1^{(s)}, \ldots, e_s^{(s)}$ are the canonical unit s-tuples of $\mathcal{P}^s$. In order to introduce the term ordering concept in $\mathcal{P}^s$, it is sufficient to order only its subset $T^{(s)}$ by a module term ordering. I need in the following a weighted module term ordering defined by weights $w_1, \ldots, w_s \in T$ and an arbitrary ordering $<_1$ of $\{1, \ldots, s\}$

$$t_i e_i^{(s)} <_M t_j e_j^{(s)} \iff w_i t_i <_T w_j t_j \ \text{ or } \ (w_i t_i = w_j t_j, i <_1 j).$$

A special instance of this module term ordering was used by Furukawa, Sasaki, and Kobayashi (1986), namely $w_1 = \ldots = w_s = 1$, for $<_1$ the natural ordering of $\{1, \ldots, s\}$, and for $<_T$ a specified ordering of T. (An alternative module term ordering was used by H. Kredel, see Armbruster (1985),

$$t_i e_i^{(s)} <_K t_j e_j^{(s)} \iff i < j \ \text{ or } \ (i = j, t_i <_T t_j).$$

By prefering the first components of module elements, this ordering has similarities to the lexicographical term ordering, where the preference of lower indexed variables leads to a strong degree increase in the higher indexed variables. Therefore it is to be expected that the ordering $<_K$ has similar disadvantageous properties like the lexicographical term ordering.) In the following, a module term ordering $<_M$ with weights $w_1, \ldots, w_s$ and ordering $<_1$ is assumed to be fixed.

Every $0 \neq F \in \mathcal{P}^s$ has a unique representation $\sum_{i=1}^{l} c_i u_i$ with $u_i \in T^{(s)}$, $c_i \in K \setminus \{0\}$. Let $u_1 <_M \ldots <_M u_l$. Then $Hterm(F) := u_l$ and $Hcoef(F) := c_l$. In addition let $lt(F) := t_l$ if $u_l = t_l \cdot e_i^{(s)}$. For convenience I define also for $0 \in \mathcal{P}^s$ $Hterm(0) := 0$ and $0 <_M t$ for all $t \in T^{(s)}$.

Let $\Phi := \{F_1, \ldots, F_m\} \subset \mathcal{P}^s \setminus \{0\}$ and let $0 \neq F = \sum_{i=1}^{l} c_i u_i$ as before. Then F reduces to F' modulo Φ, briefly $F \xrightarrow{\Phi} F'$, if there are an $i \in \{1, \ldots, l\}$, an $F_j \in \Phi$, and a $t \in T$ such that $u_i = t \cdot Hterm(F_j)$, $F' = F - \frac{c_i}{Hcoef(F_j)} t \cdot F_j$. This is the direct translation of the usual reduction procedure for ideals to modules. F is called *irreducible modulo* Φ if no such F' exists. $\xrightarrow{\Phi}^*$ is the reflexive transitive closure of $\xrightarrow{\Phi}$.

Let $F, G \in \mathcal{P}^s \setminus \{0\}$, $Hterm(F) = t_1 e_i^{(s)}$, $Hterm(G) = t_2 e_j^{(s)}$. Then $Hterm(F)$ and $Hterm(G)$ have a least common multiple in $T^{(s)}$ iff $i = j$. In that case, I write briefly $lt(F, G) \cdot e_i^{(s)}$ for the least common multiple and define the S-tuple of F and G by

$$S(F, G) := \frac{lt(F, G)}{lt(F) \cdot Hcoef(F)} F - \frac{lt(F, G)}{lt(G) \cdot Hcoef(G)} G.$$

Definition 1: Let U be a submodule of $\mathcal{P}^s$. The finite set $\Gamma := \{G_1, \ldots, G_r\}$ is called a *Gröbner basis* of U, if $0 \notin \Gamma \subset U$ and if every $F \in U$ has a G-*representation*

$$F = \sum_{i=1}^{r} b_i G_i, \quad \text{with } Hterm(b_i G_i) \leq_M Hterm(F).$$

Such Gröbner basis depends also on $<_M$, but every submodule U for each of the module term orderings in consideration has a Gröbner basis. There are several equivalent conditions characterizing Gröbner bases for modules. Here I quote only three of them.

Theorem 1. *Let $\Gamma := \{G_1, \ldots, G_r\}$ generate a submodule U of $\mathcal{P}^s$ and let $0 \notin \Gamma$. Then the following conditions are equivalent.*
(1) $\{G_1, \ldots, G_r\}$ is a Gröbner basis of U.
(2) $F \xrightarrow{\Gamma}^* 0$ for all $F \in U$.

(3) $S(G_i, G_j) \xrightarrow{\ *\ }_{\Gamma} 0$ for all $G_i, G_j \in \Gamma$, such that $S(G_i, G_j)$ is defined.

If a module U is generated by $\Phi := \{F_1, \ldots, F_m\}$, then a modification of Buchberger's algorithm to modules computes a Gröbner basis for U, see alg. 4.2 of Möller and Mora (1986). This computation consists in the reduction of every (defined) $S(F_i, F_j)$ modulo the actual set Φ to an irreducible H_{ij}. If $H_{ij} \neq 0$, then it is appended to Φ as a new F_k. For practical reasons redundant elements, i. e. those with a G-representation in terms of the other elements in Φ, may be canceled in Φ. At termination, Φ is a Gröbner basis of U by condition (3) of theorem 1. Not all $S(F_i, F_j)$ have to be reduced because some of the criteria developed for the case of ideals U can be adopted to the module case. This can be seen by inspecting the arguments of Gebauer and Möller (1988).

Having a Gröbner basis Γ of U, the test $F \in U$ can be performed easily by condition (2). It amounts in testing finitely many terms $u_i \in T^{(s)}$ for being multiples of certain $Hterm(G_j)$, $G_j \in \Gamma$, and subtracting by appropriate multiples of some G_j. By collecting these multiples of the G_j's, the G-representation of F is found provided the reduction procedure terminates with 0, i. e. provided $F \in U$.

If $\Phi := \{G_1, \ldots, G_r\}$ is a Gröbner basis of U, then each defined $S(G_i, G_j)$ is in U and has therefore a G-representation $\sum_k s_{ijk} G_k$. Hence the r-tuple

$$ S_{ij} := \frac{lt(G_i, G_j)}{lt(G_i) \cdot Hcoef(G_i)} e_i^{(r)} - \frac{lt(G_i, G_j)}{lt(G_j) \cdot Hcoef(G_j)} e_j^{(r)} - \sum_{k=1}^{r} s_{ijk} e_k^{(r)} $$

belongs to $Syz(G_1, \ldots, G_r)$. In addition, when the module term ordering in $\mathcal{P}^r$ has weights $w_i = lt(G_i), i = 1, \ldots, r$, (and an arbitrary ordering of $\{1, \ldots, r\}$), then the S_{ij} corresponding to defined $S(G_i, G_j)$ constitute a Gröbner basis of $Syz(G_1, \ldots, G_r)$, see 4.2 and lemma 7.8 of Möller and Mora (1986).

3   Changing the module basis.

Let $G_1, \ldots, G_r$ generate the module $U \subseteq \mathcal{P}^s$, let $d_1, \ldots, d_K$ be a basis of $Syz(G_1, \ldots, G_r)$ and let

$$ F_i := \sum_{i=1}^{r} b_{ij} G_j, \quad b_{ij} \in \mathcal{P}. $$

Then I show how to find the basis of $Syz(F_1, \ldots, F_m)$, answer the question whether $F_1, \ldots, F_m$ also generates U, and if so, I will find the representations $G_i = \sum a_{ij} F_j$. This will solve the problems 1 and 2.

Let $e_i^{(m)}$ denote the canonical i-th unit vector of $\mathcal{P}^m$ and let $b_i := (b_{i1}, \ldots, b_{ir})$, $i = 1, \ldots, m$. Then the following result holds.

Proposition 1.   *The $(r + m)$-tuples*

$$(b_1, e_1^{(m)}), \ldots, (b_m, e_m^{(m)}), (d_1, 0), \ldots, (d_K, 0)$$

generate

$$W := \{(u, v) \mid u \in \mathcal{P}^r, v \in \mathcal{P}^m, \sum_{i=1}^{r} u_i G_i = \sum_{i=1}^{m} v_i F_i\}.$$

Proof: By construction $(b_i, e_i^{(m)}) \in W$ and $(d_i, 0) \in W$. Let $(u, v) \in W$. Then

$$(u, v) - \sum_{i=1}^{m} v_i(b_i, e_i^{(m)}) = (u - \sum_{i=1}^{m} v_i b_i, 0) \in W.$$

Hence $u - \sum_{i=1}^{m} v_i b_i \in Syz(G_1, \ldots, G_r)$. Therefore there are $q_1, \ldots, q_K \in \mathcal{P}$ such that $u - \sum v_i b_i = \sum q_i d_i$. This gives

$$(u, v) = \sum_{i=1}^{m} v_i(b_i, e_i^{(m)}) + \sum_{i=1}^{K} q_i(d_i, 0)$$

Q. E. D.

Proposition 2. $a_1, \ldots, a_r \in \mathcal{P}^m$ *exist with* $(e_i^{(r)}, a_i) \in W$, *if and only if* $U = Mod(F_1, \ldots, F_m)$.

Proof: By construction we have $Mod(F_1, \ldots, F_m) \subseteq U$. Therefore $G_i \in Mod(F_1, \ldots, F_m)$ iff $Mod(F_1, \ldots, F_m) = U$. But $G_i \in Mod(F_1, \ldots, F_m)$ iff there are $a_i = (a_{i1}, \ldots, a_{im}) \in \mathcal{P}^m$ with $G_i = \sum a_{ij} F_j$. This means $(e_i^{(r)}, a_i) \in W$. Q. E. D.

In order to detect distinct elements $(e_i^{(r)}, a_i) \in W$, I introduce an ordering $<_o$ in $\mathcal{P}^r \times \mathcal{P}^m$ by using module orderings $<_M$ in $\mathcal{P}^r$ and $<_{M'}$ in $\mathcal{P}^m$ and composition

$$(u_1, v_1) <_o (u_2, v_2) :\Longleftrightarrow u_1 <_M u_2 \text{ or } (u_1 = u_2, v_1 <_{M'} v_2).$$

Then the following algorithm solves problem 1 and problem 2.

Algorithm:

Input: $(b_1, e_1^{(m)}), \ldots, (b_m, e_m^{(m)})$ with $b_i = (b_{i1}, \ldots, b_{ir}) \in \mathcal{P}^r$, $i = 1, \ldots, m$ and $(d_1, 0), \ldots, (d_K, 0) \in \mathcal{P}^r \times \mathcal{P}^m$, such that $\{d_1, \ldots, d_K\}$ generates $Syz(G_1, \ldots, G_r)$ and $F_i := \sum_{j=1}^{r} b_{ij} G_j, i = 1, \ldots, m$.

Output: A pair (A, C) of sets. $Mod(G_1, \ldots, G_r) = Mod(F_1, \ldots, G_m)$ holds, iff A is not empty. In that case A consists in $a_i = (a_{i1}, \ldots, a_{im}) \in \mathcal{P}^m$, such that $G_i = \sum_{j=1}^m a_{ij} F_j, i = 1, \ldots, r$. The set C consists in $c_1, \ldots, c_L$ generating $Syz(F_1, \ldots, F_r)$.

Step 1:   For the module generated by $(b_1, e_1^{(m)}), \ldots, (b_m, e_m^{(m)}), (d_1, 0), \ldots,$ $(d_K, 0)$ compute a Gröbner basis w. r. to $<_o$.

Step 2:   Reduce each element of the Gröbner basis modulo the other Gröbner basis elements and normalize it to $Hcoef = 1$. Let GB be this Gröbner basis. Collect in C all m-tuples c with $(0, c) \in GB$ and in A all a_i with $(e_i^{(r)}, a_i) \in GB$. If $length(A) \neq r$, then replace A by the empty set. Return (A, C).

The correctness of the algorithm is easily shown.

Theorem 2.   *Let the $G_i, F_i, b_i,$ and d_i be as in the input of the algorithm. Then the Gröbner basis computed in step 1 consists after normalization and reduction in elements $(0, c_i), i = 1, \ldots, L$, such that $c_1, \ldots, c_L$ generate $Syz(F_1, \ldots, F_m)$ and in elements (u, v) with $u \neq 0$. If and only if $Mod(G_1, \ldots, G_r) = Mod(F_1, \ldots, F_m)$, the latter elements are $(e_i^{(r)}, a_i), i = 1, \ldots, r$, with suitable $a_i = (a_{i1}, \ldots, a_{im})$ and*

$$G_i = \sum_{j=1}^m a_{ij} F_j.$$

Proof: The Gröbner basis in question generates the module W as defined in proposition 1. Therefore, $(0, c) \in W$ iff $c \in Syz(F_1, \ldots, F_m)$. By construction of $<_o$ the submodule $\{(u, v) \in W \mid u = 0\}$ is generated by $(0, c_1), \ldots, (0, c_L)$. Hence $c_1, \ldots, c_L$ generate $Syz(F_1, \ldots, F_m)$ and all other basis elements are of type (u, v) with $u \neq 0$.

By proposition 2, $(e_i^{(r)}, q_i)$ exist in W for $i = 1, \ldots, r$ if and only if $Mod(G_1, \ldots, G_r) = Mod(F_1, \ldots, F_m)$. I will show, that $(e_i^{(r)}, q_i)$ is in W iff an $(e_i^{(r)}, a_i)$ is in the (normalized and reduced) Gröbner basis. Assume in the contrary $(e_i^{(r)}, q_i) \in W$ but no $(e_i^{(r)}, a_i)$ in the reduced Gröbner basis. By condition (2) of theorem 1, $(e_i^{(r)}, q_i)$ is reducible to 0. Thus there is a Gröbner basis element H_k with $Hterm(H_k) = Hterm(e_i^{(r)}, q_i)$ $(= (e_i^{(r)}, 0))$, but $H_k - (e_i^{(r)}, 0) = (u, h)$ with $0 \neq u \in \mathcal{P}^r$. Therefore $(e_i^{(r)}, q_i) - H_k = (-u, -h + q_i) \in W$ is still reducible, which means there is an other Gröbner basis element H_j with the same $Hterm$ as $(-u, -h + q_i)$. This $Hterm$ depends only on u, such that H_k is reducible by H_j, a contradiction.

Any $(u, v) \in W$ with $u \neq 0$ and $u \neq e_i^{(r)}$ for $i = 1, \ldots, r$ is reducible by means of the $(e_i^{(r)}, a_i)$. Thus all elements of the reduced Gröbner basis are of the required type, if $Mod(G_1, \ldots, G_r) = Mod(F_1, \ldots, F_m)$.

$(e_i^{(r)}, a_i) \in W$ means $G_i = \sum a_{ij} F_j$.   Q. E. D.

The output of the algorithm produces more than what is required in the two problems. For solving problem 2 it is sufficient to store only the first r components of the input $(r+m)$-tuples, because the first r components of the output tuples contain the relevant information whether for every $i \in \{1, \ldots, r\}$ there is an output element with first r components equal to $e_i^{(r)}$. The second part $<_{M'}$ of the ordering $<_o$ is of no relevance for this simplified version of the algorithm.

The ordering $<_{M'}$ guarantees that the output $(0, c_i)$ constitute a Gröbner basis w. r. to $<_o$ for $\{(u, v) \in W \mid u = 0\}$ in case $Mod(G_1, \ldots, G_r) = Mod(F_1, \ldots, F_m)$. This means $c_1, \ldots, c_L$ is a Gröbner basis w. r. to $<_{M'}$ for $Syz(F_1, \ldots, F_m)$. But if as required in problem 1 only a basis for $Syz(F_1, \ldots, F_m)$ is needed, then manipulations are superfluous which use tuples $(0, v)$. Therefore, in that case a so called *incomplete* Buchberger algorithm can be employed which at most reduces $S((u_i, v_i), (u_j, v_j))$, where both u_i and u_j are not 0, and for reductions only tuples (u_k, v_k) with $u_k \neq 0$ are used.

The algorithm simplifies a bit when $G_1, \ldots, G_r$ is a Gröbner basis. In this case I may take the S_{ij} as defined at the end of section 2 for the basis of $Syz(G_1, \ldots, G_r)$ and take for $<_M$ the module ordering of $\mathcal{P}^r$ with weights $w_i := lt(G_i)$, $i = 1, \ldots, r$, and an arbitrary ordering $<_1$ of $\{1, \ldots, r\}$. Then $\{(d_1, 0), \ldots, (d_K, 0)\}$ is a Gröbner basis. Hence all $S((d_i, 0), (d_j, 0))$ reduce to 0 modulo the subset $\{(d_1, 0), \ldots, (d_K, 0)\}$ of the input set. Therefore such S-tuples may be omitted in the algorithm giving a slightly faster performance.

If $G_1, \ldots, G_r$ is a Gröbner basis, and if the b_{ij} needed for the input are not known in advance, then they can be calculated easily, because these b_{ij} can be read off from the easily obtainable G-representation of $F_i \in Mod(G_1, \ldots, G_r)$.

4   An example.

For illustrating how the algorithm works I present a simple example which uses as input a Gröbner basis. This example is derived from an example of Lazard (1986).

Let $\mathcal{P} = \mathbf{Q}[x, y]$, $s := 1$, and

$$\begin{aligned}
G_1 &:= y^4 - 6x^2y^2 + x^4 + 1, \\
G_2 &:= xy^3 - x^3y, \\
G_3 &:= -5x^3y^2 + x^5 - x, \\
G_4 &:= 4x^5y - xy, \\
G_5 &:= 4x^7 - 5xy^2 + 4x^3.
\end{aligned}$$

(2)

This is a Gröbner basis with respect to the graduated lexicographical ordering $<_{GL}$ defined by $x^i y^j <_{GL} x^k y^l$ iff $i+j < k+l$ or $i+j = k+l, j < l$. For the F_i I take

(3)
$$
\begin{aligned}
F_1 &:= y^4 - 6x^2 y^2 + x^4 + 1 &&= G_1, \\
F_2 &:= 5xy^2 - 4x^7 - 4x^3 &&= -G_5, \\
F_3 &:= 4x^5 y - xy &&= G_4, \\
F_4 &:= 4x^9 + 3x^5 - x &&= -G_3 + x^2 G_5.
\end{aligned}
$$

Hence $b_1 = (1,0,0,0,0)$, $b_2 = (0,0,0,0,-1)$, $b_3 = (0,0,0,1,0)$. The 4-tuple $b_4 = (0,0,-1,0,x^2)$ is obtained by reduction:

$$
F_4 \longrightarrow F_4 - x^2 G_5 \longrightarrow F_4 - x^2 G_5 + G_3 = 0.
$$

The module $Syz(G_1, \ldots, G_5)$ is already generated by

(4)
$$
\begin{aligned}
d_1 &:= S_{12} = (\; -x, & y, & \quad 1, & \quad 0, & \quad 0), \\
d_2 &:= S_{23} = (\; 0, & 5x^2, & \quad y, & \quad 1, & \quad 0), \\
d_3 &:= S_{34} = (\; 0, & 0, & \quad 4x^2, & \quad 5y, & \quad -1), \\
d_4 &:= S_{45} = (\; 0, & 5, & \quad 0, & \quad -x^2, & \quad y).
\end{aligned}
$$

(In order to have integer coefficients, I multiplied here the tuples with their least common multiples.) Now the incomplete Buchberger algorithm is started with the weighted ordering $<_M$ for $\mathcal{P}^5$ defined by $<_T = <_{GL}$, weights $w_1 = lt(G_1) = y^4$, $w_2 = lt(G_2) = xy^3$, $w_3 = lt(G_3) = x^3 y^2$, $w_4 = lt(G_4) = x^5 y$, $w_5 = lt(G_5) = x^7$, and $<_1$ the natural ordering of $\{1,2,3,4,5\}$. Initially, the input rows are $(b_1, e_1^{(4)})$, $(b_2, e_2^{(4)})$, $(b_3, e_3^{(4)})$, $(b_4, e_4^{(4)})$, $(d_1, 0)$, $(d_2, 0)$, $(d_3, 0)$, $(d_4, 0)$, which I renumber by $r_1, \ldots, r_8$. Underlined are the positions which determine the $Hterm$ of the respective row.

(5)
$$
\begin{pmatrix}
\underline{1} & 0 & 0 & 0 & 0 & 1 & 0 & 0 & 0) \\
0 & 0 & 0 & 0 & \underline{-1} & 0 & 1 & 0 & 0) \\
0 & 0 & 0 & \underline{1} & 0 & 0 & 0 & 1 & 0) \\
0 & 0 & -1 & 0 & \underline{x^2} & 0 & 0 & 0 & 1) \\
\underline{-x} & y & 1 & 0 & 0 & 0 & 0 & 0 & 0) \\
0 & \underline{5x^2} & y & 1 & 0 & 0 & 0 & 0 & 0) \\
0 & 0 & \underline{4x^2} & 5y & -1 & 0 & 0 & 0 & 0) \\
0 & 5 & 0 & \underline{-x^2} & y & 0 & 0 & 0 & 0)
\end{pmatrix}
$$

Only the S-tuples $S(r_1, r_5)$, $S(r_3, r_8)$, and $S(r_2, r_4)$ exist. After any of these S-tuple computation, the higher indexed row may be replaced by the

S-tuple, because it is redundant. These operations give

(6)
$$
\begin{pmatrix}
1 & 0 & 0 & 0 & 0 & 1 & 0 & 0 & 0 \\
0 & 0 & 0 & 0 & -1 & 0 & 1 & 0 & 0 \\
0 & 0 & 0 & 1 & 0 & 0 & 0 & 1 & 0 \\
0 & 0 & -1 & 0 & 0 & 0 & x^2 & 0 & 1 \\
0 & y & 1 & 0 & 0 & x & 0 & 0 & 0 \\
0 & 5x^2 & y & 1 & 0 & 0 & 0 & 0 & 0 \\
0 & 0 & 4x^2 & 5y & -1 & 0 & 0 & 0 & 0 \\
0 & 5 & 0 & 0 & y & 0 & 0 & x^2 & 0
\end{pmatrix}
$$

Now only the S-tuples $S(r_5, r_6), S(r_4, r_7), S(r_2, r_8)$ are defined. The corresponding rows r_7 and r_8 are then redundant and may be replaced by the last two S-tuples. The computation of $S(r_5, r_6)$ will be delayed. Later in the computation it turns out, that it has not to be calculated because both r_5 and r_6 will be redundant.

(7)
$$
\begin{pmatrix}
1 & 0 & 0 & 0 & 0 & 1 & 0 & 0 & 0 \\
0 & 0 & 0 & 0 & -1 & 0 & 1 & 0 & 0 \\
0 & 0 & 0 & 1 & 0 & 0 & 0 & 1 & 0 \\
0 & 0 & -1 & 0 & 0 & 0 & x^2 & 0 & 1 \\
0 & y & 1 & 0 & 0 & x & 0 & 0 & 0 \\
0 & 5x^2 & y & 1 & 0 & 0 & 0 & 0 & 0 \\
0 & 0 & 0 & 5y & -1 & 0 & 4x^4 & 0 & 4x^2 \\
0 & 5 & 0 & 0 & 0 & 0 & y & x^2 & 0
\end{pmatrix}
$$

At this point we see that $Mod(F_1, F_2, F_3, F_4) = Mod(G_1, \ldots, G_5)$, because the rows $r_1, r_8/5, -r_4, r_3, -r_2$ have for first five components the five unit vectors of $\mathcal{P}^5$. The continuation of theses row transformations give finally after row normalizations

(8)
$$
\begin{pmatrix}
1 & 0 & 0 & 0 & 0 & 1 & 0 & 0 & 0 \\
0 & 0 & 0 & 0 & 1 & 0 & -1 & 0 & 0 \\
0 & 0 & 0 & 1 & 0 & 0 & 0 & 1 & 0 \\
0 & 0 & 1 & 0 & 0 & 0 & -x^2 & 0 & -1 \\
0 & 0 & 0 & 0 & 0 & x & x^2 - y^2/5 & -x^2y/5 & 1 \\
0 & 0 & 0 & 0 & 0 & 0 & 0 & x^4 + 1 & -y \\
0 & 0 & 0 & 0 & 0 & 0 & 4x^4 - 1 & -5y & 4x^2 \\
0 & 1 & 0 & 0 & 0 & 0 & y/5 & x^2/5 & 0
\end{pmatrix}
$$

This means, that $c_1 := (x, x^2 - y^2/5, -x^2y/5, 1)$, $c_2 := (0, 0, x^4 + 1, -y)$ and $c_3 := (0, 4x^4 - 1, -5y, 4x^2)$ generate $Syz(F_1, F_2, F_3, F_4)$, and

(9)
$$
\begin{aligned}
G_1 &= F_1 \\
G_2 &= yF_2/5 + x^2 F_3/5, \\
G_3 &= -x^2 F_2 - F_4 \\
G_4 &= F_3 \\
G_5 &= -F_2.
\end{aligned}
$$

Other examples can be presented in a similar way showing how to solve linear systems of equations and how to verify that a reconstructed Gröbner basis is in fact a Gröbner basis of a module given by a generating system.

REFERENCES

[1] Armbruster, D., *Bifurcation theory and Computer Algebra: An initial approach*, in "EUROCAL '85," Springer Lecture Notes in Comp. Sci. 204, 1985, pp. 126 – 137.

[2] Furukawa, A., Sasaki, T., Kobayashi, H., *Gröbner basis of a module over $K[x_1, \ldots, x_n]$ and polynomial solutions of a system of linear equations*, in "SYMSAC '86," proceedings, 1986, pp. 222 – 224.

[3] Gebauer, R., Möller, H. M., *On an installation of Buchberger's algorithm*, J. Symb. Comp. 6 (1988), 275 – 286.

[4] Lazard, D., *Ideal bases and primary decomposition: case of two variables*, J. Symb. Comp. 1 (1985), 261 – 270.

[5] Maeß, G., "Vorlesungen über numerische Mathematik," Vol. 1, section 2.2.2, Akademie-Verlag Berlin, 1984.

[6] Melenk, H., Möller, H.M., Neun, W., *On Gröbner basis computation on a supercomputer using REDUCE*, Preprint SC 88-2 of Konrad-Zuse-Zentrum für Informationstechnik, Berlin 1988.

[7] Möller, H.M., Mora, F., *New constructive methods in classical ideal theory*, J. Algebra 100 (1986), 138 – 178.

[8] Traverso, C., *Gröbner trace algorithms*, in "ISSAC '88," Springer Lecture Notes in Comp. Sci., 1988.

[9] Zacharias, G., *Generalized Gröbner bases in commutative polynomial rings*, Thesis at M.I.T. Dept. Comp. Sci. 1978.

H. Michael Möller
FB Mathematik und Informatik der FernUniversität
D-5800 Hagen 1

The non-scalar Model of Complexity
in Computational Geometry

JOSE L. MONTAÑA   LUIS M. PARDO   TOMAS RECIO

Abstract. An outline on the relation between algebraic complexity theories and semialgebraic sets is presented. First we discuss the concepts of total and non-scalar complexities both for polynomials and semialgebraic sets observing that they are "geometric complexities" verifying the "semialgebraic" version of the Benedetti-Risler conjecture [4]. Moreover, we remark that total and non-scalar complexities of semialgebraic sets are decidable theories. Finally, using non-scalar complexity and intersection numbers of semialgebraic sets we get new lower bounds for several problems in computational geometry, generalizing the results obtained by M. Ben-Or using total complexity and number of connected components. An expanded version of the ideas sketched here is [10]

0. Introduction

A semialgebraic set (c.f. Definition 1.2 below) is a subset of $\mathbf{R}^n$ defined by a finite collection of polynomial equalities and inequalities (c.f. [1] for a throughly treatise on the subject). Due to this open definition, semialgebraic sets arise quite naturally in many topics; in particular, they appear in Computational Geometry and Algebraic Computational Complexity: the book of F. Preparata and M.I. Shamos [12] and the survey of J. von zur Gathen [7] are references on each topic that quite explicitly contain mentions to the theory of semialgebraic sets, although it is exclusively represented by Thom-Milnor's Theorem as extended by M. Ben-Or in [3].

On the other hand, this theory of semialgebraic sets has been subject to a systematic development in the last twenty years (more or less the period starting after the publication of Thom-Milnor's Theorem); but, curiously, most of the results attained during this time have not been explored enough regarding to their application to computational geometry or algebraic computational complexity. The aim of this contribution is to go further in this direction presenting some results based on techniques from real algebraic

Partially Supported by D.G.C.Y T. 0062/86

geometry, but of potential interest in various fields of computational mathematics. We just outline the main ideas and results here, leaving proofs and extended comments aside, c.f. [10].

We shall consider in the following the field **R** of real numbers as ground field; but clearly, using Tarski Principle, most of our results are also true for any real closed field.

1. Complexities of semialgebraic sets in Real Algebraic Geometry and Computational Geometry

Several authors have dealt explicitely with the problem of measuring the complexity of *one given representation* of a semialgebraic set: c.f. [17], [14]. The book of R. Benedetti and J.J. Risler [4] contains the first systematic approach to an abstract definition of complexity in the following terms:

Definition 1.1: A *geometric complexity* **C** on **R** is a family of mappings

$$\{C_n : \mathbf{R}[X_1,\ldots,X_n] \longrightarrow \mathbf{N} \mid n \in \mathbf{N}\}$$

verifying the following axioms:

I.- If p is a constant, $C_n(p) = 0$, and $C_n(X_i) = 1$, $i = 1,\ldots,n$.

II.- $C_n(p + q) \bigtriangledown (C_n(p), C_n(q))$ and $C_n(p \times q) \bigtriangledown (C_n(p), C_n(q))$.

III.- $C_n(p_{X_i}) \bigtriangledown (n, C_n(p))$, where p_{X_i} is the partial derivative of p with respect to the variable X_i.

IV.- For $p_1,\ldots,p_n \in \mathbf{R}[X_1,\ldots,X_n]$, the number $N(S)$ of non-degenerated solutions of the system S, given by $p_1 = 0,\ldots,p_n = 0$, verifies

$$N(S) \bigtriangledown (n, C_n(p_1),\ldots,C_n(p_n))$$

V.- Let ϕ be any affine transformation of $\mathbf{R}^n$, then for any $p \in \mathbf{R}[X_1,\ldots,X_n]$ $C_n(p \circ \phi) \bigtriangledown (C_n(p), n)$.

The $\bigtriangledown$ notation means that the left term is bounded by an effective function of the elements in the rigth term. Moreover, axiom IV can be replaced without changing the theory, by the following

IV'.- For every $p \in \mathbf{R}[X_1,\ldots,X_n]$, $B_0(p^{-1}(0)) \bigtriangledown (n, C_n(p))$, where $B_0(p^{-1}(0))$ is the number of connected components of the algebraic set $p^{-1}(\{0\})$.

Examples of geometric complexities are the *total degree*, and the *additive complexity* of polynomials (c. f. [4]).

Definition 1.2: A subset W of $\mathbf{R}^n$ is a semialgebraic set if there are polynomials in n variables $f_{i,j}, g_{i,j}, i \in I, j \in J, I, J$ finite, such that $W = \cup_{i \in I} W_i$, where $W_i = \{x \in \mathbf{R}^n \mid f_{i,j}(x) = 0, g_{i,j}(x) > 0, j \in J\}$.

In [4] the two definitions above are merged to provide the concept of complexity of one representation of a semialgebraic set S, as follows:

Definition 1.3: Let $\mathbf{C} = \{C_n \mid n \in \mathbf{N}\}$ be any geometric complexity and let W be a semialgebraic set described as in the above definition. Then the **C**- *complexity of this representation* is the triple

$$(n, Card(I \times J), \max\{C_n(f_{i,j}), C_n(g_{i,j})\})$$

Given a triple of natural numbers (n, r, s), a semialgebraic set W is said to be in the class $S_{\mathbf{C}}(n, r, s)$ for a complexity $\mathbf{C}$, if and only if there is a representation of S such that the **C**-complexity of this representation is (n, r', s'), with $r' \leq r$ and $s' \leq s$.

Besides classical measures of complexity of polynomials (total degree, the additive one), computational mathematics provide a different measure:

Definition 1.4: The *total complexity* of a polynomial $p \in \mathbf{R}[X_1, \ldots, X_n]$ is the natural number $Tot(p)$ given as the minimum height of any algebraic computation tree evaluating the polynomial p (c.f. [7], [16] or [3]).

We shall assume the following constraints to the standard concept of algebraic computation tree:

Division along the tree is preceeded by a test node (on the sign condition of the divisor) allowing only division in the branch that corresponds to a non-zero answer to the test. Also, division by constants is substituted by a product (of their inverses). Thus, division by constants does not appear in these algebraic computation trees.

Similarly, we consider the non-scalar model of complexity for polynomials introduced by A.M. Ostrowski, and studied by V.Y. Pan, V. Strassen and others (c.f. [2], [7], [8] or [16] for more detailed references). Essentially it measures the complexity of a polynomial as an invariant under linear affine transformations of the ambient space.

Definition 1.5: Given an algebraic computation tree $T(X_1, \ldots, X_n)$ we say that a node v is *non-scalar* if and only if it is either a branching node or a computation node of the form: $f_v := f_{v_1} \circ_v f_{v_2}$ where $[\circ_v \in \{\times\}$ and $f_{v_1}, f_{v_2} \notin \mathbf{R}]$ or $[\circ_v \in \{:\}$ and $f_{v_2} \notin \mathbf{R}]$.

Non-scalar nodes are those that make increase the degree of the polynomials evaluated along a tree. In this sense, we measure:

Definition 1.6: In the non-scalar model, the *height of an oriented path* T in a tree T, $h_{N.S.}(\Gamma)$, is the number of non-scalar nodes it contains.

Therefore, the *non-scalar height* of an algebraic computation tree T, $h_{N.S.}(T)$ is the maximum (non-scalar) height of all the oriented paths in T.

Finally, the *non-scalar complexity* of a polynomial $p \in \mathbf{R}[X_1, \ldots, X_n]$ is the natural number $N.S.(p)$ given as the minimum non-scalar height of any algebraic computation tree evaluating p at any input in $\mathbf{R}^n$.

Proposition 1.7. *Both the non-scalar and the total complexity of polynomials are geometric complexities in the sense of Benedetti-Risler (definition 1.1) that also verify the following property:*

Axiom III'.- $C_n(p_{X_i}) \bigtriangledown C_n(p)$, where p_{X_i} is the partial derivative of p with respect to the variable X_i.

Remarks 1.8:

i) It can also be observed that, taking as complexity the total degree or the additive one, the effective function in Axiom III of definition 1.1 does not depend upon the number of variables.

ii) One point of intererest regarding the concept of C-complexity of a description of a semialgebraic set is the analysis of the following question in [4]:

Benedetti-Risler Conjecture. *Let $\mathbf{C}$ be a geometric complexity. There is an effective mapping $\eta : \mathbf{N} \times \mathbf{N} \times \mathbf{N} \longrightarrow \mathbf{N}$ such that, for every $(n, r, s) \in \mathbf{N}^3$, there is a finite collection of semialgebraic sets $W_1, \ldots, W_k$ with $k \leq \eta(n, r, s)$ such that:*

1) W_i is not homeomorphic to W_j, for all pairs $i \neq j$.

2) For every W in $S_{\mathbf{C}}(n, r, s)$, there is $j \in \{1, \ldots, k\}$ such that W and W_j are homeomorphic.

3) $W_1, \ldots, W_k$ are in $S_{\mathbf{C}}(n, r, s)$.

This conjecture is also known as the *"finiteness of topological types of semialgebraic sets of bounded (representation) complexity "*. As stated in [4], the geometric complexity defined by the total degree of a polynomial Deg: $\mathbf{R}[X_1, \ldots, X_n] \longrightarrow \mathbf{N}$, verifies a somehow stronger fact than the above conjecture, replacing in this case 'homeomorphic' by 'semialgebraically homeomorphic' (c.f. also [1] for basic definitions of semialgebraic homeomorphism). Now we state that the same holds when considering Tot-complexities and non-scalar complexities of polynomials:

Proposition 1.9. *The stronger "semialgebraic" version of the Benedetti-Risler conjecture also holds when considering Tot-complexities (respectively N.S.- complexities) of representations of semialgebraic sets.*

As observed in [2], into the context of computational algebra and geometry the word complexity has to be understood as a measure of what makes functions and sets 'to be hard to compute'. Then, complexity of a set has to be understood as the complexity of the evaluation of the characteristic function defined by that set. This is the underlying idea in the works [3],[11], [12], [13]. Next, we introduce formally such a definition of complexity of a semialgebraic set (and not of a specific representation) and we observe that it also follows the structure demanded in [4] for a geometric complexity.

Definition 1.10: Let W be a semialgebraic subset of $\mathbf{R}^n$. We define the *total complexity* of W, $C_{Tot}(W)$, as the height of any optimal algebraic computation tree over $\mathbf{R}$ computing the characteristic function defined by W in $\mathbf{R}^n$, $\chi_W : \mathbf{R}^n \longrightarrow \{0,1\}$.

Using a more classical terminology, the $C_{Tot}(W)$ is the minimum height of any *algebraic computation tree solving the membership problem to the semialgebraic set* W (c.f. [3],[11] or [12]).

Examples and remarks 1.11:

i) It could be interesting to compare Tot complexities of polynomials (Def. 1.4) $p(X) \in \mathbf{R}[X_1, \ldots, X_n]$ and C_{Tot} complexities of the semialgebraic sets $A_1 = \{p > 0\}$, $A_2 = \{p = 0\}$ and $A_3 = \{p \geq 0\}$. Clearly $C_{Tot}(A_i) \leq Tot(p)$ for i=1,2,3. In fact, $\max\{C_{Tot}(A_i)\}$ is the complexity of the *evaluation*, not of the whole polynomial, but only of the *sign* of $p(X)$ at any input.

ii) *The above inequality is not an equality in general.* First observe that the Total complexity of any semialgebraic subset W of the affine line $\mathbf{R}$ is a function of the same class than the base two logarithm of the number of its connected components $\log B_0(W)$. This can be easily deduced considering the structure of semialgebraic sets in $\mathbf{R}$ ([1]) and using both [3], for the lower bounds and a procedure of 'Divide and Conquer' as the optimal algorithm.

Following this method we get that an optimal tree for semialgebraic sets on the affine line exists involving only linear (degree 1) polynomials. Moreover, the total number of these polynomials is also linear in the number of connected components, although for every input $\alpha \in \mathbf{R}$ we only have to evaluate $\log B_0(W)$ polynomials in order to decide whether $\alpha \in \mathbf{R}$ is or not in W.

For instance, given any proper algebraic subset $V = \{a_1, \ldots, a_t\}$ of $\mathbf{R}$, there is a completely determined polynomial of degree t generating $I(V)$. Let $p(X) \in \mathbf{R}[X]$ be such a polynomial. In most cases (c.f. [2] or [16] for more detailed descriptions) $Tot(p) \geq t$. However, we have remarked that $C_{Tot}(V) \in \theta(log_2 t)$.

Similarly we can introduce the non-scalar complexity of a semialgebraic set by taking into account just the non-scalar nodes of the trees:

Definition 1.12: We define the non-scalar complexity of a semialgebraic subset W of $\mathbf{R}^n$, $C_{N.S.}(W)$, as the minimum of the non-scalar heights of all those algebraic computation trees T computing the characteristic mapping defined by W:

$$C_{N.S.}(W) := min\{h_{N.S.}(T) \mid W(T) = W\}$$

Examples of upper bounds and remarks 1.13:

i) $C_{N.S.}(W) \leq C_{Tot}(W)$ for every semialgebraic set W.

ii) For every linear affine variety A of $\mathbf{R}^n$, the non-scalar complexity of A is $C_{N.S.}(A) \leq n - d$, where d is the dimension of A.

iii) A quadratic algebraic variety in $\mathbf{R}^n$ is an algebraic set defined by $f(X_1, \ldots, X_n) = 0$ where f is a polynomial of degree 2.

Clearly, for every quadratic algebraic variety $E = \{f(X_1, \ldots, X_n) = 0\}$ of $\mathbf{R}^n$ there is a tree T evaluating the characteristic mapping of E with non-scalar cost in $O(s)$, where $s(\leq n)$ is the maximum between the positive terms and the negative terms in any diagonalized expression of the degree two homogeneous component of f.

iv) Since every algebraic subset V of $\mathbf{R}^n$ can be given as the set of solutions of a single equation, if the degree of a single polynomial, defining V as the set of its zeroes, is bounded by k, the non-scalar complexity of V is in $O(n^{k-1})$.

Proposition 1.14 (structural properties of total and non-scalar complexities). *Let W, V be two semialgebraic subsets of $\mathbf{R}^n$. Let $\pi : \mathbf{R}^n \longrightarrow \mathbf{R}^{n-1}$ be the projection that forgets the last variable and, respectively, let $Int(W)$, $Adh(W)$ and $Bd(W)$ be the interior, the closure and the boundary of the semialgebraic set W. Then, we have:*

i) $C_{Tot}(\emptyset) = 0, C_{Tot}(\mathbf{R}^n) = 0$.

ii) $C_{Tot}(V \cap W) \leq C_{Tot}(V) + C_{Tot}(W)$

iii) $C_{Tot}(V \cup W) \leq C_{Tot}(V) + C_{Tot}(W)$

iv) $C_{Tot}(V \times W) \leq C_{Tot}(V) + C_{Tot}(W)$

v) For every affine linear transformation of the space $\Phi : \mathbf{R}^n \longrightarrow \mathbf{R}^n$, $C_{Tot}(\Phi(W)) \leq C_{Tot}(W) + n^2$.

vi) $C_{Tot}(\pi(W)) \bigtriangledown C_{Tot}(W)$.

vii) $C_{Tot}(Int(W)) \bigtriangledown C_{Tot}(W), C_{Tot}(Adh(W)) \bigtriangledown C_{Tot}(W)$ *and* $C_{Tot}(Bd(W)) \bigtriangledown C_{Tot}(W)$.

Analogous statements can be shown when substituting total complexities, C_{Tot}, by non-scalar complexities, $C_{N.S.}$, of semialgebraic sets, excepting the v) and vii) ones. Property v) will be substituted by

v') For every affine linear transformation of the space $\Phi : \mathbf{R}^n \longrightarrow \mathbf{R}^n$,

$$C_{N.S.}(\Phi(W)) = C_{N.S.}(W)$$

As for property vii), we have to include the dimension of the ambient space, n, on the rigth hand of all the inequalities.

Sketch of proofs: The statements from i) to v) and v') can be obtained by joining suitably the respective algebraic computational trees. As for vi) and vii) we just have to follow carefully the algorithmic methods proving Tarski's Principle.

Under these notations it is also posible to offer a similar statement to the one claimed at the Benedetti-Risler conjecture:

Theorem 1.15. *The invariant C_{Tot} is a sufficient parameter for effectively bounding and classifying the number of topological types, i.e. there is an effective function $\eta : \mathsf{N} \longrightarrow \mathsf{N}$ such that, for every $h \in \mathsf{N}$ there is a finite number $k \leq \eta(h)$ of semialgebraic sets $W_1, \ldots, W_k$, verifying:*

i) W_i is not semialgebraically homeomorphic to W_j, where $i \neq j$.

ii) For every semialgebraic set W, if $C_{Tot}(W) \leq h$, there is $j \in \{1, \ldots, k\}$ and $n \in \mathsf{N}$ such that W and $W_j \times \mathsf{R}^n$ are semialgebraically homeomorphic.

iii) $C_{Tot}(W_j) \leq h$

An analogous statement for non-scalar complexity can be introduced (but, in this case, we have to include also the number of variables as a parameter for bounding the number of topological types).

In the above Remark 1.11 ii) we have sketched a method for computing the complexity of a semialgebraic set in the real, one dimensional, affine space, asuming the set is given in the usual way: a collection of polynomial equalities and inequalities. One can compute from this presentation the number of connected components (this is essentially Sturm's theorem) and then apply the remark. The following theorem states that this feature of the notion of complexity generalizes to any dimension.

Theorem 1.16. *The mappings Tot and C_{Tot} are computable, i.e. there are algorithms computing $Tot(p)$ (resp. $C_{Tot}(W)$) for every polynomial p (resp. for every semialgebraic set W). The same holds when considering N.S. complexities of polynomials and non-scalar complexities of semialgebraic sets.*

Remarks 1.17:

i) The proof of the theorem above is based on the existence of quantifier elimination over real closed fields, therefore it also produces for every polynomial and every semialgebraic set the respective optimal trees.

ii) We can also define, along the same terms, the concept of intrinsic complexity of constructible subsets of affine spaces K^n, where K is any field. Here the computation schemes are the same (algebraic computation trees) as for the reals; due to the lack of order over the field, the branching nodes just decide if an element $a \in K$ is $a \neq 0$ or $a = 0$.

Within these schemes we can also define the Total complexity of a polynomial $p \in K[X_1, \ldots, X_n]$ and the Total complexity of a constructible subset W of K^n. If K is algebraically closed these mappings can also be computed, just following quantifier elimination for algebraically closed fields.

iii) Observe that for computable real closed fields R (i.e. those countable fields where the arithmetic operations and the sign are Turing-computable mappings, for instance R_{alg} or the field of real algebraic rational Puiseux series, (c.f.[5] or [6])) the corresponding (replacing the reals by these fields in the definitions) complexity mappings C_{Tot} and Tot are "Turing-computable" mappings.

iv) Since quantifier elimination over real closed fields is doubly exponential the procedure that one can deduce from the theorem is, in fact, intractable. Thus, in order to find the complexity of concrete sets or polynomials we have to follow different strategies, and this will be the task of the next section.

2. Finding lower bounds for the non-scalar model of computation

In [3] a method for obtaining lower bounds of problems in Computational Geometry has been developped as an application of Thom-Milnor's theorem. The method consists on associating to a given problem a suitable semialgebraic set whose total complexity is a lower bound of the complexity of the problem. Then, the main result in this work states that the total complexity of a semialgebraic set W is bounded below by the logarithm in base two of the number of connected components of W. This method becomes very useful when the associated semialgebraic set has many connected components. Actually, when the associated set is connected, Ben-Or's method is inefficient as we can see in the following examples:

Example 2.1: Largest Empty Circle,Maximum GAP, Even Distribution Problem

Within the context of proximity problems, [12] shows an algorithm that solves the LARGEST EMPTY CIRCLE problem for N points in $O(N \cdot \log N)$. Actually, these authors mention the work of [9] who have obtained some results on lower bounds of the complexity of this problem, exhibiting a $\Omega(N \cdot \log N)$ bound when considering only LINEAR trees, i.e., trees evaluating and testing only polynomials of degree at most 1.

Thus [12] asked for a more complete answer to the complexity of this problem in the general model of algebraic computation trees. Let us recall that the L.E.C. for N points is stated as follows:

LARGEST EMPTY CIRCLE: (**L.E.C.**)
Given N points in the real plane, $\mathbf{R}^2$, find the largest circle, centered in the convex hull of these points, and without containing any of the points in the open disk.

Lower bounds for the complexity of this problem can be deduced — a standard procedure in computational geometry — from lower bounds of the following problem:

MAXIMUM GAP: (**MAX GAP**)

Given a line r in $\mathbf{R}^2$ and N points lying on r, solve the LARGEST EMPTY CIRCLE Problem for them.

Moreover, lower bounds for MAXIMUM GAP can be obtained from lower bounds of the following decisional problem:

EVEN DISTRIBUTION PROBLEM: (**E.D.P.**)

Given N real numbers, $x_1, \ldots, x_N \in \mathbf{R}$, decide whether the diameter of a maximal empty open interval inside the convex hull of $\{X_1, \ldots, X_N\}$ is ≤ 1 or not.

This last problem can be stated as the membership problem to the semialgebraic subset of $\mathbf{R}^N$:

$$W_N = \bigcup_{\sigma \in \Sigma_N} \{(x_1, \ldots, x_N) \in \mathbf{R}^N \mid 0 \leq x_{\sigma(i+1)} - x_{\sigma(i)} \leq 1, i = 1, \ldots, N-1\}$$

where Σ_N is the permutation group of order N.

It can be easily shown that W_N is a *connected* semialgebraic set and by Ben-Or's method we can only get $C_{Tot}(W_N) \in \Omega(1)$.

Example 2.2 - Some membership problems defined by connected semialgebraic sets:

2.2.1.- Consider the algebraic set in $\mathbf{R}^2$ given as the union of N meridians in S^2, i.e. the union of N maximal circles in S^2, passing all of them through two fixed antipodal points (c.f. Fig. 1). An algorithm that solves the membership problem defined by this set in $O(\log N)$ arithmetic operations can be easily designed.

2.2.2.- Consider a family of N tangent closed disks in $\mathbf{R}^2$ "following" a cusp (c.f. Fig. 3) and the membership problem they define.

2.2.3.- A convex polygon with N vertices in $\mathbf{R}^2$ defines a membership problem that can be solved in $O(\log N)$ arithmetical operations. This algorithm can be obtained by developing a Divide and Conquer method from a point in the kernel of the set.

2.2.4.- Similarly, for every starshaped set with N vertices in $\mathbf{R}^2$, we can easily show an algebraic computation tree that solves the membership problem it defines in $O(\log N)$ arithmetic operations.

In all these examples the parameter N has been chosen as an intuitive indication of the size of the problem. In fact, we observe all of these problems can be solved by means of an algebraic computation tree in $O(\log N)$. Unfortunately, all of them are connected semialgebraic sets and [3] yields to the trivial lower bound $\Omega(1)$. In order to solve this gap we introduce the following:

Lemma and Definition 2.3.

For every semialgebraic subset W of $\mathbf{R}^n$ there is a smallest natural number $In(W) \in \mathbf{N}$, such that for every linear affine subvariety A of $\mathbf{R}^n$, the number of connected components $B_0(W \cap A) \le In(W)$.

We shall call $In(W)$ the *first intersection number* of the semialgebraic set W.

Likewise we can define *the second intersection number* of W, $In^2(W)$, replacing in the definition the linear affine varieties by quadratic algebraic varieties.

Analogously the *k-th intersection number*, $In^k(W)$ is defined when one considers the intersection with algebraic varieties given by a single polynomial of degree at most k.

Sketch of proof: Thom-Milnor's Theorem shows an upper bound for the number of connected components obtained when intersecting a fixed semialgebraic set with semialgebraic sets of bounded Deg-complexity.

Under these notations we are able to show some useful lower bounds both for the total and the non-scalar complexity of semialgebraic sets:

Theorem 2.4. *For every semialgebraic subset W of $\mathbf{R}^n$ we have:*
  i) $C_{Tot}(W) \ge C_{N.S.}(W) \in \Omega(\log B_0(W) - n)$.
  ii) $C_{Tot}(W) \ge C_{N.S.}(W) \in \Omega(\log In(W) - n)$.
  iii) $C_{Tot}(W) \ge C_{N.S.}(W) \in \Omega(\log(\max\{In^2(W), In^2(W^c)\}) - n)$.
  iv) For $k \le 2$ we also have

$$C_{Tot}(W) \ge C_{N.S.}(W) \in \Omega(\log(\max\{In^k(W), In^k(W^c)\}) - n^{k-1})$$

where W^c is the complement of W in $\mathbf{R}^n$.

Sketch of proof: A suitable modification of the main proof in [3] gives property i). As for ii) and iii) we observed in 1.13 that the non-scalar complexity of linear affine or quadratic varieties is bounded by the dimension of the space. Thus, by means of the good behaviour of the non-scalar complexity with respect to intersections, for every linear affine or quadratic algebraic variety A we get

$$n + C_{N.S.}(W) \ge C_{N.S.}(W \cap A) \in \Omega(\log(B_0(W \cap A) - n)$$

which leads to the wanted property. Property iv) follows from a similar argument.

Remarks 2.5:

i) This theorem extends the main one in [3] since $C_{Tot}(W) \geq C_{N.S.}(W) \in \Omega(log(In(W))) - n)$ and as $\mathbf{R}^n$ is an affine subvariety of itself, $In(W) \geq B_0(W)$.

ii) As $In(W) > B_0(W)$ in many cases, the above theorem may improve previously obtained lower bounds. For instance, consider in $\mathbf{R}^3$ the closed semialgebraic subset W_N given in Example 2.2.1 (c.f. Fig. 1).

In fact, since $In(W_N) = 2 \times N$, by cutting with the plane $\{Z = 0\}$, we get from Theorem 2.4 that the complexity of the example in 2.2.1 is the logarithm in base two of the number of meridians.

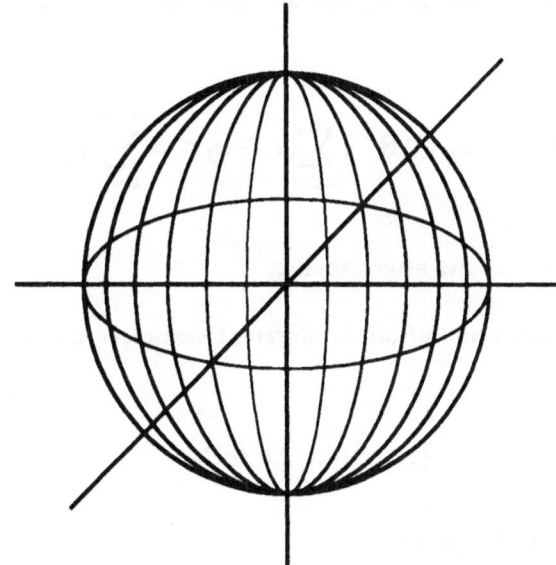

Fig.1.- A connected semialgebraic set of high first intersection number.

Similar procedures of intersecting semialgebraic subsets with algebraic subvarieties of $\mathbf{R}^n$ turn out to be very accurate to give answers to the above examples.

2.6.- Solving problem 2.1: Using the techniques of Theorem 2.4 we can state the following result concerning Largest Empty Circle, Maximum Gap and Even Distribution problems:

Theorem 2.6.1. *The second intersection number of W_N, the semialgebraic set of the E.D.P. with N points (c.f. 2.1) is $In^2(W_N) \geq N!$ and consequently (by 2.4) we have:*

   i) $C_{N.S.}(W_N) \in \Omega(N \cdot \log N)$

   ii) $C_{Tot}(W_N) \in \Omega(N \cdot \log N)$ in the model of algebraic computation trees.

   Therefore, L.E.C., MAX GAP, E.D.P. $\in \Omega(N \cdot \log N)$.

Sketch of proof: The proof of Theorem 2.6.1 has a very simple structure. First, we look for a quadratic algebraic variety ϵ_N in $\mathbf{R}^N$, an ellipsoid, such that it intersects W_N in $N!$ connected semialgebraic components. For every $\sigma \in \Sigma_N$, we define the semialgebraic set:

$$W_{\sigma,N} := \{(x_1,\ldots,x_N) \in \mathbf{R}^N \mid 0 \leq x_{\sigma(i+1)} - x_{\sigma(i)} \leq 1, i = 1,\ldots,N-1\}$$

getting the decomposition $W_N = \bigcup_{\sigma \in \Sigma_N} W_{\sigma,N}$.

Moreover, we denote by ϵ_N the quadratic algebraic subvariety of $\mathbf{R}^N$ given by

$$\epsilon_N = \{(x_1,\ldots,x_N) \in \mathbf{R}^N \mid \sum_{i<j}(x_i - x_j)^2 = \sum_{k=1}^{N-1} k.(N-k)^2\}$$

The Theorem follows after proving:

Claim 1. *The collection of all the different semialgebraic connected components of $W_N \cap \epsilon_N$ is*

$$\{W_{\sigma,N} \cap \epsilon_N \mid \sigma \in \Sigma_N\}$$

and therefore $I_n{}^2(W_N) \geq N!$.

The proof of this claim is a consequence of the following more technical assertion:

Claim 2. *Consider $k_1,\ldots,k_{N-1} \in [0,1]$ a collection of N-1 real numbers. Define*

$$H := \{(\sum_{i=s}^{t} k_i)^2 \mid 1 \leq s \leq t \leq N-1\}$$

Then, $\sum_{k \in H} k \leq \sum_{r=1}^{N-1} r.(N-r)^2$
and equality holds if and only if $k_1 = \cdots = k_{N-1} = 1$.

2.7.- Solving the other examples in 2.2.

2.7.1.- The example in 2.2.2 can be seen as in the following picture:

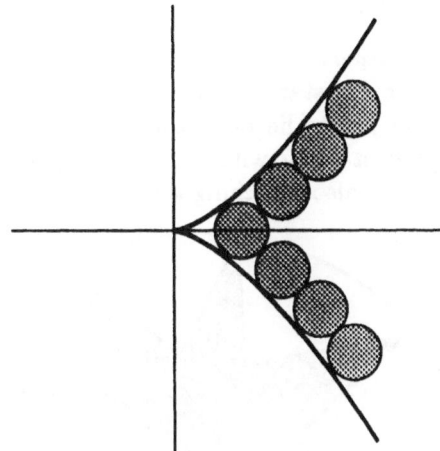

Fig. 2.- A connected semialgebraic set of high third intersection number

Since it is a connected semialgebraic set of high third order degree, this is an example of how to apply Theorem 2.4, iv): both the number of variables and the degree of the polynomial defining the curve are constant, and we get as lower bound of its complexity the logarithm in base two of the number of disks.

2.7.2.- Pseudo-regular polygons By a pseudo-regular polygon in $\mathbf{R}^2$ we mean a convex polygon whose vertices lie in a certain conic.

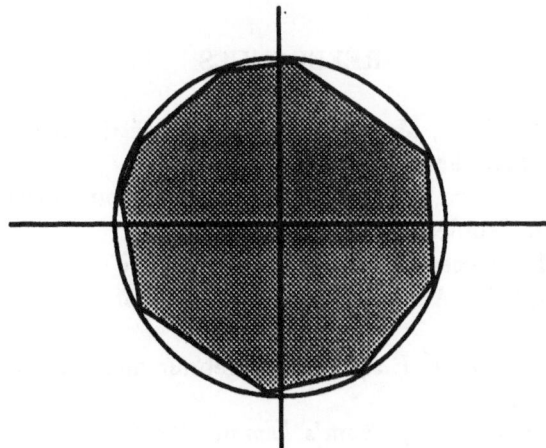

Fig. 3.- A convex polygon of high second order intersection number.

For every pseudo-regular polygon P_N of N vertices, we observe that

$In^2(P_N) \geq N$, therefore implying that $C_{N.S.}(P_N) \in \Omega(\log N)$. Because of 2.2.3 we match the complexity of the problem.

2.7.3.- Pseudo-Regular Starshaped sets

By a pseudo-regular star we mean those starshaped sets in $\mathbf{R}^2$, such that all its extreme points lie on a conic. Again, we conclude that for every pseudo-regular star S_N with N extreme points, $In^2(S_N) \geq N$ and $C_{N.S.}(S_N) \in \Omega(log N)$, also matching the wanted complexity in 2.2.4

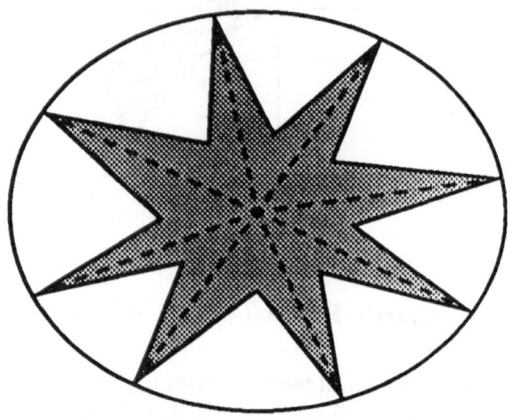

Fig. 4.- A connected star-shaped set of high second intersection number.

REFERENCES

[1] Bochnack, J., Coste, M., Roy, M. F., "Géométrie Algébrique Réelle," Springer-Verlag, 1988.

[2] Borodin, A., Munro, I., "The Computational Complexity of Algebraic and Numeric Problems," Elsevier, 1975.

[3] Ben-Or, M., *Lower Bounds for Algebraic Computation Trees*, in "Proc.15-th ACM Annual Symp. on Theory of Comput.," 1982., pp. 80–86.

[4] Benedetti, R., Risler, J. J., "Real Algebraic and Semialgebraic Sets," Hermann, 1990.

[5] Coste, M., Roy, M. F., *Thom's Lemma, the Coding of Real Algebraic Numbers and the Computation of Real Semialgebraic Sets*, Journal of Symbolic Computation 5 (1988), 121–129,.

[6] Cucker, F., Pardo, L. M., Raimondo, M. Recio, T., Roy, M.F., *On the Computation of the Local and Global Analytic Structure of a Real*

Algebraic Curve, in "Proc. of A.A.E.C.C.-5," Springer Lect. Notes in Comp. Science.356, 1989, pp. 161–182.

[7] von zur Gathen, J., *Algebraic Complexity Theory*, Technical Report of the Department of Computer Science. University of Toronto, (1988).

[8] Grigor'ev, D. Yu., *Lower Bounds in Algebraic Complexity Computational Complexity*, Journal of Soviet Math. **29** (1986), 1388–1425.

[9] Manber, U., Tompa, M., *Probabilistic, Nondeterministic and Alternating Decision Trees*, Tech. Report No. 82–03–01, Univ. of Washington., (1982).

[10] Montaña, J. L. Pardo, L. M. Recio, T., *The Complexity of Semialgebraic Sets*, Preprint 3/1990 - Dept. de Matemàticas - Univ. de Cantabria, (1990).

[11] Pardo, L. M. Recio, T., *Rabin's Width of a Complete Proof and the Width of a Semialgebraic Set*, in "Proc. EUROCAL'87," Springer Lect. Notes in Comp. Science 378, 1989.

[12] Preparata, F. P. Shamos, M. I., "Computational Geometry, An Introduction," Texts and Monographs in Computer Science, Springer - Verlag, 1985..

[13] Rabin, M. O., *Proving Simultaneous Positivity of Linear forms*, J. of Comp. and Sys. Sciences **6** (1972), 639–650.

[14] Schürmann, F., *Gradunabhängigen Stanken für die Topologie Semi-Algebraischer Mengen*, Inaugural-dissertation zur Erlangun des Doktorgrades, Münster Universität, (1988).

[15] Steele, J. N. Yao, A. C., *Lower Bounds for Algebraic Decision Trees*, J. Algorithms **3** (1982), 1–8.

[16] Strassen, V., *Polynomials with rational coefficients which are hard to Compute*, SIAM J. on Comp. **3** (1974), 128–149.

[17] Yomdin, J., *Metric Properties of semialgebraic sets and mappings and their applications in smooth analysis*, in "Géométrie Algébrique et Applications III," Travaux en cours 24, Hermann, 1987.

José L. Montaña, Luis M. Pardo, Tomás Recio
Departamento de Matemáticas, Estadística y Computación
Facultad de Ciencias - UNIVERSIDAD DE CANTABRIA
Avda. de los Castros, s/n
39071- SANTANDER - SPAIN.

Géométrie et Interpretations Génériques un Algorithme

BERNARD MOURRAIN [*]

Abstract. The aim of this paper is to treat geometric configurations of points in a way which is as "invariant" as possible, in order to be able to interpret calculus for instance when one want to prove geometric properties of a figure. The algebraic background of this problem consists in selecting good irreducible components of the variety defined by the hypotheses and in reducing the conclusion on these sets. Instead of dealing directly with polynomials as in Ritt-Wu method, we look for geometric objects and limite ourself to relations of colinearity. A natural space where such objects can be treated easily is $\wedge E$ (the exterior algebra of E). Extending it to $\wedge E \otimes S(\wedge E)$ (symmetric algebra of $\wedge E$) allows us to transform two conditions on a vector to a single one, equivalent to the first. We translate this method in the case of "formal" vectors and show that an irreducible component of the corresponding variety plays a special part for the geometric configurations. It is called the generic component. This leads to an algorithm which reduces an element of the "formal" space $\wedge F \otimes S(\wedge F)$ to zero if and only if the corresponding geometric property is true on the generic component.

1 Introduction.

En géométrie "classique", deux démarches s'opposent depuis maintenant quelques siècles. Faut-il croire en la puissance des objets géométriques ou au contraire aux vertus de l'algèbre? Suivant cette dernière idée, la géométrie analytique traduit toute propriété d'une configuration de points en termes d'équations polynômiales par rapport aux coordonnées de ces points: $A_1(x_1, x_2), A_2(x_3, x_4), \ldots, A_n(x_{2n-1}, x_{2n})$. Cette méthode se prête bien aux démarches algébriques d'éliminations, par exemple dans le cadre de vérifications automatiques de propriétés géométriques.

---

[*]Je voudrais remercier tout particulèrement M. Demazure pour ses conseils et remarques propices à un formalisme précis et rigoureux, me montrant par la même occasion que, lorsque certains objets mathématiques deviennent nécessaires, il suffit de les appeler par leur nom pour éclaircir la situation.

Ainsi les hypothèses d'un théorème traduites sous forme d'éléments $h_1 = 0$, ..., $h_l = 0$ de $k[x_1,\ldots,x_{2n}]$ (k étant le corps de base en général algébriquement clos) définissent dans k^{2n} une variété algébrique V. Cette variété se décompose de façon minimale en composantes irréductibles

$$V = V_1 \cup \ldots \cup V_p.$$

Soit P_i l'idéal premier de $k[x_1,\ldots,x_{2n}]$ associé à la composante V_i et I l'idéal engendré par $h_1,\ldots h_l$. Une décomposition primaire minimale de la racine de l'idéal I est de la forme

$$\sqrt{I} = P_1 \cap \ldots \cap P_p.$$

Dans ce cadre-là, à quoi correspond la démonstration d'un théorème g? (voir [5])

La propriété g est vraie

si et seulement si sous les hypothèses $h_1 = 0,\ldots,h_l = 0$ et sous certaines conditions de non-dégénérescence $s \neq 0$, g est nul.

si et seulement si $s.g \in \sqrt{I}$ mais $s \notin \sqrt{I}$.

si et seulement si g est nul sur une des composantes de V.

Ainsi une telle variété se décompose en composantes de dégénérescence ($s = 0$) et en composantes où les propriétés cherchées sont vraies. Mais étant donnée une propriété $p = 0$, comment être sûr qu'elle soit ou qu'elle ne soit pas vraie sur une composante non dégénérée? En effet une condition de dégénérescence $s \neq 0$ pour un théorème $g = 0$ peut se transformer en un théorème $s = 0$ sous la condition $g \neq 0$. Prenons un exemple: Soient A, B, C trois points du plan et D un point de (A, C) et tel que (D, B) orthogonale à (A, C). Choisissons comme coordonnées $A(0,0)$, $B(0,1)$, $C(x_1, x_2)$, $D(x_3, x_4)$. Les hypothèses se traduisent par $x_4 x_1 - x_2 x_3 = h_1 = 0$ et $x_4 x_2 + x_1 x_3 - x_2 = h_2 = 0$. La conclusion $AB = BC$ se traduit elle par $c = x_1^2 + (x_2 - 1)^2 - 1$. On a alors

$$(2x_3^2 + 2x_4^2 - 2x_4).c = (2h_1 - 2x_1 + 4x_3)h_1 + (2h_2 + 2x_2 - 4x_4)h_2$$

(voir [5]). Donc, $AB = BC$ est vraie à condition que $(2x_3^2 + 2x_4^2 - 2x_4) \neq 0$. (C'est-à-dire $AD^2 + DB^2 \neq AB^2$. Dans ce cas $A = B$).

Le problème consiste donc à séparer les composantes dégénérées des autres en respectant les positions générales que doivent vérifier les points et qui découlent implicitement de l'énoncé des hypothèses. Ici par exemple A, B, C sont choisis arbitrairement et donc $A \neq B$ (sauf cas de dégénérescence), la propriété cherchée n'est plus vérifiée.

Une solution consiste à localiser l'idéal I par une partie multiplicative S qui rencontre les composantes dégénérées.

$$S^{-1}.\sqrt{I} = \cap_{P_i \cap S = \emptyset} S^{-1}.P_i$$

Dans la pratique, il suffit par exemple de prendre certaines coordonnées de points comme des paramètres u_i pouvant être fixés presque librement. La partie multiplicative S sera alors $k[u_1, \ldots, u_d] - \{0\}$. En localisant par S, on ne s'autorise aucune relation de dégénéréscence entres les paramètres $u_1, \ldots, u_d$. (voir la méthode de Wu [6]).

L'approche analytique permet donc de se placer dans un cadre correct où l'on ne risque pas de démontrer des propriétés dégénérées d'une figure si les paramètres sont convenablement choisis. Pourtant la structure des objets géométriques disparaît dès les premiers calculs [5].

Dans les paragraphes suivants, nous allons voir comment on peut manipuler directement ces objets géométriques en essayant de faire apparaître naturellement les paramètres de libertés d'une figure, ceci afin d'en déduire les propriétés géométriques non dégénérées qu'elle nous cacherait.

2   Elimination géométrique.

Un environnement naturel.

Soit E un k-espace vectoriel de dimension n $(n > 0)$. L'espace projectif associé est $\mathbf{P}^{n-1} = \mathbf{P}(E)$.

Les configurations que nous nous proposons d'étudier sont celles définies par des relations de colinéarités. Une manière naturelle d'exprimer qu'un vecteur $v \in E$ est lié avec $v_1, \ldots, v_d$ est d'écrire que $v \wedge v_1 \wedge \ldots \wedge v_d = 0$. On se place dans $\wedge E$. A un sous-espace F de E de base $(v_1, \ldots, v_d)$ on associe l'élément $v_1 \wedge \ldots \wedge v_d$ de $\wedge E$. (Pour être plus précis, on devrait associer à F un élément de $\mathbf{P}(\wedge E)$. Ceci sera pris en compte implicitement par la suite en manipulant des polynômes homogènes). Ce qui nous intéresse ici est plutôt lié aux fonctions définies sur les éléments de $\wedge E$ qu'aux éléments eux-mêmes.

Nous nous placerons donc dans $M = \wedge E \otimes_k C$ où C est une k-algèbre commutative telle que $\wedge E$ s'injecte dans C. Nous noterons $(v_1, \ldots, v_d)$, l'image de $v_1 \wedge \ldots \wedge v_d$ par cette injection. Donnons quelques exemples:

- On peut prendre pour C l'algèbre des fonctions $\wedge E$ dans k. Elle contient le dual $\wedge E^*$ qui s'identifie à $\wedge E$ par l'isomorphisme suivant:

$$\begin{array}{cccc} \psi & \oplus_l \wedge^l E & \to & \oplus_l (\wedge^{n-l} E)^* \\ & v_1 \wedge \ldots \wedge v_l & \mapsto & (t_{l+1} \wedge \ldots \wedge t_n \mapsto |v_1, \ldots, v_l, t_{l+1}, \ldots, t_n|). \end{array}$$

(Ceci dépendant du déterminant $|\ |$ choisi).

- Comme autre exemple, nous prenons pour C l'algèbre symétrique $\mathbf{S}(\wedge E)$ de $\wedge E$ (c'est-à-dire l'agèbre tensorielle de $\wedge E$ quotientée par l'idéal des relations $x \otimes y - y \otimes x$). Cet exemple est très proche du précédent.

- Un dernier exemple: $C = \mathbf{S}(\wedge E) \otimes \mathbf{S}(\wedge E) \otimes \ldots \otimes \mathbf{S}(\wedge E)$. Ici, il existe plusieurs injections de $\wedge E$ dans C correspondant à chaque choix d'un facteur $\mathbf{S}(\wedge E)$.

Pour les différentier, nous les noterons $(v_1, \ldots, v_d)_1, (v_1, \ldots, v_d)_2, \ldots$ suivant l'ordre du facteur choisi.

Dérivations.

Dans ce qui suit, nous prenons pour C n'importe quelle algèbre commutative et pour L un C-module libre de type fini. Les éléments de C seront notés entre parenthèses. (Nous prendrons ultérieurement les notations $L = E \otimes C$ et $M = \wedge_C L = \wedge E \otimes C$).

Considérons l'application suivante:

$$\begin{array}{ll} \delta & L \times L \to L \times L \\ & (x, y) \mapsto (x, x + y). \end{array}$$

L'espace $L \times L(\sim L \oplus L)$ se plonge dans $\wedge_C(L \oplus L)$ qui est isomorphe au produit tensoriel gauche $\wedge_C L \otimes_C^g \wedge_C L$ (voir [1]). Le produit de deux éléments $x \otimes y$ et $x' \otimes y'$ de $\wedge_C L \otimes_C^g \wedge_C L$ est:

$$(x \otimes y) \wedge (x' \otimes y') = (-1)^{deg(y) \cdot deg(x')} (x \wedge x') \otimes (y \wedge y')$$

L'application δ s'étend canoniquement en un morphisme d'algèbres extérieures:

$$\begin{array}{ll} D & \wedge_C L \otimes_C^g \wedge_C L \to \wedge_C L \otimes_C^g \wedge_C L \\ & x \otimes 1 + 1 \otimes y \mapsto x \otimes 1 + 1 \otimes x + 1 \otimes y \end{array}$$

On a ainsi:

$$\begin{aligned} D(x_1 &\wedge \ldots \wedge x_p \otimes y_1 \wedge \ldots y_q) \\ =\ & D((x_1 \otimes 1) \wedge \ldots \wedge (x_p \otimes 1) \wedge (1 \otimes y_1) \wedge \ldots \wedge (1 \otimes y_q)) \\ =\ & x_1 \wedge \ldots \wedge x_p \otimes y_1 \wedge \ldots \wedge y_q \\ &+ x_1 \wedge \ldots \wedge x_{p-1} \otimes x_p \wedge y_1 \ldots \wedge y_q \\ &\qquad - \ldots \pm x_2 \wedge \ldots \wedge x_p \otimes x_1 \wedge y_1 \wedge \ldots \wedge y_q \\ &+ x_1 \wedge \ldots \wedge x_{p-2} \otimes x_{p-1} \wedge x_p \wedge y_1 \ldots \wedge y_q \\ &\qquad - \ldots \pm x_3 \wedge \ldots \wedge x_p \otimes x_1 \wedge x_2 \wedge y_1 \ldots \wedge y_q \\ &\ \ \vdots \\ &+ 1 \otimes x_1 \wedge \ldots \wedge x_p \wedge y_1 \wedge \ldots \wedge y_q \end{aligned}$$

De façon générale pour tout élément Z de $\wedge_C L \otimes_C^g \wedge_C L$ de bi-degré (p, q), on notera suivant le degré de la dernière composante

$$D(Z) = D^0(Z) + D^1(Z) + \ldots + D^p(Z)$$

Quitte à introduire un nouveau paramètre t on notera

$$D_t(Z) = D^0(Z) + t.D^1(Z) + \ldots + t^p.D^p(Z)$$

qui est l'application obtenue en étendant $(x, y) \to (x, y + t.x)$ aux $C[t]$-modules correspondants.

Nous noterons $\check{D}^i(Z) = D^{p-i}(Z)$ si Z de bi-degré (p, q). Ce terme est lui de bi-degré $(i, p + q - i)$.

Règle de Cramer.

Lemme 2.1 *Soient* $v, v_1, \ldots, v_d \in E$ *tels que* $v_1 \wedge \ldots \wedge v_d \neq 0$
Le terme de $\wedge E \otimes \wedge E$

(1)
$$\check{D}^1(v_1 \wedge \ldots \wedge v_d \otimes v) =$$
$$v_1 \otimes v \wedge v_2 \wedge \ldots \wedge v_d + \ldots + v_d \otimes v_1 \wedge \ldots \wedge v_{d-1} \wedge v$$

se factorise en $[\sum_{i=1}^d \alpha_i.v_i] \otimes v_1 \wedge \ldots \wedge v_d$ *avec* $\alpha_i \in k$ *si et seulement si* $v \wedge v_1 \wedge \ldots \wedge v_d = 0$. *Dans ce cas* $v = \alpha_1.v_1 + \ldots + \alpha_d.v_d$.

Démonstration: Si $v \wedge v_1 \wedge \ldots \wedge v_d = 0$, le vecteur v s'exprime de manière unique en fonction des (v_i) sous la forme

$$v = \alpha_1.v_1 + \ldots + \alpha_d.v_d.$$

En remplacant v dans (1) par l'expression ci-dessus, $\check{D}^1(v_1 \wedge \ldots \wedge v_d \otimes v)$ se factorise en $[\sum_{i=1}^d \alpha_i.v_i] \otimes v_1 \wedge \ldots \wedge v_d$. Si $\check{D}^1(v_1 \wedge \ldots \wedge v_d \otimes v) = [\sum_{i=1}^d \alpha_i.v_i] \otimes v_1 \wedge \ldots \wedge v_d$, en multipliant par $1 \otimes v_1$ (qui n'est pas nul) la relation précédente on obtient

$$v_1 \otimes v \wedge v_2 \wedge \ldots \wedge v_d \wedge v_1 = 0$$

donc $v \wedge v_2 \wedge \ldots \wedge v_d \wedge v_1 = 0$.

En notant la relation (1) sous la forme: $v_1(v \wedge v_2 \wedge \ldots \wedge v_d) + \ldots + v_d(v_1 \wedge \ldots \wedge v_{d-1} \wedge v)$, les termes entres paranthèses jouent les roles des (α_i).

Intersection

L'opération D permet de plus de définir l'intersection et la somme d'espaces vectoriels et de traiter le problème suivant:

Etant donnés deux sous-espaces vectoriels $\vec{E}_1, \vec{E}_2$ de $\vec{E}$, comment caractériser par une seule équation les hypothèses $v \in \vec{E}_1, v \in \vec{E}_2$?

Nous prenons maintenant $L = E \otimes_k C$ et $M = \wedge E \otimes_k C$ où C est une k-algèbre commutative et intègre contenant $\wedge E$.

L'application D est définie de $M \otimes_C M$ dans $M \otimes_C M$. Cet espace $M \otimes_C M = \wedge_C L \otimes_C \wedge_C L$ est isomorphe à $\wedge E \otimes_k \wedge E \otimes_k C$. Comme $\wedge E$

s'injecte dans C, D induit une application C-linéaire de $M \otimes_C M$ dans M, en multipliant dans C les deux derniers facteurs de $\wedge E \otimes_k \wedge E \otimes_k C$. Elle est encore notée D. par abus de notation.

Définition 2.2 Notons Δ l'ensemble des éléments m de M qui se "factorise" sous la forme suivante $m = u_1 \wedge \ldots \wedge u_l \otimes c$ avec $u_i \in E$, $c \in C \backslash \{0\}$ et $\Delta^* = \Delta \backslash \{0\}$. Ce sont les éléments diagonaux de M.

A chacun de ces éléments m de degré l on associe l'espace vectoriel engendré par $u_1, \ldots, u_l$. Cet espace vectoriel ne dépend pas de la représentation choisie. En effet si $u_1 \wedge \ldots \wedge u_l \otimes c = u'_1 \wedge \ldots \wedge u'_{l'} \otimes c'$ et si $\langle u_1, \ldots, u_l \rangle \neq \langle u'_1, \ldots, u'_{l'} \rangle$ il existe un vecteur v de E tel que $v \wedge u_1 \wedge \ldots \wedge u_l \neq 0$ et $v \wedge u'_1 \wedge \ldots \wedge u'_{l'} = 0$ donc $v \wedge u_1 \wedge \ldots \wedge u_l \otimes c = 0$, ce qui contredit l'hypothèse $c \neq 0$. Cet espace vectoriel est noté $\vec{m}$.

Nous allons démontrer le théorème suivant:

Théorème 2.3 *Soient E_1, E_2 deux éléments de Δ^* de degré p, q. Soient $\vec{E}_1$, $\vec{E}_2$ les deux espaces vectoriels associés à ces éléments.*

$$D_t(E_1 \otimes E_2) = D^0(E_1 \otimes E_2) + \ldots + t^{p-d}.D^{p-d}(E_1 \otimes E_2)$$

Le coefficient de la plus grande puissance de t (noté $\Gamma(E_1, E_2)$) est dans Δ^* et est associé à $\vec{E}_1 \cap \vec{E}_2$.
De plus

$$v \wedge E_1 = 0, \quad v \wedge E_2 = 0$$

est équivalent dans M à

$$v \wedge \Gamma(E_1, E_2) = 0.$$

Démonstration: Les éléments E_1, E_2 de Δ^* s'écrivent sous la forme $E_1 = e_1^1 \wedge \ldots \wedge e_p^1 \otimes c_1$, $E_2 = e_1^2 \wedge \ldots \wedge e_p^2 \otimes c_2$ avec $c_1, c_2 \in C^*$.

Notons $E'_1 = e_1^1 \wedge \ldots \wedge e_p^1$, $E'_2 = e_1^2 \wedge \ldots \wedge e_p^2$. Les deux espaces vectoriels $\vec{E}_1$, $\vec{E}_2$ ont donc pour bases $(e_1^1, \ldots, e_p^1)$ et $(e_1^2, \ldots, e_p^2)$ Quitte à choisir convenablement ces bases on peut supposer que

$$e_1^1 = e_{q-d+1}^2 = i_1, \ e_2^1 = e_{q-d+2}^2 = i_2, \ldots, e_d^1 = e_q^2 = i_d$$

avec $(i_1, \ldots, i_d)$ une base de $\vec{E}_1 \cap \vec{E}_2$. Posons $F_1 = e_{d+1}^1 \wedge \ldots \wedge e_p^1$, $F_2 = e_1^2 \wedge \ldots \wedge e_{q-d}^2$. Les éléments $I = i_1 \wedge \ldots \wedge i_d$ et $S = I \wedge F_1 \wedge F_2$ sont donc associés à $\vec{E}_1 \cap \vec{E}_2$ (resp $\vec{E}_1 + \vec{E}_2$). Calculons $D_t(E'_1 \otimes E'_2)$:

$$
\begin{aligned}
D_t(E'_1 \otimes E'_2) = \ & e_1^1 \wedge \ldots \wedge e_p^1 \otimes (E'_2) \\
+ \ & t \sum \pm e_1^1 \wedge \ldots \wedge e_{p-1}^1 \otimes (e_p^1 \wedge E'_2) \\
+ \ & \vdots \\
+ \ & t^p \, (e_1^1 \wedge \ldots \wedge e_p^1 \wedge E'_2).
\end{aligned}
$$

(les éléments entre paranthèses sont à considérer comme des coefficients).
Comme le produit $e_{i_1}^1 \wedge \ldots \wedge e_{i_k}^1 \wedge E_2'$ est nul dès que $1 \le i_j \le d$

$$
\begin{aligned}
D_t(E_1 \otimes E_2) = \ & I \wedge F_1 \otimes (F_2 \wedge I) \\
& + \ t \sum \pm I \wedge e_{d+1}^1 \wedge \ldots \wedge e_{p-1}^1 \otimes (e_p^1 \wedge F_2 \wedge I) \\
& \vdots \\
& + \ t^{p-d} I \otimes (F_1 \wedge F_2 \wedge I) \\
= \ & E_1 \otimes (E_2) + \ldots \pm t^{p-d} I \otimes (S)
\end{aligned}
$$

Par linéarité, le terme de plus haut degré de $D(E_1 \otimes E_2)$ noté $\Gamma(E_1, E_2)$
est donc $I \otimes [(S)c_1 c_2]$ avec $(S)c_1 c_2 \ne 0$ car C intègre. Ce terme est associé
à l'intersection des espaces vectoriels $\vec{E_1}$, $\vec{E_2}$.

Nous avons donc les équivalences suivantes

$$
\left. \begin{aligned} v \wedge E_1 &= 0 \\ v \wedge E_2 &= 0 \end{aligned} \right\} \Leftrightarrow \left. \begin{aligned} v \wedge E_1' &= 0 \\ v \wedge E_2' &= 0 \end{aligned} \right\} \Leftrightarrow v \wedge I = 0
$$

$$
\Leftrightarrow v \wedge I \otimes [(S)\ c_1\ c_2] = v \wedge \Gamma(E_1, E_2) = 0.
$$

D'où le théorème.

Cette méthode se généralise à l'intersection de plusieurs espaces vectoriels E_1, E_2, E_3, ... Pour cela, il suffit de recommencer le même calcul avec
$\Gamma(E_1, E_2)$ et E_3 qui sont dans Δ^*. En itérant ce procédé, nous avons donc
la propriété suivante:

Proposition 2.4

$$
\left. \begin{aligned} v \wedge E_1 &= 0 \\ v \wedge E_2 &= 0 \\ &\vdots \\ v \wedge E_l &= 0 \end{aligned} \right\} \text{ est équivalent dans } \wedge E \otimes C \text{ à } v \wedge \Gamma_l = 0
$$

avec $\Gamma_2 = \Gamma(E_1, E_2)$, $\Gamma_3 = \Gamma(\Gamma_2, E_3)$, ..., $\Gamma_l = \Gamma(\Gamma_{l-1}, E_l)$. *De plus*
$\forall i \in [1, \ldots, l]$, $\Gamma_i \in \Delta^*$.

Ce formalisme traitant directement d'objets géométriques dans l'algèbre extérieure introduit par Cramer en 1754 permet de définir ce qui correspond aujourd'hui, au produit régressif (voir [1]A.3 p 200, ex 16). Cet
outil permettant de représenter l'intersection d'espaces vectoriels trouve
une nouvelle présentation dans un article clé sur ce sujet [4]. L'algèbre de
Cayley munie des opérations "Intersection" et "Somme" y apparait, four-
nissant ainsi un langage simple pour décrire les relations géométriques d'in-
cidences. Le problème consistant à traduire dans ce langage les relations
polynômiales entre déterminants est abordé dans [3] (voir aussi [2]).

Ici, nous essayons d'étendre le produit régressif qui se calcule seulement
quand la somme des dimensions des espaces vectoriels est supérieure à la
dimension de l'espace ambiant E. Ceci peut être fait en étudiant les fonctions définies sur les éléments de $\wedge E$, plutôt que les éléments eux-mêmes
et en prenant une algèbre de scalaires contenant $S(\wedge E)$.

3 L'idéal associé aux configurations.

Soit k un corps algébriquement clos. Plaçons nous dans un espace vectoriel E de dimension n (par exemple k^n avec $n > 0$). Considérons m vecteurs génériques $X_1, \ldots, X_m$ de coordonnées $X_i = \begin{pmatrix} x_{1,i} \\ \vdots \\ x_{n,i} \end{pmatrix}$.

On supposera que $m < n$ (sinon on étend k^n à k^m).

Soit $R = k[x_{i,j}]_{1 \leq i \leq n, 1 \leq j \leq m} = k[X_1, \ldots, X_m]$.

Désignons par F le R-module $F = E \otimes_k R$.

Pour relier le caractère géométrique des sections précédentes à un environnement algébrique, nous nous plaçons maintenant dans $\wedge_R F \otimes S(\wedge_R F)$ ou dans un espace où il se représente facilement.

Pour cela, prenons une base de vecteurs génériques $\mathbf{T} = (T_1, \ldots, T_n)$.

On notera ces nouveaux vecteurs $T_i = \begin{pmatrix} t_{1,i} \\ \vdots \\ t_{n,i} \end{pmatrix}$, les $t_{i,j}$ étant de nouvelles

variables. On notera $R[\mathbf{T}] = R[t_{i,j}]$.

Tout élément $(X_1, \ldots, X_l)$ de $\wedge_R F$ s'identifie à l'élément

$$|X_1, \ldots, X_l, T_{l+1}, \ldots, T_n|$$

de $R[\mathbf{T}]$. En utilisant les propriétés universelles de l'algèbre symétrique, à chaque élément de $S(\wedge_R F)$ correspond un élément de $R[\mathbf{T}]$. On prendra cette algèbre comme algèbre des coefficients et on travaillera dans $\wedge_R F \otimes S(\wedge_R F)$ plongé dans $\wedge_R F \otimes R[\mathbf{T}]$. (Les coefficients sont notés entre parenthèses).

Par abus de langage, l'idéal de R engendré par un élément de $\wedge_R F \otimes S(\wedge_R F)$ désignera celui engendré par ses coordonnées dans la base canonique. (Exemple: L'élément $X_1 \wedge \ldots \wedge X_k$ de $\wedge_R F$ a pour coordonnées dans la base canonique les mineurs d'ordre k des colonnes $X_1, \ldots, X_k$).

Nous allons nous intéresser aux points $a_1, \ldots, a_m$ correspondant à des substitutions qui annulent certains éléments de $\wedge_R F \otimes S(\wedge_R F)$. En particulier nous nous restreindrons aux variétés de $(k^n)^m$ définies par des relations de la forme $X_{i_1} \wedge \ldots \wedge X_{i_k} = 0$. On peut supposer que les vecteurs X_{i_j} sont rangés par ordre décroissant des indices ($i_j > i_l$ si $j < l$).

Dans ce cas, les polynômes qui seront manipulés pourront être supposés homogènes par rapport à chaque vecteur (configuration projective de points).

Considérons un ensemble d'hypothèses de la forme ci-dessus et notons H_i l'ensemble formé par celles dont le plus grand vecteur est X_i.

A de telles hypothèses, correspond un ensemble de configurations géométriques de points $a_1, \ldots, a_m$. Le problème que l'on se propose de regarder est de dégager la ou les composantes associées à des figures non-dégénérées.

4   La composante générique

La variété $V_m = V[H_1 = 0, \ldots, H_m = 0]$ de E^m se décompose en plusieurs composantes irréductibles:

$$V_m = \cup_{1 \leq i \leq s} V_m^i.$$

Pourtant une de ces composantes joue un rôle particulier pour l'énoncé géométrique correspondant aux hypothèses. Nous allons la caractériser de la façon suivante:

Désignons par p_j la projection p_j $E^m \to E^j$ sur les j premières composantes. Pour toute composante irréductible V_m^i, désignons par $d_j = dim[p_j(V_m^i)]$ (dim: dimension)

Théorème 4.1 *Il existe une seule composante irréductible parmi les (V_m^i) telle que $(d_1, d_2, \ldots, d_m)$ soit maximal pour l'ordre lexicographique.*
Nous l'appellerons la composante générique.

Pour cela nous allons utiliser quelques lemmes:

Lemme 4.2 *Soit W une composante irréductible de la variété $V_{m-1} = V[H_1 = 0, \ldots, H_{m-1} = 0]$. Il existe un ouvert non-vide U de W et un élément $\Gamma \in \wedge_R F \otimes S(\wedge_R F)$ ne dépendant que de $X_1, \ldots, X_{m-1}$ tel que sur U,*

$$\{H_m = 0\} \Leftrightarrow X_m \wedge \Gamma = 0$$

De plus

$$\forall (a_1, \ldots, a_{m-1}, a_m) \in U, \ \Gamma(a_1, \ldots, a_{m-1}) \in \Delta^*.$$

Démonstration: La composante W est associée à un idéal premier $P = I(W)$ de R. Les hypothèses de H_m sont:

$$\begin{cases} X_m \wedge Z_1 = 0 \\ X_m \wedge Z_2 = 0 \\ \vdots \\ X_m \wedge Z_l = 0 \end{cases}$$

avec $Z_i = X_{i,1} \wedge \ldots \wedge X_{i,k_i}$. Les vecteurs $X_{i,j}$ sont rangés par ordre décroissant. Plaçons nous dans W.

Parmi ces hypothèses, celles dont les Z_i appartiennent à P (c'est-à-dire celles dont tous les coefficients des Z_i sont dans P) sont inutiles car nulles sur W. On peut donc supposer qu'elles n'apparaissent pas ci-dessus.

Pour tout élément Y de $\wedge_R F \otimes S(\wedge_R F)$ et tout élément Z de $\wedge_R F$, désignons par $\Gamma(Y, Z)$ le plus grand terme du développement suivant:

$$D(Y \otimes Z) = Y \otimes Z + \ldots + D^l(Y \otimes Z) + \ldots$$

dont les coordonnées ne sont pas toutes dans P.

D'après le théorème 3.3, pour toute configuration $(a_1, \ldots, a_m)$ appartenant à l'ouvert non-vide de W tel que $\Gamma(Z_1, Z_2) \neq 0$ on a

$$\left.\begin{array}{l} a_m \wedge Z_1(a_1, \ldots, a_{m-1}) = 0 \\ a_m \wedge Z_2(a_1, \ldots, a_{m-1}) = 0 \end{array}\right\} \Leftrightarrow a_m \wedge \Gamma(Z_1, Z_2)(a_1, \ldots, a_m) = 0$$

avec en plus $\Gamma(Z_1, Z_2)(a_1, \ldots, a_m)$ diagonal sur cet ouvert ($\in \Delta^*$).

Ceci nous permet de construire comme pour la proprosition 3.4, une condition équivalente aux hypothèses $\{H_m = 0\}$ sur un ouvert de W.

Posons $\Gamma_2 = \Gamma(Z_2, Z_1), \Gamma_3 = \Gamma(Z_3, \Gamma_2), \ldots, \Gamma_l = \Gamma(Z_l, \Gamma_{l-1})$. Notons U l'ouvert de W tel que $\Gamma_2 \neq 0, \ldots, \Gamma_l = \Gamma \neq 0$. Cet ouvert est de la forme $U' \times E$ car les polynômes Γ_i ne dépendent pas de X_m.

Sur U, le système $\{H_m = 0\}$ est donc équivalent à $X_m \wedge \Gamma = 0$ (proposition 3.4) et l'élément Γ de $\wedge_R F \otimes \mathbf{S}(\wedge_R F)$ se substitue en un élément diagonal non-nul de la forme

$$r_1 \wedge \ldots \wedge r_d \otimes c$$

avec $r_1 \wedge \ldots \wedge r_d \neq 0$ et $c \in \mathbf{S}(\wedge_R F) \backslash \{0\}$

Notons $U^\circ = U \cap V(X_m \wedge \Gamma = 0)$. Considérons la substitution

$$\begin{array}{rccc} s_m : & R[\mathbf{T}] & \to & R[\mathbf{T}] \\ & X_m & \mapsto & \check{D}^1(\Gamma \otimes X_m) \end{array}$$

Lemme 4.3 *Soit*

$$P^\circ = \{g \in R = k[X_1, \ldots, X_m] \ g(X_1, \ldots, X_{m-1}, s_m(X_m)) \in P.R[\mathbf{T}]\}$$

L'idéal $I(U^\circ)$ des polynômes nuls sur U° est l'idéal premier P°.

Démonstration: Soit $g \in I(U^\circ)$ homogène par rapport à chaque vecteur X_i. On peut s'y ramener car il s'agit de configurations projectives de points.

Notons $h(X_1, \ldots, X_m) = g(X_1, \ldots, X_{m-1}, s_m(X_m)) \in R[\mathbf{T}]$

Soit $(a_1, \ldots, a_{m-1})$ un élément quelconque de $p_{m-1}(U) = U'$. En utilisant les notations du lemme 5.2,

$$\Gamma(a_1, \ldots, a_{m-1}) \otimes X_m = (r_1 \wedge \ldots \wedge r_d \otimes X_m).c$$

On a alors

$$\begin{aligned} s_m(X_m) &= \check{D}^1(\Gamma(a_1, \ldots, a_{m-1}) \otimes X_m) \\ &= r_1.(X_m, r_2, \ldots, r_d)c + \ldots + r_d.(r_1, \ldots, r_{d-1}, X_m)c \end{aligned}$$

Notons

$$\begin{aligned} \alpha_i(X_m) &= (r_1, \ldots, r_{i-1}, X_m, r_{i+1}, \ldots, r_d), \\ a'_m(X_m) &= \textstyle\sum_{i=1}^d \alpha_i(X_m).r_i. \end{aligned}$$

Rappelons-nous qu'un terme (...) s'identifie à un élément du dual de $\wedge E$. Pour toute valeur de $a_m \in E$ et pour toute valeur $\overline{\alpha_i(a_m)}$ des fonctions $\alpha_i(a_m) \in \wedge E^*$, on pose $\overline{a'_m} = \sum_{i=1}^{d} \overline{\alpha_i(a_m)}.r_i$ qui est un élément de E. Comme

$$\overline{a'_m} \wedge r_1 \wedge \ldots \wedge r_d \otimes c = \overline{a'_m} \wedge \Gamma(a_1, \ldots, a_{m-1}) = 0$$

le lemme 5.2 nous dit que $(a_1, \ldots, a_{m-1}, \overline{a'_m}) \in U^\circ$.

Donc $g(a_1, \ldots, a_{m-1}, \overline{a'_m}) = 0$. Cette égalité étant vraie pour toute valeur des fonctions $(\alpha_i(a_m))_i$,

$$g(a_1, \ldots, a_{m-1}, \sum_{i=1}^{d} \alpha_i(X_m).r_i) = 0 \Leftrightarrow g(a_1, \ldots, a_{m-1}, \sum_{i=1}^{d} \alpha_i(X_m).r_i.c) = 0$$

(par homogénéité suivant le dernier vecteur).

$$g(a_1, \ldots, a_{m-1}, s_m(X_m)) = h(a_1, \ldots, a_{m-1}, X_m) = 0$$

Nous voyons donc que les coefficients de h dans $R[\mathbf{T}]$ s'annulent pour toute configuration $(a_1, \ldots, a_{m-1}, a_m) \in U' \times E = U$ (U ouvert non-vide donc dense de la composante W). Comme $P = I(W)$, $h \in P.R[\mathbf{T}] \Leftrightarrow g \in P^\circ$.

$$I(U^\circ) \subset P^\circ.$$

Réciproquement, prenons $g \in P^\circ$ homogène par rapport à chaque vecteur X_i (de degré p en X_m).

$$g(X_1, \ldots, X_{m-1}, s_m(X_m)) \in P.R[\mathbf{T}].$$

Pour tout $(a_1, \ldots, a_{m-1}, a_m) \in U^\circ$,
$g(a_1, \ldots, a_{m-1}, s_m(a_m))$
    $= 0$ car $(a_1, \ldots, a_{m-1}, a_m) \in U^\circ \subset W = V[P]$ et
        $g(X_1, \ldots, X_{m-1}, s_m(X_m)) \in P.R[\mathbf{T}]$
    $= g(a_1, \ldots, a_{m-1}, a_m.(r_1, \ldots, r_d)c)$ d'après le lemme 3.1
        $(a_m \wedge r_1 \wedge \ldots \wedge r_d = 0)$
    $= g(a_1, \ldots, a_{m-1}, a_m)(r_1, \ldots, r_d)^p c^p$ par homogénéité.
d'où $g(a_1, \ldots, a_{m-1}, a_m) = 0$ car $(r_1, \ldots, r_d)c \neq 0$ et $g \in I(U^\circ)$.

$$P^\circ = I(U^\circ).$$

Cet idéal est de plus premier: Soient $g, f \in k[X_1, \ldots, X_m]$ tels que $f.g \in P^\circ$. C'est-à-dire $f(X_1, \ldots, X_{m-1}, s_m(X_m)).g(X_1, \ldots, X_{m-1}, s_m(X_m)) \in P.R[\mathbf{T}]$. Comme $P.R[\mathbf{T}]$ est premier,

- soit $f(X_1, \ldots, X_{m-1}, s_m(X_m)) \in P.R[\mathbf{T}]$

- soit $g(X_1, \ldots, X_{m-1}, s_m(X_m)) \in P.R[\mathbf{T}]$.

En d'autres termes, l'un des deux polynômes f ou g est dans P^o. L'idéal P^o est donc associé à une composante irréductible W^o de la variété $W \cap V[H_m = 0]$ Q. E. D.

Corollaire 4.4 *Etant donnée une composante irréductible W_{m-1} de la variété $V_{m-1} = V[H_1 = 0, \ldots, H_{m-1} = 0]$, il existe une seule composante W_m de $W_{m-1} \cap V[H_m = 0]$ dominante sur $p_{m-1}(W_{m-1})$.*

Démonstration: Reprenant les notations des lemmes précédents, U_{m-1} désigne l'ouvert U de la composante $W = W_{m-1}$ défini au lemme 5.2 et W_m la composante générique associée à $P^o = I(U^o)$ (lemme 5.3). La composante W_m se projette sur une partie de E^{m-1} qui contient $p_{m-1}(U_{m-1})$ car pour tout $(a_1, \ldots, a_{m-1}) \in p_{m-1}(U_{m-1})$, on peut trouver un point a_m vérifiant $a_m \wedge r_1 \wedge \ldots \wedge r_d = 0$ donc tel que $(a_1, \ldots, a_{m-1}, a_m) \in W_m$. La projection de $p_{m-1}(W_m)$ comme $p_{m-1}(U_{m-1})$ est donc dense dans $p_{m-1}(W_{m-1})$. (On a $U_{m-1} = U'_{m-1} \times E$ et $W_{m-1} = W'_{m-1} \times E$ avec $p_{m-1}(U_{m-1}) = U'_{m-1}$ et $p_{m-1}(W_{m-1}) = W'_{m-1}$).

Le reste des composantes de $W_{m-1} \cap V[H_m = 0]$ se trouve d'après le lemme 5.2 dans

$$\{V[\Gamma_2.\Gamma_3 \ldots \Gamma_l = 0] \cap W_{m-1}\} \cap V[H_m = 0].$$

Toute composante irréductible de $V[\Gamma_2.\Gamma_3 \ldots \Gamma_l = 0] \cap W_{m-1}$ est de dimension strictement plus petite que $dim(W_{m-1})$ car $\Gamma_i \notin P$ donc la projection de toute composante de $\{V[\Gamma_2.\Gamma_3 \ldots \Gamma_l = 0] \cap W_{m-1}\} \cap V[H_m = 0]$ ne peut être dense dans $p_{m-1}(W_{m-1})$.

Nous appellerons W_m la composante générique au-dessus de $p_{m-1}(W_{m-1})$ Q. E. D.

Ceci nous permet maintenant de démontrer le théorème 5.1. Construisons d'abord la composante en question. On peut supposer que $H_1 = \emptyset$. (sinon $H_1 = X_1$ et l'hypothèse $X_1 = 0$ irait à l'encontre du bon sens d'une description projective de points).

Prenons pour $W_1 = E = V[H_1 = 0]$. Supposons construit W_i comme une composante de $V[H_1 = 0, \ldots, H_i = 0]$ dans E^i. On prend alors pour W_{i+1} la composante dans E^{i+1} de $V[H_1 = 0, \ldots, H_i = 0, H_{i+1} = 0]$ qui est dominante sur W_i (Corollaire 5.4). Notons $d_i^+ = dim(W_i)$.

Soit Y une variété irréductible de $V[H_1 = 0, \ldots, H_m = 0]$. On notera $Y_i = p_i(Y)$ et $d_i = dim(Y_i)$.

Nous allons montrer par récurrence que $(d_1, \ldots, d_i) \leq (d_1^+, \ldots, d_i^+)$ (pour l'ordre lexicographique) avec égalité si et seulement si $Y_i = W_i$. La variété Y_1 est dans $E = W_1$. On a donc $d_1 \leq d_1^+ = n$ avec égalité si $Y_1 = E$. Supposons que la propriété soit vraie pour i. Si $(d_1, \ldots, d_i) < (d_1^+, \ldots, d_i^+)$ alors $(d_1, \ldots, d_{i+1}) < (d_1^+, \ldots, d_{i+1}^+)$. Sinon $(d_1, \ldots, d_i) = (d_1^+, \ldots, d_i^+)$ et

$Y_i = W_i$ par hypothèse de récurrence. La variété irréductible Y_{i+1} de E^{i+1} est incluse dans une des composantes de $V[H_1 = 0, \ldots, H_{i+1} = 0]$. Comme $p_i(Y_{i+1}) = Y_i$, Y_{i+1} est donc dominante sur $W_i = Y_i$. D'après le corollaire 5.4 Y_{i+1} est donc incluse dans l'unique composante W_{i+1} dominante sur W_i. Donc $d_{i+1} \leq d_{i+1}^+$ et il y a égalité si $Y_{i+1} = W_{i+1}$. Cette propriété récurrente nous montre donc que W est l'unique composante de $V[H_1 = 0, \ldots, H_m = 0]$ tel que $(d_1, \ldots, d_m)$ soit maximal pour l'ordre lexicographique Q. E. D.

5   Algorithme.

Nous allons utiliser les résultats de la section précédente pour construire un algorithme permettant de démontrer des propriétés multilinéaires d'une figure géométrique. Dans un énoncé géométrique par nature triangulaire, la construction d'un point se fait à partir des points déjà construits, ce dernier étant choisi le plus génériquement possible. La composante de la variété sur laquelle on se place est donc celle qui est non-dégénérée au-dessus de celle correspondant aux points antérieurs (aucune condition supplémentaire). C'est sur cette composante générique que nous vérifierons les propriétés géométriques proposées.

Données:
1) La liste des points $(x_1, x_2, \ldots, x_m)$ de la figure correspondant à l'ordre dans lequel ils sont construits.
2) Les hypothèses traduisant pour chaque point sa position par rapport aux autres, de la forme: $\forall i \in [1, \ldots, m]$

$$\left\{ \begin{array}{l} x_i \wedge Z_0^i = 0 \\ \vdots \\ x_i \wedge Z_{k_i}^i = 0 \end{array} \right.$$

les $(Z_l^i)_l$ étant de la forme $x_{l_1} \wedge \ldots \wedge x_{l_d}$ avec $l_j < i$.

Résultat:
Des règles de substitutions de la forme $x_i \to \sum_{j<i} \alpha_j^i . x_j$ avec $\alpha_j^i \in S(\wedge_R F)$ vérifiant la propriété suivante:
Tout élément de $\wedge_R F \otimes S(\wedge_R F)$ nul sur la composante générique de la variété se réduit à 0 en appliquant successivement la règle (si elle existe) concernant x_{m-1} puis celle concernant $x_{m-2}, \ldots$ jusqu'à x_1. Réciproquement, tout polynôme se réduisant à zéro est nul sur la composante générique.

Construction: :
1. *Ranger les hypothèses par ordre décroissant des variables.* 2. *Regrouper les hypothèses suivant leur variable de tête en ensembles (éventuellement vides)* $H_1, \ldots, H_m$.

On notera $H_{i \leq p} = \cup_{i \leq p} H_i$.

3. *Traiter le cas* $H_1, H_{i \leq 2}, \ldots, H_{i \leq m-1}$.

Si aucune hypothèse n'est faite sur x_m, passer au cas suivant.

Sinon, nous supposons construites les règles de substitutions pour l'ensemble d'hypothèses $H_{i \leq m-1}$. Il faut construire celle correspondant au point x_m à partir des hypothèses

$$\{x_m \wedge Z_0^m = 0, \ldots, x_m \wedge Z_{k_m}^m = 0\}$$

Certaines hypothèses sont nulles au dessus de la composante générique et donc inutiles. Par construction, elles se réduisent à 0 par rapport aux règles concernant les points $x_1, \ldots, x_{m-1}$.

Réduire les hypothèses par rapport au système de règles existant, et ne garder que celles qui ne se réduisent pas à 0.

Dans le cas où le système restant est vide, on passe à x_{m+1}. Sinon on le note encore sous la forme 3.1.

Pour tout élément Y_1,(resp Y_2) de $\wedge_R F$ (resp $\wedge_R F \otimes S(\wedge_R F)$) ne dépendant que des variables $x_1, \ldots, x_{m-1}$, posons $\Gamma(Y_1 \otimes Y_2)$ le plus grand terme du développement suivant

$$D(Y_1 \otimes Y_2) = D^0(Y_1 \otimes Y_2) + \ldots + D^l(Y_1 \otimes Y_2) + \ldots$$

qui ne se réduit pas à 0 en appliquant les règles de substitutions déjà construites. Par construction, c'est le dernier terme qui ne s'annule pas au-dessus de la composante générique sur laquelle nous nous plaçons. C'est le même terme que celui défini au paragraphe précédent.

Calculer $\Gamma_2 = \Gamma(Z_2 \otimes Z_1)$, $\Gamma_3 = \Gamma(Z_3 \otimes \Gamma_2), \ldots,$ $\Gamma_{k_m} = \Gamma(Z_{k_m} \otimes \Gamma_{k_m-1})$.

Sur un ouvert de la composante en question, le système de départ (3) est équivalent (d'après le paragraphe précédent) à l'équation:

$$x_m \wedge \Gamma_{k_m} = 0.$$

Dériver une dernière fois pour obtenir une relation de la forme:

$$x_m.(\Gamma_{k_m}) = \check{D}^1(\Gamma_{k_m} \otimes x_m).$$

(le degré de Γ_{k_m} est d_m.) Le paragraphe précédent nous dit que la substitution $s_m : x_m \to \check{D}^1(\Gamma_{k_m} \otimes x_m)$ transforme un polynôme de $\wedge_R F \otimes S(\wedge_R F)$ nul sur la composante générique concernant les points $x_1, \ldots, x_m$ en un polynôme dont tous les coefficients dans la base canonique de $\wedge_R F \otimes S(\wedge_R F)$ sont nuls sur la composante générique pour les points $x_1, \ldots, x_{m-1}$ quel que soit x_m. Ce polynôme substitué se réduit donc à 0 par rapport aux règles correspondant aux points $x_1, \ldots, x_{m-1}$.

D'où un ensemble de règles répondant à la question pour $H_{i \leq m}$. Ce qui termine la description des différentes étapes de calculs   Q. E. D.

Remarque: Dans la pratique, on traite souvent le cas où les hypothèses précisant la position d'un point x_i de la figure de la forme $x_i \wedge Z_0^i = 0, \ldots, x_i \wedge Z_{k_i}^i = 0$, sont telles que chaque nouvelle condition ne fait descendre que de 1 la dimension de l'espace précédent. Il suffit alors de ne considérer que la première dérivation:

$$D(Z_j^i \otimes \Gamma_{j-1}^i) = D^0(Z_j^i \otimes \Gamma_{j-1}^i) + D^1(Z_j^i \otimes \Gamma_{j-1}^i).$$

ce qui simplifie beaucoup la longueur des calculs.

REFERENCES

[1] N. Bourbaki, "Eléments de Mathématiques, Algèbre," Ch 3, Hermann, Paris, 1970.

[2] Bokovski J., Sturmfels B., "Computational Synthetic Geometry," Lect. Notes Math. 1355, 1989.

[3] Crapo H., Havel T. F., Sturmfels B., Whiteley W., White N. L., *Symbolic computations in geometry*, IMA preprint series 389, Univ. of Minnesota (1988).

[4] Doubilet P., Rota G.-C., Stein J.,, *On the fondations of combinatorial theory: IX, Combinatorial methods in invariant theory*, Studies in Applied Mathematics 53 (1974), 185–216.

[5] Chou Shang Shing, "Automatic prover in Geometry," Reidel, 1988.

[6] Wu Wen-tsun, *Some Recent Advances in Mechanical Theorem Proving: After 25 years*, Contemporary Mathematics 29 (1984), 235–242, American Mathematical Society.

Bernard Mourrain
Centre de Mathématiques
Ecole Polytechnique
F 91128 PALAISEAU Cedex (France)

Canonical Bases: Relations with Standard Bases, Finiteness Conditions and Application to Tame Automorphisms

FRANÇOIS OLLIVIER

Abstract. Canonical bases for k-subalgeras of $k[x_1, \ldots, x_n]$ are analogs of standard bases for ideals. They form a set of generators, which allows to answer the membership problem by a reduction process. Unfortunately, they may be infinite even for finitely generated subalgeras. We redefine canonical bases, and for that we recall some properties of monoids, k-algebras of monoids and "binomial" ideals, which play an essential role in our presentation and the implementation we made in the IBM computer algebra system Scratchpad II. We complete the already known relations between standard bases and canonical bases by generalizing the notion of standard bases for ideals of any k-subalgebra admitting a finite canonical basis. We also have a way of finding a set of generators of the ideal of relations between elements of a canonical basis, which is a standard basis for some ordering.

We then turn to finiteness conditions, and investigate the case of integrally closed subalgebras. We show that if some integral extension B of a subalgebra A admits a finite canonical basis, we have an algorithm to solve the membership problem for A, by computing the generalized standard basis of a B-ideal. We conjecture that any integrally closed subalgebra admits a finite canonical basis, and provide partial results.

There is a simple case, but of special interest, where the complexity of computing a canonical basis is known: the case where $k[f_1, \ldots, f_n] = k[x_1, \ldots, x_n]$. We show that the canonical bases procedure give more information than previously known methods and may provide a tool for the tame generators conjecture.

Introduction

Standard bases first appeared in the work of Hironaka and became one of the main tools in computer algebra for solving systems of algebraic equations. This notion is very natural, and Janet in 1920, working on

---
Partially supported by GDR G0060 *Calcul Formel, Algorithmes, Langages et Systèmes* and PRC *Mathématiques et Informatique*.

partial differential equations, described particular sets of generators of an ideal, which are not far from standard bases theory. Canonical bases seem to have a much shorter history.

Indeed, previously known methods like that of Shannon and Sweedler (see [SS]) for solving the membership problem in the case of an k-subalgebra $k[f_1, \ldots, f_m]$, uses the ideal defining the graph of the polynomial map f associated to $f_1, \ldots, f_m$, and standard bases. Nearly at the same time, in 1986, Kapur and Madlener discovered a direct approach, introducing canonical bases (see [KM]). Independently, Robbiano and Sweedler defined the same objects, which they called SAGBI, standing for subalgebra analog of Groebner bases of ideals (see [RS]). They have shown that it is possible to translate many properties of standard bases in the vocabulary of canonical bases.

We will complete those works, by a description of a first implementation of canonical bases, which is an essential step to have a precise idea of their efficiency, and new relations with standard bases, who belong to the folklore but also have practical consequences. The main difference is that k-subalgebras of $k[n]$ do not satisfy the ascending chain condition. As a consequence, there exist finitely generated k-subalgebras with infinite canonical basis already in 2 variables, as proved by Robbiano. This serious drawback apparently discouraged Kapur and Madlener to publish their work earlier.

Anyway, it is easy to provide examples where the canonical basis computation is almost free whereas it is impossible to compute the standard basis of Shannon and Sweedler's method on any existing computer. In the worst cases, standard bases calculations may have a double exponential complexity, and for a practical point of view, it is the same as if it would never stop. It is one of the major issues in that field to determine "good cases", with reasonable complexity, starting with the works of Lazard on -1 or 0-dimensional cases (see [L] and [G]). We may think that there is a strong relation with good complexity and "nice" algebraic properties of the ideal.

For canonical bases, nothing is known yet. If we analyse the examples given by Robbiano of finitely generated k-subalgebras with infinite canonical bases, we can remark that they are not integrally closed. This situation led us to conjecture that the canonical basis is finite for any integrally closed k-subalgebra (see [O2]). At least, we can prove a general relation between the canonical basis of a subalgebra and that of its integral closure. We will provide partial results and illustrate the practical consequences of a positive answer to our conjecture.

Investigating the simple case $k[f_1, \ldots, f_n] = k[x_1, \ldots, x_n]$, we show that we have a bound on the complexity of the standard basis computation, which is of the same order—simple exponential—as for the graph method, but yet smaller (see [O1]). If $f_1, \ldots, f_n$ determine a "generic" tame automorphism, the complexity is even better for canonical bases. This gives new interest to the tame generator conjecture, and canonical bases are shown to split any generic tame automorphism into a composition of elementary generators.

1. Monoids and Standard Bases

If not stated otherwise, k will denote a field of arbitrary characteristic, $k[n]$ the k-algebra of polynomials in n variables $k[x_1, \ldots, x_n]$ and M an abelian monoid with additive law. If E is a subset of M, $\mathrm{Mon}\, E$ will denote the submonoid generated by E.

1.1. Abelian Monoids and Algebras of Monoids.

Before coming to canonical bases, we need first some results about abelian monoids. Although they are "well known", it is best to introduce them explicitly. The reader can refer to the work of Jouanolou in [Jo] for a complete exposition. As we will only consider abelian monoids, we will denote them simply by monoids, and monoideals will be both right and left monoideals.

We recall that an abelian monoid M been given, any subset I of M such that $x \in I$ implies $yx \in I$ for all $y \in M$ is called a monoideal. Any monoid has a natural structure of poset, with a partial ordering defined by $x \leq y$ if there exists z such that $x + z = y$. This ordering is admissible for the monoid structure, i.e. $x \leq y$ implies that $zx \leq zy$. For any admissible partial ordering $\prec$, we may define the *e-set* generated by a subset S of M to be the set $\mathcal{E}(E) = \{x \in M \,|\, \exists y \in S\; x \succeq y\}$. For $\leq$, e-sets are monoideals.

It is known that we can associate to any abelian monoid M an abelian k-algebra $k[M]$; the polynomial algebra $k[n]$ is $k[\mathsf{N}^n]$. If M is a submonoid of N^n, we can consider $k[M]$ as a k-subalgebra of $k[n]$. In general, denoting by $\mathsf{N}^{(S)}$ the free abelian monoid generated by a set S, the polynomial algebra $k[S]$ is the monoid algebra $k[\mathsf{N}^{(S)}]$. There are close relations between properties of the monoids and properties of their k-algebras. For example, any finitely generated monoid is coherent for the ordering coming from its monoid structure, which means that every e-set, or monoideal in this case, is finitely generated. This property implies that for any finitely generated monoid M, the ring $k[M]$ is noetherian as it is the case for $k[n]$. The situation is not so good when submonoids of N^n are considered. They may be of infinite type, except for $n = 1$. As a consequence, there exist k-subalgebras of $k[n]$ of infinite type.

There is a natural bijection between admissible orderings on monomials of $k[n]$ and admissible orderings on N^n. An admissible ordering $\prec$ being chosen, we can associate to any non-zero polynomial $P = cx_1^{\alpha_1} \cdots x_n^{\alpha_n} + \cdots$ in $k[n]$ its multidegree $\mathrm{mdeg}\, P = (\alpha_1, \ldots, \alpha_n) \in \mathsf{N}^n$. Then, for any ideal I (resp k-subalgebra A) of $k[n]$, the set $\{\mathrm{mdeg}\, P | P \in I\}$ (resp. $\{\mathrm{mdeg}\, P | P \in A\}$) is a monoideal (resp. submonoid) of N^n.

Proposition 1. *Let $\prec$ be an admissible ordering on $\mathbf{N}^n$, then any chain*

$$x_0 \succ x_1 \succ \cdots \succ x_k \succ x_{k+1} \succ \cdots$$

is finite.

Corollary 2. *Any submonoid M of $\mathbf{N}^n$ admits a minimal set of generators.*

Proof: We only have to consider the set of minimal elements for the natural ordering $<$, which is the wanted set.   QED.

1.2. The Graph Method for Monomial Ideals.

We will give relations between congruences on a monoid M and binomial ideals of $k[M]$. A binomial ideal of $k[M]$ is an ideal generated by polynomials of the form $m - m'$ where m and m' are primitive monomials. A congruence on a monoid M is an equivalence relation $\equiv$ between elements of M such that $\forall (x, y, z) \in M^3 \ x \equiv y \Rightarrow zx \equiv zy$. The monoid structure of M induces then a unique monoid structure on the set of equivalence classes $M/\equiv$.

Proposition 1. *An equivalence relation $\equiv \subset M \times M$ on a monoid M is a congruence iff $\equiv$ is a submonoid of $M \times M$.*

Proof: See [Jo 1.4.1 p. 14].   QED.

Proposition 2. *Let R be an integral domain and $\equiv$ a congruence on the monoid M, we associate to it the binomial ideal of $R[M]$ generated by the polynomials of the form $m - m'$ such that $m \equiv m'$. Then, for any two elements m, m' of M, $m \equiv m' \Leftrightarrow m - m' \in I$.*

Proof: See [MM lemmas 1 and 2 p. 311], where a proof is given for $R = \mathbf{Z}$, which generalizes to any integral domain.   QED.

We will need the following proposition, which allows to build a "standard basis" for a congruence by computing a standard basis for the associated ideal.

Proposition 3. *Let $\equiv$ be a congruence on $\mathbf{N}^n$ and I the binomial ideal associated to $\equiv$ as in the previous proposition, then for any total compatible ordering on monomials of $k[n]$ the polynomials in the standard basis G of I are differences of monomials and the set $\{(m, m')|m - m' \in G\}$ generates the congruence.*

Proof: It is easy to see that the polynomials in G are differences of monomials, for I is binomial and any S-polynomial coming from a syzygy between

$x - y$ and $z - t$ is of the form $my - m't$, so that it is still a difference of monomials.

The last part is true by the proof of [Jo cor. 1.6.6.2. p. 34]. Q. E. D.

Corollary 4. *(Theorem of Redei) Every congruence in $\mathbf{N}^n$ is finitely generated.*

Corollary 5. *Let $\phi : \mathbf{N}^n \mapsto \mathbf{N}^m$ and $\psi : \mathbf{N}^\ell \mapsto \mathbf{N}^m$ be two morphisms of monoids and M the subset of $\mathbf{N}^n \times \mathbf{N}^\ell$ defined by $M = \{(x, y) | \phi(x) = \psi(y)\}$, then M is a finitely generated monoid.*

Remark 6: It is known (see [R1]) that every admissible preordering in $\mathbf{N}^n$ is induced by a morphism of monoid $\phi : \mathbf{N}^n \mapsto \mathbf{R}^m$, where $\mathbf{R}^m$ is ordered by the pure lexicographic ordering. Such orderings were already used by Riquier and Janet. Robbiano has given a complete description of those orderings and shown that we can take $m \leq n$. If we consider a subset S of $k[n]$, and the k-algebra morphism $\psi : k[\mathbf{N}^{(S)}] \mapsto k[n]$ defined by $\psi(R) = R(S)$, any admissible ordering on $k[n]$ induces an admissible preordering of $k[\mathbf{N}^{(S)}]$, and so a graduation of $k[\mathbf{N}^{(S)}]$.

Theorem 7. *Let $A = k[P_1, \ldots, P_m]$ be a k-subalgebra of $k[x_1, \ldots, x_n]$ and G be the standard basis of the ideal $I = (P_i - y_i)_{k[x_1, \ldots, x_n, y_1, \ldots, y_m]}$ for an admissible ordering which eliminates the x_i, then $Q \in A$ iff $Q \xrightarrow{G^*} R(y)$.*

Furthermore the subset of G of polynomials which do not involve any x_i is a standard basis of $I \cap k[y] = \{R | R(P) = 0\}$.

Proof: See [SS].   Q. E. D.

Corollary 8. *Let M be the submonoid of $\mathbf{N}^n$ generated by the finite set $\{\alpha_i; \ i \in [1, m]\}$. We define on $\mathbf{N}^n \times \mathbf{N}^m$ a congruence, associated to the morphism of monoid defined by $\phi(e_i) = m_i$, where e_i stands for the ith elementary generator of $\mathbf{N}^m$, and $\phi(x) = x$ for any x in $\mathbf{N}^n$. We also denote by ϕ the associated morphism of k-algebra. Let $\prec$ denote an admissible ordering on $\mathbf{N}^n$, we extend it to $\mathbf{N}^n \times \mathbf{N}^m$ using ϕ, and complete it to an total ordering $\ll$, eliminating the n first variables. Let G be the standard basis for $\ll$ of the ideal $I = (x^{m_i} - y_i)_{k[x_1, \ldots, x_n, y_1, \ldots, y_m]}$. Then β belongs to M iff $x^\beta \xrightarrow{G} y^\gamma$, and this may be tested by computing only critical pairs up to the rank of β according to the graduation defined by $\prec$.*

The elements of $G \cap k[y]$ generate the congruence induced on $\mathbf{N}^m$ by ϕ.

Proof: These are simple consequences of the last theorem and prop 3. The bound on the standard basis computation comes from the remark made above and classical considerations on homogeneous ideals, I being homogeneous for the graduation defined by $\prec$.   Q. E. D.

2. Canonical Bases

We will denote by $k[n]$ the algebra of polynomials in n variables $x_1, \ldots, x_n$. We give ourselves an admissible ordering $\prec$ on monomials of $k[n]$. The leading coefficient of a polynomial P will be denoted by $\mathrm{lc}P$ and the leading primitive monomial of P by $\mathrm{lpm}P$. A will denote a k-subalgebra of $k[n]$. To avoid unuseful complications, we will suppose all polynomials to be monic, if not stated otherwise. It will be easy to think of the necessary modifications if it is not the case.

2.1. Definition.

Definition 1: Let A be a k-subalgebra of $k[n]$ and E a subset of A, we denote by $\mathrm{Mon}E$ the submonoid of $\mathbf{N}^n$ generated by $\{\mathrm{mdeg}P | P \in A\}$. A subset E of A is said to be a canonical basis of A if $\mathrm{Mon}E = \mathrm{Mon}A$.

Obviously, we have a similar definition for standard bases by replacing k-subalgebra by ideal and submonoid by e-set—or monoideal.

An admissible ordering being given, we can associate to any subset of a k-subalgebra a reduction relation, in the following way.

Definition 2: Let Q, and Q' be two polynomials of $k[n]$ and C a subset of $k[n]$, then we say that Q is reduced to Q' by C if $\mathrm{mdeg}Q \in \mathrm{Mon}C$ and

$$Q' = Q - \prod_{i=1}^{k} R_i^{\alpha_i},$$

where the α_i and R_i are integers and elements of C such that $\mathrm{mdeg}Q = \sum_{i=1}^{k} \alpha_i \mathrm{mdeg}R_i$. This relation will be written

$$Q \xrightarrow{P} Q'.$$

We will denote by $\xrightarrow{C^*}$ the inductive limit of the relation $\xrightarrow{C}$.

We say that P is reduced with respect to C if there is no Q such that $P \xrightarrow{C} Q$, and that P is strongly reduced if P is reduced and the reductum of P is strongly reduced, which means that no monomial of P belongs to $\mathrm{Mon}C$.

Definition 3: We say that C is a reduced canonical basis of A if C is a canonical basis, the polynomials in C are monic and each polynomial $P \in C$ is strongly reduced with respect to $C \setminus \{P\}$.

As k is a field, any k-subalgebra A admits a unique reduced canonical basis, which is finite iff A admits a finite canonical basis. We will refer to the reduced canonical basis as *the* canonical basis of A.

Lemma 4. *If P belongs to a k-subalgebra A of $k[n]$ and if C is a subset of A, then any polynomial Q such that $P \xrightarrow{C^*} Q$ belongs to A.*

Lemma 5. *Any chain of reduction*

$$Q_0 \xrightarrow{C} Q_1 \ldots Q_{k-1} \xrightarrow{C} Q_k \xrightarrow{C} \ldots$$

has to be finite.

Proof: This is only a translation of prop 1.1.1. Q. E. D.

Definition 6: If C is a subset of $k[n]$ we can extend any admissible ordering $\prec$ on monomials of $k[n]$ to a preordering on $k[\mathbf{N}^{(C)} \times \mathbf{N}^n]$ by setting $m \prec m' \Leftrightarrow m(C, x) \prec m'(C, x)$. That preordering will be used each time we will deal with polynomials in $k[\mathbf{N}^{(C)} \times \mathbf{N}^n]$ or $k[\mathbf{N}^{(C)}]$. The multidegree of a polynomial R will be then the maximal multidegree of $m(C)$, for all monomials m of R.

Lemma 7. *If $P \xrightarrow{C^*} 0$, then there exists a polynomial $R \in [\mathbf{N}^{(C)}]$, of multidegree not greater than P, such that $R(C) = P$.*

Proof: We can build R by reducing P, each step of reduction giving a monomial. The monomials appear then in strictly decreasing order according to $\prec$. Q. E. D.

The following notion is an analog of syzygies in the case of standard bases.

Definition 8: Let C be a subset of $k[n]$, $\{P_1, \ldots, P_\ell\}$ and $\{Q_1, \ldots, Q_m\}$ two finite subsets of C, whose elements are all different, let M be the submonoid of $\mathbf{N}^\ell \times \mathbf{N}^m$ whose elements $((\alpha_1, \ldots, \alpha_\ell), (\beta_1, \ldots, \beta_m))$ satisfy

$$\sum_{i=1}^{\ell} \alpha_i \mathrm{mdeg} P_i = \sum_{i=1}^{m} \beta_i \mathrm{mdeg} Q_i.$$

Then, we call a superposition between elements of C a 4-uple

$$((P_1, \ldots, P_\ell), (Q_1, \ldots, Q_m), \alpha, \beta)$$

such that (α, β) belongs to the minimal set of generators of M.

The polynomial

$$\prod_{i=1}^{\ell} P_i^{\alpha_i} - \prod_{i=1}^{m} Q_i^{\beta_i}$$

is called the S-polynomial associated to the superposition. The multidegree of the superposition is the common multidegree of both products in the formula above.

Remark 9: With the same notations, the 2-uples of exponents $((\alpha_P), (\beta_Q))$ associated to all superpositions between elements of C, generate the congruence defined by

$$\sum_{P \in C} \alpha_P \text{mdeg} P = \sum_{Q \in C} \beta_P \text{mdeg} Q.$$

In this case, minimal sets of generators do not exist, but if C is finite, the construction of cor. 1.2.7. provide a finite set of superposition, generating the congruence, which is in general smaller than the set of all superpositions.

Definition 10: If S is a set of superpositions generating the congruence defined above, it is said to be a generating set of superpositions. It is said to be confluent if all the corresponding S-polynomials are reduced to 0 by C.

Lemma 11. *If C is a subset of $k[n]$, m and m' two monomials of $k[\mathbf{N}^{(C)}]$ such that $m(C)$ and $m'(C)$ have the same leading monomial and S a generating set of superpositions, then there exist ℓ monomials M_i of $k[\mathbf{N}^{(C)}]$ and S-polynomials R_i associated to superpositions of S such that*

$$m(C) - m'(C) = \sum_{i=1}^{\ell} M_i(C) R_i.$$

We have then a fundamental theorem, which also has an analog in the case of standard bases.

Theorem 12. *Let A be a k-subalgebra of $k[n]$ and C a subset of A, then the three following propositions are equivalent:*
A) C is a canonical basis,
B) $\forall P \in A$ $P \xrightarrow{C^*} 0$,
C) C generates A and there exists a generating confluent set S of superpositions between elements of C.

Proof: A) $\Rightarrow$ B) For any element P of A, mdeg$P$ is in Mon$C$ so that P needs to be reduced by C if P is not 0. By lemmas 4 and 5, $P \xrightarrow{C^*} 0$.

B) $\Rightarrow$ C) As any polynomial in A is reduced to 0, it is obviously the case of any S-polynomial. This also implies that C generates A.

C) $\Rightarrow$ A) That will be the consequence of a more precise result.

Proposition 13. *If $P = T(C)$ is a polynomial in A and if all superpositions between elements of C of multidegree not greater than the multidegree of T are reduced to 0 by C, then P is reduced to 0 by C.*

Proof: We recall that we have extended $\prec$ to an ordering on monomials of $k[\mathbf{N}^{(C)}]$. We suppose the result is false and search a contradiction. Let us consider the non reducible $P = R(C)$ such that R is minimal according to $\prec$, we have $R \preceq T$. Then we can choose some P among them such that R has minimal number of monomials.

Let m be a maximal monomial of R, $m(C)$ is obviously reducible, and $R(C) - m(C)$ is reducible too, for its maximal monomials are not greater than m and $R - m$ has smaller number of monomials than R. $m(C)$ and $R(C) - m(C)$ have the same leading monomial and opposite leading coefficients, if not P would be reducible. Then $R(C) - m(C) \xrightarrow{C} Q(C)$, with $Q(C)$ reducible so that Q is smaller than $R - m$ according to lemma 7. Now $R(C) - m(C)$ is equal to $m'(C) + Q(C)$ where m' is a monomial greater than Q. $m(C)$ and $m'(C)$ have obviously opposite leading monomials and by lemma 11, $m(C) - m'(C)$ is of the form $\sum m_i S_i$, where the S_i are S-polynomials associated to superpositions in C and the $m_i S_i$ are smaller than R. We can then use the hypothesis on S-polynomials and apply again lemma 7 on each $m_i S_i$. So $m(C) + m'(C) = Q'(C)$ with Q' smaller than R.

The conclusion of this construction is that $P = Q(C) + Q'(C)$ and $Q + Q'$ is smaller than R, a contradiction. Q. E. D.

Remark 14: We have no need in this proof to suppose that A is of finite type, nor that C is finite. Of course, we shall have to restrict ourselves to that case for effective applications.

2.2. Completion Procedure. Implementation.

Using cor. 1.2.7, we can solve the membership problem for $\text{Mon}C$, and it is then easy to build a reduction procedure. The same standard basis construction will give a generating set of superposition, so that the construction of superpositions is also effective (see also [H]).

Definition 1: We say that a completion procedure is fair if all S-polynomials which are not discarded using some criteria have to be considered and reduced during the computation.

For example, if we sort S-polynomials according to the multidegree of corresponding superposition the procedure is fair iff the ordering is archimedean.

We have then the following result.

Theorem 2. *Let A be $k[P_1, \ldots, P_n]$, then if A admits a finite canonical basis, any fair procedure of the following form will stop and return a canonical basis:*

    C := [P1,...,Pn]
    (1) LS := *List-of-S-polynomials-not-considered-yet*(C)
    if LS = [] *then* output C *fi*
    Sp := *Choose*(LS); LS := LS - [Sp]
    if Red(Sp) ≠ 0 *then* C := cons(Red(Sp),C) *fi*
    goto (1).

If A admits no finite standard basis, the sets of polynomials C_i, returned at each loop are such that $\bigcup_{i=1}^{\infty} C_i$ is a standard basis.

Proof: See [KM].   Q. E. D.

Remark: This way of computing a finite standard basis if there exists one, in a situation where finite standard bases do not exist in general, has already been intruced by F. Mora in [Mor] for a different situation, viz. non-commutative standard bases.

Remark 3: We did not implement exactly a procedure of that type. Superpositions are determined, using a standard computation as described in 1.2.8. Each time a new element corresponding to a superposition is appended to the standard basis, its computation is suspended after returning the superposition to the canonical basis process. It computes the S-polynomial, reduces it, updates the list C as above and call the standard basis algorithm again. In this way, not all superpositions are found, but we still secure a generating set, which is enough, and better for efficiency. If a superposition corresponds to the reduction of a polynomial in C, we can discard it.

This algorithm is fair iff $\prec$ is archimedean. This is the case for the degree ordering, implemented in Scratchpad II. It would have been too complicated and inefficient to use the standard basis algorithm of the public system (implemented by Gebauer and Moeller), so that we have rewritten it in the case of binomial ideals and made it incremental. We use $\prec$, refined by the inverse lexicographical ordering on variables, sorted by "order of appearance". Indeed, for each element appended to C, a new variable appear in the standard basis computation. With such an ordering, we will never have to consider superpositions involving a polynomial which has been removed.

Two packages have been implemented, STANDMON computes standard bases for binomial ideals, monomials with suitable ordering been implemented in the domain MOFAM. The last package, BASECAN implements the canonical bases process.

Remark 4: During the standard basis computation, some superpositions may be found, coming from the reduction of a syzygy between two superpositions—as in 1.2.8, superpositions are identified with binomials. In such a case, this superposition needs not to be considered, for it is generated by superpositions already treated and reduced. It seems that with the chosen ordering such a situation never occurs.

Remark 5: Reducing to a generating set of superpositions is the canonical bases analog of the criterion of Moeller allowing to reduced the set of syzygies to a generating set of the module of relations between leading monomials (see [Mo]).

3. Relations with Standard Bases

We will consider here a k-subalgebra A of $k[n]$ with a finite canonical basis C, according $\prec$. M will denote the submonoid MonA.

3.1. A Generalization of Standard Bases.

The generalized standard bases presented here are special cases of those described by Sweedler in [S] and Robbiano in [R2]. The connection made with canonical bases allows simpler definitions, and a more effective presentation. Moreover, canonical bases could be extended too, in the same way as Sweedler did for standard bases.

We first remark that if A is of finite type—it is obviously the case if A admits a finite canonical basis—then A is noetherian. So we may hope to generalize standard bases to A without much trouble. We will see it is indeed the case.

Definition 1: Let I be an ideal of A, M the submonoid MonA and E the e-set $\{\mathrm{mdeg}P|P \in I\}$ of M. Then we say that a subset G of I is a standard basis of I if the set $\{\mathrm{mdeg}P|P \in G\}$ generates E.

Remark 2: We have to notice that we must use the same ordering to define the canonical basis and the standard basis. In the case of $k[n]$, we do not have such a trouble for $\{x_1, \ldots, x_n\}$ is a canonical basis for all orderings. As shown in [RS], other algebras share this property, for example the elementary symmetrical polynomials form a standard basis of the subalgebra they generate, for all orderings.

Proposition 3. *All ideals of a k-subalgebra A admitting a finite canonical basis, admit a finite standard basis.*

Proof: With the same notations as in the definition, M is of finite type so that it is coherent and E is of finite type.   Q. E. D.

We will now generalize the notion of syzygy.

Definition 4: Let $S = \{R_1, \ldots, R_q\}$ be a finite subset of a A, which admits a finite canonical basis $C = \{P_1, \ldots, P_\ell\}$, Q and R two elements of S, and E the set of 2-uple of monomials $(m, m') \in k[\ell] \times k[\ell]$ such that

$$\text{mdeg}(m(P)Q) = \text{mdeg}(m'(P)R).$$

Denoting by M the submodule generated by E, we call syzygy between Q and R a 4-uple (R, S, m, m'), such that (m, m') belongs to the minimal subset of E which generates M.

Remark 5: Such minimal elements are in finite number and we can again restrict ourselves to a generating set of syzygies, obtained in the following way. We consider the polynomial algebra $k[w, x, u, y]$, with 1 variable w, n variables x, q variables u associated to the polynomials R, and ℓ variables y associated to the polynomials P. We define weights on variables such that the weight of w and the u is 1, and the weight of the other variables 0. The binomial ideal $(\text{lpm}P_i - y_i, w\text{lpm}R_j - u_j)$ of $k[w, x, u, y]$, is homogeneous for this weight—this is why we need the extra variable w. Then, we compute the standard basis of this ideal up to weight 1, for an ordering which eliminates w and the x and then the u.

The elements of weight 1 in this basis whose leading monomial depends only of the variables u and y are of the form $\prod y^{\alpha_i} u_j - \prod y^{\beta_i} u_{j'}$. They are associated to a set of syzygies, generating the module of relations between leading monomials. As pointed out by P. Conti and C. Traverso in [CT], an efficient algorithm for standard bases of modules can be derived from an algorithm for ideals if we forget syzygies of weight 2 and more.

The considerations of remark 2.2.5 also apply in this case.

Remark 6: We have seen that in the case of canonical bases, superpositions involve in general more than two polynomials. Here, syzygies involve only two polynomials, but there can be more than just one syzygy between two given polynomials (see [S]).

We can define a notion of reduction with respect to a subset G of A in an obvious way and we get the usual theorem.

Theorem 7. *If A is a k-subalgebra of $k[n]$, I an ideal of A and G a subset of I, then the following properties are equivalent:*
  A) G is a standard basis of I,
  B) all elements of I are reduced to 0 by G,
  C) G generates I and there exists a generating confluent set of syzygies between elements of G.

Proof: We can adapt the proof of th. 2.1.12, or any proof for "usual" standard bases (see [Bu]).   Q. E. D.

Again, we will have a completion procedure, relying on successive reduction of S-polynomials.

3.2. Ideal of Relations.

Definition 1: Let A be a k-subalgebra of $k[n]$ admitting a finite canonical basis $C = \{P_1, \ldots, P_m\}$, then we can define an ideal of relations between polynomials of C by $I = \{R \in k[m] | R(P) = 0\}$.

Definition 2: Let S be a superposition between elements of a finite canonical basis $C = \{f_1, \ldots, f_m\}$, P the S-polynomial associated to S. Reducing $P(f)$ to zero by C, we secure a polynomial $R(f)$, of smaller multidegree than P, such that $P - R \in I$. We denote $P - R$ by $R(S)$.

Theorem 3. *With the same notations, if we consider the whole generating set of superpositions G determined by a standard basis computation, using some total ordering $\ll$ compatible with $\prec$ as described in cor. 1.2.7, then the the set of polynomials $R(G)$ associated by the previous construction form a standard basis of the ideal of relations I according to $\ll$.*

Proof: It is easily seen using cor 1.28 and lemma 2.1.11 that all polynomials in I are reduced to 0 by $R(G)$.   Q. E. D.

4. Finiteness Conditions

4.1. Examples.
We will begin by two examples of Robbiano, which show that the canonical basis of a finitely generated k-subalgebra may be infinite.

Example 1: Let $A = k[x, xy - x^2, xy^2] \in k[x, y]$. If k is of characteristic 0 and if we consider some ordering with $x > y$, then the reduced canonical basis of A is is

$$\{x, xy - y^2, xy^2, xy^3 - \frac{1}{2}y^4, xy^4, xy^5 - \frac{1}{3}y^6, \ldots\},$$

so that A admits no finite canonical basis. If we consider some ordering with $y > x$, then the canonical basis is finite.

If k is of positive characteristic p, then A admits a finite canonical basis for all orders, for then $y^{2p} \in A$.

It takes 11 s to compute the standard basis with $x > y$ up to degree 7, using Scratchpad II. Only two S-polynomials are reduced to 0 during this computation. As the degree increases, more and more unuseful and undetected superpositions are considered, coming from the particular structure of the algebra; $d - 3$ well chosen superpositions would be enough to go up to degree d.

Example 2: Let A be $k[x + y, xy, xy^2]$, where k is an arbitrary field, then the canonical basis of A for some ordering with $x > y$ is

$$\{x + y, xy, xy^2, xy^3, xy^4, \ldots\}.$$

If we take $y > x$ then the canonical basis is also infinite by symmetry.

Remark 3: We can remark on those two examples that A is not integrally closed and that its integral closure is $k[x, y]$, which has a finite canonical basis.

In example 1, the extension $A[y^2]$ is an integral extension of A with finite canonical basis. Indeed, $y^2 = xy^2/x$ is in the integral closure, so that $I = xA$ is both a A ideal and a $A[y^2]$ ideal. Now, if we want to test that a polynomial P is in A, this can be done by computing a generalized standard basis for I in $A[y^2]$ and then test if xP belongs to I. In example 2, we can take $A[y] = k[x, y]$, and remarking that $y = xy^2/xy$ is in the integral closure, consider the ideal $xy A = xy A[y]$.

This method generalizes each time we know (by its generators) an integral extension B=$A[P_i/Q_i]$ of A in its fraction field, with finite canonical basis. The ideal $I = (\prod Q_i^{a_i-1}) A$, where a_i is the degree of a monic polynomial $R_i \in A[z]$ such that $A_i(P_i/Q_i) = 0$, is equal to $(\prod Q_i^{a_i-1}) A[P_i/Q_i]$. This allows to reduce the membership problem for A to the membership problem for a the B ideal I, generated by a single element.

We can easily apply to those two examples the method of Shannon and Sweedler, but we can give some example where this method fails whereas the canonical basis method have a pretty good complexity.

Example 4: If we consider the k-subalgebra A of $k[n]$ generated by the n polynomials

$$P_1 = x_1 + \cdots + x_n$$
$$P_2 = x_1x_2 + x_2x_3 + \cdots + x_nx_1$$
$$\vdots$$
$$P_n = x_1x_2 \cdots x_n,$$

the standard basis of Shannon and Sweedler's method cannot be computed with the program Macaulay of Bayer and Stillman, already for $n = 7$. But the canonical basis of A for the degree ordering is $\{P_1, \ldots, P_n\}$. Indeed, there is no superposition between those polynomials, for their multidegrees are linearly independent. We can remark that the computation of a canonical basis for the ideal $(P_1, \ldots, P_{n-1}, P_n - 1)$ of $k[n]$ is itself a difficult problem, known as the Arnborg–Davenport problem. For the best of our knowledge it has been done only up to $n \leq 7$, using Macaulay, and $n = 8$

using the program of J. C. Faugère. It takes more than a week on ALLIANT FX40.

We could give many other examples of this kind, e.g. the polynomials of the Mayr–Meyer examples ([MM]), form a canonical basis for some ordering.

4.2. A Conjecture and Related Results.

We have stated in [O2] the following conjecture, to which the remark 4.1.3 gives a particular interest.

Conjecture. *If A is a finitely generated integrally closed k-subalgebra of $k[n]$, then its canonical basis for any admissible ordering is finite.*

Remark 1: The hypothesis that A is finitely generated is essential, for there exist integrally closed k-subalgebra of infinite type (consider for example $k[x, xy, xy^2, \ldots] \subset k[x, y]$).

We will give some partial results relating the standard basis of A and that of its integral closure $\overline{A}$.

Definition 2: Let A be any k-subalgebra of $k[n]$, we call cone of A, the convex cone CA generated in $\mathbf{R}_+^n$ by Mon$A \in \mathbf{N}^n$, with vertex at the origin.

Lemma 3. *If $P \in k[n]$ belongs to the integral closure $\overline{A}$ of A, then* mdeg$P \in \overline{CA}$, *which stands for the topological closure of CA.*

Proof: P belongs to $\overline{A}$ so that $P = R/Q$ with $R \in A$ and $Q \in A$, and P satisfies some polynomial equation $P^k + a_1 P^{k-1} + \cdots + a_k = 0$ where the a_i belong to A. Now, multiplying this equation by Q^k, we get $R^k + a_1 Q R^{k-1} + \cdots + a_k Q^k = 0$, so that R^{k+1}/Q belongs to A. We can now prove by induction that $R^k P^i = R^{k+i}/Q^i$ belongs to A for all positive integer i. The mutidegree of $R^k P^i$ is $k$mdeg$R + i$mdeg$P$, hence the wanted result. Q. E. D.

Theorem 4. *Let A be any k-subalgebra of $k[n]$, then*

$$CA \subset \overline{CA} \subset \overline{CA}.$$

Proof: The first inclusion is obvious and the second is a mere consequence of the lemma. Q. E. D.

Remark 5: Our conjecture would imply that if A is finitely generated, $C\overline{A} = \overline{CA}$, for the canonical basis would be finite, so that its cone would be closed and generated by a finite number of points with integral coefficients. Of this, we would deduce that $\overline{CA}$ is generated by a finite number of integral points for any k-subalgebra. We will see that this result can be proved for graded k-algebras of dimension 2.

4.3. Special Results for 2-dimensional Graded k-Algebras.
We will first introduce some results, valid in general case.

Proposition 1. *Let* $A = k[P_1, \ldots, P_n]$ *be a finitely generated graded k-subalgebra of $k[n]$ of dimension μ, $I \in k[m]$ be the ideal of relations between polynomials P_i, $\Delta = \mathrm{lcm}(\deg P_i)$, $\delta = \gcd(\deg P_i)$, then if we denote by $H(d)$ the number of elements of degree d in $\mathrm{Mon}A$, there exist polynomials $R_i \in \mathbf{Q}[x]$ of common degree equal to $\mu - 1$, such that*

$$H(j\Delta + i\delta) = R_i(j),$$

for j great enough. Furthermore $H(j\delta + k) = 0$ for $0 < k < \delta$.

Proof: The last part is obvious. Now, if we define a degree $\deg_p$ in $k[y_1, \ldots, y_m]$ by $\deg_p(y_i) = \deg P_i$, we can remark that the number of elements of degree $\deg_p = d$ in $k[y_1, \ldots, y_m]$ satisfies the wanted property. The ideal of relations I is obviously $\deg_p$-homogeneous. This implies our result, for we have a finite free resolution of $A = k[m]/I$, which preserves the graduation $\deg_p$. Q. E. D.

Corollary 2. *If A is a finitely generated k-algebra of dimension μ and $h(d)$ the number of points of degree less or equal to d in $\mathrm{Mon}A$, then there exists some polynomial $R \in \mathbf{Q}[x]$ of degree μ such that $h(d) \geq R(d)$.*

Definition 3: Let A be a k-subalgebra, we call dimension of $\mathcal{C}A$, the maximal number of linearly independent points in $\mathcal{C}A$.

Proposition 4. *If A is a finitely generated k-subalgebra, the dimension of A is equal to the dimension of $\mathcal{C}A$.*

Proof: The dimension of $\mathcal{C}A$ is the maximal number ℓ of linearly independent points in $\mathrm{Mon}A$. If $P_1, \ldots, P_\ell$ are polynomials of A such that their multidegrees are linearly independent, then $k[P]$ is isomorphical to $k[\ell]$, so that $\dim A \geq \ell$. We also have $\dim A \leq \ell$ by cor. 2, hence the result. Q. E. D.

We will need the following simple lemma about submonoids of $\mathbf{N}^n$.

Lemma 5. *If M is a submonoid of $\mathbf{N}^n$ and $p_1, \ldots, p_m$ points in $\mathcal{C}M$, then if we denote by G the subgroup of $\mathbf{Z}^n$ generated by M, there exist a point $q \in \mathcal{C}M$ such that the cone C' of vertex q generated by the points $p_i + q$ satisfies $M \cap C' = G \cap C'$.*

Proposition 6. *If A is a 2-dimensional graded finitely generated k-subalgebra of $k[n]$, then for any ordering $\prec$, $\overline{\mathcal{C}A}$ is generated by 2 points in $\mathbf{N}^n$.*

Proof: We will prove this result in $k[x, y]$, but the argument also applies in $k[n]$. We can remark that at most 2 canonical bases exist for A, one for orderings such that $x > y$ the other for $y > x$. We can consider only one of these cases, say $x > y$. Let $P_1, \ldots, P_m$ be homogeneous generators of A, $(\alpha_1, \beta_1), \ldots, (\alpha_m, \beta_m)$ their multidegrees, we choose P_j such that α_j/β_j is maximal—we consider it is the case if $\beta_j = 0$. It is easily seen that the S-polynomials coming from a superposition between the P_i have smaller slope than P_j. This implies that $p = (\alpha_j, \beta_j)$ generates the right border of CA.

If the left border of CA is vertical, we have our result, if not we have to prove that its slope σ is rational. We denote by D the lcm of the degrees of P_i. By lemma 5, for any point $p' = (1, \sigma - \varepsilon) \in CA$, the number $\mu(aD)$ of points of degree aD in $C\{p, p'\} \cap \mathrm{Mon} A$ is asymptotically equivalent to the number of points in $G \cap C'$. We denote by $\nu(aD)$ the number of points of degree aD in $C\{p, (1, 1 + \varepsilon)\} \cap G$. We can remark then that the number of points of degree aD in $G \cap \mathbf{R}_+^n$ is equivalent to aD/r for some integer r, so that

$$\frac{aD}{r}\left(\frac{\sigma + \varepsilon}{1 + \sigma + \varepsilon} - \frac{\beta}{\alpha + \beta}\right) \sim \nu(aD) \geq H(aD)$$

$$\geq \mu(aD) \sim \frac{aD}{r}\left(\frac{\sigma - \varepsilon}{1 + \sigma - \varepsilon} - \frac{\beta}{\alpha + \beta}\right).$$

Now, by prop. 1, σ must be rational.   Q. E. D.

This result is not sufficient to conclude, but it is still encouraging to prove—even in a special case—a consequence of the conjecture. Assume we can prove that the topological closure of the cone is finitely generated for any finitely generated algebra. An idea to go ahead would be to prove then that for any generator of the cone $(a_1, \ldots, a_n) \in \mathbf{N}^n$, one of the two following propositions is true:

 i) there exists a polynomial in A, which multidegree is a multiple of $(a_1, \ldots, a_n)$,

 ii) there exist a polynomial $P \in k[x_1, \ldots, x_n]$, with multidegree a multiple of $(a_1, \ldots, a_n)$, and a polynomial $R \in A$ such that $RP^p \in A$ $p \in \mathbf{N}$.

5. Application to Morphisms of $k[n]$

5.1. Complexity. If we consider an endomorphism of $k[n]$ defined by polynomials $f_1, \ldots, f_n$, it is an automorphism iff $k[f] = k[n]$, so that is can be tested using canonical bases. But, we need to secure a bound in order to stop the computation if $k[P]$ has an infinite canonical basis. That will be a consequence of a theorem of Gabber.

Definition 1: Let f be an endomorphism of $k[n]$ defined by polynomials f_i, we will call degree of f the maximum degree of the f_i.

Theorem 2. *If $f \in \text{Aut}_k k[n]$ is of degree d, then the degree of f^{-1} is bounded by d^{n-1}.*

Proof: See [BCW].   Q. E. D.

Theorem 3. *If $A = k[f_1, \ldots, f_n] = k[n]$ and the maximal degree of polynomials f_i is d, then the canonical basis of A with respect to the degree ordering is $\{x_1, \ldots, x_n\}$ and may be computed by considering only superpositions of degree less or equal to d^n.*

Proof: The first part is obvious, and the second is a simple consequence of prop. 2.1.13, using the theorem of Gabber.   Q. E. D.

Remark 4: Of that result, we can deduce a bound on the complexity of the canonical basis computation. It will be of the same order as the bound we can obtain for Shannon and Sweedler's method[1], but yet smaller. Indeed the computation of a canonical or standard basis may be considered as a linear algebra problem, once we have secured a bound on the degree of superpositions or syzygies. For the ideal of the graph the bound d^n has been proved in [O1]. For canonical basis, we have a system of $O(d^{n(n-1)})$ equations in $O(d^{n^2})$ variables; for the other method a system of $O(d^{2n^2})$ equations in $O(d^{2n^2})$ variables. Of this, we easily deduce a bound polynomial in d^{n^2} for both methods.

Remark 5: If we consider the automorphism f of $k[n]$ defined by polynomials $x_1, x_2 + x_1^d, \ldots, x_n + x_{n-1}^d$, then $\deg f^{-1} = d^{n-1}$. This shows that our bound is sharp, and that we will have to climb up to degree d^{n-1} at least using Shannon and Sweedler's method. But the canonical basis of $k[f]$ may be computed in degree d at most. We can obviously build examples where the canonical basis requires to consider superpositions of degree greater than d, but it seems difficult to reach d^n.

5.2. Tame Automorphism. We will now consider tame automorphisms of $k[n]$.

Definition 1: We say that an automorphism of $k[n]$ is tame if it is in the subgroup generated by elementary automorphisms which are:
   A) the automorphisms generated by the permutations of the variables,
   B) de Jonquières' automorphisms:

$$f(x_1, \ldots, x_n) = (x_1, \ldots, x_{n-1}, cx_n + P(x_1, \ldots, x_{n-1})) \text{ with } c \neq 0.$$

---

[1]In this special case the method has been introduced earlier by A. van den Essen in [E].

It is known that all automorphisms of $k[2]$ are tame (see [Ju] and [Ku]). It is only a conjecture in more variables, see [BCW] and [N] for further details on the subject. We will see that we have a good bound on the degree of canonical bases for automorphisms of $k[2]$.

Proposition 2. *If f is an automorphism of $k[2]$, we can be in the two following situations:*

A) there exists some integer a such that $\mathrm{mdeg}f_1 = a\,\mathrm{mdeg}f_2$ or $\mathrm{mdeg}f_2 = a\,\mathrm{mdeg}f_1$,

B) $\{f_1, f_2\}$ is a canonical basis of $k[2]$.

Proof: Using the fact that f is tame we have $f = g_h \circ \cdots \circ g_1$ where the g_i are elementary. It is then easy to prove the result by induction on h.   Q. E. D.

Corollary 3. *With the same notations, the canonical basis may be computed without considering any superposition of multidegree greater than* $\max(\mathrm{mdeg}f_1, \mathrm{mdeg}f_2)$.

Proof: If we are in situation A), we can remark that the first superposition will be for example a reduction of $f_1 \xrightarrow{\ f_2\ } f_3$ of multidegree $\mathrm{mdeg}f_1$, so that we can delete f_1 and continue with f_2 and f_3. As the reduction corresponds to a de Jonquières' automorphism $k[f_1, f_2] = k[f_2, f_3]$ and we can iterate the argument untill we are in case B). Then we have secured a canonical basis, and the bound holds for the multidegrees of $f_1, f_2, \ldots$ are decreasing.   Q. E. D.

Remark 4: By the same proof, we see that the canonical basis algorithm will split f as a composition of elementary automorphisms.

It would be tempting to try to generalize prop 2. This can be done in the following way.

Problem. *Let f be a tame automorphism of $k[n]$, does it exist $i \in [1, n]$ such that*

$$\mathrm{mdeg}f_i \in \mathrm{Mon}k[f_1, \ldots, \widehat{f_i}, \ldots, f_n]?$$

If we had a positive answer to that problem, we would be able to split f using canonical bases computations. But we would not have any more the bound of cor. 3, for we do not even know if the canonical basis of $k[f_1, \ldots, \widehat{f_i}, \ldots, f_n]$ is finite—as it is integrally closed, it would be a consequence of our conjecture.

The study of this problem has a special interest, for there is an automorphism of $k[x, y, z]$, given by Nagata in [N], which does not match its conclusion, so that if the result holds anyway, the tame generators conjecture would be false in 3 variables.

Example 5: (Nagata 1972) If we consider the automorphism

$$f : \begin{array}{ccc} x & \mapsto & x - 2y(y^2 + xz) - z(y^2 + xz)^2 \\ y & \mapsto & y + z(y^2 + xz) \\ z & \mapsto & z, \end{array}$$

we can see that for all orderings, we cannot have mdeg$f_i \in$ Mon$k[f_j, f_k]$ with all different indices. The consideration of this example convinced Nagata that the tame generators conjecture is false.

We will conclude by giving a class of tame automorphism, for which the answer to our problem is yes.

Definition 6: We say that f is a generic tame automorphism of $k[n]$ if $f = g_h \circ \cdots \circ g_1$, where the g_i are elementary automorphisms such that:
    – g_{2j+1} is de Jonquières and the polynomial P is a dense polynomial of degree a least 2,
    – all coefficients are algebraically independent on the ground field of k,
    – g_{2j} is a permutation which do not leave x_n invariant.

Proposition 7. *If f is a generic tame automorphism of $k[n]$, then the f_i form a canonical basis or there exist $i \in [1, n]$ such that* mdeg$f_i \in$ Mon$\{$mdeg$f_j | j \neq i\}$.

Proof: If f is defined as in def. 6, this is easily proved by induction on h.   Q. E. D.

Corollary 8. *If f is a generic tame automorphism, then it can be split into a composition of elementary automorphism by a canonical basis algorithm where no superposition of multidegree greater than* max$\{$mdeg$f_i\}$ *needs to be considered.*

Proof: The proposition implies that if the f_i do not form themselves a canonical basis, then the canonical basis may be computed by successive reductions.   Q. E. D.

Of course, in practice we will consider automorphism defined by polynomials in $Q[n]$. But it seems, by trying many examples, that the "average" complexity will be the same, the computational time being of the same order than the time needed to build f as a composition of elementary automorphisms.

Example 9: Consider the set of polynomials $\{x, y + x^{10}, z + y^{10}, t + z^{10}\}$. It determines a tame automorphism of $k[x, y, z, t]$ and that can be tested in 1.1s using Scratchpad on a IBM 4381. The computation of the standard basis of Shannon and Sweedler method takes 496.9 seconds using the pure lexicographical ordering.

Of course, in such an example where the inverse is of degree 1000, a method which determines it needs to get in some troubles. In cases where f and f^{-1} have the same degree, standard bases are more efficient in small examples, but canonical bases are better when the degree increases.

REFERENCES

[BCW] H. Bass, E. H. Connell and D. Wright, *The Jacobian Conjecture: Reduction of Degree and Formal Expansion of the Inverse*, Bull. of the A.M.S. **7** (1982), 287–331.

[Bu] B. Buchberger, *Groebner bases: An algorithmic method in polynomial ideal theory*, in "Multidimensional Systems Theory," N.K. Bose (ed.),, Reidel, 1985, pp. 184–232.

[CT] P. Conti and C. Traverso, *Computing the conductor of an integral extension*, Discr. Appl. Math. (to appear). Proceedings AAECC, 1989

[E] A. van den Essen, *A Criterion to Decide if a Polynomial Map is Invertible and to Compute the Inverse*, Cath. Univ. Nijmegen, rep. 8653, 1986, Communications in Algebra (to appear).

[G] M. Giusti, *On the Castelnuovo regularity for curves*, in "Proc. of ISSAC 1989, Portland," ACM press, pp. 250–253.

[Ja] M. Janet, *Systèmes d'équations aux dérivées partielles*, J. de Math., 8ᵉ série, tome III (1920).

[Jo] J.-P. Jouanolou, *Monoïdes*, publ. IRMA 297/P-162, Strasbourg, 1984.

[Ju] H. W. E. Jung, *Ueber ganze birationale Transformationen der Ebene*, J. Reine Agew. Math. **184** (1942).

[H] G. Huet, *An algorith to generate the basis of solutions to homogeneous linear diophantine equations*, Information Processing letters **7**, (1978), 144–147.

[KM] D. Kapur and K. Madlener, *A Completion Procedure for Computing a Canonical Basis of a k-Subalgebra*, in "Computers and Mathematics," E. Kaltofen and S. M. Watt editors, Springer, 1989.

[Ku] W. van der Kulk, *On polynomial rings in two variables*, Nieuw Arch. Wiskunde **1** (1953), 33-41.

[L] D. Lazard, *Résolution des systèmes d'équations algébriques*, Theoretical Computer Science **15** (1981), 77–110.

[MM] E. W. Mayr and A. Meyer, *The Complexity of the Word Problems for Commutative Semigroups and Polynomial Ideals*, Advances in Math. **46** (1982), 305–329.

[Mo] H. M. Moeller, *A reduction strategy for the Taylor resolution*, in "proceedings of EUROCAL'85," Linz, 1985, LNCS, Springer.

[Mor] F. Mora, *Groebner Bases for Non-Commutative Polynomial Rings*, in "Proceeding of AAECC 3," Grenoble, 1985, Lecture notes in Comp. Science 229, Springer.

[N1] M. Nagata, *On the automorphism group of k[x,y]*, Lect. Math. Kyoto University 1972.

[O1] F. Ollivier, *Inversibility of Rational Mappings and Structural Indentifiability in Automatics*, in "Proceedings of ISSAC 1989," Portland, ACM press.

[O2] F. Ollivier, *Inversibility of Rational Mappings and Structural Indentifiability in Control Theory*, prépublication du Centre de Mathématiques de l'École Polytechnique, expanded version of the previous paper, presented at AAECC 7, Toulouse, 1989.

[R1] L. Robbiano, *Terms ordering on the polynomial ring*, in "Proc. of EUROCAL 1985," LNCS, Springer, pp. 513–517.

[R2] L. Robbiano, *On the Theory of Graded Structures*, J. Symb. Comp. **2** (1986).

[RS] L. Robbiano and M. Sweedler, *Subalgebra Bases*, in "Proc. of the Salvador conference on Commutative Algebra," Salvador, 1988,, Springer LNM (to appear).

[S] M. Sweedler, *Ideals bases and valuation rings*, preprint, 1986.

[SS] D. Shannon and M. Sweedler, *Using Groebner bases to determine algebra membership, split surjective algebra homomorphisms and determine birational equivalence*, J. Symb. Comp. **6** (1988), 267–273.

François Ollivier
Laboratoire d'Informatique de l'X (LIX)
École Polytechnique
F-91128 Palaiseau Cedex (France)
cffoll@frpoly11.BITNET
ollivier@cmep.polytechnique.fr

The tangent cone algorithm
and some applications to local algebraic geometry

GERHARD PFISTER

1. About two conjectures

Our interest to use computers to investigate problems of algebraic geometry started because we couldn't solve two conjectures we have been interested for a long time.

The first conjecture goes back to a result of K. Saito cf. [S] who proved the following theorem.

Theorem. *Let* $f \in \mathbf{C}[[x_1, \ldots, x_n]]$ *defining an isolated singularity* $(X,0) \subseteq (\mathbf{C}^n, 0)$ *(i.e.* $\dim_{\mathbf{C}} \mathbf{C}[[x_1, \ldots, x_n]]/(f, \frac{\partial f}{\partial x_1}, \ldots, \frac{\partial f}{\partial x_n}) =: \tau(f) < \infty$).
The following conditions are equivalent:
(1) $f \in (\frac{\partial f}{\partial x_1}, \ldots, \frac{\partial f}{\partial x_n})$, *i.e.* $\tau(f) = \dim_{\mathbf{C}} \mathbf{C}[[x_1, \ldots, x_n]]/(\frac{\partial f}{\partial x_1}, \ldots, \frac{\partial f}{\partial x_n}) =:$
$\mu(f)$
(2) There is an automorphism $\varphi : \mathbf{C}[[x_1, \ldots, x_n]] \to \mathbf{C}[[x_1, \ldots, x_n]]$ *such that* $\varphi(f) = p$ *is a weighted homogeneous polynomial (i.e. there are* $w_1, \ldots, w_n, d$ *positive integers such that* $p(\lambda^{w_1} x_1, \ldots, \lambda^{w_n} x_n) = \lambda^d p$*);* f *is called quasi-homogeneous.*
(3) The Poincaré-complex

$$0 \to \mathbf{C} \to \mathcal{O}_{X,0} \to \Omega^1_{X,0} \to \ldots \to \Omega^n_{X,0} \to 0$$

is exact.

Now everything has still a meaning if instead of hypersurface-singularities we consider complete intersection singularities resp. curve singularities in general (cf.[GMP]), because μ, the Milnor number, is a topological invariant of the singularity and τ, the Tjurina number, is the dimension of the miniversal deformation. One can prove that for complete intersections, resp. Gorenstein curve singularities, (1) and (2) are equivalent. We didn't succeed with (3) being equivalent to (2). With the help of the tangent cone algorithm and an Atari we found a counterexample (cf.[PS]).

Let us consider $f, g \in \mathbb{C}[[x, y, z]]$ defining an isolated curve singularity $(X, 0)$. In this case

$$\tau(f, g) = \dim \mathbb{C}[[x, y, z]]/(f, g, M_1, M_2, M_3),$$

M_i denoting the 2-minors of the Jacobian matrix $\frac{\partial(f, g)}{\partial(x, y, z)}$.

In the general case, one defines

$$\tau = \dim_K (K[[x_1, \ldots, x_n]]/(f_1, \ldots, f_m))^m/J$$

$$J = \left(\begin{pmatrix} \partial f_1/\partial x_1 \\ \vdots \\ \partial f_m/\partial x_1 \end{pmatrix}, \ldots, \begin{pmatrix} \partial f_1/\partial x_n \\ \vdots \\ \partial f_m/\partial x_n \end{pmatrix} \right)$$

Again in the space curve case, one has

$$\mu(f, g) = \dim_{\mathbb{C}} \mathbb{C}[[x, y, z]]/(f, M_1, M_2, M_3) -$$
$$\dim_{\mathbb{C}} \mathbb{C}[[x, y, z]]/(\partial f/\partial x, \partial f/\partial y, \partial f/\partial z)$$

assuming that both dimensions are finite; in the general case,

$$\mu = \sum_{k=1}^{m} (-1)^{m-k} \dim K[[x_1 \ldots x_n]]/(f_1, \ldots, f_{k-1}, \frac{\partial(f_1 \ldots f_k)}{\partial(x_{v_1}, \ldots, x_{v_k})}),$$

$1 \leq v_1 < \ldots < v_k \leq n$.

The Poincaré-complex is given by

$$0 \to \mathbb{C} \to \mathcal{O}_{X,0} \to \Omega^1_{X,0} \to \Omega^2_{X,0} \to \Omega^3_{X,0} \to 0;$$

here

$$\Omega^i_{X,0} = \bigwedge^i \Omega^1_{X,0}$$
$$\Omega^1_{X,0} = \Omega^1_{\mathbb{C}^3,0}/f\Omega^1_{\mathbb{C}^3,0} + g\Omega^1_{\mathbb{C}^3,0} + df\mathcal{O}_{\mathbb{C}^3,0} + dg\mathcal{O}_{\mathbb{C}^3,0}$$
$$\Omega^1_{\mathbb{C}^3,0} = \mathcal{O}_{\mathbb{C}^3,0}dx + \mathcal{O}_{\mathbb{C}^3,0}dy + \mathcal{O}_{\mathbb{C}^3,0}dz.$$

With the identification $\Omega^1_{\mathbb{C}^3,0} = \mathbb{C}[[x, y, z]]^3$ we get

$$\Omega^3_{X,0} = \mathbb{C}[[x, y, z]]/(\frac{\partial f}{\partial x}, \ldots, \frac{\partial g}{\partial z})$$
$$\Omega^2_{X,0} = \mathbb{C}[[x, y, z]]^3/\mathcal{U}$$

$\mathcal{U}$ generated by $\begin{pmatrix} f \\ 0 \\ 0 \end{pmatrix}, \begin{pmatrix} 0 \\ f \\ 0 \end{pmatrix}, \begin{pmatrix} 0 \\ 0 \\ f \end{pmatrix}, \begin{pmatrix} g \\ 0 \\ 0 \end{pmatrix}, \begin{pmatrix} 0 \\ g \\ 0 \end{pmatrix}, \begin{pmatrix} 0 \\ 0 \\ g \end{pmatrix}, \begin{pmatrix} f_y \\ f_z \\ 0 \end{pmatrix},$

$\begin{pmatrix} f_x \\ 0 \\ -f_z \end{pmatrix}, \begin{pmatrix} 0 \\ f_x \\ f_y \end{pmatrix}, \begin{pmatrix} g_y \\ g_z \\ 0 \end{pmatrix}, \begin{pmatrix} g_x \\ 0 \\ -g_z \end{pmatrix}, \begin{pmatrix} 0 \\ g_x \\ g_y \end{pmatrix}.$

For a large class of examples, one can prove that the Poincaré-complex is exact iff $\mu = \dim \Omega^2_{X,0} - \dim \Omega^3_{X,0}$.

On the other hand $(X, 0)$ is quasihomogeneous iff $\mu = \tau$.

The following (unimodular) singularities have exact Poincaré complex but they are not quasihomogeneous:

$$f = xy + z^{l-1}, \quad g = xz + y^{k-1} + yz^2 \quad 4 \le l \le k, \; 5 \le k$$

$$f = xy, \quad g = xz + yz^2 + z^3 + z^{3+i} \quad i > 1.$$

The second conjecture was a conjecture of Mumford concerning the structure of moduli spaces of curve singularities.

If you start with any (irreducible) plane curve singularity defined by $f(x, y) = 0$, $f \in \mathbf{C}[[x, y]]$, then one is interested in the (local) moduli space of it, i.e. a space which classifies all plane curve singularities with the same topological type. The approach to construct it is the following (cf.[**LaP**], [**Me**]): we construct a family $F(x, y, t)$, $\mathbf{t} \in \mathbf{C}^N$ which contains all singular curves we are interested in (the miniversal μ-constant deformation). This is usually not so difficult. For instance, if $f = x^p + y^q$ and $gcd(p, q) = 1$ then

$$F = f + \sum_{\substack{qi+pj>pq \\ i \le p-2 \\ j \le q-2}} t_{ij} x^i y^j$$

is such a minimal family.

But this family still contains trivial subfamilies, i.e. the underlying parameter space is not a moduli space in the sense that different points correspond to different (non-isomorphic) singularities.

Let $\mathbf{C}^N$ be the parameter space, the moduli space will be $\mathbf{C}^N / \sim =: \mathfrak{A}$. How to study $\sim$?

The trivial subfamilies of the family above are given as integral manifolds of a Lie algebra L, the kernel of the Kodaira-Spencer map,

$$\mathrm{Der}_{\mathbf{C}} \mathbf{C}[t] \to \mathbf{C}[t][[x, y]] / (F, \partial F / \partial x, \partial F / \partial y)$$
$$\delta \rightsquigarrow [\delta F]$$

amd $\mathfrak{A} = \mathbf{C}^N / L$.

There is no chance in general for $\mathfrak{A}$ having a nice structure, because the orbits under the action of L will have different dimensions. This implies that $\mathfrak{A}$ is not Hausdorff.

We stratify $\mathbf{C}^N$ by the orbit dimension: $\mathbf{C}^N = \coprod S_i$. The conjecture was now that S_i / L exists as a geometric quotient i.e. especially it has the structure of an algebraic varity. This we could only prove for the stratum of maximal orbit dimension (cf. [**LP**]).

Again with the help of the tangent cone algorithms and the Atari we could compute the kernel of the Kodaira-Spencer map and give an example that S_i/L is not an algebraic variety (cf. [LP]).

How to compute the kernel of the Kodaira-spencer map for a family $F = f + \sum_t \alpha X^\alpha$ of hypersurface singularities with constant Milnor number μ?

Generators can be computed as follows:

Choose a monomial base $\{X^\alpha\}$ of $\mathbf{C}[[X]]/(\frac{\partial f}{\partial x_1}, \ldots, \frac{\partial f}{\partial x_n})$ and assume it is also a base of the free $\mathbf{C}[t]$-module $\mathbf{C}[t][[x_1, \ldots, x_n]]/(\frac{\partial F}{\partial x_1}, \ldots, \frac{\partial F}{\partial x_n})$ which is often the case. Consider $X^\alpha F = \sum h_{\alpha\beta}(t) X^\beta \mod (\frac{\partial F}{\partial x_1}, \ldots, \frac{\partial F}{\partial x_n})$. Then $\{\delta_\alpha\}$, $\delta_\alpha = \sum h_{\alpha\beta} \frac{\partial}{\partial T_\alpha}$ generate the kernel of the Kodaira-Spencer map.

Notice that in this situation often the standard base (cf. §2) is constant and we know in advance that computations can be stopped at a certain degree with respect to $\mathbf{X}$. This is important because the set of parameters t is usually large.

2. The tangent cone algorithm

What is the tangent cone of an algebric variety X at some point $P \in X$? It is the set of all tangents (i.e. lines intersecting X at P with a higher multiplicity then the multiplicity of X at P).

In terms of the local ring $\mathcal{O}_{X,P}$ resp. $\hat{\mathcal{O}}_{X,P}$ it is just given by the graded ring $gr_\mathfrak{m}\mathcal{O}_{X,P}$.

Let $\mathcal{O}_{X,P} = K[x_1, \ldots, x_n]_{(x_1, \ldots, x_n)}/\mathfrak{a}$, resp. $\hat{\mathcal{O}}_{X,P} = K[[x_1, \ldots, x_n]]/\mathfrak{a}$, $\mathfrak{a} \subset K[x_1, \ldots, x_n]$ the ideal defining X. Then $gr_\mathfrak{m}\mathcal{O}_{X,P} = gr_\mathfrak{m}R/gr_\mathfrak{m}\mathfrak{a}R \simeq K[x_1, \ldots, x_n]/I(\mathfrak{a})$, $R = K[x_1, \ldots, x_n]_{(x_1, \ldots, x_n)}$ resp. $K[[x_1, \ldots, x_n]]$. $I(\mathfrak{a})$ is the ideal generated by $\{I(f), f \in \mathfrak{a} \setminus \{0\}\}$. If we consider $f \in R$ as powerseries, $f = \sum_{m \geq 0} f_m$, f_m homogeneous of degree m, then $I(f) = f_k$, $k = ord(f) = \min\{m, f_m \neq 0\}$.

One is interested now to compute generators of $I(\mathfrak{a})$ because the tangent cone contains already a lot of information about X and is much easier to handle. One example is the Hilbert polynomial:

Let $F(l) := \dim_K \mathcal{O}_{X,P}/\mathfrak{m}^l$. It is well known that there is a polynomial P such that $F(l) = P(l)$ if $l \gg 0$, $P(l) = \frac{m(\mathcal{O}_{X,P})}{d!}l^d + \ldots$, $d = \dim \mathcal{O}_{X,P}$ and $m(\mathcal{O}_{X,P})$ the multiplicity. Now $F(l) = \sum_{i=0}^{l-1} \dim_k(K[x_1, \ldots x_n]/I(\mathfrak{a}))_i$ and the Hilbert-polynomial of $K[x_1, \ldots, x_n]/I(\mathfrak{a})$ can be computed.

Especially if $\dim_K \mathcal{O}_{X,P} < \infty$ than

$$\dim_K \mathcal{O}_{X,P} = \dim_K K[x_1, \ldots, x_n]/I(\mathfrak{a})$$

and the computation of the dimension is reduced to a combinatorial problem.

How to compute generators of $I(\mathfrak{a})$?

To do this with a computer we need some ordering between the monomials in R.

Let $T = \{x_1^{\alpha_1} \ldots x_n^{\alpha_n} \ , \ \alpha_i \geq 0\}$, T is a semigroup.

We consider a total ordering $<$ of T compatible with the semigroup structure. It is known (Robbiano, cf. [R]) that such an ordering can be realized by a suitable matrix $\mathcal{U} \in M_{n,s}(\mathbf{R})$, i.e. if u_j is the j-th row of $\mathcal{U}$ then $X^\alpha < X^\beta$ iff there is an i such that $u_j \cdot \alpha = u_j \cdot \beta$ for $j < i$ and $u_i \cdot \alpha < u_i \cdot \beta$ (here $\cdot$ means just the ordinary scalar product).

If such an ordering is fixed we may speak about $L(f)$, $C(f)$ the leading monomial, resp. the leading coefficient for $f \in K[x_1, \ldots, x_n]$.

Let $\mathfrak{a} \subseteq K[x_1 \ldots x_n]$ be an ideal and $L(\mathfrak{a})$ the ideal generated by $\{L(f), \ f \in \mathfrak{a} \smallsetminus \{0\}\}$ than we have the notion of a standard base: $f_1, \ldots, f_m$ is a standard base of $\mathfrak{a}$ if $L(\mathfrak{a})$ is generated by $L(f_1), \ldots, L(f_m)$.

If $<$ is a well ordering than a standard base is also called a Gröbner base.

Now let us consider the ordering given by the matrix

$$\begin{pmatrix} -1 & -1 & \ldots & -1 & -1 \\ 0 & -1 & \ldots & -1 & -1 \\ \vdots & & \ddots & & \vdots \\ 0 & 0 & \ldots & 0 & -1 \end{pmatrix}$$

the "dual" of the graded lexicographic ordering.

Let us assume we have $g_1, \ldots, g_t$ a standard base of $\mathfrak{a}$ with respect to that ordering; then $I(g_1), \ldots, I(g_t)$ generate $I(a)$. So we got generators of $I(\mathfrak{a})$ and computed the tangent cone. The problem of this ordering is that it is not a well ordering. We cannot apply Buchberger's algorithm (cf. [B]) wich lives on well-orderings.

There is one way to avoid this problem:

We choose generators of $\mathfrak{a}$: $f_1, \ldots, f_k$ and make them homogeneous by adding one variable z:

$$f_i^h = z^{d_i} f_i(x_1/z, \ldots, x_n/z), \qquad d_i = \deg f_i.$$

We considers the ordering

$$\begin{pmatrix} 1 & 0 & 0 & \ldots & 0 & 0 \\ 0 & 1 & 1 & \ldots & 1 & 1 \\ \vdots & & & \ddots & & \vdots \\ 0 & 0 & 0 & \ldots & 0 & 1 \end{pmatrix}$$

Using Buchberger's algorithm we get a Gröbner base of $(f_1^h, \ldots, f_k^h)$ $G_1, \ldots, G_t$ and $g_1 = G_1(z = 1), \ldots, g_t = G_t(z = 1)$ is a standard base of $\mathfrak{a}$.

The problem is that this way is very time and space consuming. Let us consider an example:

let $\mathfrak{a} = (\partial f/\partial x, \partial f/\partial y)$ and $f = x^5 + y^{11} + x^3 y^3 + x y^9$.

$\mathfrak{a}$ has a standard base of 5 elements with respect to the ordering defined by $\begin{pmatrix} -1 & -1 \\ -1 & 0 \end{pmatrix}$ and it took in our implementation 0.1 seconds to compute it.

The corresponding homogenization has a Gröbner base of 19 elements and it took 2.4 seconds to compute it.

T. Mora (cf. [M]) developed the tangent cone algorithm which is a modification of Buchberger's algorithm and works for a large class of orderings wich are not well-orderings.

Let us recall the idea of Buchberger's algorithm (cf. [B]; [MPT]): we need the notion of an S-polynomial and of a normal form.

(1) Let $f_1, f_2 \in K[x_1, \ldots, x_n]$, $G = \operatorname{lcm}(L(f_1), L(f_2))$; then $Spoly(f_1, f_2) = (C(f_2)G/L(f_1))f_1 - (C(f_1)G/L(f_2))f_2$.

(2) Let G be a finite set of polynomials. We define $NF(G, h) := h'$ computed by the algorithm
$h' := h$

while exists $f \in G$ such that $L(f)/L(h')$ **do**
    choose any $f \in G$ such that $L(f)/L(h')$
    $h' := h' - \frac{C(h')}{C(f)}\frac{L(h')}{L(f)}f$.

This algorithm terminates because of the well ordering.

Buchberger's algorithm runs as follows:

We start with a set F of generators of an ideal $\mathfrak{a}$ and get a Gröbner base G using the following algorithm:

$G := F$
$B := \{(f_1, f_2) \mid f_1, f_2 \in G \,,\ f_1 \neq f_2\}$
while $B \neq \emptyset$ **do**
    $(f_1, f_2) :=$ a pair in B
    $B := B \smallsetminus \{(f_1, f_2)\}$
    $h :=$ Spoly (f_1, f_2)
    $h' := NF(G, h)$
    if $h' \neq 0$ **then**
        $B := B \cup \{(g, h'),\ g \in G\}$
        $G := G \cup \{h'\}$.

The point in which we need a well ordering in this algorithm is in the NF-algorithm.

Let us consider an example:

Consider $K[x]$ with the ordering $1 > x > x^2 > \ldots$.

Let $G = \{x + x^2\}$ and $h = x$. Then $NF(G, h)$ will produce $-x^2, x^3, -x^4, \ldots$ and this is perfectly o.k. because we know that in $K[[X]]$

$$x = \sum_{\nu=0}^{\infty} (-1)^{\nu} x^{\nu}(x + x^2).$$

But NF does not terminate.

In $K[[x]]$ resp. $K[X]_{(X)}$ it is not important whether we reduce some h or a multiple of h with a unit.

If we had started with $(1 + X) \cdot X$ instead of X everything would have been O.K.

How to produce this unit with the computer?

We consider the reduction of x by $x + x^2$ to x^2. After this step we extend the set of elements which are used for the reduction process by x and the algorithm terminates. This is the general concept for the local normal form algorithm (NFL):

$h' := NFL(G, h)$
$h' := h$
$T := \emptyset$
while there exists $f \in G \cup T$ **such that** $L(f)/L(h')$ **do**
    choose a suitable (what is suitable will be explained later)
        $f \in G \cup T$ such that $L(f)|L(h')$
    $T := T \cup \{h'\}$
    $h' := h' - \frac{C(h')}{C(f)} \frac{L(h')}{L(f)} f$

This means expecially $NFL(G, h) = NF(G, e \cdot h)$ for a suitable $e \in K[x_1, \ldots, x_n]$ which is a unit in $K[[x_1, \ldots, x_n]]$.

What means suitable?

This will depende on the order (cf. [**MPT**]).

If we have the ordering given by

$$\begin{pmatrix} -1 & -1 & \ldots & -1 & -1 \\ 0 & -1 & \ldots & -1 & -1 \\ \vdots & & \ddots & & \vdots \\ 0 & 0 & \ldots & 0 & -1 \end{pmatrix}$$

and we define for any $f \in K[x_1, \ldots, x_n]$ $d(f) := \deg(f) - \text{ord}(f)$ (degree of f as a polynomial, order of f as a power series), we have to choose $f \in G \cup T$ such that $L(f)/L(h')$ and $\max\{d(h'), d(f)\}$ is minimal.

The basic idea of T. Mora tangent cone algorithm is to replace the procedure NF in Buchberger's algorithm by NFL.

The algorithm works for orderings with the following properties (tangent cone orderings (cf. [**MPT**])): let the ordering be defined by the matrix

$$\mathcal{U} = \begin{pmatrix} u_1 \\ \vdots \\ u_s \end{pmatrix} ; \text{ then there is a } k \text{ such that}$$

a) $u_1, \ldots, u_k \in \mathbf{Z}^n$.
b) For all $d_1, \ldots, d_k \in \mathbf{Z}$ the set $\{\alpha \in T \mid \alpha u_1 = d_1, \ldots, \alpha u_k = d_k\}$ is well ordered with respect to the order defined by $\mathcal{U}$.

These properties are always satisfied if $\mathcal{U} \in M_{n,s}(\mathbf{Q})$ or $u_1 = (-1, \ldots, -1)$ or $\mathcal{U}$ defines a well ordering.

The algorithm produces for a given ideal $\mathfrak{a} \subset K[x_1 \ldots x_n]$ a standard base of $\mathfrak{a}\text{Loc}(K[x_1, \ldots, x_n])$, $\text{Loc}(K[x_1, \ldots, x_n]) = \{(1 + g)^{-1} f, \; f, g \in K[x_1, \ldots, x_n], \; L(g) < 1\}$.

Notice that if $1 > x^\alpha$ for all $\alpha \neq 0$ then $\mathrm{Loc}(K[x_1,\ldots,x_n]) = K[x_1,\ldots,x_n]_{(x_1,\ldots,x_n)}$.

If additionally the order satisfies

a) $1 > x^\alpha$

b) if x^β and a sequence $x^{\alpha_1} > \ldots > x^{\alpha_s} > \ldots$ are given, then there is a k such that $x^\beta > x^{\alpha_k}$

then it produces a standard base of $aK[[x_1,\ldots,x_n]]$.

The additional properties are satisfied if the entries in $\mathcal{U}$ are all negative integers.

A slight modification of the algorithm produces also a standard base of a submodule $U \subseteq R^s$.

Especially if $\dim_k R^s/\mathcal{U} < \infty$ we get

$$\dim_k (gr_m R)^s/L(\mathcal{U}) = \dim_k R^s/U.$$

As well as Buchberger's algorithm, the algorithm of Mora is much more sophisticated (cf. [MPT]):

- The criteria can be used as in Buchberger's algorithm.
- In NFL not all h' have to be added to T: if $f \in G \cup T$ and $L(f)/L(h')$ and $d(f) \leq d(h')$ then h' need not to be added to T. If there is a $g \in T$ and $L(h')/L(g)$ and $d(h') \leq d(g)$ then g can be removed by adding h'.
- There is an early termination test if the ideal is zero dimensional: if we have $g_1,\ldots,g_n \in G$ such that $L(g_i) = x_i^{\alpha_i}$ then one can compute a β such that all $x^\alpha < x^\beta$ are already in $aK[x]_{(x)}$. Knowing this we can use NF instead of NFL and stop the reductions if we come to x^β. In some applications we even know x^β in advance and can immediately switch to NF.
- There is a "lazy" version of the algorithm which is usually faster than the "classical" one because it makes after every single reduction step a decision to postpone further reductions to the normal form and starts reducings other elements which look at that level more efficient to reduce.

3. Furhter applications

In the applications it is useful to work first in characteristic p (181 or 32003 are nice primes to work with) which turns to be often 100 times faster, and work later with the interesting examples in characteristic 0. Besides the two applications mentioned already we often use (as already many others did),

(1) the test $\mu = \tau$ to decide whether a singularity is quasihomogeneous.

(2) for a family $F(\mathbf{x},t)$ the test $\partial F/\partial t \in (F, \partial F/\partial x_1,\ldots,\partial F/\partial x_n)$ to decide whether it is analytically trivial.

(3) to compare $\dim K[[x_1,\ldots,x_n]]/(\partial F/\partial x_1,\ldots,\partial F/\partial x_n) = \mu(t)$ with $\dim K[x_1,\ldots,x_n]/(\partial F/\partial x_1,\ldots,\partial F/\partial x_n) = \mu_{\mathrm{glob}}(t)$ to decide whether the hypersurface has just one singular point.

(4) We found an example against the conjecture that τ-constant in a family of special singularities implies μ-constant (cf. [LaP]):

$$F(x, y, t) = x^{11} + y^5 + x^7 y^2 + 2tx^2 y^4 + t^2 x^4 y^3$$
$$\tau = 34, \quad \mu(F(t = 0)) = 40, \quad \mu(F(t \neq 0)) = 39.$$

(5) The way in which the algorithm is organized could be used to simplify theoretical proofs (cf. [LuP]).

(6) We are working from time to time on the Zariski conjecture: μ-constant in a family of hypersurfaces implies multiplicity is constant.

REFERENCES

[B]   Buchberger B., *Gröbner bases: an algorithmic method in polynomial ideal theory*, in "Recent trends in multidimensional systems theory," N. K. Bose, ed., Reidel, 1985.

[GMP] Greuel, G.M., Martin, B., Pfister G., *Numerische Charackterisierung quasihomogener Gorenstein Kurven singularitäten*, Math. Nach. **124** (1985), 123–131.

[LaP]   Laudal A., Pfister G., "Local moduli problems and singularities," Lecture Notes 1310, 1989.

[LuP]   Luengo I., Pfister G., *Normal forms and moduli spaces of curve singularities with semigroup* $< 2p, 2q, 2pq + d >$, Composiito Math. **76** (1990).

[ME]   Merle M. Preprint Ecole Polytechnique.

[M]   Mora, T., *An algorithm to compute the equations of tangent cones*, in "Eurocam 82," Lect. N. Comp. 144, 1982, pp. 158–165.

[MPT] Mora T., Pfister G., Traverso C., *An introduction to the tangent cone algorithm*, in "Issues in Robotics" (to appear).

[PS]   Pfister G., Schönemann H., *Singularities with exact Poincaré complex but not quasihomogeneous*, Revista Mathematica **2** (1989), 161–171.

[R]   Robbiano L, *Terms orderings on the polynomial ring*, in "Eurocal 85," Lect. N. Comp. 204, 1985, pp. 513–517.

[S]   Saito K., *Quasihomogene isolierte Singularitäten von Hyperflächen*, Inv. Math. **14** (1971), 123–142.

Gerhard Pfister
Humboldt-Universität zu Berlin
Sektion Mathematik
108 Berlin, Unter den Linden 6
Deutschland

Effective Methods for Systems of Algebraic Partial Differential Equations

JEAN-FRANÇOIS POMMARET AKLI HADDAK

> *"God created integers, men made the remaining"*
> *(KRONECKER)*
> *... but it is surely the Devil who let them conceive*
> *partial differential equations!*

As a matter of fact, despite the fast progress of computer algebra during the last ten years, only a few steps have been done towards the use of symbolic computers for studying systems of partial differential equations (PDE) ([2], [13]). In particular, one must notice a few modern tentatives for dealing with algebraic PDE through differential algebraic techniques [19] or differential elimination techniques [16], [18] or exterior calculus ([1], Novosibirsk school), but these methods, being absolutely dependent on the coordinate system as they rely on old works ([6], [14], [18]), do not seem to go far inside the intrinsic structure of the system. Despite this point, these methods have been applied with success to control theory during the last five years ([3], [12]).

On another side, that is without any reference to applications, the *formal theory* of PDE has been developped during the period 1965-1975 by a few americans along the pioneering work of D. C. Spencer [17] and H. Goldschmidt [4]. These new techniques (homological algebra, commutative algebra, diagram chasing, jet theory,...), mixing together differential geometric and algebraic arguments, have been applied to the theory of *Lie pseudogroups*, namely groups of transformations solutions of systems of PDE, again algebraic in many useful situations (volume preserving, complex analytic or holomorphic, contact, symplectic transformations,...) ([8], [9]). However, this domain became no longer in fashion and disappeared from the mathematical scenery (See Mathematical Reviews 81 f:58046). As a byproduct, almost no applied mathematician took the risk to use these ideas as they appear quite difficult and sophisticated at first sight ([4], [7]).

After a short historical survey, the purpose of this communication is to convince the reader that *the formal theory of PDE is just the missing tool for studying linear and nonlinear systems of PDE by means of computer algebra*. In particular, it will provide new intrinsic guide-lines (δ-cohomology, formal integrability) along which the computational aspect must follow patterns similar to the "old" ones ([6], [14]), though many results (the main theorem below is a good example) cannot even be proved by using classical methods.

As a general introduction to this theory is out of the scope of this short paper while details can be found in [17] and in the three books of the first-named author [9], [10], [11], we shall rather illustrate the algorithms it may provide along a *mathematical tale* (not far from a true story!) where the "puzzle" has been computed through MACSYMA on MULTICS and SUN by the second-named author. This example, taken from [6] and adapted from [9], has been presented for the first time under this form during an intensive course (april 17, 18, 19, 1989) at "Institut National de Recherche en Informatique et Automatique (INRIA, France)". Not only is this example showing how tricky PDE can be, but also it points out clearly two typical formal problems that can be asked about PDE. We also exhibit the modern solution in a rather intuitive and almost self-contained way, while comparing it with the one first proposed by M. Janet in 1920 [6]. A few similar examples are proposed to the reader at the end, as exercises. Finally, we present a few applications showing that the formal theory will fast become of increasing use in many branches of mathematical physics.

ONCE UPON A TIME... a french professor of mathematics had two excellent students, far above the other ones and so good that whenever one was not the first at an examination, the second was and vice-versa. However, the minds of these two students were quite different: one was more attracted by the possibility to pass from explicit calculus on a white sheet of paper to high level abstract ideas, while the second hated to compute and was more concerned with logic through the use of computers. Of course, the professor was desperately dreaming to know who was the most intelligent student. It happened that, during a dark night when he was not able to sleep, the Devil spoke to him and offered (freely!) a way to select the best student on the basis of the following problem ([6], [9]).

DEVIL'S PROBLEM: Let u, v, y be 3 functions of the cartesian coordinates x^1, x^2, x^3 on euclidean space, related by the following system of 2 PDE where $\partial_{33} y = \partial^2 y / \partial_{x^3} \partial_{x^3}, \ldots$:

$$\partial_{33} y - x^2 \partial_{11} y - u = 0, \quad \partial_{22} y - v = 0$$

1) If $u = v = 0$ the space of solutions of the resulting linear system of PDE for y is a vector space over the constants. What is its dimension?

2) Otherwise, what kind of compatibility conditions must be satisfied by u and v in order to insure the existence of solutions for y?

3) Does there exist a "general" way to solve such problems?

Though this problem seemed easy to the professor at first sight, apart from the last question, he called the two students and gave them a few days for providing back the solution of the two first questions at least. After a week, the answers were as follows:

FIRST WORK:

For the first question, the system that must be solved is:

$$\partial_{33}y - x^2\partial_{11}y = 0, \partial_{22}y = 0$$

The general solution of the second PDE is $y = A(x^1, x^3)x^2 + B(x^1, x^3)$. Bringing it into the first, one gets:

$$(x^2)^2\partial_{11}A + x^2(\partial_{11}B - \partial_{33}A) - \partial_{33}B = 0$$

leading to: $\partial_{11}A = 0$, $\partial_{33}A - \partial_{11}B = 0$, $\partial_{33}B = 0$ that is: $A = a(x^3)x^1 + b(x^3)$, $B = \alpha(x^1)x^3 + \beta(x^1)$
with: $x^1\partial_{33}a(x^3) + \partial_{33}b(x^3) - x^3\partial_{11}\alpha(x^1) - \partial_{11}\beta(x^1) = 0$

Differentiating with respect to x^1, I obtain:

$$\partial_{33}a(x^3) - x^3\partial_{111}\alpha(x^1) - \partial_{111}\beta(x^1) = 0$$

while differentiating again with respect to x^3, I get:

$$\partial_{333}a(x^3) - \partial_{111}\alpha(x^1) = 0$$

that is: $\partial_{111}\alpha(x^1) = \partial_{333}a(x^3) = p = cst$
Similarly: $\partial_{111}\beta(x^1) = q = cst$ with $\partial_{33}a(x^3) - px^3 - q = 0$. Integrating, I have:

$$
\begin{aligned}
a(x^3) &= (p/6)(x^3)^3 + (q/2)(x^3)^2 + rx^3 + s \\
\alpha(x^1) &= (p/6)(x^1)^3 + (1/2)(x^1)^2 + mx^1 + n \\
\beta(x^1) &= (q/6)(x^1)^3 + (u/2)(x^1)^2 + vx^1 + w
\end{aligned}
$$

Hence: $\partial_{33}b = x^3\partial_{11}\alpha + \partial_{11}\beta - x^1\partial_{33}a = lx^3 + u$, and finally:

$$b(x^3) = (l/6)(x^3)^3 + (u/2)(x^3)^2 + fx^3 + g$$

As $y = a(x^3)x^1x^2 + b(x^3)x^2 + \alpha(x^1)x^3 + \beta(x^1)$, the general solution linearly depends on the 12 constants $p, q, r, s, l, m, n, u, v, w, f, g$ and the space of solutions is a 12-dimensional space over the constants, a fact not evident at all at first sight, including the property that all the 6th order derivatives of y are zero because y is a polynomial of degree at most 5 in x^1, x^2, x^3. One can also check that all the 5th order derivatives can be known from lower order ones.

For the second question, it is easy to exhibit a 3rd order compatibility condition as follows:

$$\partial_{233}y - x^2\partial_{112}y - \partial_{11}y = \partial_2 u$$

$$\partial_{2233}y - x^2\partial_{1122}y - 2\partial_{112}y = \partial_{22}u$$

$$\partial_{22233}y - x^2\partial_{11222}y - 3\partial_{1122}y = \partial_{222}u$$

that is to say, *necessarily*:

$$\partial_{233}v - x^2\partial_{112}v - 3\partial_{11}v - \partial_{222}u = 0$$

It is much more difficult to exhibit another one of higher order, not linearly dependent on the partial derivatives of the preceding one. I can proceed as follows:

Let $\partial_{112}y = (1/2)(\partial_{33}v - x^2\partial_{11}v - \partial_{22}u) = w$.

The preceding compatibility condition becomes $(1/2)A \equiv \partial_2 w - \partial_{11}v = 0$ but we have successively:

$$\begin{aligned}
\partial_{11233}y &= \partial_1(\partial_{1233}y - \partial_{111}y) + \partial_{1111}y \\
&= \partial_{11}(\partial_{233}y - \partial_{11}y) + \partial_{1111}y \\
&= \partial_{11}(\partial_2(\partial_{33}y - x^2\partial_{11}y) + x^2\partial_{112}y) + \partial_{1111}y
\end{aligned}$$

leading to $\partial_{1111}y = \partial_{33}w - \partial_{112}u - x^2\partial_{11}w$ and:

$$\begin{aligned}
\partial_{111133}y &= \partial_{3333}w - \partial_{11233}u - x^2\partial_{1133}w \\
&= \partial_{1111}u + x^2\partial_{111111}y \\
&= \partial_{1111}u + x^2\partial_{1133}w - x^2\partial_{11112}u - (x^2)^2\partial_{1111}w
\end{aligned}$$

Hence I obtain the following 6th order compatibility condition:

$$B \equiv \partial_{3333}w - 2x^2\partial_{1133}w + (x^2)^2\partial_{1111}w - \partial_{11233}u + x^2\partial_{11112}u - \partial_{1111}u = 0$$

The various derivatives of u appearing in B are:

$$\partial_{223333}u, \partial_{112233}u, \partial_{111122}u, \partial_{11233}u, \partial_{11112}u, \partial_{1111}u$$

and it follows that B cannot be factorized through derivatives of A, that is B *is effectively a new 6th order compatibility condition for u and v.* I have not been able to find another one, but if there are other ones of order greater than 6, the computations must be extremely tedious and I did not succeed. I am also aware of the fact that, had a linear combination of x^1, x^2, x^3 been effected, no general argument could be used and even the first integration should be impossible in general!

Complement: Though I do not know if such a result can be useful or not, I have obtained the following compatibility condition between A and B:

$$\partial_{3333}A - 2x^2\partial_{1133}A + (x^2)^2\partial_{1111}A - 2\partial_2 B = 0$$

which involves the 7th order derivatives of u and v.

SECOND WORK:

After trying a few calculations, it became so tedious that I gave it up and tried to use a computer according to the standard rules of differentiations. My general idea has been to differentiate successively the two given

PDE in order to compute term after term the various derivatives of y at the origin $(0,0,0)$. I obtained the following board:

order	8	7	6	5	4	3	2	1	0	T	N	O
number	45	36	28	21	15	10	6	3	1	O	U	R
total n.	165	120	84	56	35	20	10	4	1	T	M	D
symbol rk.	41	32	24	17	11	6	2	system		A	B	E
								rank		L	E	R

74   43

58

103

79

N.

									2	2	2	0
									8	8	6	1
								20		20	12	2
							39			40	20	3
						66				70	30	4
					102					112	42	5
				148−1=147						168	56	6

FULL "PUZZLE"

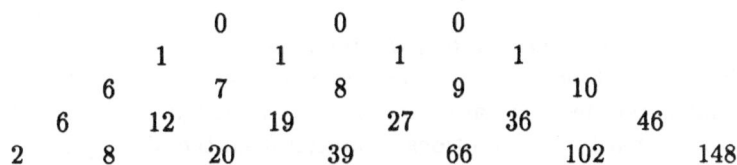

VANISHING DIFFERENCE TRIANGLE

On the horizontal lines are indicated, successively, the order q of derivatives of y, the number of derivatives at order q, the total number of derivatives of y up to order q and the rank of the derived equations with respect to the top order derivatives (*symbol* of order q), eventually the rank of the system with respect to derivatives of order 6 and 7, or 7 and 8, or 5, 6 and 7, or 6, 7 and 8. On the vertical lines are indicated, successively, the number of times one differentiates the equations, the number of equations thus obtained and the total number of equations already produced. Along the diagonal, one can read the ranks of the various sets of equations with respect to the total number of derivatives appearing.

Everything went on easily up to deriving the given PDE 5 times by using MACSYMA on MULTICS. Then the computations became much longer and I had to use SUN during 3 hours in order to obtain the unexpected total rank 147 (*that will prove to be crucial!*). Though I don't know if it is useful or not, I got some extra money to derive 7 times and got the total rank 202 in 9 hours! Any calculus involving higher order derivatives should be extremely expensive though almost impossible to obtain by hand, even if the matrices involved are quite empty, and one should use very big computers. However, the advantage of my method is that it can be used on any example with slight modifications, independently of the coefficients or coordinates.

Concerning the first question, the number of arbitrary derivatives at each order is 1 for order 0, 3 for order 1, then $6-2 = 10-6 = 15-11 = 21-17 = 28-24 = 36-32 = \ldots = 4$ at any order greater or equal to 2. In fact, one has to know at the origin y, $\partial_1 y$, $\partial_2 y$, $\partial_3 y$ and, more generally, $\partial_{1\ldots1} y$, $\partial_{1\ldots12} y$, $\partial_{1\ldots13} y$, $\partial_{1\ldots123} y$. Hence one must evaluate an infinite number of derivatives (4 at each order ≥ 2) and the space of solutions is surely an infinite dimensional vector space.

It is much more delicate to answer to the second question. Indeed, as $2-2 = 8-8 = 20-20 = 0$, there is no compatibility condition of order $0,1,2$ in u and v but there is $40-39 = 1$ compatibility condition of order 3, namely:

$$A \equiv \partial_{233}v - x^2\partial_{112}v - 3\partial_{222}u = 0$$

At order 4 there are $70-66 = 4$ compatibility conditions, that is the given one and its 3 first order derivatives, hence *essentially* no new compatibility condition. A similar comment is valid at order 5 because $112-102 = 10 = 1+3+6$.

It is extremely surprising that the things change at order 6.

Indeed, according to the well known construction of vanishing difference triangle, the awaited rank should be 148 though it is 147. As the number of compatibility conditions produced from the already known one should be $1+3+6+10 = 20$, the rank should be effectively $168-20 = 148$. The fact that it is 147 surely proves that there is one additional compatibility condition, say $B \equiv 0$, that could be known from the computer with extra work, like the preceding one (see before). However, I have not been able to stop the argument at higher order. For example, at order 7 in u and v, one should have $(1+3+6+10+15) + (1+3) = 39$ compatibility conditions and therefore a total rank of $240-39 = 201$ *if there are no new compatibility conditions and no compatibility conditions among the compatibility conditions!* In the present situation, one only find $202-201 = 1$ differential identity between A and B (see before). However, I have no way to conclude because new compatibility conditions may appear after many differentiations as we saw for A and then B. I hope that progresses will be made in the future in order to answer question 3 by finding an algorithm for solving such problems about PDE. In a certain sense, I have the feeling that the latter board is a kind of *"puzzle"* furnishing most of the desired informations... for somebody knowing how to manage with them through a convenient algorithm!

It is only after reading the two works that the professor understood why the Devil had given the challenge freely. He passed through a very sad period of his life,did not teach any longer and decided to leave for USA where he had heard about certain recent new methods for studying systems of PDE. After a year at Princeton University, almost as a student, he finally became able to solve the problem by himself and to discover that both students were, at the same time, right...and wrong!

Before sketching the modern solution of the problem proposed by M. Janet [6], we present a short historical survey of the work done on PDE since the beginning of the century (see [9], p. 1 and [11], p. 59 for a synthetic exposition).

The first ideas for applying algebraic type methods for systems of PDE are essentially contained in the works of a few frenchmen:

1) C. Riquier introduced, at the beginning of this century, the concept of a *"cut"* determined by a system of PDE on the formal power series solutions in the analytic case. Roughly, the given PDE and all their derivatives could be used (successively, if possible) in order to compute certain derivatives at a point, called *"principal"*, as functions of the n independent variables and the other derivatives, called *"parametric"*. The trick was therefore to obtain a unique way of looking at that cut (two different ways for computing the same derivative must produce the same result) while "packing" the parametric derivatives through arbitrary formal power series of $1, 2, \ldots, n$ independent variables in order to know about the *"degree of generality"* of the solution space.

2) Conversely, if one has an infinite number of PDE, one could eventually think that they are in fact coming from a finite set among them or *"basis"*, through differentiations. Such a result, coming from noetherian arguments, has been first suggested and sketched by F. Tresse.

3) In 1920, Maurice Janet, after a few months with D. Hilbert and under the impulse of J. Hadamard, understood for the first time the concept of *"involution"* through the PDE point of view, a concept that had already been proposed in a computational way by Elie Cartan within exterior calculus. One must notice that Cartan did not refer to the work of Janet in 1930, in his letters to A. Einstein, for personal reasons (see [10] for details about this story). As a consequence almost nobody paid attention to the work of Janet who turned to mechanics.

However, the purpose of Janet was also to study differential elimination theory by noticing that, whenever one is differentiating PDE, the highest order derivatives are appearing linearly. The idea was thus to introduce linear algebra techniques for dealing with orders higher than a certain one, say q, while bringing down to non-differential business, involving only a finite number of indeterminates, the way of dealing with orders smaller than q. This approach gave rise to *"differential algebra"* along the work of J. F. Ritt [14] who used the work of Janet without quoting him (see

[10] for details). Of course, one must notice that both Janet and Ritt were essentially interested by the possibility to have constructive methods and finite algorithms that could be performed at least in a manual way.

These ideas on differential algebra, namely the possibility to add the word *"differential"* in front of any algebraic concept (ring, ideal, field, extension, group, invariant,...), was then extended to the Galois theory for polynomials and culminated in the Picard-Vessiot theory for linear ordinary differential equations, brought in [7] to an up to date form by E.R. Kolchin after twenty years of work (see [10] for a critical survey).

Another approach to PDE is to consider the construction of "differential identities" and "differential sequences" in this framework, in order to get a better understanding of mathematical physics (gauge theory, general relativity,...). By a differential sequence, we mean a chain of functional spaces and operators between them such that the composition of two successive operators is zero. Vector calculus ($grad, rot, div$), exterior calculus (exteror derivative d) and tensor calculus provide good examples where the need for an intrinsic understanding is clear. Now, the compatibility conditions for strain in continuum mechanics are second order and the general problem is to know how to solve linear inhomogeneous systems of PDE (with second members) like $\mathcal{D}\xi = \eta$. In first place, Janet got the idea of a *finite length differential sequence* by constructing the *compatibility conditions* $\mathcal{D}_1\eta = 0$ for η, then adding a second member to study $\mathcal{D}_1\eta = \zeta$ while forgetting now about $\mathcal{D}$, and so on. His main result is that no more than n such successive operators $\mathcal{D}_1, \mathcal{D}_2, \ldots, \mathcal{D}_n$ could be constructed, where n is the number of independent variables (see [9] and [11] for details). However, all the arguments of Janet, Ritt,..., considered more extensively by J.M. Thomas [18] and recently by Wu Wen-Tsun [19] or F. Schwarz [16] are absolutely depending on the coordinate system. This means that the difficulty of (manual or automatic) computation can change a lot while the dimensions of spaces or the numbers of equations can change completely if one changes local coordinates.

It was therefore a challenge to modify the previous approach in such a way as to obtain only intrinsic methods and results. Such an improvement has been done during the period 1960-1975 by D. C. Spencer in USA and coworkers ([4], [8], [17]). Nevertheless, they did not mind about the constructive aspect (they did not know the work of Janet) and did not get the link with differential algebra or differential Galois theory which is achieved in our book [10] with promising applications to most of the domains of mathematical physics ([11]).

We now sketch the modern way of dealing with the given problem. We suppose, for this, that the reader is familiar with the basic definitions of differential geometry and jet theory ([5], [9], [16]), though we shall use an intuitive approach by dealing only with linear systems (For nonlinear systems and details, see [4], [9], [10], [11]).

Let X be a manifold of dimension n with local coordinates $(x^1, \cdots, x^n)$ and E be a vector bundle over X with fiber dimension m, local coordinates (x^i, y^k) and changes of charts of the form $\bar{x} = \varphi(x)$, $\bar{y} = A(x)y$.

The q-jet bundle of E is a vector bundle $J_q(E)$ with local coordinates (x^i, y^k_μ) where $\mu = (\mu_1, \ldots, \mu_n)$ is a multi-index and $0 \leq |\mu| = \mu_1 + \cdots + \mu_n \leq q$. For $\mu = 0$ we use to set $y^k_0 = y^k$ and we define

$$\mu + 1_i = (\mu_1, \ldots, \mu_{i-1}, \mu_i + 1, \mu_{i+1}, \ldots, \mu_n).$$

The modern way to deal with a linear system of PDE is to use jet coordinates instead of derivatives in order to define a vector subbundle $R_q \subset J_q(E)$ by a local system of linear equations: $A^{\tau\mu}_k(x)y^k_\mu = 0$ where the Einstein summation is done on $0 \leq |\mu| \leq q$. The prolongations $R_{q+r} \subset J_{q+r}(E)$ will be obtained by substituting derivatives instead of jets coordinates, deriving r times in the usual way and substituting again jets coordinates. For example we have

$$R_{q+1} \begin{cases} A^{\tau\mu}_k(x)y^k_\mu & = 0 \\ A^{\tau\mu}_k(x)y^k_{\mu+1_i} + \partial_i A^{\tau\mu}_k(x)y^k_\mu & = 0 \end{cases}$$

The projections $\pi^{q+r+s}_{q+r} : J_{q+r+s}(E) \longrightarrow J_{q+r}(E)$ induce maps $\pi^{q+r+s}_{q+r} : R_{q+r+s} \longrightarrow R_{q+r}$ wich are not in general surjective and we may introduce $R^{(s)}_{q+r} = \pi^{q+r+s}_{q+r}(R_{q+r+s}) \subseteq R_{q+r}$. We shall assume for simplification that all the $R^{(s)}_{q+r}$ are vector bundles.

Now the symbol M_q of R_q is defined by the linear equations:

$$M_q \ \{A^{\tau\mu}_k(x)v^k_\mu = 0 \ | \mu | = q.$$

It follows that we have:

$$M_{q+r} \ \{A^{\tau\mu}_k(x)v^k_{\mu+\nu} = 0 \ | \mu | = q, | \nu | = r.$$

and M_{q+r} is said to be the r-prolongation of M_q as we can see that M_{q+r} only depends on M_q.

Let us now introduce the cotangent bundle $T^* = T^*(X)$ of X with corresponding tensor, exterior and symmetric products respectively denoted by $\otimes, \Lambda, S$. We may define the Spencer map:

$$\delta : \Lambda^s T^* \otimes S_{q+r+1} T^* \otimes E \longrightarrow \Lambda^{s+1} T^* \otimes S_{q+r} T^* \otimes E$$

by the local formula $(\delta v)^k_\mu = dx^i \wedge v^k_{\mu+1_i}$ on families of forms.

As $A^{\tau\mu}_k(x)(\delta v)^k_\mu = dx^i \wedge A^{\tau\mu}_k(x)v^k_{\mu+1_i}$, we may obtain the restriction map:

$$\delta : \Lambda^s T^* \otimes M_{q+r+1} \longrightarrow \Lambda^{s+1} T^* \otimes M_{q+r}$$

to exterior forms with value in the symbols. One easily checks that $((\delta \circ \delta)v)^k_\mu = dx^i \wedge dx^j \wedge v^k_{\mu+1_i+1_j} \equiv 0$ and thus that $\delta \circ \delta = 0$. The cohomology at $\Lambda^s T^* \otimes M_{q+r}$ of the corresponding sequences is denoted by $H^s_{q+r}(M_q)$ as it only depends on M_q which is said to be *s-acyclic* if $H^1_{q+r} = \ldots = H^s_{q+r} = 0, \forall r \geq 0$ or *involutive* if it is *n*-acyclic.

It does not seem possible to study acyclicity in general by means of a constructive procedure though this can be done for a few specific examples (9) which happen to be extremely useful in mathematical physics [11]. However, the following key definition provides a constructive way to study involution and comes from [9] where we have adapted for this purpose the techniques of Janet [6]. The reader knowing his work will check that an involutive symbol gives rise to a "*complete*" set of "*principal*" monomials at each order but that the contrary is not true.

For giving it, let us order the v^k_μ according to the lexicographic order on (μ, k) and let us solve in cascade the various linear equations defining M_q with respect to the highest v^k_μ each time.

Let us associate with each such solved equation the multiplicative variables $(x^1, \ldots, x^i)$ if the highest v^k_μ is of class i, that is to say is such that $\mu_1 = \ldots = \mu_{i-1} = 0$ with $\mu_i \neq 0$. We shall state:

Definition 1 M_q is said to be involutive if there exists a coordinate system, called δ-regular, such that the first prolongation with respect to the multiplicative variables is producing M_{q+1}.

One can prove that this local definition of involution is equivalent to the previous intrinsic one through the vanishing of the Spencer δ-cohomology. In particular, after choosing a δ-regular coordinate system, the dimensions of the various sbspaces just computed are describing intrinsic properties of the symbols through the Hilbert polynomial $dim M_{q+r}$ for $r \geq 0$.

The key theorem for studying systems of PDE is the following delicate theorem that cannot be proved without diagram chasing and the formal theory of PDE ([9], [17]) as 2-acyclicity has no analogue in any classical work and its use in differential algebra is superseding the classical approach (compare [10], p. 246 with [7], p.167).

Theorem 1 *If M_q is 2-acyclic (involutive), then $(R^{(1)}_q)_{+r} = R^{(1)}_{q+r}$.*

Corollary 1 *If M_q is 2-acyclic (involutive) and $R^{(1)}_q = R_q$ then $R^{(1)}_{q+r} = R_{q+r}$, $\forall r \geq 0$ (formally integrable system).*

Corollary 2 *With any R_q one can associate a formally integrable $R^{(s)}_{q+r}$ through a finite algorithm for finding the integers $r, s \geq 0$.*

The second corollary which can be proved by inductive use of the first corollary as a criterion of formal integrability is essential for knowing about the formal power series solutions in a constructive way. The induction crucially uses the fact that, even if M_q is not involutive, then M_{q+r} becomes involutive for big enough r. We must notice that the main theorem mixes properties related to explicit derivations (system) with properties related to linear algebra and module theory (symbol): this is the main novelty of the american work.

The remainder of the paper illustrates all the preceding definitions and constructive results by showing out that the diagram obtained by means of a computer is sufficient in order to obtain $r = 3, s = 2$ with $M_5^{(2)} = 0$ in the case of our initial problem.

For this, if $\Phi : J_q(E) \longrightarrow F$ is a morphism between vector bundles we shall define the r-prolongation $\rho_r(\Phi) : J_{q+r}(E) \longrightarrow J_r(F)$ by means of the local formula: $\Phi : A_k^{\tau\mu}(x)y_\mu^k = u^\tau \Longrightarrow \rho_r(\Phi) : A_k^{\tau\mu}(x)y_{\mu+\nu}^k + \ldots + \partial_\nu A_k^{\tau\mu}(x)y_\mu^k = u_\nu^\tau$ for $0 \leq |\nu| \leq r$, according to the standard rules of derivations.

Now if $\Phi : E \longrightarrow F$ is a morphism, between vector bundle we may introduce $K = ker\Phi$, $Q = coker\Phi = F/im\Phi$ in the exact sequence of vector bundles:

$$0 \longrightarrow K \longrightarrow E \longrightarrow F \longrightarrow Q \longrightarrow 0$$

where one checks: $dimK - dimE + dimF - dimQ = 0$ because $dimE - dimK = dim\,im\Phi = dimF - dimQ$.

Finally, introducing $J_{q+r}^q(E) = ker\pi_q^{q+r}$ it is easy to check that $J_q^{q-1}(E) = S_qT^* \otimes E$ where $T^* = T^*(X)$ is the cotangent bundle of X and S_q the symmetric product. Also we recall these results and definitions in the following commutative diagram $diag(q,r,s)$, where $F = J_q(E)/R_q$.

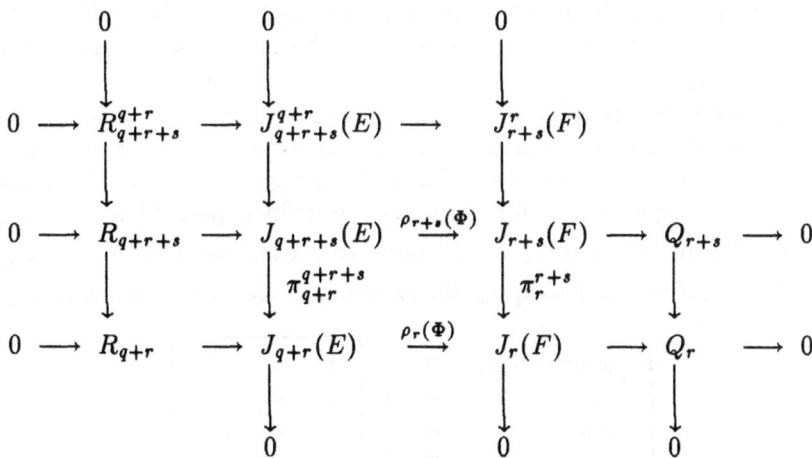

with $M_{q+r} = R_{q+r}^{q+r-1}$ and the short exact sequences $seq(q,r,s)$:

$$0 \longrightarrow R_{q+r+s}^{q+r} \longrightarrow R_{q+r+s} \longrightarrow R_{q+r}^{(s)} \longrightarrow 0$$

In our example, $n = 3, dimE = m = 1, dimF = 2, q = 2$. The numbers in the horizontal lines are successively q, $dimS_qT^* \otimes E$, $dimJ_q(E)$,

$dim S_q T^* \otimes E - dim M_q$. The numbers in the vertical columns are successively r, $dim S_r T^* \otimes F$, $dim J_r(F)$ while the diagonal numbers are just $dim\ im\rho_r(\Phi) = dim J_{q+r}(E) - dim R_{q+r}$, for $r = 0, 1, \cdots, 6$.

We are ready for applying inductively the theorem and the criterion to R_2 until we reach $R_5^{(2)}$ and prove that it is formally integrable with involutive symbol or simply *involutive*. Let us consider R_2 with $dim R_2 = 10 - 2 = 8$. Then $seq(2, 0, 1)$ gives: $dim R_2^{(1)} = dim R_3 - dim M_3 = 12 - (10 - 6) = 8 = dim R_2$ and thus $R_2^{(1)} = R_2$.

However we have M_2:

$$\begin{cases} v_{33} - x^2 v_{11} &= 0 \\ v_{22} &= 0 \end{cases} \quad \begin{array}{|ccc|} \hline x^1 & x^2 & x^3 \\ x^1 & x^2 & . \\ \hline \end{array}$$

and M_2 is not involutive because M_3:

$$\begin{cases} v_{333} - x^2 v_{113} &= 0 \\ v_{233} - x^2 v_{112} &= 0 \\ v_{223} &= 0 \\ v_{222} &= 0 \\ v_{133} - x^2 v_{111} &= 0 \\ v_{122} &= 0 \end{cases} \quad \begin{array}{|ccc|} \hline x^1 & x^2 & x^3 \\ x^1 & x^2 & . \\ x^1 & x^2 & . \\ x^1 & x^2 & . \\ x^1 & . & . \\ x^1 & . & . \\ \hline \end{array}$$

and $dim M_3 = 10 - 6 = 4$ instead of $10 - 5 = 5$. Hence R_2 is not formally integrable. Indeed $R_3^{(1)} \subset R_3$ because $seq(2, 1, 1)$ gives: $dim R_3^{(1)} = dim R_4 - dim M_4 = (35 - 20) - (15 - 11) = 15 - 4 = 11 < 12$ and we may start afresh with $R_3^{(1)}$.

Now we notice that M_3 is involutive because $dim M_4 = 15 - 11 = 4 = 15 - nb$ of multiplicative variables in M_3. However, we must also notice that, because (x^1, x^2, x^3) is δ-regular while (x^2, x^1, x^3) is not, the framed board on the right is therefore intrinsically defined (see [9] for technical details). We may thus apply the theorem to R_3 and get $(R_3^{(1)})_{+r} = R_{3+r}^{(1)}$. In particular, if we want to apply the criterion to $R_3^{(1)}$ we must study $M_3^{(1)}$:

$$\begin{cases} v_{333} - x^2 v_{113} &= 0 \\ v_{233} &= 0 \\ v_{223} &= 0 \\ v_{222} &= 0 \\ v_{133} - x^2 v_{111} &= 0 \\ v_{122} &= 0 \\ v_{112} &= 0 \end{cases} \quad \begin{array}{|ccc|} \hline x^1 & x^2 & x^3 \\ x^1 & x^2 & . \\ x^1 & x^2 & . \\ x^1 & x^2 & . \\ x^1 & . & . \\ x^1 & . & . \\ x^1 & . & . \\ \hline \end{array}$$

$M_3^{(1)}$ is not involutive because of the non multiplicative variable x^3 for $v_{112} = 0$ but its first prolongation $M_4^{(1)}$ is involutive (exercise). Hence, if we want to check the criterion we have:

$$\pi_4^5((R_4^{(1)})_{+1}) = \pi_4^5((R_3^{(1)})_{+2}) = \pi_4^5(R_5^{(1)}) = R_4^{(2)} \subseteq R_4^{(1)}$$

$$dim R_4^{(1)} = dim R_5 - dim M_5 = (56 - 39) - (21 - 17) = 13.$$

Now the reader may check by himself that $R_4^{(2)}$ is obtained from $R_4^{(1)}$ by adding the equation $y_{1111} = 0$, and $dim R_4^{(2)} = dim R_4^{(1)} - 1 = 12$.

We may thus start afresh with $R_4^{(2)}$. Its symbol $M_4^{(2)}$ is:

$$\left\{ \begin{array}{lcl}
v_{3333} & = 0 & \boxed{\begin{array}{ccc} x^1 & x^2 & x^3 \end{array}} \\
\end{array} \right.$$

		x^1	x^2	x^3
v_{3333}	$= 0$	x^1	x^2	x^3
v_{2333}	$= 0$	x^1	x^2	.
v_{2233}	$= 0$	x^1	x^2	.
v_{2223}	$= 0$	x^1	x^2	.
v_{2222}	$= 0$	x^1	x^2	.
$v_{1333} - x^2 v_{1113}$	$= 0$	x^1	.	.
v_{1233}	$= 0$	x^1	.	.
v_{1223}	$= 0$	x^1	.	.
v_{1222}	$= 0$	x^1	.	.
v_{1133}	$= 0$	x^1	.	.
v_{1123}	$= 0$	x^1	.	.
v_{1122}	$= 0$	x^1	.	.
v_{1112}	$= 0$	x^1	.	.
v_{1111}	$= 0$	x^1	.	.

$M_4^{(2)}$ is not involutive because of the non multiplicative variable x^3 for $v_{1111} = 0$. Its first prolongation is zero and thus trivially involutive. However, we have:

$$(R_4^{(2)})_{+1} = ((R_4^{(1)})^{(1)})_{+1} = ((R_4^{(1)})_{+1})^{(1)} = (R_5^{(1)})^{(1)} = R_5^{(2)}$$

because $M_4^{(1)}$ and M_3 are involutive. But we have $\pi_4^5(R_5^{(2)}) = R_4^{(3)}$ and from $diag(2, 3, 2)$ we deduce:

$$dim R_5^{(2)} = dim R_7 - dim R_7^5 = (120 - 102) - (64 - 58) = 18 - 6 = 12$$

but from $diag(2, 2, 3)$ we get:

$$dim R_4^{(3)} = dim R_7 - dim R_7^4 = (120 - 102) - (85 - 79) = 18 - 6 = 12$$

Hence $dim M_5^{(2)} = dim R_5^{(2)} - dim R_4^{(3)} = 12 - 12 = 0$ and $M_5^{(2)} = 0$ is trivially involutive. However, we have:

$$(R_5^{(2)})_{+1} = (R_4^{(2)})_{+2} = ((R_4^{(1)})^{(1)})_{+2} = (R_6^{(1)})^{(1)} = R_6^{(2)}$$

and $\pi_5^6(R_6^{(2)}) = R_5^{(3)}$. Using similarly $diag(2,3,3)$ we get:

$$dimR_5^{(3)} = dimR_8 - dimR_8^5 = (165 - 147) - (109 - 103) = 18 - 6 = 12$$

and it follows that $R_5^{(2)}$ is involutive with zero symbol. In fact, using $diag(2,4,2)$ we get: $dimR_6^{(2)} = dimR_8 - dimR_8^6 = (165-147) - (81-75) = 18 - 6 = 12$ and we check that $R_6^{(2)} \simeq R_5^{(2)}$ while $R_5^{(3)} = R_5^{(2)}$.

We have thus proved that the space of solutions of R_2 is a 12-dimensional vector space over the constants as it coincides with the space of solutions of the involutive system $R_5^{(2)}$ which has a zero symbol.

This result will be used in order to solve the second question. Indeed, introducing u and v as second members of R_2 we may add therefore to $R_5^{(2)}$ second members involving the derivatives of u and v up to order $7 - 2 = 5$. Now, as $R_5^{(2)}$ is involutive, the compatibility conditions for $R_5^{(2)}$ becomes first order in the second members of $R_5^{(2)}$ because the criterion only involves one differentiation. Accordingly, the compatibility conditions for R_2 only involves the derivatives of u and v up to order 6. Accordingly $Q_1 = Q_2 = 0$, $dimQ_3 = 1, \cdots, dimQ_6 = 21 = dimJ_3(Q_3) + 1$.

This result proves that the common answer given by the two students, being necessary, is also sufficient. However, the mistakes done by the second student (not done by the first student, of course, as he made an explicit integration) is that R_2 is not formally integrable and that R_4 provides a third order equation $y_{112} = 0$ which is not among the equations R_3. Hence the taylor series cannot be known term by term just by differentiating the given equations.

For the reader aware of the methods proposed by Janet and successors, we explain the essential differences.

The main point, discovered by the americans through the δ-sequence and cohomology, was that only involution (or 2-acyclicity) could lead to intrinsic results. This fact has been missed by Janet though he had all the material needed in hands [9]. This is the reason for which the methods $seem$ similar. One must notice that involution and acyclicity are also lacking in differential algebra (compare [7] and [10]).

As a byproduct, the main difference is that, in order to stay with intrinsic operations, one cannot mix together jet coordinates of different orders and this is the reason for which the properties of the symbols become essential in the american work. Hence, in practice, as a result of fifty years of progress, $completeness$ and $passivity$ in the sense of Janet are respectively superseded by $involutivity$ and $formal\ integrability$ in the sense of Spencer, though, of course, the second concepts imply the first ones. As for the determination of compatibility conditions, exactly like in the purely linear situation, it was proved by Spencer to be related to the inductive construction of $cokernels$ in long exact sequences.

More precisely, in dealing with the above example, y_{33}, y_{22} are the principal derivatives with associated monomials $(\chi_3)^2, (\chi_2)^2$ for a covector $\chi = \chi_i dx^i$. Such a system of monomials is not complete, the completion beeing $(\chi_3)^2, (\chi_2)^2, (\chi_2)^2\chi_3$ but *now the orders are mixed up*. Also, after a permutation $(1, 2, 3) \longrightarrow (2, 1, 3)$ for numbering the coordinates, the principal monomials become $(\chi_3)^2, (\chi_1)^2$ and the complete set becomes $(\chi_3)^2, (\chi_1)^2, (\chi_1)^2\chi_2, (\chi_1)^2\chi_3$. This result explains why, on the same example, the method of Janet does provide the same intrinsic results, whenever there are (like the dimension 12 of the solution space) but the details are quite different. In any case, we just ask the reader to consider that, if we had effected a tricky (even linear) change of coordinates before providing the example, the Janet algorithm could not be handled easily, though there is no change in the operations and results of the general algorithm proposed in this paper.

To conclude this paper, we indicate a major domain of application, namely *control theory* [3]. Indeed, up to now *classical control theory* is studying the input/output relations defined by systems of ordinary differential equations (ODE). In view of the preceding example where (u, v) may be considered as the input (source term) and y as the output, we understand that more generally, *partial differential control theory* will study the input/output relations defined by systems of partial differential equations (PDE). In the langage of control theory, the first question is nothing else than the definition of *state* as a jet space (which may be finite or infinite dimensional but is perfectly determined on the condition to know the corresponding formally integrable system having the same formal solutions). It follows from our work ([11], [12]) that the formal theory of systems of PDE will become an essential tool for adapting computer algebra to partial differential control theory.

We finally list a few examples that can be easily solved by means of computer algebra.

1. $)x^1, x^2; u, v; y$
$Dy \equiv x^1\partial_1 y + x^2\partial_2 y = u$, $\Delta y \equiv \partial_{11}y + \partial_{22}y = v$
(Hint:If $u = v = 0$ the space of solutions has dimension 2 over the constants, otherwise one needs $\Delta u - Dv - 2v = 0$.)

2. $)x^1, x^2, x^3; u, v, w; y$
$\partial_{33}y = u$, $\partial_{22}y = v$, $\partial_{112}y = w$ (Hint:If $\partial_{22}u - \partial_{33}v = 0$, $\partial_{11}v - \partial_2 w = 0$, $\partial_{112}u - \partial_{33}w = 0$.)

3. $)x^1, x^2; u; y$
$\partial_2 y - \partial_1 y - y = 0$, $\partial_{111}y + \partial_1 y - \partial_1 u = 0$, (Hint:If $u = 0$, the space of solutions has dimension 3 over the constants, otherwise one needs $\partial_{12}u - \partial_{11}u - \partial_1 u = 0$.)

REFERENCES

[1] E. A. Arais, V. P. Sapeev, N. N. Janenko, *Realization of Cartan's method of exterior forms on an electronic computer*, Dokl. Akad. Nauk. SSSR **214** (1974). (Soviet. Math. Dokl., **15** pages 203–205)

[2] J. Davenport, Y. Siret, E. Tournier, "Calcul Formel," Masson, Paris, 1986.

[3] M. Fliess, *Automatique et corps différentiels*, Forum Math. 1 (1989).

[4] H. Goldschmidt, *Integrability criteria for systems of non-linear partial differential equations*, J. Diff. Geometry 1 (1969), 269–307.

[5] M. Golubitsky, V. Guillemin, "Stable mappings and their singularities," Springer-Verlag, 1973.

[6] M. Janet, "Leçons sur les systèmes d'équations aux dérivées partielles," Cahiers scientifiques, Fasc. IV, Gauthier-Villars, Paris, 1929.

[7] E. R. Kolchin, "Differential algebra and algebraic groups," Academic Press, New-York, 1973.

[8] A. Kumpera, D. C. Spencer, "Lie equations," Annals of Math. Studies 73, Princeton University Press, 1972.

[9] J. F. Pommaret, "Systems of partial differential equations and Lie pseudogroups," New York, 1978.

[10] J. F. Pommaret, "Differential Galois theory," Gordon and Breach, New York, 1983.

[11] J. F. Pommaret, "Lie pseudogroups and mechanics," Gordon and Breach, New York, 1988.

[12] J. F. Pommaret, *Problèmes formels en théorie du contrôle aux dérivées partielles*, C. R. Acad. Sc. Paris **308**, **I** (1989), 457–460.

[13] R. H. Rand, "Computer algebra in applied mathematics: an introduction to MACSYMA," Pitman, London, 1984.

[14] J. F. Ritt, "Differential algebra,," Dover, 1950, 1966.

[15] D. J. Saunders, "The geometry of jet bundles," London Mayhematical Society Lecture Note Series, 142, Cambridge University Press, 1989.

[16] F. Schwarz, *The Riquier-Janet theory and its application to non-linear evolution equations*, Physica **110** (1984), 243–251.

[17] D. C. Spencer, *Overdetermined systems of partial differential equations*, Bull. Amer. Math. Soc. **75** (1965), 1-114.

[18] J. M. Thomas, "Systems and roots," W. Byrd Press, 1962.

[19] Wu Wen-Tsun, *On the foundation of algebraic differential geometry*, Mathematics-Mechanization research preprints 3 (1989), 1–26, Institute of Systems Science, Academia Sinica, Pekin.

Jean-François Pommaret, Akli Haddak
Centre d'Enseignement et de Recherche
en Mathématiques Appliquées (CERMA),
Ecole Nationale des Ponts et Chaussées (ENPC),
La Courtine, 93167 Noisy-le-Grand Cedex, France

Finding roots of equations involving functions defined by first order algebraic differential equations

DANIEL RICHARDSON

Abstract. A method is given for approximating solutions of $f(x) = 0$, for f in a certain class $\mathcal{F}$ of real valued analytic functions of one real variable. The method depends on being able to decide the sign of $f(r)$ for given f in $\mathcal{F}$ and rational r. That is, approximation of roots is Turing reduced to the constant problem. The last root problem is evaded: essentially solutions are only found in given finite intervals.

The class of functions $\mathcal{F}$ contains x, $\exp(x)$, $\log(x)$ for $x > 0$, $\sin(x)$ for x in $(-\pi/2, \pi/2)$, and is closed under field operations, differentiation and integration. $\mathcal{F}$ is built up by successively adding solutions g of first order differential equations $q(x, g, g') = 0$, where q is a polynomial whose coefficients may involve previously defined functions.

The main technique used is construction of a false derivative. A false derivative of $f(x)$ is a function $f*(x)$ which is continuous and has the same sign as $f'(x)$ whenever $f(x) = 0$.

I) Introduction

This paper is about finding solutions of equations, $f = 0$, where f is a real valued function of one real variable. The function f is supposed to be a polynomial in the variable x and in some monomials applied to x, and the monomials are assumed to be solutions of non singular first order algebraic differential equations. An example would be a polynomial in x, $\exp(x)$ and $\log(x)$.

The main tool used is cylindrical decomposition, together with the idea of a false derivative. In general, a false derivative of f is a simplification of the true derivative which is obtained by using $f = 0$. The false derivative $f*(x)$ is required to be continuous and so that $f*(x)$ and $f'(x)$ have the same sign (i.e. are either both less than zero, both greater than zero or both equal to zero) whenever $f(x) = 0$. Let I be an interval on which $f(x)$ and $f*(x)$ are defined, and let u and v be two roots of $f(x)$ so that there is no root of $f(x)$ between u and v. If either u or v is a double root of $f(x)$, then it is also a root of $f*(x)$. Suppose on the other hand that u

and v are both simple roots. Then the signs of $f'(u)$ and $f'(v)$ must differ. Since $f*(x)$ and $f'(x)$ have the same signs whenever $f(x) = 0$, the signs of $f*(u)$ and $f*(v)$ must differ; therefore since $f*(x)$ is continuous, $f*(x)$ has a root between u and v. Thus roots of $f*(x)$ separate roots of $f(x)$.

Suppose, for example, we wish to solve $p(x, g(x)) = 0$, where $g(x)$ is the most complicated monomial in p. Replace $g(x)$ by a new variable y to obtain $p(x, y)$. Now find a cylindrical decomposition of the x, y plane which is sign invariant for $p(x, y)$. Above each interval on the x axis, there are finitely many implicitly defined functions of x which give possible values for y. Consider a particular interval. Suppose the functions above this interval are $\alpha_1, ..., \alpha_k$. We look for solutions of $g(x) = \alpha_i(x)$. (These are also solutions of $p(x, g(x)) = 0$; there may be other solutions but these are at singularities of the cylindrical decomposition.) Assume that the derivative of $g(x)$ is $\mathcal{G}(x, g(x))$, where $\mathcal{G}$ is simpler than g. (This will be true if g satisfies an algebraic differential equation.)

A false derivative of $g(x)$ would be $\mathcal{G}(x, 0)$. What we want however is a false derivative of $g(x) - \alpha_i(x)$. We begin by replacing $g(x)$ by $\alpha_i(x)$ in $\mathcal{G}$, to obtain $\mathcal{G}(x, \alpha_i(x))$. Then between two roots of $g(x) - \alpha_i(x)$ there is either a singularity or a root of $\mathcal{G}(x, \alpha_i(x)) - \alpha_i(x)'$, which is called the false derivative of $g(x) - \alpha_i(x)$.

It is not necessary actually to construct the implicitly defined $\alpha_i(x)$. That is, cylindrical decomposition is used in the proof, and to motivate the construction, but is not needed to find the roots. We want roots of $\mathcal{G}(x, \alpha_i) - \alpha_i'$. These will be x components of the solution of the pair of equations

$$p(x, y) = 0$$
$$\mathcal{G}(x, y)p_y(x, y) + p_x(x, y) = 0$$

If $R(x)$ is the result of eliminating y between these two equations, then we have that the roots of $g(x) - \alpha_i(x)$ are separated by roots (or possibly singularities) of $R(x)$ and singularities of the cylindrical decomposition; and $R(x)$ does not contain the difficult monomial $g(x)$, and the cylindrical decompositon does not involve $g(x)$ at all.

Between two roots of $p(x, g(x))$, the curve $y = g(x)$ either crosses over a singularity of the cylindrical decomposition of $p(x, y)$, or it crosses from one implicitly defined function $\alpha_i(x)$ to another one, $\alpha_j(x)$, with i different from j; or it crosses some $\alpha_i(x)$ twice. Thus the roots of $p(x, g(x))$ are separated by singularities of the cylindrical decomposition, roots of $p_y(x, g(x))$ (which has lower degree in g than $p(x, g)$), and roots and singularities of $R(x)$.

Application of the false derivative technique allows us to separate the roots of $p(x, g(x))$ by roots of equations which are simpler.

For the rest of this paper, some measure M on the real line will be assumed. The assumptions about M are that intervals are measurable, that M is computable at intervals with rational endpoints, and is continuous;

and that it gives finite measure to the whole real line. An example of such a measure would be one which was the same as the usual Lesbegue measure in a finite interval and was 0 outside the interval; another example would be a probability measure defined by the integral of a continuous function.

To say that we have solved the equation $f(x) = 0$ will, in the context of this paper, mean that for any $\epsilon > 0$, we can find a set on the x axis which is a finite union of intervals with endpoints which are either rational numbers or plus or minus infinity and which has total measure $< \epsilon$ and which contains all the roots. This concept of solvability is just a way of focusing attention on the finite intervals and evading the last root problem.

The problem of solving the equation $f(x) = 0$ will also be called the problem of approximating the roots of $f(x)$.

According to this notion of solvability, if we know $f(x)$ has at most one root in an interval I, which may be infinite, and if we can decide whether or not $f(x) = 0$ at rational x, then we can solve $f(x) = 0$, using the bisection algorithm. The difficulty of solving $f(x) = 0$ will be measured by the number of times $f(r) = 0$ has to be tested, for rational r. Any possible difficulties about computing the measure will be ignored.

If we know $f(x)$ has at most one root in interval I, the difficulty of approximating the root, if any, to within ϵ increases only in $O(\log(1/\epsilon))$. Also,

Approximation Theorem. *Let $f(x)$ and $h(x)$ be two real valued continuous functions of a real variable x, and assume they both have the same domain of definition, which is a single open interval I, whose endpoints are either plus or minus infinity or computable real numbers. If we know that the roots of $f(x)$ are separated by the roots of $h(x)$ and we can solve $h(x) = 0$, and we have a way of testing the sign of $f(x)$ for rational x, then we can also solve $f(x) = 0$.*

Proof: Given ϵ, we wish to approximate the roots of $f(x)$ to within ϵ. First approximate the roots of $h(x)$ to within $\epsilon/2$. This gives us a finite set of intervals with rational endpoints which contain all the roots of $h(x)$. The measure of this set is below $\epsilon/2$. Let the intervals, in order, be (a_i, b_i), $i < n$. In any of the gaps, $[b_i, a_{i+1}]$, there can be at most one root of f, and we can find out for sure whether or not there is such a root by testing $f(x) = 0$ at b_i and at a_{i+1}. If there is no root, we discard the gap, and if there is a root we approximate it, using bisection to within measure $\epsilon/2(n + 2)$. The two possible long gaps to the left of a_1 and to the right of b_n can be dealt with in the same way, except that there is never any way of knowing for sure whether or not there is a root in the gap. To do the long gap to the left, find point b_0 to the left of a_1 so that the measure of I to the left of b_0 is less than $\epsilon/2(n + 2)$. We can test whether or not there is a root of f between b_0 and a_1. If there is a root we can approximate it by bisection as before. If not, we take the long interval to the left of b_0, which has small measure. The long gap to the right of b_n is dealt with in the same way. The result of all this is a set of intervals with rational endpoints which contain all the roots of $f(x)$, and, incidentally, all the roots of $h(x)$, and which has measure below ϵ.

All the decidability results here depend upon an oracle which will determine the sign of a given function at a given rational constant. That is, it is only relative to this oracle that the equations are solved.

Dependence on a nonexistant oracle is unsatisfactory in theory. However, no very difficult examples of the constant problem are known (to me). There might be two possible types of difficult example. There could be a surprising equality, which would occur, for example, if $\exp(\exp(2/3))$ were rational. There might also be a surprising near equality: two short constants which are so close to representing the same number that they could not be distinguished by any reasonable amount of computing. This might happen, for example, if $\exp(\exp(7/8))$ were surprisingly close to some small rational. Any difficult example would be a very interesting mathematical fact and, as far as I am aware, none such has been discovered. There are definitely not any theoretical undicidability results about identity of constants which are obtained by substituting rationals into functions defined by algebraic differential equations; and there are no number theoretic or logical grounds for thinking that such problems might be undecidable.

In section II below, some examples are done, giving positive real solutions for equations, $f(x) = 0$, of three types:

1) $f(x) = c_1 x^{r_1} + c_2 x^{r_2} + ... + c_k x^{r_k}$, where $r_1, ..., r_k$ are non negative and real,

2) $f(x)$ is a polynomial in x and $\log(x)$, or in x and $\exp(x)$,

3) $f(x)$ is a polynomial in x and $\sin(x)$, and x is restricted to a bounded interval.

In the last section, the main decidability result is proved. This solves the root approximation problem and the zero equivalence problem for functions which are polynomials in x and any monomial which is defined by a non singular algebraic first order differential equation; or more generally, in any finite sequence of monomials where each is defined by a non singular differential equation which is algebraic in the previous monomials.

II) Examples involving x^k, $\exp(x)$, $\log(x)$ and $\sin(x)$

False derivative for polynomials, and related functions

Of course there are many different ways in which false derivatives could be defined for polynomials. The approach taken below relates to the Descartes rule of signs.

Suppose we wish to solve a polynomial equation $p(x) = 0$, and one of the terms of $p(x)$, say $c_k x^k$, causes difficulty. (The difficulty might be caused by either the coefficient or the exponent.) Think of this difficult term as a monomial $g(x)$ which satisfies the differential equation $g' = kg/x$. Since there is a singularity at the origin, our root finding technique will only work on the positive real axis.

Write $p(x)$ as $q(x)+g(x)$. Form the false derivative $p*(x) = xq'(x)-kq$.

Let $p(x) = c_1 x^{r_1} + c_2 x^{r_2} + ... + c_k x^{r_k}$. We assume that $r_1 > r_2 > ... > r_k \geq 0$. It is not necessary to assume that the r_i numbers are rational or even algebraic. The number of changes of sign in the sequence of coefficients will be called the Descartes index. It can be proved that this notion of index is well defined but this will not be done here.

Define $p * (x)$, the false derivative of p, to be 0 if the Descartes index of $p(x)$ is 0. Otherwise, suppose the last change of sign occurs between c_i and c_{i+1}. Take $g(x)$ to be $c_i x^{r_i}$. Then $p * (x)$ is defined to be $xq' - r_i q$, where $q(x)$ is $p(x) - g(x)$. It turns out that $p*(x)$ is $c_1(r_1 - r_i)x^{r_1} + c_2(r_2 - r_i)x^{r_2} + ... + c_{i-1}(r_{i-1} - r_i)x^{r_{i-1}} + c_{i+1}(r_{i+1} - r_i)x^{r_{i+1}} + ...c_k(r_k - r_i)x^{r_k}$. Note that the Descartes index of $p*(x)$ is one less than the Descartes index of $p(x)$.

This proves that the Descartes index of $p(x)$ is an upper bound for the number of real positive roots, counting multiplicities. It also gives a way of approximating positive real roots which has only quadratic complexity in the Descartes index.

A version of Thom's lemma can also be proved, using false derivatives, in a straightforward manner.

Cylindrical decompositions

The polynomial example given above is especially simple because the awkward monomial $g(x)$ entered only linearly into $p(x)$. The next step, necessary to help understand more complex situations, is to state a definition of cylindrical decomposition of R^n.

A non empty connected subset of R^n is termed a region.

Suppose we are given finitely many continuous functions $f_1, ..., f_k$ from a region A into R with $f_i < f_{i+1}$ throughout A. Then we say that $f_1, ..., f_k$ determine a stack over A, which is a partition of $A \times R$ into the $2k + 1$ regions $\{(a,x) : a \in A, x < f_1(a)\}$, $\{(a,x) : a \in A, nx = f_1(a)\}$, $\{(a,x) : a \in A, f_1(a) < x < f_2(a)\}$, ..., $\{(a,x) : a \in A, x = f_k(a)\}$, $\{(a,x) : a \in A, f_k(a) < x\}$.

A cylindrical decomposition of R^n is defined recursively as a set of stacks: one over each region comprising a cylindrical decomposition of R^{n-1}. A cylindrical decomposition of R^0 is one point.

A decomposition is said to be algebraic if the defining functions are all algebraic, i.e. solutions of polynomial equations with no singular points.

In general we will say that a function f from region A to R is algebraic if there is a polynomial with integral coefficients $p(a, y)$ so that 1) when $y = f(a)$, $p(a, y) = 0$ and $p_y(a, y)$ is non zero, and 2) $f(a)$ is differentiable in A. That is, the function must be defined by an equation with no singularities.

So, for example, in this sense of the word algebraic, there is no algebraic square root function defined at 0.

Let F be a set of functions defined in R^n. We say that a decomposition is sign invariant for F if each element of F has the same sign throughout each component of the decomposition.

It is known that given any polynomial with integral coefficients, an algebraic cylindrical decomposition of R^n which is sign invariant for the polynomial can be constructed, for any order of the variables.

The algebraic case is important. Note, however, that, according to the definition given above, the stack functions do not have to be algebraic, only continuous. The definition is possibly, therefore, somewhat unfamiliar. In subsequent papers, it is hoped that non algebraic cylindrical decompositions will be systematically constructed which are sign invariant for transcendental functions of several real variables, such as polynomials in x, y, $\exp(x)$, and $\exp(y)$. Some of the techniques intended to be used in this way may be seen in Richardson [Bath, 89-29].

Polynomials in x and $\log(x)$

Let $p(x, y)$ be a polynomial with integral coefficients. Let $p_y(x, y)$ mean the partial derivative of p with respect to y. Suppose the polynomial $p(x, y)$ is irreducible (this means as a polynomial in y with coefficients in the field of rational functions in x). We wish to find positive roots of $p(x, \log(x))$. Using the technique mentioned above, it is sufficient to show that it is possible effectively to find algebraic numbers, and roots of $p_y(x, \log(x))$ which separate the roots of $p(x, \log(x))$.

Find a cylindrical algebraic decomposition of R^2 such that $p(x, y)$ does not change sign on any component. The separating set for the roots of $p(x, \log(x))$ will include all the endpoints of the intervals in the decomposition, and will also have some other numbers in it.

Let I be any positive interval in the decomposition. Suppose u and v are two distinct adjacent roots of $p(x, \log(x))$ in I. Consider the stack over I. There are two cases.

a) $y = \log(x)$ crosses two different curves at u and v. Suppose the curves are $y = f_i(x)$ and $y = f_{i+1}(x)$. These two curves are separated over I by at least one curve on which $p_y(x, y)$, the derivative of p with respect to y is zero. Therefore u and v are separated by a root of $p_y(x, \log(x))$.

b) $y = \log(x)$ crosses the same curve at u and v. Let this curve be $y = f(x)$. Define $G(x) = \log(x) - f(x)$. $G'(x) = (1/x) - f'(x)$. $G'(x)$ is algebraic and has a root between u and v. $G'(x)$ can not be identically zero since if it were $f(x)$ would be $\log(x)$, and $f(x)$ is algebraic. Observe that $f'(x) = -p_x/p_y$. So the roots of $G'(x)$ are x components of solutions of the pair of equations

$$p(x, y) = 0$$

and

$$p_y + x p_x = 0,$$

which can be found directly by taking the resultant with respect to y.

Note that the intervals in the decomposition also have endpoints which are roots of a resultant, namely the resultant of $p(x, y)$ and $p_y(x, y)$. So finially we have the roots of $p(x, \log(x))$ separated by the roots of two resultants, each of these being polynomials in x, and the roots of $p_y(x, \log(x))$, which, as a polynomial in $\log(x)$, has lower degree than $p(x, \log(x))$.

The special simplicity of this example comes from the fact that the derivative of $\log(x)$ is algebraic, and therefore the true derivative can be used.

Polynomials in x and $\exp(x)$, or x and $\sin(x)$

Polynomials in x and $\exp(x)$ can be solved as above after substitution, subject to the condition $x > 0$.

They can also be done as follows. The only problem is case b) above. As before we define $G(x) = \exp(x) - f(x)$. Then $G'(x) = \exp(x) - f'(x)$. Now define $h(x) = f(x) - f'(x)$. The function $h(x)$ is equal to $G'(x)$ whenever $G(x) = 0$. Therefore $h(x)$ also has a root between u and v, and of course $h(x)$ is algebraic. In terms of the ideas above, $h(x)$ is a false derivative of $g(x)$. Note that $h(x)$ can not be identically zero since $f(x)$ is not an exponential function.

The special simplicity of this example comes from the simplicity of the differential equation for $\exp(x)$.

Consider now a polynomial $p(x, \sin(x))$ over a bounded interval I. It is sufficient to solve the problem where I is $(-\pi/2, \pi/2)$. The argument is the same as above until case b). At this stage we set $G(x) = \sin(x) - f(x)$. $G'(x) = \cos(x) - f'(x)$. Now set false derivative $h(x) = (1 - f(x)^2)^{1/2} - f'(x)$.

It is necessary to avoid points at which $f(x) = 1$ or $f(x) = -1$, since at these points the differential equation has a singularity. First find these points. They are the roots of $p(x, 1)$ and $p(x, -1)$. We need only consider x intervals between these roots on which $f(x)$ is bounded between -1 and 1.

Since $f(x)$ is bounded between -1 and 1, $h(x)$ is algebraic with no singularities; also $h(x) = G'(x)$ whenever $g(x) = 0$. So $h(x)$ must have a root between u and v. As before, $h(x)$ can not be identically zero. If it were f would be a solution of the differential equation. But the equation has a unique solution, which is $\sin(x)$, and $f(x)$ is algebraic.

It is claimed that the roots of $p(x, 1), p(x, -1)$, the roots of $p_y(x, \sin(x))$ and the roots of the various $h(x)$ and the singularities of the cylindrical decomposition separate the roots of $p(x, \sin(x))$. Unless $p(x, y)$ can be factorized as a polynomial in y, the roots of the various $h(x)$ are included in the roots of $R(x)$, where $R(x)$ is the resultant of $p(x, y)$ and $(y^2 + (p_x/p_y)^2 - 1)(p_y)^3$. Thus if $p(x, y)$ is irreducible, all the roots of $p(x, sin(x))$ in $(-\pi/2, \pi/2)$ are separated by roots of $p_y(x, sin(x))$ and roots of $R(x)$, and the roots of $p(x, 1)$ and $p(x, -1)$.

This example shows why three different resultants are needed in the proof of the general theorem.

III) Main Decidability Result

In this section a general result is obtained which reduces the zero equivalence problem and the approximation of roots problem for a certain class of functions to the problem of determining the sign of a function at a rational point. The class of functions considered includes solutions of first order algebraic differential equations and, recursively, functions which satisfy first order differential equations which are algebraic in previously defined functions. The class of functions includes x, $\exp(x)$, $\log(x)$, $\sin(x)$ restricted to a finite interval, and is closed under field operations, composition, differentiation and integration and, probably, many other interesting operations. The purpose of the sequence of definitions below is to support a clear definition of the class of functions.

Let (g_i), $i = 1, \ldots, n$ be a sequence of real valued analytic functions of one real variable, all defined on a common interval I, which may be infinite. In the following the g_i, and their derivatives g_i' will all be called monomials.

Let $\mathcal{F}_i$ be the collection of functions which are obtained by substituting monomials $g_1, g_1', \ldots, g_i, g_i'$, (the first i from the sequence), for variables in a polynomial $p(x, y_1, w_1, \ldots, y_i, w_i)$ with $2i + 1$ variables $x, y_1, w_1, \ldots, y_i, w_i$, and integral coefficients. Each function in $\mathcal{F}_i$ will be represented by a large number of equivalent expressions of this type. (Two expressions are considered equivalent here if their interpretations as functions are the same on the interval I.)

Let $g(x)$ be another real valued analytic function defined on the common interval I. We will say that $g(x)$ satisfies a differential equation algebraic in $\mathcal{F}_i$ if there is a polynomial $q(x, y, z, w_1, \ldots, w_i)$ with integral coefficients so that the following two conditions hold.

C1) $q(x, g(x), g'(x), g_1, \ldots, g_i)$ identically 0 on I, and

C2) $Res_z(q(x, g(x), z), q_z(x, g(x), z))$, the resultant with respect to z of $q(x, g(x), z)$ and $q_z(x, g(x), z)$, has no root on I.

The second condition insures that the differential equation is non singular in this sense: for any value of x and any value of z, if $q(x, y, z) = 0$, with $y = g(x)$, then z is locally defined implicitly as a function of x and y. So the function $g(x)$ is uniquely defined by the first equation, together with values of $g(x)$ and $g'(x)$ at one point in I.

In case $g(x)$ satisfies a differential equation of the above sort we will say that $g(x)$ is obtained by a simple extension of $\mathcal{F}_i$. Simple extensions may be either algebraic or transcendental.

It will be assumed in the following that when we are given a simple extension, $g(x)$, we are also given the polynomial $q(x, y, z)$, defining the derivative as mentioned above. It will also be assumed in each case that together with q we are given $CAD(q)$, a cylindrical algebraic decomposition of (x, y, z) space which is sign invariant for q, projecting the variables out in reverse alphabetical order. The points on the x axis which define the bottom stack will be called the critical points of $CAD(q)$. Above one of the intervals on the x axis between these critical points, the stack functions are defined by $Res_z(q(x, y, z), q_z(x, y, z)) = 0$, as usual.

Now and for the rest of this section we assume that the sequence $(g_1, ..., g_n)$ is fixed, and for each j $\mathcal{F}_j$ is defined from $(g_1, g_1' ..., g_j, g_j')$, as discussed above, and each g_{j+1} is obtained by a simple extension of $\mathcal{F}_j$. So each g_{j+1} satisfies a non singular differential equation algebraic in the previous ones.

If monomial $g(x)$ is defined by $q(x, g, g') = 0$, then g'' is $-(q_x + q_y g')/q_z$. Note that this implies that all the derivatives of g_i are rational combinations of g_i, g_i', and other terms in $\mathcal{F}_i$. The quotient field of each $\mathcal{F}_i$ is closed under differentiation.

We can define a normal form for expressions for functions in $\mathcal{F}_n$. Write each expression as a polynomial in g_n' with coefficients which are polynomials in g_n, these having coefficients in $\mathcal{F}_{n-1}$ and then, recursively, put the coefficients in normal form. All the extensions are not transcendental, so this normal form is not a canonical form.

For an expression in normal form, the highest or most complex monomial is the latest one in the sequence $(g_1, g_1', ..., g_n, g_n')$ which occurs. We then get a partial ordering of expressions by complexity as follows. For two expressions E and F in normal form, we will say that E is less complex than F if either the highest monomial of E is below the highest monomial of F, or if they have the same highest monomial, say g, but the degree of this highest monomial in E is less than the degree in F.

Using induction on complexity, it is possible to prove that functions definable by sequences of extensions as described above are closed under field operations, differentiation, and integration.

The main result below will also be proved by induction on complexity of expressions.

One variable Decidability Theorem.   *Suppose given, as discussed above, sequence $(g_i), i < n$ of monomials, where each g_{i+1} is defined by an algebraic differential equation over $\mathcal{F}_i$, and also assume given an interval I, possibly infinite, where all of the monomials are defined, and a measure M so that $M(I)$ is finite. Assume we have an oracle which will decide, given any f in $\mathcal{F}_n$, and rational r in I, whether $f(r) < 0$, $f(r) = 0$, or $f(r) > 0$.*

Then, given any $f(x)$ in $\mathcal{F}_n$, we can in a finite number of steps involving reference to the oracle, decide whether or not $f(x)$ is identically 0 on I. If $f(x)$ is not identically 0 on I, we can, in a finite number of steps involving calls to the oracle, solve $f(x) = 0$ on I, in the sense defined previously.

Proof.

The first part of this theorem, deciding zero equivalence, is similar to results of Shackell[1989], who use similar methods.

We proceed by induction on the complexity of the expression given for f in normal form in $\mathcal{F}_n$.

We suppose that we are given an expression for $f(x)$ and, as induction hypothesis, that the two problems above, zero equivalence and root approximation have been shown to be Turing reducible to the Oracle for expressions below the given expression for $f(x)$.

The highest monomial which occurs in the expression for f is either of the form g or of the form g', for some g in the sequence. We suppose g is g_{i+1} and g is defined by differential equation $q(x, g, g') = 0$, over $\mathcal{F}_i$. There are two cases, A) in which the highest monomial is of the form g, and B) in which it is of the form g'. Note that everything under case A) is simpler than everything under case B), according to our inductive scheme. These two cases are dealt with below.

Case A) The highest monomial in the expression for f is of the form g.

The main steps of the construction below are labeled.

A1) Put the expression for $f(x)$ in normal form, i.e. write it as a polynomial in its highest monomial, which is called g. The coefficients of this polynomial do not involve g and therefore we can assume, by induction hypothesis, that we can decide zero equivalence of these coefficients.

A2) Test the leading coefficient of the expression for $f(x)$ for zero equivalence. If it is identically zero, reduce the expression for $f(x)$ accordingly and go on to solve the problem, as inductively assumed. Suppose, on the other hand, that the leading coefficient is not identically 0.

Let $f(x) = p(x, g)$ for polynomial $p(x, y)$ with coefficients in $\mathcal{F}_i$. Let p_y and p_x be $p_y(x, g)$ and $p_x(x, g)$ respectively. Note that if the coefficients of $p(x, y)$ are in $\mathcal{F}_i$, then the coefficients of p_x are in the quotient field of $\mathcal{F}_i$.

Suppose that each g_i is defined by $q_i(x, g_i, g_i') = 0$, and the last monomial g is defined by $q(x, g, g') = 0$. Let the degree of z in $q(x, y, z)$ be k.

A3) Define R_g to be the resultant, with respect to g, of $p(x, g)$ and $q(x, g, (-p_x/p_y))(p_y)^{k+1}$.

This second expression is a polynomial in g with coefficients in the quotient field of $\mathcal{F}_i$.

The expression for R_g does not involve g or g' and is therefore simpler in our inductive scheme than the original expression for $f(x)$.

A4) Test whether or not R_g is identically 0 on I. There are now two subcases, 5a) and 5b).

A5a) Suppose R_g identically 0. Then use the Euclidean algorithm to find the HCF of the two expressions. It will be necessary to do some tests for zero equivalence in $\mathcal{F}_i$ along the way here in order to avoid dividing by zero.

Let $H(x,g)$ be the HCF found. Assume this is in normal form and has non zero leading coefficient. If the degree (as a polynomial in g of course) of $H(x,g)$ is less than the degree of $p(x,g)$, we can factorize $p(x,g)$ into a product of simpler expressions, and continue by induction. Suppose, on the other hand, that the degrees of H and p are the same. In this case $p(x,g)$ is a factor of

$$q(x,g,(-p_x/p_y))(p_y)^{k+1}.$$

This subcase requires special examination, since it may happen that f is identically 0.

Let R_y be the resultant of $p(x,g)$ and $p_y(x,g)$. Recall that the leading coefficient of $p(x,g)$ was non zero.

If R_y were identically 0, we could use the Euclidean algorithm to factorize $p(x,g)$, and continue by induction.

Suppose, on the other hand, that R_y is not identically 0. In this case, solve $R_y = 0$ on I. Let (a,b) be a rational subinterval of I on which R_y has no root. The claim now is that $f(x)$ is either never equal to zero on (a,b) or it is identically 0. Since R_y has no root on (a,b), the equation $p(x,y) = 0$ together with any point $(x0, y0)$ so that $p(x0, y0) = 0$, implicitly defines y as a function of x in (a,b). Also $y' = (-p_x/p_y)$.

Suppose $f(x) = 0$ for some x in (a,b). Let $y(x)$ be the function implicitly defined. Since $p(x,g)$ divides $q(x,g,(-p_x/p_y))(p_y)^{k+1}$, $y(x)$ satisfies the same differential equation as $g(x)$, and thus if $y(x)$ and $g(x)$ are equal at one point they are identically equal. But we supposed they were equal at one point. Thus $f(x)$ is identically zero on (a,b) and on I.

Therefore $f(x)$ is identically zero on I if and only if $f(r) = 0$ for some rational r such that $p_y(x,g)$ is not zero. The test rational can be found either by solving $R_y = 0$ on I and then picking r so that $R_y(r)$ is not zero; or simply by guessing at rationals and referring them to the oracle until one is found with R_y non zero.

Having got the test rational, if we find that $f(r)$ is not zero, it follows that $f(x)$ has no roots on I except possibly at points which are also roots of R_y. In this case the solution of $f=0$ is contained in the solution of $R_y = 0$, and an approximation to the roots of R_y will also do as an approximation to the roots of f.

A5b) We now return to the main line of the proof. Suppose R_g is not identically zero.

Yet another resultant now has to be defined. Let R_q be

$$Res_y(p(x,y), Res_z(q(x,y,z),q_z(x,y,z))).$$

The inner resultant $Res_z(q, q_z)$ is a polynomial in y with coefficients which are polynomials in x. If we substitute g in for y, this polynomial in g has no root on I, by assumption C2). So the inner resultant is not identically 0.

It could happen, though, that R_q was identically 0. In this case, find the HCF of $p(x, y)$ and the inner resultant. If this HCF divides $p(x, y)$, factorize $p(x, y)$ and continue, by induction. The other possibility is that $p(x, y)$ itself divides the inner resultant. This means that when $p = 0$, both $q = 0$ and $q_z = 0$ for some z. We know, by assumption C2) that $q = 0$ and $q_z = 0$ never happens when $y = g$. Therefore in this case $p(x, g)$ is never 0, and $f(x)$ has no roots.

Next assume that R_q is not identically 0.

In this case it is claimed that f is not identically zero and that the roots of $R_g(x)$ and $R_q(x)$, together with the critical points of $CAD(q)$, separate the roots of $f(x)$. Once this has been shown, the proof of part A) is finished, using Theorem 1) from the Introduction.

Let R_y be defined as above. The roots of R_y are contained in the roots of R_g. In particular, R_y can't be identically zero.

First suppose that u is a double root of $f(x)$. It is claimed that R_g must be 0 at u. Suppose not. Then R_y is not 0 at u. We have $f'(u) = p_x(u, g(u)) + p_y(u, g(u))g'(u)$. Therefore $g' = -p_x/p_y$ at u and so $q(x, g, -p_x/p_y)) = 0$ at u and $R_g = 0$ at u. Contradiction.

Let u and v be two adjacent roots of $f(x)$. Suppose that there is no root of R_y between u and v, and neither u nor v are roots of R_y, and that there is no critical point of $CAD(q)$ between u and v. In this case, above (u, v), $p(x, y) = 0$ defines y implicitly as a stack of functions of x, say $y_1(x) < y_2(x) < ... < y_n(x)$. The function $y = g(x)$ is also defined in (u, v). Between u and v, the graph of $y = g(x)$ either crosses between two different stack functions or it crosses the same stack function twice. In the first case, it would cross a solution of $p_y(x, g) = 0$, but this would imply $R_y = 0$, which is impossible. So it must be that some stack function $y(x)$ is crossed at u and again at v.

Suppose there is also no root of R_q between u and v. It is not possible then to have $q(x, y(x), z) = 0$ and $q_z(x, y(x), z) = 0$. Between u and v, then, we can define $z(x)$ implicitly by

$$q(x, y(x), z) = 0$$

and $z = g'(x)$, when $x = u$, i.e. $z(u) = g'(u)$. As we move x from u to v, z is defined implicitly. When we get to v, $y(x)$ is again equal to $g(x)$. The curves $y(x)$ and $g(x)$ make a cycle between u and v, and the part of the xy plane between them is contained in a region of $CAD(q)$, since there are no critical points of $CAD(q)$ between u and v, and since $Res_z(q, q_z) = 0$ does not intersect with either $y = g(x)$ or $y = y(x)$. The z values which are implicitly defined above these curves are therefore on the same sheet in the cylindrical decomposition and are all determined by their projections.

We can assume then that $z(v) = g'(v)$. Now consider the false derivative $z(x) + y'(x)$. If this is zero at u or v, then there is a root of R_g at u or v. Otherwise, since the curves cross, the sign of $z(x) + y'(x)$ changes between u and v. Since both are continuous, there must be a point between u and v at which $z(x) + y'(x) = 0$. At this point $z(x) = -p_x/p_y$. So not only $p(x, y(x)) = 0$ but also $q(x, y(x), -p_x/p_y = 0$. Thus $R_g(x) = 0$, and since x is between u and v, this line of the proof is finished.

Case B) In this case we assume that $f(x)$ is of the form $p(x, g, g')$, and our induction hypothesis is that the zero equivalence problem and the root approximation problem can be solved for any simpler expression. In particular, they can be solved for any polynomial in g with coefficients in $\mathcal{F}_i$, or in a polynomial in g and g' with degree of g' lower than the g' degree of $p(x, g, g')$.

Assume the leading coefficient of $p(x, g, g')$ is not identically zero.

Let $R(x, g)$ be the resultant of $p(x, g, g')$ and $q(x, g, g')$, the defining polynomial for the derivative of g.

There are two possibilities:

B1) $R(x, g)$ is identically zero. Find the HCF of $p(x, g, g')$ and $q(x, g, g')$, treating these as polynomials in g'.

If the degree of the HCF is less than the degree of $p(x, g, g')$, factorize $p(x, g, g')$ and continue by induction.

In the other case $p(x, g, g')$ divides $q(x, g, g')$. It may happen in this case that f is identically 0. Define R_z to be the resultant of $p(x, g, g')$ and $p_z(x, g, g')$, where $p_z(x, g, g')$ means the result of first taking $p_z(x, y, z)$ and then substituting g for y and g' for z. Since the original differential equation is non singular, there are no roots of R_z or $p_z(x, g, g')$ in I. Therefore $p(x, h, h') = 0$ is a valid definition of a new monomial h. Since p divides q, h is either never equal to g or it is identically equal to g. Therefore $f(x)$ is either identically zero or never zero, and it suffices to test $f(r) = 0$ at one rational point.

B2) On the other hand, assume $R(x, g)$ is not identically zero. Since q is identically zero by definition of g, the roots of f are contained in the roots of $R(x, g)$. But by induction, we can solve $R(x, g) = 0$ in I.

That completes the proof of the main theorem.

Remarks

The problem of approximating solutions of equations in one real variable has also been solved by Lombardi [1989], who assumes only that the function to be dealt with is known to be not identically zero, and can be effectively uniformly approximated by a series of polynomials. The Lombardi method would in some cases be better than the one described above, and vica versa. It seems likely that a hybrid method would be better than both. Certainly the Lombardi techniques can be used to improve what is done above. For example, in the proof of the approximation theorem, a Lombardi approximation can be used to throw out intervals which might contain roots of $h(x)$ but could not possibly contain roots of $f(x)$.

In view of the results of Khovanskii [1980] and van den Dries [1988], the above decidability result is not surprising. In fact it seems likely that almost all local problems in real analysis, that is problems involving bounded quantifiers and analytic functions built up by solving differential equations, will turn out to be decidable.

REFERENCES

[1] Arnon, D. S., and B. Buchberger, (Editors),, "Algorithms in real algebraic geometry," Academic Press, 1988.

[2] Davenport, J. H., *Computer Algebra for Cylindrical Decomposition*, University of Bath computer science technical report 88-10.

[3] van den Dries, L., *On the elementary theory of restricted elementary functions*, The Journal of Symbolic Logic **53** (1988), 769–808.

[4] van den Dries, L. and J. Denef, *P-adic and real subanalytic sets*, Annals of Mathematics **128** (1988), 79–138.

[5] Khovanskii, A. G., *Fewnomials and Pfaff manifolds*, in "Proc. Int. Congress of Mathematicians," Warsaw 1983, pp. 549–564.

[6] Khovanskii,A.G., *On a class of systems of transcendental equations*, Soviet Math. Dokl. **22** (1980), 762–765.

[7] Lombardi, H., *Algebre Elementaire en Temps Polynomial*, Thesis, University of Nice, 1989.

[8] Richardson, D., *Wu's method and the Khovanskii finiteness theorem*, Bath Computer Science, technical report 89-29.

[9] Shackell, J., *Zero-Equivalence in function fields defined by algebraic differential equations*.

[10] Shackell, J., *A Differential-equations approach to functional equivalence*.

Daniel Richardson
Department of Mathematics
University of Bath

Some Effective Methods in the Openness of Loci for Cohen-Macaulay and Gorenstein Properties

FABIO ROSSI   WALTER SPANGHER

Introduction

In this paper we present some effective methods in order to compute the $\mathcal{P}$-locus of a quotient A of a polynomial ring $S = k[\mathbf{X}]$, where $\mathcal{P}$ is any of the following properties: "Cohen-Macaulay (C. M.)", "Gorenstein", "C. M. type $\leq r$", "Complete intersection (C. I.)". It is well known that the previous loci are open being A an excellent ring (see, for instance, [**GM**]).

We are able to compute, by applying our techniques, the equations of their (closed) complements in $\mathrm{Spec}(A)$.

As a special case, suitable computational tests on A in order to be C. M., Gorenstein, etc., are also found.

If A is an *affine ring*, to compute the previous loci we can also work, after suitable homogenization, in the homogeneous setting, so that it is possible to use the *Macaulay* system (see [**BSS**]), and then dehomogenize the ideals obtained in this way (see, Section 1).

We propose two different approaches to the previous problems. The first one is via finite free resolutions (FFR) and homological properties (see Section 1). In this way we can also find some computational criteria about the properties S_n, T_n, G_n, in addition to the computation of the previous loci.

On the other hand, when I is a prime ideal, it is well known (see [**Va**], Section 2.3),that a nice criterion of Macaulayness of $A = S/I$ can be obtained by any Noether Normalization of A.

Moreover, our homological approach needs the computation of FFR's which has an high complexity (see [**MM**]).

Hence, using some properties of A as a finite module over its Noether normalizations (see Sections 3 and 4), we also present a second approach for the computation of the $\mathcal{P}$-locus of A, where $\mathcal{P}$ is any of the following properties: "C. M.", "Gorenstein", "C. M. type $\leq r$".

We wish to thank M. Manaresi for her encouragement.

Research with the contribution of M.U.R.S.T., 40% fund

1. Effective computation of loci via homological methods

(1.1) Let $S = k[X_1, \ldots, X_n] = k[\mathbf{X}]$ be a polynomial ring over any field k; let $I \neq 0$ be any ideal of S and let $A = S/I$. It is well known that: A is C. M. iff $A_{\mathfrak{m}}$ is C. M. for every $\mathfrak{m} \in \operatorname{Specm}(A)$ (= maximal spectrum). Let us denote by $\mathfrak{m}$ a maximal ideal of S such that $\mathfrak{m} \supseteq I$ and let $A' = A_{\mathfrak{m}}$.

$A' = S_{\mathfrak{m}}/I_{\mathfrak{m}}$ is obviously a quotient of a Gorenstein ring $S' = S_{\mathfrak{m}}$; moreover it is trivial that dim $S' = n$ and that $ht\ I_{\mathfrak{m}} = h' \geq ht\ I = h$. Then, by [Sa], Proposition 5.1, it follows that A' is C. M. iff $\operatorname{Ext}^{h'}_{S'}(A', S') \neq 0$ and $\operatorname{Ext}^j_{S'}(A', S') = 0$ for every $j \neq h'$.

(1.2) We want to find the (closed) locus C of $\operatorname{Spec}(A)$ given by:

$$C := \{\mathfrak{p} \in \operatorname{Spec}(S), \mathfrak{p} \supseteq I \mid S_{\mathfrak{p}}/I_{\mathfrak{p}} \text{ is not C. M. }\}.$$

Proposition 1.2. *With the previous notation it follows:* $C = V(\mathfrak{a})$, *where* $\mathfrak{a} \subseteq S$ *is the ideal given by*

$$\mathfrak{a} = \sum_{j=h}^{n} (\prod_{\substack{i=h \\ i \neq j}}^{n} \operatorname{Ann}_S(\operatorname{Ext}^i_S(A, S))).$$

Proof: Let

$$D := \cup_{i=h}^{n-1} [\operatorname{Supp}_S \operatorname{Ext}^i_S(A, S) \cap (\cup_{j=i+1}^{n} \operatorname{Supp}_S \operatorname{Ext}^j_S(A, S))].$$

First of all we claim that $C = D$.

Being A a Jacobson ring, it is enough to prove the previous claim for all $\mathfrak{m} \in V(I) \cap \operatorname{Specm}(S)$. Let $\mathfrak{m} \in C$, then $A_{\mathfrak{m}}$ is not a C. M. ring. If $ht\ I_{\mathfrak{m}} = s \geq h$, it follows that $\operatorname{Ext}^j_{S_{\mathfrak{m}}}(A_{\mathfrak{m}}, S_{\mathfrak{m}}) = 0$ if $j = 0, 1, \ldots, s-1$ and $\operatorname{Ext}^s_{S_{\mathfrak{m}}}(A_{\mathfrak{m}}, S_{\mathfrak{m}}) \neq 0$. But $A_{\mathfrak{m}}$ is not C. M. , so there is a $\bar{j} \in \{s+1, \ldots, n\}$ such that $\operatorname{Ext}^{\bar{j}}_{S_{\mathfrak{m}}}(A_{\mathfrak{m}}, S_{\mathfrak{m}}) \neq 0$; therefore $\mathfrak{m} \in \operatorname{Supp}_S \operatorname{Ext}^{\bar{j}}_S(A, S)$ and finally, $\mathfrak{m} \in D$.

Conversely, if $\mathfrak{m} \in D$, then there exists r, $(h \leq r \leq n-1)$, such that $\mathfrak{m} \in \operatorname{Supp}_S \operatorname{Ext}^r_S(A, S) \cap (\cup_{j=r+1}^{n} \operatorname{Supp}_S \operatorname{Ext}^j_S(A, S))$. In this case there are two possibilities :

a) $r = \min_{h \leq r' \leq n-1} \{r' \mid \mathfrak{m} \in \operatorname{Supp}_S \operatorname{Ext}^{r'}_S(A, S)\}$; in this case, $ht\ I_{\mathfrak{m}} = \operatorname{grade}(I_{\mathfrak{m}}) = r$, but $\mathfrak{m} \in \cup_{j=r+1}^{n} \operatorname{Supp}_S \operatorname{Ext}^j_S(A, S)$, so $A_{\mathfrak{m}}$ is not C. M. by [Sa], Proposition 5.1; therefore $\mathfrak{m} \in C$.

b) There exists r', $(h \leq r' < r)$, such that $\mathfrak{m} \in \operatorname{Supp}_S \operatorname{Ext}^{r'}_S(A, S)$. In this situation it is obvious that $\mathfrak{m} \in C$ by applying the above mentioned [Sa] Proposition 5.1. If we put

$$\mathfrak{b} := \prod_{i=h}^{n-1} (\operatorname{Ann}_S \operatorname{Ext}^i_S(A, S), \prod_{j=i+1}^{n} \operatorname{Ann}_S \operatorname{Ext}^j_S(A, S))$$

it is easy to see that $C = V(\mathfrak{b})$.

Finally the thesis follows being $\text{rad}(\mathfrak{a}) = \text{rad}(\mathfrak{b})$. \hfill Q. E. D.

(1.3) *Some remarks about the computability of C:* (For a more effective method, see also the remark (1.11)). In order to compute C we can:

(i) compute a FFR of S/I over S truncated at n-th module;

(ii) compute a presentation

$$S^{m_j} \to S^{n_j} \to \text{Ext}_S^j(S/I, S) \to 0$$

using the previously computed FFR;

(iii) compute $F_0(\text{Ext}_S^j(S/I, S))$, where $F_0(-)$ is the 0-th Fitting's invariant of $\text{Ext}_S^j(S/I)$, using the previous presentation. Since the Proposition 1.2 implies that

$$C = V(\sum_{j=h}^{n}(\prod_{\substack{i=h \\ i \neq j}}^{n} F_0(\text{Ext}_S^i(A, S))))),$$

so we can compute C.

A remark about notation: Let $R^m \xrightarrow{M} R^n \to N \to 0$ be a finite presentation of an R-module N, and let M be the associated presentation matrix (with R-entries). Given such a matrix M, let $F_i(N)$ denote the ideal of R generated by all $(n - i)$-rowed subdeterminants of the matrix M ($i = 0, \ldots, n - 1$), and let $F_i(N) = R$ for $i \geq n$; put $F_i(N) = (0)$ for $i < 0$. $F_i(N)$ is called i-th Fitting's invariant of N (see [No] for the main properties of Fitting's invariants).

(1.4) *Remarks*:

(i) It is obvious that: A is C. M. iff $C = \emptyset$ iff

$$1 \in \sum_{j=h}^{n}(\prod_{\substack{i=h \\ i \neq j}}^{n}(F_0(\text{Ext}_S^i(A, S))))).$$

Hence, the computation of C gives also a Macaulayness criterion.

(ii) The previous computations can be considerably simplified if all the minimal prime divisors of I have the same height in S (for instance, if I is prime). In this case we say that A is an *equidimensional ring*.

Now, *let us suppose A to be equidimensional*. In this situation it follows that, for every $\mathfrak{m} \supseteq I$ and $\mathfrak{m} \in \text{Specm}(S)$, $\text{Ext}_{S_\mathfrak{m}}^h(A_\mathfrak{m}, S_\mathfrak{m}) \neq 0$ ($h = \text{ht } I$); therefore

$$C = V(\prod_{j=h+1}^{n} F_0(\text{Ext}_S^j(A, S)))$$

so A is C. M. iff $1 \in \prod_{j=h+1}^{n} F_0(\operatorname{Ext}_S^j(A,S))$ iff $\operatorname{Ext}_S^j(A,S) = 0$ for every $j = h+1, \dots, n$.

(iii) (*Equidimensionality test*). If $h = ht\ I$, then A is equidimensional iff $\operatorname{Supp}_S \operatorname{Ext}_S^h(A,S) = V(I)$ iff $\operatorname{rad}(I) \supseteq F_0(\operatorname{Ext}_S^h(A,S))$.

Proof: First of all, remark that $\operatorname{Supp}_S \operatorname{Ext}_S^h(A,S) \subseteq V(I)$. If A is equidimensional, then $ht\ I_{\mathfrak{m}} = ht\ I = h$ for every maximal ideal $\mathfrak{m} \supseteq I$; hence $\operatorname{Ext}_{S_{\mathfrak{m}}}^h(A_{\mathfrak{m}}, S_{\mathfrak{m}}) \neq 0$. On the other hand, if A is not equidimensional, there is a minimal prime divisor $\mathfrak{p}$ of I such that $ht\ \mathfrak{p} > ht\ I = h$. If we denote by $\mathfrak{p}_1, \dots, \mathfrak{p}_s$ the other minimal prime divisors of I, it is easy to see, being S a Jacobson ring, that there is a maximal ideal $\mathfrak{n}$ such that : $\mathfrak{n} \supseteq \mathfrak{p}$ and $\mathfrak{n} \not\supseteq \cap_{i=1}^{s} \mathfrak{p}_i$; hence $\operatorname{Ext}_{S_{\mathfrak{n}}}^h(A_{\mathfrak{n}}, S_{\mathfrak{n}})$ must be zero, because $ht\ I_{\mathfrak{n}} > h$. Q. E. D.

(1.5) *Some known results about the canonical module:*

Let $(R, \mathfrak{m})$ be a Gorenstein local ring and let J be an ideal of R with $ht\ J = h$. The following properties are essentially contained in [HK].

(i) The canonical module $\Omega(R/J)$ of R/J exists and moreover: $\Omega(R/J) = \operatorname{Ext}_R^h(R/J, R)$.

(ii) If in addition, R/J is a C. M. ring, then $r(R/J) = $ C. M. type of $R/J = \mu(\Omega(R/J))$, where $\mu(-)$ denotes the cardinality of a minimal basis.

(1.6) Let us go back to our previous setting and notation. We want to compute now the equations of the (closed) locus $C_r(r \geq 1)$ defined by:

$$C_r := \{ p \in \operatorname{Spec}(S), \mathfrak{p} \supseteq I \mid (S/I)_{\mathfrak{p}} \text{ is C. M. with } r(S_{\mathfrak{p}}/I_{\mathfrak{p}}) > r$$
$$\text{or } (S_{\mathfrak{p}}/I_{\mathfrak{p}}) \text{ is not C.M}\}.$$

Note that C_1 is the "not Gorenstein" locus of $A = S/I$.

Proposition 1.6. *If we denote by C the not C. M. locus of A, then*

$$C_r = C \cup \bigcup_{i=h}^{n} \{ \mathfrak{p} \mid \mu(\mathfrak{p}; \operatorname{Ext}_S^i(A,S)) > r \}$$

where $\mu(\mathfrak{p}; M) = \mu(M_{\mathfrak{p}})$ and $h = ht\ I$.

Proof: Let us put $D := \cup_{i=h}^{n} \{ \mathfrak{p} \mid \mu(\mathfrak{p}; \operatorname{Ext}_S^i(A,S)) > r \}$, and let $\mathfrak{p} \in C_r$. If $A_{\mathfrak{p}}$ is not C. M. then there is nothing to prove, being $\mathfrak{p} \in C$. Otherwise, let us suppose that $\mathfrak{p} \in C_r$ and $A_{\mathfrak{p}}$ is C. M. Hence, if $ht\ I_{\mathfrak{p}} = s \geq h$, it follows that, $\operatorname{Ext}_{S_{\mathfrak{p}}}^s(A_{\mathfrak{p}}, S_{\mathfrak{p}}) \neq 0$; $\operatorname{Ext}_{S_{\mathfrak{p}}}^j(A_{\mathfrak{p}}, S_{\mathfrak{p}}) = 0$ if $j \neq s$ and $\mu(\Omega(A_{\mathfrak{p}})) > r$. Therefore $\mathfrak{p} \in D$. Conversely, let $\mathfrak{p} \in D$. If $A_{\mathfrak{p}}$ is C. M. then $\operatorname{Ext}_{S_{\mathfrak{p}}}^s(A_{\mathfrak{p}}, S_{\mathfrak{p}})$ is the unique Ext different from zero; moreover necessarily $\mu(\mathfrak{p}; \operatorname{Ext}_S^s(A,S)) > r$, hence $\mathfrak{p} \in C_r$. The case in which $\mathfrak{p} \in D$ and $A_{\mathfrak{p}}$ is not C. M. is trivial.   Q. E. D.

(1.7) *On the computability of the locus C_r:*

Let us denote by $F_i(\mathrm{Ext}_S^j(A,S))$ the Fitting's invariants of the S-module $\mathrm{Ext}_S^j(A,S)$. It is easy to see that $C_r = C \cup \bigcup_{i=h}^n V(F_r(\mathrm{Ext}_S^i(A,S)))$. Hence, it follows:

Proposition 1.7. $C_r = C \cup V(\mathfrak{b})$ *where* $\mathfrak{b} = \prod_{j=h}^n F_r(\mathrm{Ext}_S^j(A,S))$.

(1.8) *Remark:* Let us suppose that A is an equidimensional ring (we recall that we can easily test this fact using 1.4. (iii)). In this context it follows that $C_r = C \cup \{\mathfrak{p} \mid \mu(\mathfrak{p}; \mathrm{Ext}_S^h(A,S)) > r\}$. Hence, C_r can be computed as: $C_r = C \cup V(\mathfrak{b})$ where $\mathfrak{b} = F_r(\mathrm{Ext}_S^h(A,S))$. ($h = ht\ I$).

(1.9) *Some remarks about* $(R_m), (S_m), (T_m)$, *m-Gorenstein and* (G_m), *criteria:*

We recall a few definitions about this matter.

- We say that: a ring A is (R_m) if $A_\mathfrak{p}$ is a regular ring for every $\mathfrak{p} \in \mathrm{Spec}(A)$ such that $ht\ \mathfrak{p} \le m$; A is (S_m) if $\mathrm{depth}(A_\mathfrak{p}) \ge \min\{m; ht\ \mathfrak{p}\}$ for every $\mathfrak{p} \in \mathrm{Spec}(A)$; A is (T_m) if $A_\mathfrak{p}$ is a Gorenstein ring for any $\mathfrak{p} \in \mathrm{Spec}(A)$ such that $ht\ \mathfrak{p} \le m$.

- A ring A is called: m-Gorenstein if it is S_m and (T_{m-1}); (G_m) if it is (S_m) and (T_m).

By applying the results of this Section and using Zariski's Jacobian criterion (see [S], pp. 73-76), it is easy to find some criteria about the claimed properties; indeed:

Let $C = V(\mathfrak{a})$ be the singular (respectively, not C. M. , not Gorenstein) locus of $A = S/I$, and let $ht\ \mathfrak{a} = h$. It follows that: If $h \le m$, then A is not (R_m) (respectively, not (S_m), not (T_m)); if $h > m$, then A is (R_m) (respectively, is (S_m), is (T_m)).

Obviously, similar criteria about m-Gorenstein and (G_m) can be given just as slight modifications of the previous one.

(1.10) *About the complete intersection locus:*

We recall that a local ring $(A, \mathfrak{m},)$ is called complete intersection (C. I.) if $\hat{A}$ is a quotient of a complete local ring R by an ideal generated by an R-sequence. For the polynomial case $A = S/I$ it is easy to see (for instance [M2] Th. 21.1 (iii) and also [K] Th. 129) that $A_\mathfrak{p} = S_\mathfrak{p}/I_\mathfrak{p}$ is C. I. iff $\mu(I_\mathfrak{p}) = ht\ (I_\mathfrak{p}) = \mathrm{grade}(I_\mathfrak{p})$, being $A_\mathfrak{p}$ a quotient of a regular ring.

Proposition 1.10. *Let* $C := \{\mathfrak{p} \in \mathrm{Spec}(S), \mathfrak{p} \supseteq I \mid A_\mathfrak{p}\ is\ not\ C.\ I.\}$ *and* $h = ht\ I$. *Then* $C = \cup_{i=h}^n [V(F_i(I)) \cap \mathrm{Supp}_S \mathrm{Ext}_S^i(A,S)]$, *where* $F_i(I)$ *is the i-th Fitting's invariant of I,*

Proof: Let us denote by $D = \cup_{i=h}^n [V(F_i(I)) \cap \mathrm{Supp}_S \mathrm{Ext}_S^i(A,S)]$. If $\mathfrak{p} \in C$ it follows that $A_\mathfrak{p}$ is not C. I.; if we suppose that $ht\ I_\mathfrak{p} = h' \ge h = ht\ I$, then $\mathfrak{p} \in \mathrm{Supp}_S \mathrm{Ext}_S^{h'}(A,S)$; moreover $\mu(I_\mathfrak{p}) > h'$, so $\mathfrak{p} \in V(F_{h'}(I))$

because $\mu(I_{\mathfrak{p}}) = \min\{r \mid F_r(I_{\mathfrak{p}}) = S_{\mathfrak{p}}\}$.Conversely, let $\mathfrak{p} \in D$; hence $\mathfrak{p} \in V(F_{h'}(I)) \cap \operatorname{Supp}_S \operatorname{Ext}_S^{h'}(A, S)$. But now $ht\ I_{\mathfrak{p}} = \operatorname{grade}(I_{\mathfrak{p}}) \le h'$ because $\mathfrak{p} \in \operatorname{Supp}_S \operatorname{Ext}_S^{h'}(A, S)$; on the other hand $\mathfrak{p} \in V(F_{h'}(I))$ so $\mu(I_{\mathfrak{p}}) > h'$, hence $\mathfrak{p} \in C$.   Q. E. D.

Remark: If A is equidimensional, then $C = V(F_h(I))$ where $h = ht\ I$.

(1.11) *How to compute the previous loci using the Macaulay system*:

If we want to compute the equations of non $\mathcal{P}$ locus (where $\mathcal{P}$ is any of the following properties: C. M., Gorenstein, C. M. type $\le r$, C. I.), we can use the *Macaulay* system in order to do the computations as it is pointed out in the following:

Proposition 1.11. *Let* $C := \{\mathfrak{p} \in \operatorname{Spec}(S), \mathfrak{p} \supseteq I \mid A_{\mathfrak{p}}$ *is not* $\mathcal{P}\}$. *Let us denote by* $S' = k[X_0, X_1, \dots, X_n]$ *and by* $A' = S'/^h I$ *the homogenizations of* S *and* A *with respect to* X_0.

Let $C' := \{\mathfrak{q}' \in \operatorname{Spec}(S'), \mathfrak{q}'$ *homogeneous*, $\mathfrak{q}' \supseteq {}^h I \mid A'_{\mathfrak{q}'}$ *is not* $\mathcal{P}\} = V(\mathfrak{a}')$, *where* $\mathfrak{a}'$ *is a suitable homogeneous ideal of* S'. *Then* $C = V({}^a\mathfrak{a}')$.

Proof: First of all it is easy to prove that for any $\mathfrak{q}' \in \operatorname{Proj}(S')$ it follows that $A'_{\mathfrak{q}'}$ is not $\mathcal{P}$ iff $A'_{(\mathfrak{q}')}$ is not $\mathcal{P}$. Indeed the canonical local homomorphism $A'_{(\mathfrak{q}')} \to A'_{\mathfrak{q}'}$ is faithfully flat, and the fiber over $\mathfrak{q}' A'_{(\mathfrak{q}')}$ is a field. Thus the previous claim follows immediately from [M2] – Sec. 23, and from [A] – Th. 2. Moreover if $\mathfrak{p} \in \operatorname{Spec}(S)$ and $\mathfrak{p}' = {}^h\mathfrak{p}$, then $A'_{(\mathfrak{p}')} \cong A_{\mathfrak{p}}$. Q. E. D.

Obviously, we can compute $\mathfrak{a}'$ using the techniques presented in this Section and using *Macaulay* in the homogeneous setting. Then, in order to find $\mathfrak{a}$, it is enough to dehomogenize $\mathfrak{a}'$.

Moreover in the homogeneous setting it is also very easy to test if A' is C. M. (respectively, Gorenstein, C. M. with r type). Indeed, to test the C. M. property we can compute the minimal FFR of A' over S', and then apply Auslander - Buchsbaum (i.e. $dh_{S'} A' = h$) (other techniques are given in [BSS] - A.5 p. 114). When it is known that A' is C. M., it is enough to compute $\beta_h(A')$, where β_h is the last Betti number, in order to test if A' is Gorenstein or C. M. with r type, because it is well known that $\beta_h(A') = r(A')$ ($h = ht\ I$).

2. Preliminaries about the Noether Normalization Lemma

(2.1) Let I be an ideal of $S = k[X_1, \dots, X_n]$ and let $A = S/I$, $\dim A = d$.

A subalgebra $B = k[Y_1, \dots, Y_d]$ of A is called a *Noether Normalization* of A if the following properties are satisfied:
  i) $\{Y_1, \dots, Y_d\}$ *are algebraically independent over* k;
  ii) A *is a finite extension of* B.

We showed in (1.11) that, in order to find — for instance — the locus of all prime ideals $\mathfrak{p}$ of $\mathrm{Spec}(S), \mathfrak{p} \supseteq I$ such that $A_{\mathfrak{p}}$ is not C. M., it is possible to work in the homogeneous setting, finding the same locus for ^{h}I.

Then, in this Section, we can assume to work with a homogeneous ideal.

(2.2) Let k be an *infinite field*, let $S = k[X_0, X_1, \ldots, X_n]$ and let I be any non-zero homogeneous ideal of S. It is well known that $A = S/I$ is a graded finitely generated B-module over a suitable polynomial ring B.

We need some preliminary definitions. Let $A = S/I$, $\dim A = d + 1$ ($ht\ I = n - d$).

(i) We say that an *effective Noether Normalization* of A is given if the following properties are satisfied:

a) *It is given an h-sop (i. e. a homogeneous system of parameters)* $\{g_0, g_1, \ldots, g_d\}$ *of A (see* [ZS], *Ch. VII, §7; hence* $B = k[g_0, \ldots, g_d]$ *is a polynomial ring and* $\{g_0, \ldots, g_d\}$ *is a homogeneous system of integrity)*;

b) *It is given a finite presentation of A as a B- module:*

$$B^s \xrightarrow{M} B^t \to A \to 0.$$

(ii) We say that an *effective Noether Normalization of S with respect to I* is given, if the following properties are satisfied:

a₁) *It is given an h-sop* $\{h_0, \ldots, h_n\}$ *of S such that* $I \cap R = (h_{d+1}, \ldots, h_n)$, *where* $R = k[h_0, \ldots, h_n]$.

a₂) *It is given a finite presentation of S as R-module:*

$$R^p \xrightarrow{N} R^q \to S \to 0.$$

Remark 2.3: Every effective Noether Normalization of S with respect to I implies, in a canonical way, an effective Noether Normalization of A.

Indeed, let $\{h_0, h_1, \ldots, h_n\}$ be an h-sop of S such that $I \cap R = (h_{d+1}, \ldots, h_n)$.

It is quite obvious that $\{g_0 = \bar{h}_0, \ldots, g_d = \bar{h}_0\}$ (where $\bar{h}_i = [h_i]_I$) is an h- sop of A. Moreover, if $\{f_1, \ldots, f_q\}$ is a (homogeneous) system of generators of S over R, then $\{\bar{f}_1, \ldots, \bar{f}_q\}$ is a system of generators of A over $B = R/I \cap R$. Hence the presentation $B^s \xrightarrow{M} B^q \to A \to 0$ of A, is easily obtained from the given one $R^p \xrightarrow{N} R^q \xrightarrow{\lambda} S \to 0$, as follows: consider a system of generators $\{G_1, \ldots, G_s\}$ of the sub-R-module $\lambda^{-1}(I) \subseteq R^q$; the columns of the matrix M are the images of the G_i's in $B^q = (R/I \cap R)^q$.

(2.4) *How to find an h-sop of the Noether Normalization w. r. t. I:*

Let us suppose that $h_i = \sum_{j=0}^{n} a_{ij} X_j (i = 0, 1, \ldots, n), a_{ij} \in k$, is a linear change of coordinates; so we have $S = k[h_0, \ldots, h_n]$. Moreover, since we assume k infinite, there are many $a_{ij} \in k (i = 0, 1, \ldots, d; \; j = 0, \ldots, n)$ such that $(I, h_0, \ldots, h_d)$ is an irrelevant ideal in S (about this point, see also next 3.9. - for another construction about Noether Normalizations, see, for instance, [Se]); hence there is a $\rho \in \mathbf{N}$ such that $(h_0, \ldots, h_n)^\rho \subseteq (I, h_0, \ldots, h_d)$, so $h_i^\rho = \phi_i + b_i$ where $\phi_i \in I$ and $b_i \in (h_0, \ldots, h_d)$ $(i = d+1, \ldots, n)$. Finally, $\{h_0, \ldots, h_d, \phi_{d+1}, \ldots, \phi_n\}$ is the claimed h-sop. We observe that the minimal $\bar{\rho} \in \mathbf{N}$ such that $(h_0, \ldots, h_n)^{\bar{\rho}} \subseteq (I, h_0, \ldots, h_d)$ is the regularity index of the Hilbert's function of $(I, h_0, \ldots, h_d)$; so we can find it.

(2.5) *How to find an h-sop of A:*

As a consequence of (2.3) and (2.4), we can find an h-sop of A just taking the equivalence classes $\bar{h}_0, \ldots, \bar{h}_d$ of the previous h-sop of S. Let us point out that, in this case, $h_{d+1}, \ldots, h_n, \phi_{d+1}, \ldots, \phi_n$ do not play any role, but their knowledge could be profitable in the computation of a presentation matrix (see (2.8)).

(2.6) *How to find a finite system of generators of S over R.:*

As in (2.4) we can find $\sigma \in \mathbf{N}$ such that

$$(h_0, \ldots, h_n)^\sigma \subseteq (h_0, \ldots, h_d, \phi_{d+1}, \ldots, \phi_n).$$

Then

$$\{h_{d+1}^{\alpha_{d+1}} \cdot \ldots \cdot h_n^{\alpha_n} \mid \sum_{j=d+1}^{n} \alpha_j < \sigma\}$$

is a finite (homogeneous) system of generators of S over R (see [ZS], Lemma of p. 198).

(2.7) *How to find a finite system of generators of A over B:*

We can find $\sigma \in \mathbf{N}$ such that $(\bar{h}_0, \ldots, \bar{h}_n)^\sigma \subseteq (\bar{h}_0, \ldots, \bar{h}_d)$; then

$$\{\bar{h}_{d+1}^{\alpha_{d+1}} \cdot \ldots \cdot \bar{h}_n^{\alpha_n} \mid \sum_{j=d+1}^{n} \alpha_j < \sigma\}$$

is a (homogeneous) system of generators of A over B.

(2.8) *How to find a presentation of A over B:*

Several techniques can be used.

a) First of all, let us suppose that A *is torsion free as a B-module*. This is true iff I is unmixed (i.e. *all* the prime divisors of I have the same dimension - see [ZS], Lemma 2, p.206). In this case there is an embedding of A in a suitable B^r. Starting from a system of generators $\{f_1, \ldots, f_t\}$ of A over B, we can easily compute a maximal B-free system $\{f_1, \ldots, f_r\}$, hence we can compute the embedding $A \to B^r$. By means of Buchberger's algorithm applied to $B^t \to A \to B^r$, we can compute the presentation matrix M of A over B.

b) Let us suppose now that A is not B-torsion free. In this case we can do as follows. Starting from an h-sop of S and from a system of generators of R- module S computed as in (2.4) and (2.6), we can compute (as in a)) the embedding of S in a suitable R^r. Hence we can give the presentation matrix N of the R-module S and then we can apply Remark (2.3).

c) Let us describe another general technique to compute a presentation matrix of a B-module A. Let us start from an h-sop $\{\bar{h}_0, \ldots, \bar{h}_d\}$ of A, and from a homogeneous system of generators $\{1 = \bar{f}_1, \ldots, \bar{f}_t\}$ ($f_i \in k[\mathbf{X}]$) of the B-module A. Let us suppose $z_2, z_3, \ldots, z_t$ be "tag variables" (see [Sp]) and let us define the homomorphism

$$\phi : k[z_2, \ldots, z_t, h_0, \ldots, h_d] \to k[h_0, \ldots, h_n]$$

by: $\phi(h_i) = h_i (i = 0, \ldots, d)$; $\phi(z_j) = f_j (j = 2, \ldots, t)$. Let us point out that both rings involved are polynomial ones. Then, it is well known that

$$\phi^{-1}(I) = (I, z_2 - f_2, \ldots, z_t - f_t)k[\underline{z}, h_0, \ldots, h_n] \cap k[\underline{z}, h_0, \ldots, h_d],$$

so we can compute it.

Let

$$\phi^{-1}(I)_{\leq 1} := \{ F_1(h_0, \ldots, h_d) + \sum_{i=2}^{t} F_i(h_0, \ldots, h_d)z_i \mid$$

$$F_1(h_0, \ldots, h_d) + \sum_{i=2}^{t} F_i(h_0, \ldots, h_d)f_i \in I\}$$

with $F_i(h_0, \ldots, h_d) \in k[h_0, \ldots, h_d]$

It is obvious that the B-module of the syzygies of $\{1, \bar{f}_2, \ldots, \bar{f}_t\}$ is an epimorphic image of $\phi^{-1}(I)_{\leq 1}$, so it is enough to compute a system of generators of the latter in order to give the claimed presentation; to do it, we can compute a Grobner basis of $\phi^{-1}(I)$ with respect to any termordering such that $\underline{z} > h_0, \ldots, h_d$. (It is more convenient to use a termordering such that $h_{d+1}, \ldots, h_n > \underline{z} > h_0, \ldots, h_d$ in order to compute also $\phi^{-1}(I)$ - see [Sp]).

3. On the C. M. locus via Noether Normalization

(3.1) Let us assume that $S = k[X_1, \ldots, X_n]$, I is an ideal of S, $h = ht\ I$, and $A = S/I$. If $I = \cap_{j=1}^{s} I_j$, where I_j are comaximal ideals, then the ring A is a decomposable one, so $A = \prod_{j=1}^{s} A_j$ (where $A_j = S/I_j$). It is easy to see that A is C. M. iff A_j is C. M. for every j. Indeed, the prime ideals of A are exactly of the kind $\mathfrak{p} = A_1 \oplus \ldots \oplus A_{j-1} \oplus \mathfrak{p}_j \oplus A_{j+1} \oplus \ldots \oplus A_s$, where $\mathfrak{p}_j \in \text{Spec}(A_j)$, and it is also true that $A_\mathfrak{p} \cong (A_j)_{\mathfrak{p}_j}$. Moreover, if we denote by $r(A)$ the global C. M. type of A, it follows that $r(A) = \max_{1 \le j \le s} r(A_j)$.

Then: the not C. M. locus of A is the union of the not C. M. loci of A_j. Similarly: C. M. type $> r$ locus of A is the union of the C. M. type $> r$ loci of $A_j (r \ge 1)$.

Remark: In order to find a decomposition $A = \oplus A_j$ of A such that each $\text{Spec}(A_j)$ is connected, we could compute a primary decomposition of I, but this is very expensive from a computational point of view. However, the authors do not know any other approach for this problem.

(3.2) Let us assume now that $\text{Spec}(A)$ is connected (as will be pointed out in (3.9) this is *not* a restrictive hypothesis for our aim).

Let $B = k[Y_1, \ldots, Y_{n-h}] \rightarrow A$ be a Noether Normalization of A, and let $d = n - h = \dim A$. We need some preliminary remarks.

a) A is finite B-module and the extension $B \rightarrow A$ verifies GU and INC (see [K]); moreover, for every $\mathfrak{q} \in \text{Spec}(B)$ there is just a finite number of $\mathfrak{p} \in \text{Spec}(A)$ lying over $\mathfrak{q}$ (i.e. $\mathfrak{p} \cap B = \mathfrak{q}$).

b) For any $\mathfrak{q} \in \text{Spec}(B)$, $A_\mathfrak{q}$ is a finite $B_\mathfrak{q}$-module, and the extension $B_\mathfrak{q} \rightarrow A_\mathfrak{p}$ is a quasi-finite one. Then, $(B_\mathfrak{q})\hat{} \rightarrow (A_\mathfrak{p})\hat{}$ is a finite extension. Moreover $A_\mathfrak{q}$ is a semilocal ring, and it follows that $(A_\mathfrak{q}, \mathfrak{q}A_\mathfrak{q})\hat{} = (A_\mathfrak{q}, \text{rad}(A_\mathfrak{q}))\hat{} = \prod_{\mathfrak{p} | \mathfrak{p} \cap B = \mathfrak{q}} (A_\mathfrak{p})\hat{}$.

For the sake of completeness let us prove the following Proposition in the general context (i.e. A not necessarily indecomposable).

Proposition 3.3. *Let B be a Noether Normalization of A.*
(i) If A is B-flat, then A is a C. M. ring;
(ii) Let $\mathfrak{q} \in \text{Spec}(B)$; if $A_\mathfrak{q}$ is $B_\mathfrak{q}$-flat, then $A_\mathfrak{q}$ is a C. M. ring.

Proof.: (i) Let $\mathfrak{m}$ be any maximal ideal of A; hence $\mathfrak{n} = \mathfrak{m} \cap B$ is maximal too, and $ht\ \mathfrak{m} \le ht\ \mathfrak{n}$ because GU holds (see [K], Th. 45). Moreover, $ht\ \mathfrak{n} = \text{grade}(\mathfrak{n})$ by the regularity of B, and $\text{grade}(\mathfrak{n}) \le \text{grade}(\mathfrak{m})$ by the B-flatness of A, so the thesis follows.

(ii) The proof runs exactly like the previous one.   Q. E. D.

Remark: We observe that if A is B-flat, then GD holds for the extension $B \to A$ (see [M1], (5.D)).

Proposition 3.4. *Let A be indecomposable and let B be its Noether Normalization. Then:*
(i) If A is a C. M. ring, then A is equidimensional;
(ii) If A is C. M., then A is B-flat;
(iii) If A_q is C. M., then A_q is B_q-flat, where $q \in \text{Spec}(B)$.

Proof: (i) Let us suppose that there are two minimal prime divisors $\mathfrak{p}, \mathfrak{p}'$ of I such that $ht\, \mathfrak{p} \neq ht\, \mathfrak{p}'$. Since $\text{Spec}(A)$ is connected, there is a maximal ideal $\mathfrak{m}$ such that $\mathfrak{m} \supseteq \mathfrak{p}, \mathfrak{p}'$; hence $A_\mathfrak{m}$ is not C. M.

(ii) It is enough to prove that $A_\mathfrak{n}$ is $B_\mathfrak{n}$-flat for every $\mathfrak{n} \in \text{Specm}(B)$. But $A_\mathfrak{n}$ is a semilocal equidimensional (by i) C. M. ring, so $ht\, J + \dim J = \dim A_\mathfrak{n}$ holds for every ideal J of $A_\mathfrak{n}$. Then, by [K], Th. 47, it follows that $ht\, \mathfrak{p} = ht\, \mathfrak{q}$ for all $\mathfrak{p} \in \text{Spec}(A)$ such that $\mathfrak{p} \cap B = \mathfrak{q}$. We take now a regular sop $\{z_1, \ldots, z_d\}$ in $B_\mathfrak{n}$; then z_1 is also $A_\mathfrak{n}$-regular because $A_\mathfrak{n}$ is C. M. and $ht\, (\mathfrak{p} \cap B) = 0$ (hence $\mathfrak{p} \cap B = (0)$) for every associate prime $\mathfrak{p}$ of $A_\mathfrak{n}$; then $dh_{B_\mathfrak{n}}(A_\mathfrak{n}) = dh_{B_\mathfrak{n}/(z_1)}(A_\mathfrak{n}/(z_1))$. Further, $B_\mathfrak{n}/(z_1)$ is still a regular ring, $A_\mathfrak{n}/(z_1)$ is equidimensional, C. M. , and it is an integral extension of $B_\mathfrak{n}/(z_1)$. Hence, by induction, it follows that $dh_{B_\mathfrak{n}}(A_\mathfrak{n}) = dh_{B_\mathfrak{n}/(z_1,\ldots,z_d)}(A_\mathfrak{n}/(z_1, \ldots, z_d)) = 0$.

(iii) As the previous one.   Q. E. D.

Corollary 3.5. *Let A be indecomposable.*
(j) A is C. M. iff A is B-flat;
(jj) Let $\mathfrak{q} \in \text{Spec}(B)$, then A_q is C. M. iff A_q is B_q-flat, that is, A_q is B_q-flat iff $A_\mathfrak{p}$ is C. M. for every $\mathfrak{p} \in \text{Spec}(A)$ such that $\mathfrak{p} \cap B = \mathfrak{q}$.

Proof: It is an obvious consequence of (3.3) and (3.4)   Q. E. D.

Remark - Counterexample 3.6: The hypothesis that A be indecomposable is necessary in order that (3.4) (ii) holds; indeed:
    Let $S = k[X, Y], I = (X) \cap (Y, X - 1); ht\, I = 1, \dim A = 1.$ $B = k[Y] \to k[X, Y]/(X) \cap (Y, X - 1) = A$; so A is C. M. , $A \cong k[Y] \oplus k$, so A is not B-flat; moreover, it is easy to see that the extension $B \to A$ cannot verify GD (see a previous Remark; for other examples see [GNN]).

Proposition 3.7. *Let A be indecomposable and let B be its Noether Normalization.*
    Let $\bar{C} := \{\mathfrak{p}, \mathfrak{p} \in \text{Spec}(A) \mid A_q$ is not B_q-flat where $\mathfrak{q} = \mathfrak{p} \cap B\}$. $\bar{C}$ is closed in $\text{Spec}(A)$, and $\bar{C} = V(F_j(A) \cdot A)$, where $F_j(A)$ is the first non-zero Fitting's invariant of the B-module A.

Proof: Let $0 = F_{j-1}(A) \subsetneq F_j(A) \subset \ldots \subset F_\sigma(A) = B$ be the chain of the Fitting's invariants. If $\mathfrak{q} \in \text{Spec}(B)$ it follows easily that $(F_j)_q \neq 0$ (B is a domain) and $(F_j)_q = B_q$ iff $q \notin V(F_j)$; on the other hand $(F_j)_q = B_q$ iff A_q is B_q-free (of rank j).

Hence the thesis follows.   Q. E. D.

(3.8) Let $C := \{\mathfrak{p}, \mathfrak{p} \in \text{Spec}(A) \mid A_{\mathfrak{p}} \text{ is not C. M. }\}$, then, by 3.5, $\bar{C} \supseteq C$, and this inclusion is proper, in general. Indeed,

$$C = \{\mathfrak{p} \in \bar{C} \mid (B_{\mathfrak{p} \cap B})\hat{\ } \to (A_{\mathfrak{p}})\hat{\ } \text{ is not flat}\}$$

as it is shown by the following remark: $B_q \to A_q$ is a flat homomorphism iff $(B_q)\hat{\ } \to (A_q, q A_q)\hat{\ }$ is flat too. Moreover, $(A_q, q A_q)\hat{\ } = \oplus_{\mathfrak{p} \mid \mathfrak{p} \cap B = q}(A_{\mathfrak{p}})\hat{\ }$ and $A_{\mathfrak{p}}$ is C. M. iff $(A_{\mathfrak{p}})\hat{\ }$ is too.

So, in order to compute C, we need to know, for every $q \in V(F_j(A))$, all the ideals $\mathfrak{p} \in \text{Spec}(A)$ lying over q such that $(B_q)\hat{\ } \to (A_{\mathfrak{p}})\hat{\ }$ is not flat. It is quite clear that this method is not computable at all.

(3.9) *How to find C by means of a "generic" Noether Normalization:*

Without any restriction, we can assume that I is a homogeneous ideal of $S' = k[X_0, \dots, X_n]$ (see (2.1) and (1.11)). Let $ht\, I = h$ $(0 \le h \le n)$, $d = n - h = \dim(\text{Proj } S'/I)$, and let us suppose that k is *infinite*. We remark that $\text{Spec}(S'/I)$ is connected, so S'/I is indecomposable (see also (3.1) and (3.2)).

(j) Let $(h_0, \dots, h_d)$ be the "generic" linear variety of (projective) dimension $n-d-1$, namely the "generic" point of the Grassmannian $\mathcal{G}_{n,h-1}(k)$ where $h_i = \sum_{j=0}^{n} a_{ij} X_j$ (a_{ij} are parameters). The set of all $(h'_0, \dots, h'_d)$ such that $(h'_0, \dots, h'_d, I)$ is not an irrelevant ideal, is given by the set of the solutions (a'_{ij}) of the equation $F(a_{ij}) = 0$, obtained from the Chow-Cayley form of I, $(h_0, \dots, h_d, I)k[a_{ij}, \mathbf{X}] \cap k[a_{ij}]$ (this is a multihomogeneous polynomial).

On the other hand, by [Mu], Ch. II, Th. 1, §4, p. 185, we can see that $F(a'_{ij}) = 0$ iff $V(h'_0, \dots, h'_d) \cap V(I) \ne \emptyset$ holds in $\mathbf{P}^n(\bar{k})$ (where $\bar{k}$ is the algebraic closure of k). But F is a non-zero polynomial, so, by [Bo], Ch. IV, §2.5, Prop. 8, there are infinitely many a_{ij} in k which do not satisfy $F(a_{ij}) = 0$.

(jj) The C. M. locus of A is a non-empty dense open of $\text{Spec}(A)$, so $\dim C < \dim \text{Spec}(A)$; similarly, $\dim(C \cap \text{Proj}(A)) < d$. Let $C = V(\mathfrak{a})$ ($\mathfrak{a}$ is a homogeneous ideal), and let $\mathfrak{m}$ be a homogeneous ideal in $\text{Spec}(A)$ such that $A_{\mathfrak{m}}$ is C. M. We can prove the following:

Proposition 3.9. *There are (a'_{ij}), $a'_{ij} \in k$, $h'_i = \sum_j a'_{ij} X_j$, such that, if $\mathfrak{n}' = \mathfrak{m} \cap k[h'_0, \dots, h'_d]$, then $A_{\mathfrak{p}}$ is C. M. for every $\mathfrak{p}$ lying over $\mathfrak{n}'$.*

Proof: Indeed, the previous assertion is not true, exactly for each (a_{ij}) such that $\mathfrak{n} A + \mathfrak{a}$ is not an irrelevant ideal $(\mathfrak{n} = \mathfrak{m} \cap k[h_0, \dots, h_d])$, namely for each (a_{ij}) which verifies a suitable system of equations $H_\sigma(a_{ij}) = 0$

(with coefficients in k). It is easy to see (working in $\bar{k}$) that the ideal (H_σ) is not zero; then, applying [Bo], Prop. 8, there are infinitely many (a'_{ij}) in k such that $H_\sigma(a'_{ij}) \neq 0$. Q. E. D.

The above observations justify the following method to find C.

(jjj) Let $B = k[h_0, \ldots, h_d] \rightarrow A$ a "generic" Noether Normalization of A, and let $B^s \xrightarrow{M} B^t \rightarrow A \rightarrow 0$ be its effective presentation (see §2). Let $F_j(A)$ be the first non-zero Fitting's invariant of the B-module A, and let $G_1(a_{ij}, \mathbf{X}), \ldots, G_r(a_{ij}, \mathbf{X})$ be a finite system of generators of the ideal $F_j(A)$ of B. We can think of every G_p as a polynomial with coefficients in $k[\mathbf{X}]$ and with a_{ij} as indeterminates. If a is the ideal in $k[\mathbf{X}]$ generated by the coefficients of all the G_p's, then $C = V(a)$.

4. On the C. M. type $\leq r$ locus via Noether Normalization

(4.1) Let $A = S/I$ and let B be its Noether Normalization. Moreover, let C be the not C. M. locus of A. Thinking $\mathrm{Hom}_B(A, B)$ as an A-module and applying [HK] Satz 5.12 and 6.11, it follows that

$$r(A_\mathfrak{p}) = \mu_{A_\mathfrak{p}}(\mathfrak{p}; \mathrm{Hom}_B(A, B))$$

for any $\mathfrak{p} \in \mathrm{Spec}(A), \mathfrak{p} \notin C$. Hence, if

$$C_r := \{\mathfrak{p} \in \mathrm{Spec}(A) \mid A_\mathfrak{p} \text{ is C. M. with } r(A_\mathfrak{p}) > r \text{ or } A_\mathfrak{p} \text{ is not C. M.}\},$$

then $C_r = C \cup V(F_r(\mathrm{Hom}_B(A, B)))$ where F_r is the r-th Fitting's invariant of the A-module $\mathrm{Hom}_B(A, B)$.

(4.2) *Remark*: In order to use just the B-structure of $\mathrm{Hom}_B(A, B)$ (which is easier to compute), we could try to work (as in 3.9) by a "generic" Noether Normalization.

In this case, since

$$(\mathrm{Hom}_{B_\mathfrak{q}}(A_\mathfrak{q}, B_\mathfrak{q}), \mathfrak{q}B_\mathfrak{q})\hat{} = \oplus_{\mathfrak{p}|\mathfrak{p}\cap B=\mathfrak{q}}\mathrm{Hom}_{(B_\mathfrak{q})\hat{}}((A_\mathfrak{p})\hat{}, (B_\mathfrak{q})\hat{}),$$

it follows that

$$\mu(\mathfrak{q}; \mathrm{Hom}_B(A, B)) = \sum_{\mathfrak{p}|\mathfrak{p}\cap B=\mathfrak{q}} \mu(\mathfrak{p}; \mathrm{Hom}_B(A, B)) = \sum_{\mathfrak{p}|\mathfrak{p}\cap B=\mathfrak{q}} r(A_\mathfrak{p}).$$

It is easy to find (by means of the Fitting's invariants) the locus of all ideals $\mathfrak{q} \in \mathrm{Spec}(B)$ such that $\mu_{B_\mathfrak{q}}(\mathfrak{q}; \mathrm{Hom}_B(A, B)) > s$ ($s \in \mathbf{N}$). However, without more informations about the ramification properties of the finite map $\mathrm{Spec}(A) \rightarrow \mathrm{Spec}(B)$ (see [Sh], Ch. II, §5), we cannot find all $r(A_\mathfrak{p})$ with $\mathfrak{p} \cap B = \mathfrak{q}$.

So, we have to compute the presentation of $\operatorname{Hom}_B(A, B)$ as an A-module.

(4.3) With the same notation as in §2, let $B^s \xrightarrow{\ M\ } B^t \to A \to 0$ be an effective Noether Normalization of A, and let

$$\{m_1, \ldots, m_t\} = \{\bar{h}_{d+1}^{\alpha_{d+1}} \cdot \ldots \cdot \bar{h}_n^{\alpha_n} \mid \sum_{j=d+1}^n \alpha_j < \sigma\}$$

be a system of generators of A over B (see 2.7 and 2.8). By duality, we obtain: $0 \to \operatorname{Hom}_B(A, B) \to B^t \xrightarrow{\ {}^tM\ } B^s$ and hence we can compute a system of generators $\{\phi_1, \ldots, \phi_\rho\}$ (also as an A-module) of $\operatorname{Hom}_B(A, B) \cong \ker({}^tM)$. Let:

$$\phi_1 = \begin{pmatrix} b_{11} \\ \vdots \\ b_{t1} \end{pmatrix}, \ldots, \phi_\rho = \begin{pmatrix} b_{1\rho} \\ \vdots \\ b_{t\rho} \end{pmatrix};$$

(as a B-linear transformation each ϕ_i is such that $\phi_i(m_j) = b_{ji} \in B$, $i = 1, \ldots, \rho, j = 1, \ldots, t$). Now we want to compute a system of generators of the A-module of syzygies of $\{\phi_1, \ldots, \phi_\rho\}$.

First of all, we denote by $P = (\lambda_{kr}^h), h, k, r = 1, \ldots, t, \lambda_{kr}^h \in B$, the multiplication matrix of the B-algebra A (this is easily computed; indeed $m_h m_k = \sum_{r=1}^t \lambda_{kr}^h m_r$), and we put $P_h = \begin{pmatrix} \lambda_{11}^h \ldots \lambda_{1t}^h \\ \vdots \\ \lambda_{t1}^h \ldots \lambda_{tt}^h \end{pmatrix}$. It is easy to

see that $m_h \phi_i = P_h \cdot \begin{pmatrix} b_{1i} \\ \vdots \\ b_{ti} \end{pmatrix}$, $h = 1, \ldots, t; i = 1, \ldots, \rho$, so we are able

to compute a system of generators $(\mu_{hi}^\ell)_{h=1,\ldots,t,\ i=1,\ldots,\rho}^{\ell=1,\ldots,\tau}$, $\mu_{hi}^\ell \in B$ of the B-module of the syzygies of $\{m_h \phi_i \mid h = 1, \ldots, t; i = 1, \ldots, \rho\}$, being $m_h \phi_i \in B^t$ (B is a polynomial ring). Let $a_i^\ell = \sum_{h=1}^t \mu_{hi}^\ell m_h$. Then $\{(a_1^\ell, \ldots, a_\rho^\ell) \mid \ell = 1, \ldots, \tau\}$ is a system of generators of the A-module of syzygies of $\{\phi_1, \ldots, \phi_\rho\}$, so we have the presentation matrix of the A-module $\operatorname{Hom}_B(A, B)$; hence we are able to compute the Fitting's invariants (see 4.1).

REFERENCES

[A]   L. L. Avramov, *Flat morphisms of complete intersections*, Soviet Math. Dokl. 16 (1975), 1413–1417.

[Ba]   D. A. Bayer, *The division algorithm and the Hilbert scheme*, Ph. D. Thesis, Harvard (1982).

[BSS] D. Bayer, Ma. Stillmann, Mi. Stillmann, "Macaulay user manual," 1989.

[Bo]   N. Bourbaki, "Algèbre," Ch. IV-V, Hermann, Paris, 1959.

[GM]   S. Greco, M. G. Marinari, *Nagata's criterion and openness of loci for Gorenstein and complete intersection*, Math. Z. 160 (1978), 207–216.

[GNN] R. Gilmer, B. Nashier, W. Nichols, *On the heights of prime ideals under integral extensions*, Arch. Math. 52 (1989), 47–52.

[HK]   J. Herzog, E. Kunz, "Der kanonische Modul eines Cohen-Macaulay Ringes," Lecure Notes in Math. 238, Springer, 1971.

[K]   I. Kaplansky, "Commutative rings," Allyn and Bacon, 1970.

[M1]   H. Matsumura, "Commutative algebra," (2 edn.), Benjamin, 1980.

[M2]   H. Matsumura, "Commutative ring theory," Cambridge U. Press, 1986.

[MM]   E. Mayr, A. Meyer, *The complexity of the word problems for commutative semigroups and polynomial ideals*, Adv. Math. 46 (1982), 305–329.

[Mu]   D. Mumford, "Introduction to algebraic geometry," preliminary version of first 3 chapters, Harvard Univ. mimeogr..

[No]   D.G. Northcott, "Finite free resolutions," Cambridge U. Press, 1976.

[Sa]   J. D. Sally, "Numbers of generators of ideals in local rings," Lect. Notes Pure Appl. Math. 35, Marcel Dekker, 1978.

[S]   P. Samuel, "Méthodes d'algèbre abstraite en géométrie algébrique," Springer, 1955.

[Se]   A. Seidenberg, *Contructions in algebra*, Trans. Amer. Math. Soc 197 (1974), 273–313.

[Sh]   I. R. Shafarevich, "Basic algebraic geometry," Springer, 1974.

[Sp]   D. Spear, *A constructive approach to commutative ring theory*, in "Proc. 1977 Macsyma User's conference," 1977, pp. 369–376.

[Va]   W. V. Vasconcelos, "Divisor theory in module categories," North Holland, Amsterdam, 1974.

[ZS]   O. Zariski, P. Samuel, "Commutative algebra," Vol. II, Van Nostrand, New York - London, 1968.

Fabio Rossi, Walter Spangher
Università degli Studi di Trieste, Dipartimento di Scienze Matematiche
Piazzale Europa, 1, I-34121 Trieste, ITALIA

Sign determination on zero dimensional sets

MARIE-FRANÇOISE ROY   AVIVA SZPIRGLAS

Introduction

The aim of this paper is to study the following problem:
- Let $f_1, f_2, \cdots, f_n$, be n polynomials with integer coefficients such that:

1. for $1 \leq i \leq n$, f_i is a polynomial of the variables $(X_1, \cdots, X_i)$, monic in X_i, of degree d_i in X_i;

2. for $1 \leq i \leq n$, for $1 \leq j \leq i$, f_i is of degree $\delta_j \leq d_j$ in X_j.

- We consider the following "triangular" system:

$$S : \begin{cases} f_1(X_1) = 0 \\ f_2(X_1, X_2) = 0 \\ \cdots \\ f_n(X_1, \cdots, X_n) = 0 \end{cases}$$

- Let Q be a polynomial in $(X_1, \cdots, X_n)$ with integer coefficients. We want to determine the sign taken by Q at each real solution of S. The main result is the following:

Theorem   *The computation of the sign taken by Q at each real solution of the system S is done with a number of elementary arithmetical operations polynomial in $D = \prod_{i=1}^{n} d_i$.*

Remark.

1. The parameter D for the complexity is very natural in that context for it is the dimension as a vector space of the quotient of the ring of polynomials in n indeterminates by the ideal generated by $(f_i)_{1 \leq i \leq n}$.

2. We do not know if our method gives a binary complexity polynomial in D and in the size of the coefficients.

To solve the problem, we can proceed as suggested in [1] by eliminating successively the variables, in a D5 way. We shall see (part 1) that this method gives a complexity exponential in n, but not polynomial in D.

The method we use here is based on the following idea: we compute the radical of the ideal generated by $(f_i)_{1 \leq i \leq n}$, and, by a change of coordinates, we have to solve a problem in one variable for which we can use the results of [8]. This idea is taken from [4](see in D.Lazard's paper [6] also).

In fact, we shall compute step by step the radical of this ideal and the new system of coordinates. We work at each step with equations in two variables. For, if we work with systems in n variables, we have to consider in the computation of the complexity (because of the determination of standard basis) the number of monomials of degree D at most in a polynomial of degree D, and this number is in $O(D^n)$.

In part 2, we explicitly work with systems in two variables.

In part 3, we determine a system R such that:

$$R : \begin{cases} f(Y_1) = 0 \\ X_2 = h_2(Y_1) \\ \cdots \\ X_n = h_n(Y_1) \end{cases}$$

where

- the polynomials f and, for $2 \leq i \leq n$, h_i are of degree at most D;

- $Y_1 = X_1 + \sum_{i=2}^{n} \lambda_i X_i$

Remark. If we consider the same problem for the general case where the system is not triangular, i.e for a system where each f_i is a polynomial in all the variables $(X_1, \cdots, X_n)$, the system R exists but its construction has a complexity which is not polynomial in D.

Let $\bar{S}$ be the system deduced from S by substituing each X_i by its value in function of X_1, and by eventually reducing the polynomials obtained, using the first equation of the system R.

The construction of the system R allows us to say that each solution of $\bar{S}$ is solution of R, and each solution of R is solution of $\bar{S}$.

(It is for the construction of R that we work by steps; at each step, we work with systems of two equations in two variables.)

When we have obtained the system R, we use in part 4 the methods introduced in [1] and developped in [8] on computations with real algebraic numbers, to determine the sign of $\bar{Q}$ (polynomial coming from Q by substitution and reduction) at each real solution of R.

This gives the resolution of our problem (sign of Q at each solution of S), and the complexity of this resolution.

All along this paper, we suppose that the polynomials f_i, for $1 \leq i \leq n$ verify the hypothesis of the introduction.

We denote: $D_i = \prod_{j=1}^{i} d_j$, for $1 \leq i \leq n$

The system S_i, for $1 \leq i \leq n$, is the following one:

$$S_i : \begin{cases} f_1(X_1) = 0 \\ \cdots \\ f_i(X_1, \cdots, X_i) = 0 \end{cases}$$

1   Method by successive elimination of the variables

In this part, we study the complexity of the determination of the sign of the polynomial Q at the real solutions of S whith a method using successive elimination of the variables.

1.1   Description of the method

Let, for $2 \leq i \leq n$, $(\xi_1 \cdots, \xi_i)$ be a solution of the system S_i

Let, for $2 \leq i \leq n$, $\bar{f}_i(\xi_1 \cdots, \xi_{i-1})(X_i)$ be the polynomial in one variable X_i with real coefficients, such that:

$\bar{f}_i(\xi_1 \cdots, \xi_{i-1})(X_i) = f_i(\xi_1 \cdots, \xi_{i-1}, X_i)$

Then, ξ_i is a real root of $\bar{f}_i(\xi_1 \cdots, \xi_{i-1})(X_i)$.

Hence, if we want to solve the problem, we need only to know the sign taken by $\bar{Q}(\xi_1 \cdots, \xi_{n-1})(X_n)$ $(= Q(\xi_1 \cdots, \xi_{n-1}, X_n))$ at each real root of $\bar{f}_n(\xi_1 \cdots, \xi_{n-1})(X_n)$, for each $(n-1)$-uple $(\xi_1, \cdots, \xi_{n-1})$ solution of the system S_{n-1}.

The computation is done as in [8], by replacing Sturm sequences by Sturm Habicht sequences (see [3]) as in [9].

1.2   Analysys of the complexity

We have the following steps:

1. we compute for each $(n-1)$-uple $(\xi_1, \cdots, \xi_{n-1})$ solution of S_{n-1} the number of real roots of $f_n(\xi_1, \cdots, \xi_{n-1})(X_n)$, i.e:

   - we compute the Sturm Habicht sequence of $\bar{f}_n(X_1 \cdots, X_{n-1})(X_n)$

   - we compute the signs taken by the d_n leading coefficients of the elements of that sequence (but we first reduce those coefficients modulo S_{n-1}) at $(\xi_1, \cdots, \xi_{n-1})$. Then we can apply Sturm Habicht theorem.(see [9]).

2. we compute the Sturm Habicht sequence associated to
$\bar{f}_n(X_1 \cdots, X_{n-1})(X_n)$ and $\bar{Q}(X_1 \cdots, X_{n-1})(X_n)$

3. we compute the signs taken by the d_n leading coefficients of the elements of this sequence at each $(\xi_1, \cdots, \xi_{n-1})$, solution of S_{n-1}. Then, we apply once again Sturm Habicht theorem (see [9]).

The method of [8] gives the conclusion.

Hence, we see that we have to solve a more general problem than ours; we have to know the sequence of signs taken by k polynomials $Q_1, \cdots, Q_k$ in the variables $X_1, \cdots, X_i$ at each solution of S_i, for $2 \le i \le n$.

So, we work in $\mathcal{A}_{S_{i-1}}[X_i]$, where, for $2 \le j \le n$, $\mathcal{A}_{S_j} = \mathbf{R}[X_1, \cdots, X_j]/I_j$ (I_j is the ideal generated by the (f_l) for $1 \le l \le j$). This gives a control of the degree of the polynomials.

We have then the following steps:

1. we apply k times the procedure $\mathbf{SIadd}_{(\xi_1, \cdots, \xi_{i-1})}$ (see procedure $\mathbf{SIadd}$ of [8]); for that, we need to compute at most d_i products in $\mathcal{A}_{S_i}$, with for each of these products a number of elementary arithmetical operations polynomial in D_i (see [7]).

2. we compute the Sturm Habicht sequences of $\bar{f}_i$ and each of these products; so, we make a number of elemenrary operations in $\mathcal{A}_{S_{i-1}}$ in $O(D_i^3)$, the cost in term of elementary arithmetical operations being hence polynomial in D_i (after [7]).

   For each computation of product and associated Sturm Habicht sequence,we have hence a cost polynomial in D_i^3.

3. we use 1 and 2 to evaluate the signs of the leading coefficients of the elements of those Sturm Habicht sequences at the solutions of S_{i-1}.

So, we have the following lemma:

Lemma 1.1 *The number of elementary arithmetical operations in evaluating the signs taken by k polynomials $Q_1, \cdots, Q_k$ in the variables $X_1, \cdots, X_i$ at each solution of S_i, for $2 \le i \le n$ is polynomial in D_i^i.*

Proof. we have to do i steps with, for the j th step ($1 \le j \le i$), a complexity polynomial in $\prod_{l=j}^i D_l$. So the total cost is polynomial in $\prod_{l=1}^i D_l$, i.e in D_i^i. Q. E. D.

As a corollary of lemma 1.1, we have the following result:

The complexity, in terms of number of elementary arithmetical operations, of the computation of the sign taken by Q at all the solutions of S is not polynomial in D, if we use the method of successive elimination of the variables described here.

2   First case: triangular system in two variables.

We consider the following system S_2:

$$S_2 : \begin{cases} f_1(X_1) = 0 \\ f_2(X_1, X_2) = 0 \end{cases}$$

We construct in this section a system R_2 of the following form:

$$R_2 : \begin{cases} f_{12}(Y_{12}) = 0 \\ X_2 = h_{22}(Y_{12}) \end{cases}$$

where f_{12} et h_{22} are polynomials in one variable of degree at most D_2, the solutions of R_0 being the solutions of the system $\bar{S}_2$ obtained from the system S_2 by the substitution $X_1 = Y_{12} - \lambda_2 X_2$.

From its construction, we shall deduce that the ideal generated by $f_{12}(Y_{12})$ et $X_2 - h_{22}(Y_{12})$ is radical.

2.1   Description of the system R_2

We know (see [5]) that there exists two polynomials $g_{12}(X_1)$ and $g_{22}(X_2)$ of degree at most respectively D_1 and D_2, such that the ideal generated by f_1, f_2, g_{12}, and g_{22} is the radical of the ideal $\mathcal{I}_2$ generated by f_1 and f_2. And the ideal generated by g_{12} et g_{22} itself is radical.

Precisely, we have: $g_{12} = \frac{f_1}{\gcd(f_1, f_1')}$; and, as the vector space $\mathbf{R}[X_1, X_2]/(f_1, f_2)$ is of dimension at most D_2, there exists a polynomial h in the variable X_2, of degree at most D_2 which is in the ideal (f_1, f_2).

Then, $g_{22} = \frac{h}{\gcd(h, h')}$

Remark. The radical of the ideal generated by f_1 and f_2 is in fact generated by g_{12}, f_2 and g_{22}. We consider the following system S_2':

$$S_2' : \begin{cases} g_{12}(X_1) = 0 \\ f_2(X_1, X_2) = 0 \\ g_{22}(X_2) = 0 \end{cases}$$

The set of distinct real solutions of S_2 is the set of real solutions of S_2'

The system S_2' has r_2 distinct solutions ($r_2 \leq D_2$), which are r_2 distinct points of the plan (X_1, X_2).

The projection of those r_2 distinct points on the X_1 coordinate axis is composed by r_1 distinct points ($r_1 \leq D_1$ and $r_1 \leq r_2$).

We change the direction of projection (i.e the X_2 axis) such that this projection contains exactly r_2 distinct points.

So, we make a change of variables of the type:

$$\begin{cases} Y_{12} = X_1 + \lambda_2 X_2 \\ X_2 = X_2 \end{cases}$$

which corresponds to a "good" direction of projection.

Remark (⋆). There are at most $\frac{d_2(d_2-1)}{2}$ "bad" directions of projection, included the "bad" complex directions, hence there are at most $\frac{d_2(d_2-1)}{2}$ "bad" choices for the parameter λ_2.

Let $\bar{g}_{12}(Y_{12}, X_2)$ and $\bar{f}_2(Y_{12}, X_2)$ be the polynomials deduced from g_{12} and f_2 by substitution:

- $\bar{g}_{12}(Y_{12}, X_2) = g_{12}(Y_{12} - \lambda_2 X_2)$.

- $\bar{f}_2(Y_{12}, X_2) = f_2(Y_{12} - \lambda_2 X_2, X_2)$.

As the direction of projection is "good", the standard basis of the ideal generated by $\bar{g}_{12}$, g_{22} and $\bar{f}_2$ for the order $Y_{12} > X_2$ is composed by $f_{12}(Y_{12})$ and $X_2 - h_{22}(Y_{12})$ where f_{12} and h_{22} are polynomials of degree at most D_2. Moreover, f_{12} has exactly r_2 real distinct roots.

Solving S'_2 is equivalent of solving:

$$R_2 : \begin{cases} f_{12}(Y_{12}) = 0 \\ X_2 = h_{22}(Y_{12}) \end{cases}$$

2.2   Complexity of the construction of the system R_2.

- the computation of h has a complexity in $D_2^{o(1)}$; in fact, we have to solve a linear system with a size in $D_2^{o(1)}$; from [2] we deduce the complexity of the computation.

- hence, the computation of g_{12} and g_{22} is done in $D_2^{o(1)}$

- the determination of λ_2 is done in the following way:

  1. we fix an arbitrary λ_2.

  2. we substitute, to X_1, $Y_{12} - \lambda_2 X_2$ in g_{12} and f_2.

  3. we determine the standard basis of the ideal generated by $\bar{g}_{12}$, g_{22} and $\bar{f}_2$ for the order $Y_{12} > X_2$

  4. if we obtain a polynomial of degree 1 in X_2, the choice of λ_2 is "good".

  5. if not, we change λ_2 and do step 2 again.

- the construction of R_2 is then finished.

So, we see, using remark $(\star)$, that we have at most D_2^2 computations of standard basis, and each of these computations is done in $D_2^{o(1)}$ (see [2])

Moreover, the two substitutions of step 2 need a number of elementary arithmetical operations in $O(D_2^3)$.

Hence, we have the lemma:

Lemma 2.1 *The number of elementary arithmetical operations for the construction of R_2 is in $D_2^{o(1)}$*

Remark. The construction of R_2 implies that its solutions give the solutions of S_2', which are themselves the solutions of S_2

3   Triangular system in n variables

For the construction of the system R announced in the introduction, we proceed in $n - 1$ steps. The i th step is constituted by a system R_i of the following type:

$$R_i : \begin{cases} f_{1i}(Y_{1i}) = 0 \\ X_2 = h_{2i}(Y_{1i}) \\ X_3 = h_{3i}(Y_{1i}) \\ \cdots \\ X_i = h_{ii}(Y_{1i}) \end{cases}$$

with:

1. $Y_{1i} = X_1 + \sum_{j=2}^{i} \lambda_j X_j$;

2. f_{1i} is a polynomial of degree at most D_i;

3. for $1 \leq j \leq i$, h_{ji} is a polynomial of degree at most D_i;

4. the ideal generated by $f_{1i}(Y_{1i})$, $X_2 - h_{2i}(Y_{1i})$, $\cdots$, $X_i - h_{ii}(Y_{1i})$ is radical;

5. the set of solutions of the system R_i is the set of solutions of $\bar{S}_i$, system obtained from S_i by substitution $(X_1 = Y_{1i} - \sum_{j=1}^{i} \lambda_j X_j)$ and reduction (using the equation $f_{1i}(Y_{1i}) = 0$).

Remark. R_n is the system R we want to construct, with:

$$\begin{cases} f = f_{1n} \\ h_i = h_{in} \text{ for } 2 \leq i \leq n \end{cases}.$$

3.1   From step i to step $i+1$

Let $\bar{f}_{i+1}$ be the poynomial in $Y_{1i}, X_2, \cdots, X_i, X_{i+1}$ obtained from f_{i+1} by substitution (and eventually reduction):

$\bar{f}_{i+1}(Y_{1i}, X_2, \cdots, X_i, X_{i+1}) = f_{i+1}(Y_{1i} - \sum_{j=2}^{i} \lambda_j X_j, X_2, \cdots, X_{i+1}).$
The following system T_{i+1}:

$$T_{i+1} : \begin{cases} f_{1i}(Y_{1i}) = 0 \\ X_2 = h_{2i}(Y_{1i}) \\ \cdots \\ X_i = h_{ii}(Y_{1i}) \\ \bar{f}_{i+1}(Y_{1i}, X_2, \cdots, X_i, X_{i+1}) = 0 \end{cases}$$

has r_{i+1} distinct solutions $(r_{i+1} \le D_{i+1})$.

The system T_{i+1} is such that the projection on the plan of the coordinates X_{i+1} and Y_{1i} of these r_{i+1} dictinct points is exactly r_{i+1} dictinct points, which are the solutions of the following triangular system of two equations in two variables $S_{2,i+1}$:

$$S_{2,i+1} : \begin{cases} f_{1i}(Y_{1i}) = 0 \\ \bar{f}_{i+1}(Y_{1i}, h_{2i}(Y_{1i}), \cdots, h_{ii}(Y_{1i}), X_{i+1}) = 0 \end{cases}$$

$S_{2,i+1}$ gives all the solutions of T_{i+1}.

But, we know (see part 2) that we can construct a system $R_{2,i+1}$ with $Y_{1i+1} = Y_{1i} + \lambda_{i+1} X_{i+1}$ and:

$$R_{2,i+1} : \begin{cases} f_{1i+1}(Y_{1i+1}) = 0 \\ X_{i+1} = h_{i+1,i+1}(Y_{1i+1}) \end{cases}$$

where f_{1i+1} and $h_{i+1,i+1}$ are polynomials of degree at most $D_i.d_{i+1} = D_{i+1}$, such that the set of solutions of this system $R_{2,i+1}$ gives all the solutions of $S_{2,i+1}$; moreover, the ideal generated by $f_{1i+1}(Y_{1i+1})$ and $X_{i+1} - h_{i+1,i+1}(Y_{1i+1})$ is radical.

Let $h_{2i+1}, \cdots, h_{ii+1}$ be the polynomials in Y_{1i+1} obtained by substitution (and eventually reduction) of Y_{1i} by $Y_{1i+1} - \lambda_{i+1} h_{i+1,i+1}(Y_{1i+1})$ in $h_{2i}, \cdots, h_{ii}$.

The system R_{i+1} is then constructed.

Remark.   With the choice we have done for the projection on the first coordinate axis (i.e with the choice of the direction of the axis for the coordinate X_{i+1} in the plan (Y_{1i}, X_{i+1})), the projection on the linear subspace of the coordinate axis $Y_{1i+1}, X_2, \cdots, X_i$ of the r_{i+1} distinct solutions of T_{i+1} (or of R_{i+1}) are exactly r_{i+1} dictinct points.

3.2 Complexity of the computations from step i to step $i+1$

3.2.1 Complexity of the construction of $\bar{f}_{i+1}$

We have $f_{i+1}(X_1, X_2, \cdots, X_{i+1}) = \sum_{j=0}^{d_{i+1}} A_j X_{i+1}^j$, where A_j, for $0 \leq j \leq d_{i+1}$, is a polynomial of degree at most d_k in each variable X_k, for $1 \leq k \leq i$.

Hence, A_j has at most D_i monomials of the type $X_1^{\alpha_1} X_2^{\alpha_2} \cdots X_i^{\alpha_i}$, where $\alpha_k \leq d_i$, for $1 \leq k \leq i$.

But, we have the following relations $(\star\star)$:
$$\begin{cases} X_1 = Y_{1i} - \sum_{k=2}^i \lambda_k X_k \\ X_2 = h_{2i}(Y_{1i}) \\ \cdots \\ X_i = h_{ii}(Y_{1i}) \end{cases}$$
where, for $2 \leq k \leq i$, h_{2i} is of degree at most D_i.

To construct $\bar{f}_{i+1}$, we have to compute each monomial A_j by substitution using the relations $(\star\star)$, in $\mathbf{R}[Y_{1i}]/(f_{1i})$. The computation of one of these monomials needs $\sum_{k=1}^i \alpha_k$ products between elements of $\mathbf{R}[Y_{1i}]/(f_{1i})$; then, we have to add these D_i monomials in $\mathbf{R}[Y_{1i}]/(f_{1i})$.

We know (see [7]) that an elementary operation in $\mathbf{R}[Y_{1i}]/(f_{1i})$ needs a number of elementary arithmetical operations polynomial in D_i.

So, we have the

Lemma 3.1 *The construction of $\bar{f}_{i+1}$ needs a number of elementary arithmetical operations polynomial in D_i.*

3.2.2 Complexity of the construction of f_{1i+1} and $h_{i+1,i+1}$

Lemma 2.1 gives:

Lemma 3.2 *The construction of f_{1i+1} and $h_{i+1,i+1}$ needs a number of elementary arithmetical operations in $D_{i+1}^{o(1)}$*

3.2.3 Complexity of the construction of h_{ji+1} for $2 \leq j \leq i$

The polynomial $h_{ji+1}(Y_{1i+1})$ is obtained by computing in $\mathbf{R}[Y_{1i+1}]/(f_{1i+1})$ the polynomial $h_{ji}(Y_{1i+1} - \lambda_{i+1} h_{i+1,i+1}(Y_{1i+1}))$. But, h_{ji} is a polynomial of degree at most D_i; by arguments of the same type we gave for the proof of lemma 3.1, we have the

Lemma 3.3 *The construction for $2 \leq j \leq i$ of h_{ji+1} needs a number of elementary arithmetical operations polynomial in D_{i+1}.*

3.2.4 In conclusion:

We have the

Lemma 3.4 *The computations from step i to step $i+1$ need a number of elementary arithmetical operations polynomial in D_{i+1}.*

3.3   Construction of R

We deduce from the preceeding results:

Proposition 3.5 *The system R is obtained with a number of elementary arithmetical operations polynomial in D.*

4   Where we solve the initial problem

We suppose that R and the change of variable $Y_1 = X_1 + \sum_{i=2}^{n} \lambda_i X_i$ are already computed. Let $\bar{Q}(Y_1, X_2, \cdots, X_n) = Q(Y_1 - \sum_{i=2}^{n} \lambda_i X_i, X_2, \cdots, X_n)$

Let, for $1 \leq i \leq n$, $\bar{f}_i(Y_1, X_2, \cdots, X_n) = f_i(Y_1 - \sum_{i=2}^{n} \lambda_i X_i, X_2, \cdots, X_n)$.

Let $\bar{S}$ be the system:
$$\begin{cases} \bar{f}_1(Y_1, X_2, \cdots, X_n) = 0 \\ \cdots \\ \bar{f}_n(Y_1, X_2, \cdots, X_n) = 0 \end{cases}$$

4.1   Determination of the sign of $\bar{Q}$ at each solution of $\bar{S}$

Remark. A solution of R, hence of $\bar{S}$ is given by a n-uple $(\xi, h_2(\xi), \cdots, h_n(\xi))$ where ξ is in the set of real roots of f, and then is known by the coding of ξ (given by algorithm **RAN** of [8]) as a real root of f, i.e by the sequence of signs taken by the successive derivatives of f at ξ.

Let $\tilde{Q}(Y_1) = \bar{Q}(Y_1, h_2(Y_1), \cdots, h_n(Y_1))$, reduced in $\mathbf{R}[Y_1]/(f(Y_1))$. Let ξ be a real root of f.

The sign of $\tilde{Q}(\xi)$ is the sign of $\bar{Q}(\xi, h_2(\xi), \cdots, h_n(\xi))$. Hence, it is the sign taken by $\bar{Q}$ at the solution $(\xi, h_2(\xi), \cdots, h_n(\xi))$ of R (hence of $\bar{S}$).

So, if we want to determine the sign taken by $\bar{Q}$ at each solution of R (hence of $\bar{S}$), we have to compute the sign taken by $\tilde{Q}$ at each real root of f. This computation is done by using algorithm **RANSI** of [8].

Hence, we have the proposition (using the results of [8]):

Proposition 4.1 *The computation of the sign taken by $\bar{Q}$ at each solution of $\bar{S}$ needs a number of elementary arithmetical operations polynomial in D.*

4.2   The sign determination.

In conclusion, the computation of the sign of Q at each solution of S is done in the following way:

1. compute the change of variable $Y_1 = X_1 + \sum_{i=2}^{n} \lambda_i X_i$, and the system R;

2. compute the sign taken by $\bar{Q}$ at each solution of R (i.e at each solution of $\bar{S}$); this gives the sign taken by Q at each solution of S.

Remark. When doing all these computations, we have to make some "substitutions-reductions" for some polynomials. As the number of elementary arithmetical operations as we do those "substitutions-reductions" is polynomial in D, and as the number of those "substitutions-reductions" depends only on n, the total number of elementary arithmetical operations in these "substitutions-reductions" is itself polynomial in D.

From the preceeding results, we deduce the theorem:

Theorem 4.2 *The number of elementary arithmetical operations in the computation of the sign taken by Q at each solution of S is polynomial in D.*

REFERENCES

[1] M. Coste, M-F. Roy, *Thom's lemma, the coding of real algebraic numbers and the topology of semi algebraic sets*, Journal of Symbolic Computation **5** (1988), 121–129.

[2] A. Dickenstein, N. Fitchas, M. Giusti, C. Sessa, *The membership problem for unmixed polynomials ideals is solvable in subexponential time.* Preprint

[3] L. Gonzales, H. Lombardi, T. Recio, M-F. Roy, *Sous résultants et spécialisation de la suite de Sturm*, RAIRO (to appear).

[4] J. Heintz, M-F. Roy, P. Solernò, *Sur la complexité du principe de Tarski-Seidenberg*, Bulletin de la SMF (to appear).

[5] T. Krick, A. Logar, *Membership problem, representation problem and the computation of the radical for one dimensional ideals.* (in these proceedings)

[6] D. Lazard, *Solving zero dimensional algebraic system*, Preprint LITP june 1989.

[7] H. Lombardi, *Algèbre élémentaire en temps polynomial*, Thèse, Université de Nice, Publications Mathématiques de Besançon (1989).

[8] M-F. Roy, A. Szpirglas, *Complexity of computation on real algebraic numbers*, Journal of Symbolic Computation (to appear).

[9] M-F. Roy, A. Szpirglas, *Complexity of the computation of cylindrical decomposition and topology of real algebraic curves using Thom's lemma*, in "Real algebraic and analytic geometry," Trento, Springer Lecture Notes in Math, 1988.

Marie Françoise Roy
IRMAR
Université de Rennes
Campus de Beaulieu
35 042 Rennes CEDEX

Aviva Szpirglas
CNRS U.R.A 742
Université Paris-Nord
Avenue Jean-Baptiste Clément
93 430 Villetaneuse

A Classification of
Finite-dimensional Monomial Algebras

KIYOSHI SHIRAYANAGI

Abstract. Monomial algebras are finitely presented algebras defined by monomials. This notion is considered the most fundamental of "standard" finitely presented algebras, in the sense that they have finite noncommutative Gröbner bases as defining relations.

This paper classifies and constructs finite-dimensional monomial algebras modulo an algebra isomorphism by means of trees of nonzero monomials. These trees are deeply related to partially ordered sets called LSGOP. As a corollary, it is shown that every finite-dimensional monomial algebra has a unique irredundant presentation up to a permutation of generators. This result will serve as an effective method for solving the isomorphism problem in this area.

1 Introduction

In this paper, k is a fixed field of arbitrary characteristic. Monomial algebras are non-commutative associative algebras over k generated by a finite number of variables having a finite number of their monomials as the defining relations. To avoid a redundancy of generators, we assume that each monomial of the relations has a degree larger than one. Furthermore, we suppose that the defining relations have no redundancies; in other words, they are what is called the reduced Gröbner basis. An introduction to the Gröbner basis theory can be found in [1].

Ufnarovskij([7]) defined an oriented graph for a monomial algebra to investigate its growth. The growth of an algebra means how the dimension of the linear subspace spanned by monomials with degree smaller than n increases with n. Especially, the Ufnarovskij graph gives a criterion for deciding whether a monomial algebra is finite-dimensional or not.

Gateva-Ivanova and Latyshev([2]) consider monomial algebras as one of the most fundamental of standard finitely presented algebras, which have finite Gröbner bases as the defining relations. Refer to [5] or [4] for the definition and construction of Gröbner basis in the non-commutative case. Gateva-Ivanova and Latyshev also define the Ufnarovskij graph for a stan-

dard finitely presented algebra to look into its growth. That is, the graph
is defined by the Ufnarovskij graph of the corresponding monomial algebra
constructed by taking all the leading monomials. Moreover, they study
algorithmically recognizable properties of monomial algebras in various algebraic aspects.

This paper shows that a monomial algebra determines a tree called
a *word tree* deeply related to a partially ordered set called an LSGOP(see
[3]) in a Galois cohomology theory. In the finite-dimensional case, there is
a one-to-one correspondence between the isomorphism classes of monomial
algebras and the isomorphism classes of word trees. The latter isomorphism
is a little stronger than the ordinary tree isomorphism. We can then combi-
natorially classify and construct the finite-dimensional monomial algebras
modulo an algebra isomorphism. As a corollary, we show the uniqueness
of irredundant presentation up to a permutation of generators. This corollary can offer an effective method for solving the isomorphism problem for
finite-dimensional monomial algebras.

2   LSGOP-algebras and Monomial Algebras

LSGOP (lower subtractive G orbit poset) was introduced by Haile, Larson,
and Sweedler to classify k-algebra U which are, for a Galois extension A/k
of fields with Galois group G, isomorphic to $End_k A$ as A-bimodules and
"idempotent as algebras over A"(refer to [3]). Here we consider algebras
over the ground field k associated with LSGOP, disregarding the Galois
group action. Therefore, we define the LSGOP without the terminology of
group action.

Let (G, P) be a group G having a partial order $\leq (= P)$.

Definition 1 (G, P) is an LSGOP (for G)
$\Longleftrightarrow$
(1) $1 \in G$ is the smallest element of P (P is therefore a rooted poset), and
(2) $x \leq y \leq z \Leftrightarrow x^{-1}y \leq x^{-1}z$, when $x \leq z$.

We also say that G is an LSGOP or P is an LSGOP (for G), if there is
no confusion. Now let us define a map $e_P : G \times G \to \{0, 1\}$ in the following
fashion:
$$e_P(g, h) = \begin{cases} 1 & \text{if } g \leq gh \\ 0 & \text{otherwise.} \end{cases}$$
In [6], we defined an algebra associated with LSGOP as follows:

Definition 2 Let (G, P) be an LSGOP.
A is the algebra associated with (G, P) (or LSGOP-algebra for (G, P))
$\Longleftrightarrow$
The underlying vector space is $\bigoplus_{g \in G} ku_g$ (the k vector space having a basis
$\{u_g\}_{g \in G}$) and the multiplication is given by $u_g u_h = e_P(g, h)u_{gh}$.

We write this algebra $k[G]_P$. Note that the existence of the unit element and the associativity of $k[G]_P$ is guaranteed by (1) and (2) of Definition 1 respectively. $k[G]_P$ is considered to be an extension of the group algebra $k[G]$ in a sense.

Theorem 1 *(proved in [6]) If G is a finite group and two LSGOPs (G, P) and (G, Q) are trees, then $P \simeq Q$ (as LSGOPs) if and only if $k[G]_P \simeq k[G]_Q$ (as k-algebras).*

Here $P \simeq Q$ as LSGOPs means that there exists a poset isomorphism $\sigma : P \to Q$ such that if $e_P(g, h) = 1$, then $\sigma(g)\sigma(h) = \sigma(gh)$.

Furthermore, it is easy to see but important that if (G, P) is a (possibly infinite) tree, then $k[G]_P$ is a monomial algebra.

Example 1 When $G = \mathbf{Z}_5$ (the cyclic group of degree 5) and P is described by the tree below $(G = \langle g \rangle)$, $k[G]_P$ has a presentation: $k\langle x, y | x^3 = y^2 = yx = x^2 y = 0 \rangle$.

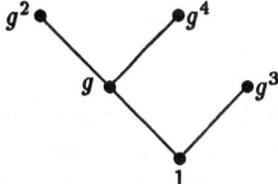

Throughout this paper, $k\langle X | R \rangle$ denotes the associative algebra generated by variables in X having R as defining relations.

Then how about the converse: can every monomial algebra be represented by an algebra associated with some tree LSGOP?

The answer is NO. As a counterexample, it can be proved that there does not exist a group G such that $k\langle x, y | x^3 = xy = yx^2 = y^2 x = y^3 = 0 \rangle$ and $k[G]_P$ are isomorphic.

3 The Word Trees

So far, we have considered the relationship between LSGOP-algebras and monomial algebras such that the tree LSGOP-algebra class is properly contained by the monomial algebra class. The difference is caused by the existence of a group G. In fact, without the introduction of groups, we can define a partially ordered set *word tree* from a monomial algebra. From now on, we say A is a monomial algebra for X if A has X as its algebra generators, i.e. $A = k\langle X | R \rangle$ for some relation R of X.

Definition 3 Let A be a monomial algebra for X.
The word tree $\tau(A)$ of A is as follows:
(1) the elements of $\tau(A)$ are $1(\in k)$ and nonzero monomials in A (then, $\tau(A)$

$\subset \langle 1, X \rangle$, where $\langle 1, X \rangle$ is the free monoid generated by the set $\{1\} \cup X$ with the unit 1) and
(2) the partial order is given by '$x \leq y \iff \exists u \in \tau(A)$ s.t. $y = xu$ in A'.

Remark 1: The assumption of monomial algebras deduces an asymmetric law required for a partial order.
Remark 2: The set $\tau(A)$ gives a basis of the underlying vector space of A.
Example 2 The word tree of the monomial algebra $k\langle x, y | x^3 = xy = yx^2 = y^2x = y^3 = 0 \rangle$ given in the above counterexample:

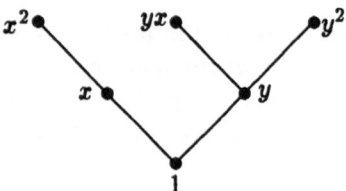

Next we define a word tree for a finite number of variables to construct monomial algebras from them.

Definition 4 Let X be a finite set of variables and 1 the unit element of k. A word tree τ for X is a set of monomials in X including 1 and X (i.e. $\{1\} \cup X \subset \tau \subset \langle 1, X \rangle$) such that τ is a rooted tree with the root 1, endowed with a partial order "being a left prefix," i.e. "$xu \leq xv \iff u \leq v$ and $xv \in \tau$".

We can easily prove the following theorem.

Theorem 2 *Let X be any finite set of variables. Then, for any monomial algebra A for X, the word tree of A is a word tree for X.*

4 Finite-dimensional Monomial Algebras

4.1 Classification and Construction

Let us consider the finite-dimensional case. First, the converse of Theorem 2 holds.

Theorem 3 *For any finite word tree τ for X, there exists a finite-dimensional monomial algebra A for X such that $\tau(A) \equiv \tau$ (identical).*

In fact, for such a tree τ, we can easily see that the following algebra gives the desired monomial algebra A:
the basis of the underlying vector space is the set τ and the multiplication is given by $a \cdot b = e_\tau(a, b)ab$, where ab is in the monoid $\langle 1, X \rangle$ and the map $e_\tau : \tau \times \tau \to \{0, 1\}$ is defined by

$$e_\tau(a,b) = \begin{cases} 1 & \text{if } ab \in \tau \\ 0 & \text{otherwise.} \end{cases}$$

Remark 1: In this way, we clearly write the algebra multiplication "·" when it is necessary to distinguish it from the monoid multiplication.

Remark 2: We can construct such an algebra A by directly giving its presentation (see the end of 4.1).

Remark 3: Theorem 3 for the infinite case does not hold. For example, $k\langle x, y, z | x^n y^n z = 0; n \geq 1\rangle$ (not monomial algebra, just because it is not a finitely presented algebra) corresponds to an infinite word tree. However, if we admit an infinite number of relations as monomial algebras, the theorem holds in general.

Now we define an isomorphism of (possibly infinite) word trees in a manner similar to LSGOPs.

Definition 5 Let X_1 and X_2 be two finite sets of variables. For two word trees τ_1 for X_1 and τ_2 for X_2,

$\tau_1 \simeq \tau_2$ as word trees

$\Leftrightarrow$

there exists a poset isomorphism $\sigma\colon \tau_1 \to \tau_2$ such that
(1) the restriction of σ to X_1 gives a bijection from X_1 into X_2 (therefore $\sharp X_1 = \sharp X_2$), and
(2) if $uv \in \tau_1$, then $\sigma(u)\sigma(v) \in \tau_2$ and $\sigma(u)\sigma(v) = \sigma(uv)$.

Our main theorem below is a correspondence theorem for finite-dimensional monomial algebras.

Theorem 4 (Correspondence Theorem) *For two finite-dimensional monomial algebras A for X and B for Y,*
$A \simeq B$ *(as k-algebras) if and only if $\tau(A) \simeq \tau(B)$ (as word trees), where $\tau(A)$ and $\tau(B)$ are word trees for X and Y respectively.*

Remark: The "if" part holds also in the infinite-dimensional case.

A complete proof of this theorem will be given in Section 5. Theorem 2, 3, and 4 give us a one-to-one correspondence between the algebra isomorphism classes of finite-dimensional monomial algebras and the isomorphism classes of finite word trees. Furthermore, theorem 5 provides a very useful technique for determining when a rooted tree T ($\subset \langle 1, X\rangle$) is a word tree for X.

Theorem 5 (Word Tree Criterion) *T is a word tree if and only if for any $x \in T$ at level 1 (i.e. immediately above the root),*
$V_x \equiv \{y \in T | x \leq y\}$ is $xS(S \subset T)$ such that
(1)S is convex (i.e. $u, v \in S \& w \in T$ with $u \leq w \leq v \Rightarrow w \in S$), and
(2)S and V_x are poset isomorphic via $u \to xu$,
where the poset structure of S is induced from T.

Intuitively, (1) and (2) say that S can be '*moved*' from the place of V_x so that it completely fits into some place of T as the root of S comes on the root of T. The proof of this theorem is essentially due to Hail, Larson, and Sweedler, who give a technique for determining when a G orbit poset is an LSGOP (see [3], Ground Level Theorem 8.7(a), p.792).

Now we can determine finite-dimensional monomial algebras for each dimension using word trees.

Example 3 The way of '*moving*' in the above intuitive description has a one-to-one correspondence with an isomorphism class of finite-dimensional monomial algebras. That is, in fact, an isomorphism of LSGOPs or of word trees means that their ways of '*moving*' are the same. In the figures of classification examples below, one tree indicates one algebra isomorphism class. A way of '*moving*' is described by an arrow as necessary. In each tree, once we assign variables to all the nodes of level 1, we can uniquely determine their monomials to put in the other nodes, (according to arrows, if they exist).

1dim:	
2dim:	
3dim:	
4dim:	
5dim:	

How to construct: It is easy to construct a presentation of finite-dimensional monomial algebras from a word tree τ. First, all nodes at level 1 in τ give algebra generators. Then for each node u in τ, if ux does not belong to τ for some generator x, then $ux = 0$ gives a defining relation. Finally, however, we need to check a redundancy of relations. It is therefore an optimum way to make a relation in order from below in τ.

4.2   Presentation Uniqueness Theorem

We can deduce the following corollary from the result of Theorem 4.

Corollary 6 (Presentation Uniqueness Theorem) *Every finite-dimensional monomial algebra has a unique irredundant presentation up to a permutation of generators.*

   That is, let A and B be two finite-dimensional monomial algebras $k\langle X_A | R_A \rangle$ and $k\langle X_B | R_B \rangle$ respectively. (By definition, R_A and R_B are irredundant relations.) Then $A \simeq B$ (as k-algebras) if and only if
(1) $\sharp X_A = \sharp X_B$, and
(2) there exists a permutation $\pi : X_A \to X_B$ such that $R_B = \pi(R_A)$, where the meaning of $\pi(R_A)$ is defined in a natural fashion.

   Obviously, we cannot expect this assertion to be true in the general universal algebra. For example, in the group theory, $\mathfrak{S}_3$ (the symmetric group of degree 3) has two group presentations $\langle x, y | x^2 = y^2 = (xy)^3 = 1 \rangle$ and $\langle x, y | x^2 = y^3 = (xy)^2 = 1 \rangle$. Incidentally, the two corresponding monomial algebras $k\langle x, y | x^2 = y^2 = (xy)^3 = 0 \rangle$ and $k\langle x, y | x^2 = y^3 = (xy)^2 = 0 \rangle$ are not isomorphic with each other.

   This corollary can offer an effective method for solving the isomorphism problem: given two finite-dimensional monomial algebras, decide whether or not they are isomorphic.

5   Proof of the Main Theorem

This section gives a complete proof of our main theorem (Theorem 4).

Proof: We choose $\tau(A)$, $\tau(B)$ as bases of the underlying k-vector space of A, B respectively. We put $e_A \equiv e_{\tau(A)}$ and $e_B \equiv e_{\tau(B)}$.
"if" part: Let σ be a word tree isomorphism: $\tau(A) \to \tau(B)$. We define φ by the natural extension of σ to A. Then $\varphi(u \cdot v) = \varphi(e_A(u,v)uv) = e_A(u,v)\sigma(uv)$ and $\varphi(u)\varphi(v) = \sigma(u) \cdot \sigma(v) = e_B(\sigma(u),\sigma(v))\sigma(u)\sigma(v)$ for $\forall u, \forall v \in \tau(A)$. On the other hand, $e_A(u,v) = 1$ implies $\sigma(uv) = \sigma(u)\sigma(v)$, and so $e_B(\sigma(u),\sigma(v)) = 1$ since σ is a poset isomorphism. Again because σ is a poset isomorphism,
$\sharp\{(u,v) \in \tau(A) \times \tau(A) | e_A(u,v) = 1\} = \sharp\{(u',v') \in \tau(B) \times \tau(B) | e_B(u',v') = 1\}$.

It follows that $e_A(u,v) = e_B(\sigma(u), \sigma(v))$; therefore, so $\varphi(u \cdot v) = \varphi(u)\varphi(v)$. The bijectivity results from that of σ.

"only-if" part:

Lemma 1 (Isomorphism Condition) *Let a k-linear map $\varphi: A \to B$ be given by $\varphi(u) = \sum_{v \in \tau(B)} a_{u,v} v$. We put $A_\varphi = (a_{u,v})_{u \in \tau(A), v \in \tau(B)}$ (a matrix with entries in k). Then we have: φ is an isomorphism of k-algebras if and only if*
(1) A_φ is a $n \times n$ square matrix $(n = \sharp\tau(A) = \sharp\tau(B))$ with $\det A_\varphi \neq 0$; and
(2) $e_A(u,v)a_{uv,w} = \sum_{s,t \in \tau(B) \text{such that } st=w} a_{u,s}a_{v,t}e_B(s,t)$ for $u,v,w \in \tau(A)$.

Proof: It suffices to compare the coefficients of each base (in $\tau(B)$) of B between $\varphi(u \cdot v)$ and $\varphi(u) \cdot \varphi(v)$ for $u, v \in \tau(A)$. In fact, $\varphi(u \cdot v) = \varphi(e_A(u,v)uv) = \sum_{w \in \tau(B)} e_A(u,v)a_{uv,w}w$ and $\varphi(u) \cdot \varphi(v) = \sum_{s \in \tau(B)} a_{u,s} \cdot \sum_{t \in \tau(B)} a_{v,t}t = \sum_{s,t \in \tau(B)} a_{u,s}a_{v,t}e_B(s,t)st$. Q. E. D.

Now let us define the levels of elements of a rooted tree T with the root ρ. For $x \in T$, we write $l_T(x) = n$ (the level of x is n) if $\rho = x_0 < x_1 < x_2 < \cdots < x_{n-1} < x_n = x$ in T and there does not exist $y_i \in T$ such that $x_i < y_i < x_{i+1}$ for any $i \in [0, n-1]$. We put $l_A(u) \equiv l_{\tau(A)}(u)$.

Lemma 2 *Let $R = R_A = \bigoplus_{u \in \tau(A)-\{1\}} ku \subset A$. Then,*
(1) R is the largest nilpotent two-sided ideal of A.
(2) $R^i = \bigoplus_{u \in V_i} ku$, where $V_i = \{u \in \tau(A)|l_A(u) \geq i\}$ for $i \in \mathbb{N} \cup \{0\}$, and $R^0 = A$. Especially if we denote $L_i^A = \{u \in \tau(A)|l_A(u) = i\}$, then we have $\sharp L_i^A = \dim R^i - \dim R^{i+1}$ for $i \in \mathbb{N} \cup \{0\}$.

Proof: (1) First, R is a two-sided ideal of A, since if $u \neq 1$, then $uv \neq 1$ and $vu \neq 1$ for all $v \in A$. Suppose $u_1 u_2 \cdots u_n \neq 0$ for $u_1, u_2, \cdots, u_n \in R$. Associate this to the left, then $(\cdots((u_1 \cdot u_2) \cdot u_3) \cdots u_n) \neq 0$. Therefore, $1 < u_1 < u_1 u_2 < u_1 u_2 u_3 < \cdots < u_1 u_2 \cdots u_n$. This gives a proper chain with $n + 1$ elements. Hence if the proper chains in $\tau(A)$ have at most n elements, then $R^n = \{0\}$. Let any nilpotent ideal of A be N. It remains to show that $N \subset R$. However, if an element $1 + a$ ($a \in R$) is in N, then $(1 + a)^n \neq 0$ for $\forall n$, which contradicts the nilpotency of N.
(2) By induction on i. If $i = 0$, it is immediate since $V_0 = \tau(A)$. Assume that it is true for $j < i$. Every element of R^i is a k-linear combination of $u \cdot v$'s, where $u \in V_{i-1}$ and $v \in V_1$, since $R^i = R^{i-1} \cdot R$. $u \cdot v = e_A(u,v)uv$ and if $e_A(u,v) = 1$, i.e. $u < uv$ (since $v \neq 1$), then $uv \in V_i$, and so $R^i \subset \bigoplus_{u \in V_i} ku$. Conversely, for any $u \in V_i$, there exist $s \in V_{i-1}$ and $t \in V_1$ such that $u = st$ and $e_A(s,t) = 1$. In fact, choose the element immediately below u as s and note that $t \neq 1$. Hence $u \in R^{i-1} \cdot R = R^i$, i.e. $\bigoplus_{u \in V_i} ku \subset R^i$. Q. E. D.

Remark: Throughout, let us call R the *radical* of A.

Lemma 3 *Let R be the radical of A. Then*

$$\#\{(u,v) \in \tau(A) \times \tau(A)|e_A(u,v) = 1\} = \sum_{i \geq 0} dim R^i.$$

Remark: The summation above is a finite sum, since R is nilpotent.

Proof: $(u, uv) = (u', u'v') \Leftrightarrow (u, v) = (u', v')$ implies that
$\#\{(u,v) \in \tau(A) \times \tau(A)|e_A(u,v) = 1\} = \#\{(u, uv) \in \tau(A) \times \tau(A)|e_A(u,v) = 1\}$.
It suffices to compute $\nu = \#\{(u,t) \in \tau(A) \times \tau(A)|u \leq t\}$. For any $t \in L_i^A$,
$\#\{u \in \tau(A)|u \leq t\} = l_A(t) + 1 = i + 1$, since $\tau(A)$ is a tree. Hence
$\nu = \sum_{i \geq 0}(i+1)\#L_i^A$, and so the equality in Lemma 2(2): $\#L_i^A = dim R^i - dim R^{i+1}$ implies that $\nu = \sum_{i \geq 0} dim R^i$.   Q. E. D.

In addition, let us show a "hereditary" property of $\tau(A)$, which will play an important role in the proof of the main theorem.

Lemma 4 (Hereditary Property of $\tau(A)$) *For any $x \in \tau(A)$, $V_x \equiv \{y \in \tau|x \leq y\}$ is xS ($S \subset \tau(A)$) such that*
(1) S is convex (i.e. $u, v \in S \& w \in \tau(A)$ with $u \leq w \leq v \Rightarrow w \in S$), and
(2) S and V_x are poset isomorphic via the map $L_x : u \to xu$, where the poset structure of S is induced from $\tau(A)$.

Proof: (1) We can show the stronger property: for $v \in S$ and $w \in \tau$, $w \leq v \Rightarrow w \in S$. In fact, because $w \leq v$ and $xv \in \tau$, we have $xw \leq xv$ since $\tau(A)$ is a word tree. Thus, $xw \in \tau$, i.e. $w \in S$.
(2) First, the surjectivity of L_x is immediate by the definition of V_x and partial order of $\tau(A)$. Second, the injectivity is shown by the fact that the assumption of monomial algebra implies $xu = xv \Rightarrow u = v$. Finally, "$u \leq v$ and $xv \in \tau(A) \Leftrightarrow xu \leq xv$" shows that both L_x and its inverse map L_x^{-1} are poset maps.   Q. E. D.

Proof of "only-if" part: Assume that there exists an isomorphism

$$\varphi : A \to B, \quad \varphi(u) = \sum_{v \in \tau(B)} a_{u,v}v, a_{u,v} \in k.$$

We put $A_\varphi = (a_{u,v})_{u \in \tau(A), v \in \tau(B)}$. Using Lemma 1,
(∗1) $det A_\varphi \neq 0$ (A_φ is a square matrix),
(∗2) $e_A(u,v)a_{uv,w} = \sum_{s,t \in \tau(B) such that \ st=w} a_{u,s}a_{v,t}e_B(s,t)$.

Claim 1 *(1)* $l_A(u) > l_B(v) \Rightarrow a_{u,v} = 0$.
(2) $u \neq 1 \Rightarrow a_{1,u} = a_{u,1} = 0$

Proof: (1) Let R_A and R_B be the radicals of A and B, respectively. Since φ is an isomorphism of k-algebras, $\varphi(R_A^i) \subset R_B^i$ (in fact, both sides are equal) for all $i \in \mathbf{N} \cup \{0\}$. By Lemma 2(2), $R_A^i = \bigoplus_{u \in V_i^A} ku$ where $V_i^A = \{u \in \tau(A) | l_A(u) \geq i\}$ and similarly for B. Hence if we put $l_A(u) = i$, then $a_{u,v} \neq 0 \Rightarrow l_B(v) \geq i = l_A(u)$.
(2) $\varphi(1_A) = 1_B$ (the isomorphism image of the unit element of A is the unit element of B) implies that $a_{1,u} = 0$ for $u \neq 1$. It is immediate from (1) that $a_{u,1} = 0$ for $u \neq 1$.   Q. E. D.

Now by the definition of matrix determinant,
$det A_\varphi = \sum_{\sigma \in \mathfrak{S}_{A,B}} sgn\sigma \prod_{u \in \tau(A)} a_{u,\sigma(u)}$, where $\mathfrak{S}_{A,B}$ denotes the set of all bijections from $\tau(A)$ to $\tau(B)$. φ induces the isomorphism $R_A^i \simeq R_B^i$ for all i, and so $dim R_A^i = dim R_B^i$ for all i. Hence by Lemma 2(2), $\sharp L_i^A = \sharp L_i^B$ for all i. Together with Claim 1(1), this implies

$$(1) \quad det A_\varphi = \sum_{\sigma \in \mathfrak{S}_{A,B} \text{ s.t. } \forall u \in \tau(A), l_A(u) = l_B(\sigma(u))} sgn\sigma \prod_{u \in \tau(A)} a_{u,\sigma(u)}.$$

Claim 2 $a_{u,u'} a_{v,v'} \neq 0, l_A(u) = l_B(u')$ and $l_A(v) = l_B(v')$ *implies* $e_A(u, v) = 0 \Rightarrow e_B(u', v') = 0$.

Proof: Assume that $e_A(u, v) = 0$ and $e_B(u', v') = 1$. Then using (∗2), and putting $w = u'v'$,

$$0 = \sum_{s, t \in \tau(B) \text{ such that } st = w} a_{u,s} a_{v,t} e_B(s, t).$$

On the right hand side, it suffices to observe only those s's such that $e_B(s, t) = 1$, that is, $1 \leq s \leq w$ in $\tau(B)$.
    For $s = 1$, $a_{u,s} = 0$ by Claim 1(2), since $e_A(u, v) = 0 \Rightarrow u \neq 1$.
For $1 < s < u'$, $a_{u,s} = 0$ by Claim 1(1), since $l_B(s) < l_B(u') = l_A(u)$.
For $u' < s < w$, the hereditary property of $\tau(B)$ (Lemma 4) implies $l_B(t) =$ "length of the path from s to $w = st$" and $l_B(v') =$ "length of the path from u' to $w = u'v'$."
Therefore $l_B(t) < l_B(v') = l_A(v)$ and by Claim 1(1), $a_{v,t} = 0$.
For $s = w$ in $\tau(B)$, $t = 1$ and $a_{v,t} = 0$ by Claim 1(2), since $e_A(u, v) = 0 \Rightarrow v \neq 1$.
    Consequently, we have $0 = a_{u,u'} a_{v,v'}$, which contradicts the first assumption $a_{u,u'} a_{v,v'} \neq 0$.   Q. E. D.

Claim 3 *If σ is an element of $\mathfrak{S}_{A,B}$ such that $a_{u,\sigma(u)} \neq 0$ and $l_A(u) = l_B(\sigma(u))$ for all $u \in \tau(A)$, then $e_A(u, v) = e_B(\sigma(u), \sigma(v))$ for all $u, v \in \tau(A)$.*

Proof: $\sigma \in \mathfrak{S}_{A,B}$ in the assumption satisfies $a_{u,\sigma(u)} a_{v,\sigma(v)} \neq 0$, $l_A(u) = l_B(\sigma(u))$, and $l_A(v) = l_B(\sigma(v))$ for all $u, v \in \tau(A)$. Hence by Claim 2, if $e_A(u,v) = 0$, then $e_B(\sigma(u), \sigma(v)) = 0$. On the other hand, by Lemma 3, $R_A^i \simeq R_B^i$ for all i implies $\#\{(u,v) \in \tau(A) \times \tau(A) | e_A(u,v) = 1\} = \#\{(u',v') \in \tau(B) \times \tau(B) | e_B(u', v') = 1\}$. Hence $e_A(u,v) = 0 \Leftrightarrow e_B(\sigma(u), \sigma(v)) = 0$, that is, $e_A(u,v) = e_B(u',v')$ for all $u, v \in \tau(A)$.   Q. E. D.

Now by equality (1) and (*1): $det A_\varphi \neq 0$, we know that there exists $\sigma \in \mathfrak{S}_{A,B}$ such that

$$(2) \qquad \prod_{u \in \tau(A)} a_{u,\sigma(u)} \neq 0 \text{ and } l_A(u) = l_B(\sigma(u)) \text{ for all } u \in \tau(A).$$

We wish to construct a poset isomorphism $\tilde{\sigma} : \tau(A) \to \tau(B)$ such that $\tilde{\sigma}(uv) = \tilde{\sigma}(u)\tilde{\sigma}(v)$ if $e_A(u,v) = 1$, starting with σ in (2).

Notation: By Lemma 4, for any $u \in L_{i+1}^A$, there exist unique $s \in L_i^A$ and $t \in L_1^A$ such that $u = st$ and $e_A(s,t) = 1$, so we write $s = u_i$, $t = u_1$ for $u \in L_{i+1}^A$.

Claim 4 (Inductive Construction) *Suppose* $\sigma_i \in \mathfrak{S}_{A,B}$ *induces a bijection* $L_i^A \to L_i^B$ *and satisfies* $a_{u,\sigma_i(u)} \neq 0$ *and* $l_A(u) = l_B(\sigma_i(u))$ *for all* $u \in \tau(A)$. *Define a new map* $\sigma_{i+1} : \tau(A) \to \tau(B)$ *as follows:*
$$\sigma_{i+1}(u) = \begin{cases} \sigma_i(u_i)\sigma_i(u_1) & \text{if } u \in L_{i+1}^A \\ \sigma_i(u) & \text{otherwise.} \end{cases}$$
Then $e_A(u,v) = e_B(\sigma_i(u), \sigma_i(v))$ *for all* $u, v \in \tau(A)$, *and* σ_{i+1} *is an element of* $\mathfrak{S}_{A,B}$ *which induces a bijection* $L_{i+1}^A \to L_{i+1}^B$ *such that* $a_{u,\sigma_{i+1}(u)} \neq 0$ *and* $l_A(u) = l_B(\sigma_{i+1}(u))$ *for all* $u \in \tau(A)$.

Proof: The first statement is immediate from Claim 3. For $u \in L_{i+1}^A$, $\sigma_i(u_i)\sigma_i(u_1) \in L_{i+1}^B$, since $e_B(\sigma_i(u_i), \sigma_i(u_1)) = e_A(u_i, u_1) = 1$, $\sigma_i(u_i) \in L_i^B$, and $\sigma_i(u_1) \in L_1^B$. We show that σ_{i+1} is an injection of L_{i+1}^A into L_{i+1}^B. Assume $u \neq v$ for $u, v \in L_{i+1}^A$. Then we have two cases because $\tau(A)$ is a tree: (1) $u_i = v_i$ and $u_1 \neq v_1$ and (2) $u_i \neq v_i$. However, in case (1), $\sigma_i(u_i) = \sigma_i(v_i)$ and $\sigma_i(u_1) \neq \sigma_i(v_1)$. In case (2), $\sigma_i(u_i) \neq \sigma_i(v_i)$ and so $\sigma_{i+1}(u) = \sigma_i(u_i)\sigma_i(u_1) \neq \sigma_i(v_i)\sigma_i(v_1) = \sigma_{i+1}(v)$ since $\tau(B)$ is a tree. In both cases, we have $\sigma_{i+1}(u) \neq \sigma_{i+1}(v)$, which indicates an injectivity of σ_{i+1} from L_{i+1}^A to L_{i+1}^B. Furthermore, $\#L_{i+1}^A = \#L_{i+1}^B$ verifies that σ_{i+1} is a bijection. Hence, by the definition of σ_{i+1}, σ_{i+1} is in $\mathfrak{S}_{A,B}$ and $l_A(u) = l_B(\sigma_{i+1}(u))$ for all $u \in \tau(A)$. For $u \in L_{i+1}^A$, (*2) of the isomorphism condition implies that

$$e_A(u_i, u_1)a_{u,\sigma_{i+1}(u)} = \sum_{\substack{s,t \in \tau(B) \text{such that } st = \sigma_{i+1}(u)}} a_{u_i,s}a_{u_1,t}e_B(s,t).$$

Then $a_{u,\sigma_{i+1}(u)} = a_{u_i,\sigma_i(u_i)} a_{u_1,\sigma_i(u_1)}$ by

$$e_A(u_i, u_1) = e_B(\sigma_i(u_i), \sigma_i(u_1)) = 1, \quad \sigma_{i+1}(u) = \sigma_i(u_i)\sigma_i(u_1)$$

and the same argument as in the proof of Claim 2. Hence $a_{u_i,\sigma_i(u_i)} \neq 0$ and $a_{u_1,\sigma_i(u_1)} \neq 0$ implies $a_{u,\sigma_{i+1}(u)} \neq 0$. For $u \notin L_{i+1}^A$, $a_{u,\sigma_{i+1}(u)} = a_{u,\sigma_i(u)} \neq 0$. Consequently $a_{u,\sigma_{i+1}(u)} \neq 0$ for all $u \in \tau(A)$.   Q. E. D.

Now we choose σ in (2) as σ_1 in Claim 4. By the inductive construction of Claim 4, we obtain σ_n, where $n = max_{u \in \tau(A)} l_A(u)$. We put $\sigma_n = \tilde{\sigma}$, then $\tilde{\sigma}(1) = 1$ and $\tilde{\sigma}|_{L_i^A} = \sigma_i$ for all $i \in [1, n]$.

Claim 5 $\tilde{\sigma}$ *is a desired isomorphism of word trees from $\tau(A)$ to $\tau(B)$; that is, $\tilde{\sigma}$ is a poset isomorphism $\tau(A) \rightarrow \tau(B)$ such that if $e_A(u,v) = 1 \Rightarrow \tilde{\sigma}(u)\tilde{\sigma}(v) = \tilde{\sigma}(uv)$.*

Proof: First, let us show that if $u \in L_i^A$ and $v \in L_{i+1}^A$, then $u < v \Leftrightarrow \tilde{\sigma}(u) < \tilde{\sigma}(v)$. If $u < v$, then $u = v_i$ and so $\tilde{\sigma}(v) = \tilde{\sigma}(u)\tilde{\sigma}(v_1)$ by the definition of $\tilde{\sigma}$. By Claim 4, $e_B(\tilde{\sigma}(u), \tilde{\sigma}(v_1)) = e_A(u, v_1) = 1$. Hence $\tilde{\sigma}(u) < \tilde{\sigma}(u)\tilde{\sigma}(v_1) = \tilde{\sigma}(v)$. Consequently, $u < v \Rightarrow \tilde{\sigma}(u) < \tilde{\sigma}(v)$, and so $\sharp\{v \in L_{i+1}^A | u < v\} \leq \sharp\{v' \in L_{i+1}^B | \tilde{\sigma}(u) < v'\}$. But $\sharp L_i^A = \sharp L_i^B$ and $L_{i+1}^A = \sharp L_{i+1}^B$ imply that the inequality above is really an equality. Therefore,

$$(3) \qquad u < v \Leftrightarrow \tilde{\sigma}(u) < \tilde{\sigma}(v) \text{ for } u \in L_i^A, v \in L_{i+1}^A.$$

In general, for $u < v$ in $\tau(A)$, if $u \in L_i^A, v \in L_j^A (i < j)$, then we have a sequence $u = u_i < u_{i+1} < \cdots < u_\nu < \cdots < u_j = v$, where $u_\nu \in L_\nu^A$ for $\nu \in [i, j]$. But by (3), $u_\nu < u_{\nu+1} \Leftrightarrow \tilde{\sigma}(u_\nu) < \tilde{\sigma}(u_{\nu+1})$ for all $\nu \in [i, j-1]$. Hence $u < v \Rightarrow \tilde{\sigma}(u) < \tilde{\sigma}(v)$. The converse, $\tilde{\sigma}(u) < \tilde{\sigma}(v) \Rightarrow u < v$ can be proved by the same argument using (3). Thus $\tilde{\sigma}$ is a poset isomorphism: $\tau(A) \rightarrow \tau(B)$.

Second, we show that if $e_A(u,v) = 1 \Rightarrow \tilde{\sigma}(u)\tilde{\sigma}(v) = \tilde{\sigma}(uv)$. Let $u \in L_i^A, v \in L_j^A$, and suppose $e_A(u,v) = 1$. Then we can write $u < uu_1 < uu_1u_2 < \cdots < uu_1u_2\cdots u_\nu < \cdots < uu_1u_2\cdots u_{j-1} < uu_1u_2\cdots u_j = uv$; that is, $e_A(uu_1u_2\cdots u_{\nu-1}, u_\nu) = 1$ and $l_A(u_\nu) = 1$ for all $\nu \in [1, j]$ $(u_0 = 1)$ by the hereditary property of $\tau(A)$ (Lemma 4). Then we have $\tilde{\sigma}(uu_1u_2\cdots u_{\nu-1} \cdot u_\nu) = \tilde{\sigma}(uu_1u_2\cdots u_{\nu-1}) \cdot \tilde{\sigma}(u_\nu)$ for all $\nu \in [1, j]$ by the definition of $\tilde{\sigma}$. Hence $\tilde{\sigma}(uv) = \tilde{\sigma}(u)\prod_{\nu \in [1,j]} \tilde{\sigma}(u_\nu)$. On the other hand, $1 < u_1 < u_1u_2 < \cdots < u_1u_2\cdots u_\nu < \cdots < u_1u_2\cdots u_{j-1} < v$ since $\tau(A)$ is a word tree. Thus, similarly, $\tilde{\sigma}(v) = \prod_{\nu \in [1,j]} \tilde{\sigma}(u_\nu)$. Consequently, $\tilde{\sigma}(uv) = \tilde{\sigma}(u)\tilde{\sigma}(v)$.   Q. E. D.

This completes the proof of the theorem.   Q. E. D.

6   Conclusion

We can completely determine finite-dimensional monomial algebras modulo an algebra isomorphism, by means of word trees of nonzero monomials. We obtain the uniqueness of irredundant presentation up to a permutation of generators. The result can be used to construct an effective algorithm for solving the isomorphism problem for finite-dimensional monomial algebras.

Acknowledgement: The author is grateful to Professor T. Mora for his constructive remark on the infinite case: the author's proof of the "only-if" part of the Correspondence Theorem implies that the result also holds for the infinite-dimensional *graded* monomial algebras and *homogeneous* (i.e. length preserving) isomorphism.

The author also would like to thank Professor M. Sweedler and Professor B. Buchberger for their encouragement.

REFERENCES

[1] Buchberger, B., *Gröbner Bases: An Algorithmic Method in Polynomial Ideal Theory*, Chapter 6, in "Multidimensional Systems Theory," (N. K. Bose ed.), D. Reidel Publishing Company, 1985, pp. 184–232.

[2] Gateva-Ivanova, T. and Latyshev, V., *On recognisable properties of associative algebras*, J. Symbolic Computation 6 (1988), 371–388.

[3] Haile, D., Larson, R., and Sweedler, M., *A new invariant for* C *over* R*: Almost invertible cohomology theory and the classification of idempotent cohomology class and algebras by partially ordered sets with a Galois group action*, Amer. J. of Math. 105 (1983), 689–814.

[4] Le Chenadec, P., "Canonical forms in finitely presented algebras," Research Notes in Theoretical Computer Science, Pitman, 1986.

[5] Mora, T., *Groebner bases for non-commutative polynomial rings*, in "Proc. AAECC3," L. N. Comp. Sci. 229, 1986, pp. 353–362.

[6] Shirayanagi, K., *Algebras associated with LSGOP*, J. Algebra 113(2) (1988), 318–338.

[7] Ufnarovskij, V., *A growth criterion for graphs and algebras defined by words*, Mat. Zametki 31 (1982), 465–472. (in Russian); English transl. in Math. Notes 31

Kiyoshi Shirayanagi
NTT Software Laboratories
3-9-11 Midori-cho, Musashino-shi
Tokyo 180, Japan
shirayan%ntt-20.ntt.jp@relay.cs.net

An Algorithm related to Compactifications of adjoint Groups

ELISABETTA STRICKLAND

Introduction

A classical problem of projective geometry is the determination of methods for the computation of the number of figures which satisfy certain properties. In the case of subspaces, a method which has been largely developed and is well known is the study of the Grassmann varieties and their Chow ring. In the case of quadrics, pairs of spaces and more in general in the case of a symmetric variety G/H, where G is a semisingle adjoint group, $H = G^\sigma$, σ an order two automorphism of G, a general method has been introduced, defining a suitable intersection ring for G/H. Such ring is difficult to calculate being the direct limit of the Chow ring of all the equivariant regular compactifications of G/H. For a given compactification, one can give a general combinatorial method for the computation of the Chow ring, using root systems, fans and sheaves over fans, which can be treated in a purely algebraic way [BDP].

In [S$_1$] a systematic method for the computation of the Poincarè series and the equivariant Poincarè series of a large class of compactifications of homogeneous spaces has been introduced. Moreover in [S$_2$] a method has been determined which gives the equivariant cohomology ring of the G–equivariant embeddings of a semisingle adjoint group. Such result was obtained on one hand using the so called Stanley–Reisner systems associated to the fans of the embeddings, on the other hand working with a sheaf over the fundamental Weyl chamber of G, independent from the choice of the embedding. In this paper we inspect closely the algorithm introduced in [S$_2$], showing its power through explicit applications.

1 Description of the algorithm

If G is a semisingle adjoint group, let $R \subset Q^h = V$ be its root system, $h =$rank G.

Moreover let

$$A = \{a_{ij}\}$$

be the Cartan matrix and choose simple roots

$$\Delta = \{\alpha_1 \ldots \alpha_n\} \subset R .$$

Consider the cone C in V^* given by all forms φ such that

$$\varphi(\alpha_i) \geq 0 \quad \forall i .$$

Such cone C in V^* is called the "fundamental Weyl chamber".

One can define a combinatorial sheaf over C in the following way.

Set a $1 - 1$ correspondence between the faces of C and the subsets of Δ, defining

$$C_\Gamma = \{\varphi \in C | \varphi(\alpha_i) = 0 \quad \alpha_i \in \Gamma\}$$

to each face or to each Γ, we associate a ring A_Γ which is obtained as follows. In $Q^h = V$ define the linear transformations

$$s_i(\alpha_j) = \alpha_j - a_{ij}\alpha_i$$

a_{ij} being the coefficients of the Cartan matrix.

One knows that such elements are reflections and they map the root system into itself, therefore they generate a finite group, the so called Weyl group W.

For any $\Gamma \subset \Delta$, consider the subgroup

$$W_\Gamma = \langle s_i | \alpha_i \in \Gamma \rangle .$$

It is clear that

$$\Gamma' \supset \Gamma \Leftrightarrow C_{\Gamma'} \subset C_\Gamma \text{ and } W_{\Gamma'} \supset W_\Gamma .$$

Set

$$A_\Gamma = \mathbf{Q}[V^*]^{W_\Gamma} .$$

If $\Gamma' \supset \Gamma$ then $A_{\Gamma'} \subset A_\Gamma$ and call such inclusions

$$\varphi_{\Gamma'}^\Gamma : A_{\Gamma'} \rightarrow A_\Gamma .$$

The set

$$\{A_\Gamma , \varphi_{\Gamma'}^\Gamma\}$$

is the sheaf one is reduced to work with.

This is done as follows.

We recall first some facts.

If H is a finite group of linear transformations of a space V generated by reflections, then (Coxeter 1951, [B]):

a) $\mathbf{Q}[V^*]^H$ is a polynomial ring.

b) If H is the Weyl group of a root system $R \subset V$, $\mathbf{Q}[V^*]^H$ is a polynomial ring because of a), with generators of degrees

$$m_1 + 1 \ldots m_s + 1 \qquad (s = \dim V)$$

where $m_1, \ldots, m_s$ are the exponents of the root system.

Such exponents are known in all cases, [B]. For example, in A_n, $(= Sl(n+1))$ they are $1, 2, \ldots n$.

Now, if we take our V and W_Γ and we consider the action of W_Γ over V, then W_Γ is generated by reflections, so

1) A_Γ is a polynomial ring

2) Take in V the space V_Γ generated by the α_i, with $\alpha_i \in \Gamma$. Such space is stable for W_Γ and W_Γ is the Weyl group of the root system $R \cap V_\Gamma$. Moreover, if we consider

$$U_\Gamma = V^{W_\Gamma}$$

then

$$V = V_\Gamma \oplus U_\Gamma$$

therefore

$$A_\Gamma = Q[V_\Gamma^*]^{W_\Gamma} \otimes Q[U_\Gamma^*] \ .$$

For b), this last fact tells us:

a) in which degrees we should look for generators;

b) the number of such generators in each degree.

Then, if one computes in such degrees the invariants, one can determine explicitly a set of polynomial generators for A_Γ and the inclusions $\varphi_{\Gamma'}^\Gamma$.

In the classical cases this can be done in a reasonable way, as we shall see in the next section.

In the exceptional cases, (which are only five), one can write a program suited for each case. We are going to perform the case of G_2, because it is the handiest.

2   Applications

Let $G = Sl(3)$. The Cartan matrix is

$$\begin{pmatrix} 2 & -1 \\ -1 & 2 \end{pmatrix}$$

there are two simple roots α_1, α_2 and six roots, α_1, α_2, $\alpha_1 + \alpha_2$, $-\alpha_1$, $-\alpha_2$, $-\alpha_1 - \alpha_2$. The Weyl group W is S_3, which acts on Q^2 as a group of symmetries of the regular hexagon.

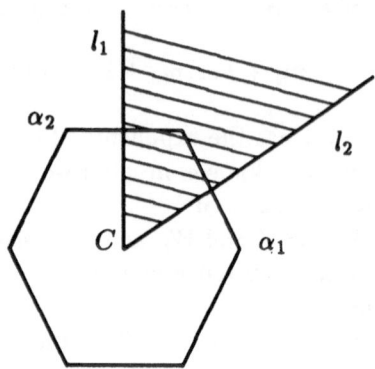

So the fundamental Weyl chamber C has 4 faces,

$$\{0\}\,,\,l_1\,,\,l_2\,,\,C$$

the $1-1$ correspondence between the faces of C and the subset of Δ is the following.

$$
\begin{aligned}
\{0\} &\leftrightarrow \{\alpha_1,\alpha_2\}\\
l_1 &\leftrightarrow \{\alpha_1\}\\
l_2 &\leftrightarrow \{\alpha_2\}\\
C &\leftrightarrow \emptyset
\end{aligned}
$$

the corresponding groups W_Γ are:

$$
\begin{aligned}
W_{\{\alpha_1,\alpha_2\}} &\cong S_3\\
W_{\{0\}} &\cong \{e\}\\
W_{\{\alpha_1\}} &\cong S_2 \subset S_3\\
W_{\{\alpha_2\}} &\cong \tilde{S}_2 \subset S_3
\end{aligned}
$$

In order to simplify the exposition, consider

$$U = V \oplus Q$$

and let W act trivially over Q, so we get an action of W on U.

It is well known that choosing a suitable basis for U, the action becomes the obvious one of S_3 over the space of dimension 3 obtained permuting coordinates.

$$B_\Gamma = Q[U^*]^{W_\Gamma} = Q[V^*]^{W_\Gamma} \otimes Q[x] = A_\Gamma[x]\ .$$

Clearly, if we know B_Γ, we know A_Γ. Moreover, if we know the inclusions

$$B_{\Gamma'} \subset B_\Gamma$$

we know the inclusions $\varphi_{\Gamma'}^{\Gamma}$.

Using the chosen basis for U

$$Q[U] = Q[x_1, x_2, x_3] .$$

We obtain

$$B_\Gamma = \begin{cases} \text{symmetric polynomials,} & \text{if } \Gamma = \{\alpha_1, \alpha_2\} \\ \text{symmetric polynomials in } x_1, x_2, & \text{if } \Gamma = \{\alpha_1\} \\ \text{symmetric polynomials in } x_2, x_3, & \text{if } \Gamma = \{\alpha_2\} \\ Q[U], & \text{if } \Gamma = \emptyset \end{cases}$$

therefore

$$\begin{aligned} B_{\{\alpha_1,\alpha_2\}} &= Q[x_1 + x_2 + x_3, x_1 x_2 + x_1 x_3 + x_2 x_3, x_1 x_2 x_3] \\ &= Q[x, t_2, t_3] \\ B_{\{\alpha_1\}} &= Q[x_1 + x_2 + x_3, x_3, x_1 x_2] = Q[x, y_1, y_2] \\ B_{\{\alpha_2\}} &= Q[x_1 + x_2 + x_3, x_1, x_2 x_3] = Q[x, z_1, z_2] \\ B_\emptyset &= Q[x_1, x_2, x_3] \end{aligned}$$

where

$$\begin{aligned} x &= x_1 + x_2 + x_3 \\ t_2 &= x_1 x_2 + x_1 x_3 + x_2 x_3 \\ t_3 &= x_1 x_2 x_3 \\ y_1 &= x_3 \\ y_2 &= x_1 x_2 \\ z_1 &= x_1 \\ z_2 &= x_2 x_3 \end{aligned}$$

the non trivial inclusions are

$$B_{\{\alpha_1,\alpha_2\}} \rightarrow B_{\{\alpha_1\}} \text{ with } \begin{cases} x &\rightarrow x \\ t_2 &\rightarrow y_2 + (x - y_1)y_1 \\ t_3 &\rightarrow y_1 y_2 \end{cases}$$

$$B_{\{\alpha_1,\alpha_2\}} \rightarrow B_{\{\alpha_2\}} \text{ with } \begin{cases} x &\rightarrow x \\ t_2 &\rightarrow z_2 + (x - z_1)z_1 \\ t_3 &\rightarrow z_1 z_2 . \end{cases}$$

Now let $G = G_2$. The picture of the root system is

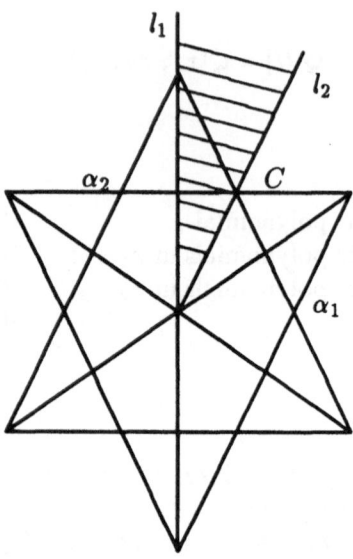

The Cartan matrix is

$$\begin{pmatrix} 2 & -3 \\ -1 & 2 \end{pmatrix}$$

The are two simple roots α_1, α_2 and twelve roots: α_1, α_2, $\alpha_1+\alpha_2$, $\alpha_2+2\alpha_1$, $\alpha_2+3\alpha_1$, $2\alpha_2+3\alpha_1$ and their negatives.

The Weyl group is $D_{2,6}$, dihedral group of the hexagon. C has 4 faces

$$\{0\}, \; l_1, \; l_2, \; C$$

the $1-1$ correspondence between the faces of C and the subsets of Γ is

$$\begin{aligned} \{0\} &\leftrightarrow \{\alpha_1, \alpha_2\} \\ l_1 &\leftrightarrow \{\alpha_1\} \\ l_2 &\leftrightarrow \{\alpha_2\} \\ C &\leftrightarrow \emptyset \end{aligned}$$

the corresponding groups W_Γ are

$$\begin{aligned} W_{\{\alpha_1,\alpha_2\}} &\cong D_{2,6} \\ W_{\{\alpha_1\}} &\cong S_2 \\ W_{\{\alpha_2\}} &\cong \tilde{S}_2 \\ W_\emptyset &\cong \{e\}. \end{aligned}$$

In this case we have

$$
\begin{aligned}
A_\emptyset &= Q[\alpha_1, \alpha_2] \\
A_{\{\alpha_1\}} &= Q[v_1, v_2] \\
A_{\{\alpha_2\}} &= Q[u_1, u_2] \\
A_{\{\alpha_1, \alpha_2\}} &= Q[z_2, z_6]
\end{aligned}
$$

where

$$
\begin{aligned}
v_1 &= 3\alpha_1 + 2\alpha_2 \\
v_2 &= \alpha_1^2 \\
u_1 &= 2\alpha_1 + \alpha_2 \\
u_2 &= \alpha_2^2 \\
z_2 &= 12\alpha_1^2 + 12\alpha_1\alpha_2 + 4\alpha_2^2 \\
z_6 &= 2^5(54\alpha_1^6 + 162\alpha_1^5\alpha_2 + 135\alpha_1^4\alpha_2^2 - 45\alpha_1^2\alpha_2^4 - 18\alpha_1\alpha_2^5 - 2\alpha_2^6)
\end{aligned}
$$

the non trivial inclusions are the following two

$$
A_{\{\alpha_1, \alpha_2\}} \to A_{\{\alpha_1\}} \quad
\begin{cases}
z_2 \;\to\; v_1^2 + v_2 \\
z_6 \;\to\; -v_1^6 + 45v_1^4 v_2 - 135v_1^2 v_2^2 - 27v_2^3
\end{cases}
$$

$$
A_{\{\alpha_1, \alpha_2\}} \to A_{\{\alpha_2\}} \quad
\begin{cases}
z_2 \to 3u_1^2 + u_2 \\
z_6 \to 27u_1^6 - 135u_1^4 u_2 + 45u_1^2 u_2^2 - u_2^3 \ .
\end{cases}
$$

REFERENCES

[B]   Bourbaki, N., "Groupes et algébres de Lie," Cap. IV, V, VI, Hermann, Paris, 1968.

[BDP] Bifet, E., De Concini, C., Procesi C., *Equivariant cohomology of regular embeddings*, Advances in Matematics (to appear).

[S1]   E. Strickland, *Equivariant Betti numbers for symmetric varieties*, J. of Algebra (to appear).

[S2]   E. Strickland, *Computing the equivariant cohomology of group compactifications*, preprint.

Elisabetta Strickland
II Università di Roma "Tor Vergata"
Roma
ITALY

Deciding Consistency of Systems of Polynomial in Exponent Inequalities in Subexponential Time

NIKOLAJ N. VOROBJOV (JR.)

Introduction

Let $h \in \mathbf{Z}[X_1, \ldots, X_n]$ be an arbitrary polynomial. In the present paper a subexponential-time algorithm is described for deciding consistency in $\mathbf{R}^n$ of a system of inequalities

$$(1) \qquad f_1 \geqq 0, \ldots, f_{k_1} \geqq 0, \; f_{k_1+1} > 0, \ldots, f_k > 0$$

where $f_i = P_i(e^{h(x_1, \ldots, x_n)}, X_1, \ldots, X_n)$ and the polynomials P_i are in $\mathbf{Z}[U, X_1, \ldots, X_n]$ for all $1 \leqq i \leqq k$. Expressions of the form f_i we shall call E-polynomials. This problem is a generalization of the problem of deciding the consistency of systems of polynomial inequalities, which was considered in [1,2,3,4,5]. It can be considered also as a contribution to the solution of Tarski's decidability problem concerning the first order theory of reals with exponentiation. Note, that algorithms from [2],[3] have exponential complexity, while procedures from [4],[5] are subexponential.

A rational function $g \in \mathbf{Q}(Y_1, \ldots, Y_s)$ can be represented as $g = g_1/g_2$ where the polynomials $g_1, g_2 \in \mathbf{Z}[Y_1, \ldots, Y_s]$ are relatively prime. Denote by $\ell(g)$ the maximum bit lengths of the integer coefficients of polynomials g_1, g_2. We suppose that the following inequalities are valid:

$$(2) \quad \deg_{u, x_1, \ldots, x_n}(P_i), \; \deg_{x_1, \ldots, x_n}(h) < d; \; \ell(P_i), \; \ell(h) \leqq M : 1 \leqq i \leqq k$$

We estimate the size of the system (1) by the value $\mathcal{L} = Mkd^n$.

To describe the concept "polynomially bounded", we use the notation $h_1 \leqq \mathcal{P}(h_2, \ldots, h_t)$ for functions $h_1, \ldots, h_t > 0$ to mean that, for suitable natural numbers p, q, the inequality $h_1 \leqq p \cdot (h_2 \cdot \ldots \cdot h_t)^q$ is true.

The running time of the algorithms for solving systems of polynomial inequalities from [2,3] has the upper bound $\mathcal{P}(M, (kd)^{2^{O(n)}})$. Algorithms from [4], [5] have the complexity $\mathcal{P}(M, (kd)^{n^2})$.

Our main result is the following theorem.

Theorem. *There is an algorithm which decides whether the system (1), satisfying bounds (2), is consistent. The running time of the algorithm is less then $\mathcal{P}(M, (nkd)^{n^4}) \leqq \mathcal{P}(\mathcal{L}^{\log^4 \mathcal{L}})$ (i.e. is subexponential in $\mathcal{L}$).*

Let us give at first a very rough sketch of the algorithm. With the help of several elementary reductions the system (1) is transformed into an equivalent (in the sense of consistency) system of the form $H_1 > 0, \ldots, H_m > 0, f = 0$ where $H_1, \ldots, H_m$ are polynomials and f is an E-polynomial, satisfying certain conditions. Then the corollary to lemma 1 (see below) is applied, which lets us, with the help of a search over elements of a finite set and linear coordinate transformation, construct the system of inequalities of the form $H_1' > 0, \ldots, H_m' > 0, \ F_1 = \ldots = F_n = 0$ which is equivalent to the system (1), where $H_1', \ldots H_m'$ are polynomials, $F_1, \ldots, F_n$ are E-polynomials and the system $F_1 = \ldots = F_n = 0$ defines in $\mathbf{R}^n$ only a finite number of points. As a result of substituting a new variable U for the term e^h in every E-polynomial $F_i (1 \leq i \leq n)$ we obtain polynomials $P_{F_i} (1 \leq i \leq n)$ such that the system $P_{F_1} = \ldots = P_{F_n} = 0$ defines an algebraic curve. Now the consistency of the system (1) is equivalent to the existence on this curve of roots of the function $G(U, X_1, \ldots, X_n) = U - e^h$ which satisfy the inequalities $H_1' > 0, \ldots, H_m' > 0$. One can construct a polynomial G' in variables $U, X_1, \ldots, X_n$, such that its roots on the curve lie between the roots of the function G. The strict formulation of this step is the content of lemma 2.

Then using [4], [5] the algorithm finds all roots of the system of equations $P_{F_1} = \ldots = P_{F_n} = G' = 0$. To determine the consistency of system (1) it is sufficient to check for every pair of roots x, y lying on the same connected component, whether the values $G(x)$ and $G(y)$ have the same sign. From the representation of roots in the algorithm from [4], [5] it follows that the problem can be reduced to deciding for two algebraic numbers α, β the sign of the number $\alpha - e^\beta$. Using [7] one can bound from below the absolute value $|\alpha - e^\beta|$ in terms of the degrees and heights of α, β, compute rational approximations to α and to β with more accuracy than this bound, find the sign of the difference of these approximations and thus the sign of $\alpha - e^\beta$. This concludes the algorithm.

The scheme so described is really only a very rough sketch. The above-mentioned reductions of the input system (1) need the introduction of additional infinitesimal elements and therefore all computations have to be performed over a certain non-archimedean real closed field. Thus, the proofs of lemmas 1,2 use some techniques of non-standard analysis [8]. For the same reason the serious difficulty is overcome when the sign of $\alpha - e^\beta$ is determined, because in the non-archimedean case the use of rational approximations is not always possible.

The more detailed but also more formal exposition of the algorithm follows.

E-varieties and their geometry

Let K be an arbitrary real closed field. A subset of K^n is called semi-algebraic if it consists of all points from K^n, satisfying some quantifier-free formula Π with atomic subformulas of the form $(g \geq 0)$, where $g \in$

$K[X_1, \ldots, X_n]$. Denote this set by $\{\Pi\} \subset K^n$. We shall use analogous notation for subsets of points in K^n, satisfying formulas with atomic sub-formulas of the form $(f \geqq 0)$, where f is an E-polynomial. We shall call the sets of the form $\{f = 0\}$ E-varieties.

For an E-polynomial f in the variables $X_1, \ldots, X_n$ denote by P_f a polynomial in the variables $U, X_1, \ldots, X_n$ such that

$$f = P_f = (e^{h(X_1, \ldots, X_n)}, X_1, \ldots, X_n).$$

The coefficients of polynomials P_f and h shall be called coefficients of the E-polynomial f. Denote by $\mathcal{D}_a(r)$ the closed ball with radius $r > 0$ and centre at the point a.

Lemma 1. *Let f, g be E-polynomials such that:*

a) $P_f, P_g \in \mathbb{Q}[U, X_1, \ldots, X_n]$ *satisfy* $\deg_{U, X_1, \ldots, X_n}(P_f)$, $\deg_{U, x_1, \ldots, X_n}$ *and* $(P_g) < d$;

b) *if* $h(x_1, \ldots, x_n) = 0$ *for a given point* $(x_1, \ldots, x_n) \in \mathbb{R}^n$, *then we have* $P_f(1, x_1, \ldots, x_n) \neq 0$;

c) $\dim_{\mathbb{R}}(\{f = g = 0\}) < \dim_{\mathbb{R}}(\{f = 0\})$;

d) *zero is a regular value of* f;

e) $\emptyset \neq \{f = 0\} \subset \mathcal{D}_0(R)$ *for a certain* $0 < R \in \mathbb{R}$.
Let $N = 2(8dn)^n$.

Then one can find integer numbers $1 \leqq \gamma_2, \ldots, \gamma_n \leqq N^2$ such that for every $2 \leqq i \leqq n$ and any connected component (in the euclidean topology) A of the E-variety $\{f = 0\}$ there exist natural numbers $\gamma_i' = \sqrt{\gamma_i}$ and

$$A \cap \{f = (\frac{\partial f}{\partial X_2})^2 - \frac{\gamma_2}{Nn}\Delta = \ldots = (\frac{\partial f}{\partial X_i})^2 - \frac{\gamma_i}{Nn}\Delta = 0\} \cap \{g \neq 0\} \neq \emptyset,$$

where $\Delta = \sum_{1 \leqq j \leqq n} (\frac{\partial f}{\partial X_j})^2$.

Lemma 1 is one of the main technical results of this work and its proof is rather complicated. It uses "exponential" analogues of an algebraic-geometrical theorem on the dimension of intersection and of Bezout's theorem. To prove these, irreducible components of E-varieties, their dimension and related notions are introduced and then some properties of those notions are established.

Corollary. *Let f be an E-polynomial satisfying the conditions of lemma 1. Let $N_1 = 2(8dn)^{2n}$ and introduce new variables $Y_1, \ldots, Y_n, Y_{n+1}$. Denote*

$$F_1 = f(\mathcal{B}^{-1} \cdot Y), \quad F_j = \frac{\partial f(\mathcal{B}^{-1} \cdot Y)}{\partial Y_j}$$

with $2 \leqq j \leqq n$, $F_{n+1} = Y_{n+1} \cdot J_1(Y) - 1$, where $Y = (Y_1, , Y_n)$ is the vector of variables, $\mathcal{B}$ is the matrix of linear transformation of coordinates corresponding to the vector

$$(\alpha_1, \ldots, \alpha_n) = (1, \frac{\gamma_2'}{\sqrt{\beta N_1 n}}, \ldots, \frac{\gamma_n'}{\sqrt{\beta N_1 n}})$$

where

$$\beta = 1 - \sum_{2 \leq j \leq n} \frac{\gamma_i}{N_1 n}$$

and is defined as a product

$$B = \prod_{0 \leq j \leq n-3} \left(\overbrace{\begin{pmatrix} 1 & \cdots & \alpha_{n-j} & & \\ & 1 & & & \mathbf{O} \\ & & \ddots & \vdots & \\ \vdots & & & 1 & \\ -\alpha_{n-j} & \cdots & & 1 & \\ \mathbf{O} & & & & \ddots \\ & & & & & 1 \end{pmatrix}}^{n-j} \left.\rule{0pt}{2em}\right\} n-j \right) \cdot \begin{pmatrix} \alpha_1 & \alpha_2 & & \mathbf{O} \\ -\alpha_2 & \alpha_1 & & \\ & & \mathbf{O} & \ddots \\ & & & 1 \end{pmatrix}$$

and $J_1(Y)$ is the Jacobian of the system $F_1 = \ldots = F_n = 0$.

a) *There exist numbers $1 \leq \gamma_2, \ldots, \gamma_n \leq N_1^2, \gamma_i = (\gamma_i')^2$ for natural numbers γ_i', $2 \leq i \leq n$ such that system of equations $F_1 = \ldots = F_n = F_{n+1} = 0$ is consistent and at each of its roots $z \in \mathbf{R}^{n+1}$ the Jacobian $J(Y_1, \ldots, Y_{n+1})$ of this system does not vanish.*

b) *Take a semi-algebraic set $U = \{g_1 > 0 \ \& \ \ldots \ \& \ g_t > 0\} \subset \mathbf{R}^{n+1}$, where $g_j \in \mathbf{Q}[Y_1, \ldots, Y_{n+1}], 1 \leq j \leq t$ such that $\{g_1 \ldots g_t = F_1 = 0\} = \emptyset$ and $\{g_1 > 0 \ \& \ \ldots \ \& \ g_t > 0 \ \& \ F_1 = 0\} \neq \emptyset$. Then*

$$\{g_1 > 0 \ \& \ \ldots \ \& \ g_t > 0 \ \& \ F_1 = 0\} \cap \{F_1 = \ldots = F_n = F_{n+1} = 0\} \neq \emptyset.$$

In the deduction of the corollary from lemma 1 the E-polynomial g is chosen so that the E-variety $\{f = g = 0\}$ contains the set of all points where the Gaussian curvature of $\{f = 0\}$ vanishes. As was noted in [9] a system of equations which fixes the direction of the gradient to the hypersurface $\{f = 0\}$ at the points with non-vanishing curvature defines a set of isolated points.

Let us call a set T a representative set for $V \subset \mathbf{R}^n$ iff for each connected component U of V the intersection $U \cap T \neq \emptyset$.

Lemma 2. *Consider a system of inequalities*

(3) $$H_1 > 0, \ldots, H_m > 0 \ , \ F_1 = \ldots = F_n = 0$$

where the polynomials $H_i \in \mathbf{Q}[X_1, \ldots, X_n](1 \leq i \leq m)$ and F_j are E-polynomials such that $P_{F_j} \in \mathbf{Q}[U, X_1, \ldots, X_n]$. Let $\{F_1 = \ldots = F_n = H_1 \cdot \ldots \cdot H_m = 0\} = \emptyset$, the Jacobian J of the system of equations does not vanish at every root of the system. Introduce a new variable X_{n+1}. Denote by $P_{n+1} = X_{n+1} \cdot (M_1^2 + \ldots + M_{n+1}^2) - 1$ where $M_1, \ldots, M_{n+1}$ are all maximal minors of the Jacobian matrix of the system $P_{F_1} = \ldots = P_{F_n} = 0$. Introduce once more a new variable X_{n+2} and suppose that for some $0 < R \in \mathbf{R}$ all roots of the system $P_{F_1} = \ldots = P_{F_n} = P_{n+1} = G = 0$ lie in a ball $\mathcal{D}_0(R)$ and the rank of the Jacobian matrix of the system $P_{F_1} = \ldots = P_{F_n} = P_{n+1} = P_{n+2} = 0$ is maximal at every root of the system, where

$$P_{n+2} = X_1^2 + \ldots + X_n^2 + X_{n+1}^2 + X_{n+2}^2 + U^2 - R \ , \ G = U - e^h.$$

Denote by T_1 some finite representative set for the semi-algebraic set

$$\{H_1 \geqq 0 \ \& \ \ldots \ \& \ H_m \geqq 0 \ \&$$

$$\& \ P_{F_1} = \ldots = P_{F_n} = P_{n+1} = P_{n+2} = \hat{J}_1 = 0\},$$

where $\hat{J}_1 \in \mathbf{Q}[U, X_1, \ldots, X_{n+2}]$ is the result of substituting variable U for the term e^h in the Jacobian J_1 of the system $P_{F_1} = \ldots = P_{F_n} = P_{n+2} = G = 0$; by T_2 some finite representative set for the semi-algebraic set

$$B = \{H_1 \geqq 0 \ \& \ \ldots \ \& \ H_m \geqq 0$$

$$\& \ P_{F_1} = \ldots = P_{F_n} = P_{n+1} = P_{n+2} = H_1 \cdots \cdots H_m = 0\}.$$

Let $A_1, \ldots, A_r$ be all connected components of the semi-algebraic set

$$\{H_1 \geqq 0 \ \& \ \ldots \ \& \ H_m \geqq 0 \ \& \ P_{F_1} = \ldots = P_{F_n} = P_{n+1} = P_{n+2} = 0\}$$

Write $V_j = A_j \cap (T_1 \cup T_2) (1 \leqq j \leqq r)$.

The system of inequalities (3) is consistent iff the E-polynomial G changes its sign on one of finite sets $V_j (1 \leqq j \leqq r)$.

The proof of the lemma essentially uses some ideas from [6], concerning analytical functions, which form Pfaffian chains.

As was noted above, we shall need to establish the formulated propositions in the case when, instead of the field $\mathbf{R}$, its "non-standard" extension, containing infinitesimals, is considered. Let us recall for this purpose some facts from non-standard analysis (see e.g. [8]).

There exists the sequence of ordered fields $\mathbf{R}_0 = \mathbf{R} \subset \mathbf{R}_1 \subset \ldots \subset \mathbf{R}_k \subset \ldots$ in which the field $\mathbf{R}_k (k \geqq 1)$ contains an element $\epsilon_k > 0$ infinitesimal relative to the elements of the field $\mathbf{R}_{k-1}$ (i.e. for every element $0 < a \in \mathbf{R}_{k-1}$ the inequality $\epsilon_k < a$ is true).

Let us consider the language $\mathcal{L}_k (k \geqq 0)$ of first order predicate calculus in which the set of functional symbols is in a bijective correspondence with the set of all functions of several arguments from $\mathbf{R}_k$ receiving values from $\mathbf{R}_k$ and the only predicate is the relation $=$. The following "transfer principle" is valid for all $0 \leqq i \leqq j$: a logically closed formula of the language $\mathcal{L}_i$ is true in $\mathbf{R}_i$ iff it is true in $\mathbf{R}_j$.

An element $z \in \mathbf{R}_k (k \geqq 1)$ is called infinitesimal (relative to the elements of the field $\mathbf{R}_{k-1}$) if for every $0 < w \in \mathbf{R}_{k-1}$ we have that $|z| < w$. An element $z \in \mathbf{R}_k$ is called infinitely large iff $z = 1/z_1$, where z_1 is infinitesimal. It is easy to prove that if an element $z \in \mathbf{R}_k (k \geqq 1)$ is not infinitely large (relative to $\mathbf{R}_{k-1}$) then there exist elements $z_1 \in \mathbf{R}_{k-1}$ and $z_2 \in \mathbf{R}_k$ where z_2 is infinitesimal, such that $z = z_1 + z_2$. In this case the element z_1 is called the standard part of the element z (relative to $\mathbf{R}_{k-1}$) and is denoted by $z_1 = st_{\mathbf{R}_{k-1}}(z)$. If the vector $x = (x_1, \ldots, x_n) \in \mathbf{R}_k^n$ and the element x_i for every $1 \leqq i \leqq n$ is not infinitely large then let $st_{\mathbf{R}_{k-1}}(x) = (st_{\mathbf{R}_{k-1}}(x_1), \ldots, st_{\mathbf{R}_{k-1}}(x_n))$.

One can prove that the field $\mathbf{R}_k (k \geqq o)$ is real closed. If an element $0 < \epsilon_{k+1} \in \mathbf{R}_{k+1}$ is infinitesimal relative to $\mathbf{R}_k$ then ϵ_{k+1} is transcendental over $\mathbf{R}_k$ and the real closure $\mathbf{Q}(\epsilon_1, \ldots, \widetilde{\epsilon_{k+1}})$ of the field $\mathbf{Q}(\epsilon_1, \ldots, \widetilde{\epsilon_{k+1}})$ is contained in $\mathbf{R}_{k+1}$. Furthermore, $\mathbf{Q}(\epsilon_1, \ldots, \widetilde{\epsilon_{k+1}})$ is contained in the field of formal Puiseux series in ϵ_{k+1}, i.e. for any $0 \neq a \in \mathbf{Q}(\epsilon_1, \ldots, \widetilde{\epsilon_{k+1}})$ the representation is true : $a = \sum_{i \geqq 0} \alpha_i \epsilon_{k+1}^{\nu_i / \mu}$ where $0 \neq \alpha_i \in \mathbf{Q}(\epsilon_1, \ldots, \epsilon_k)$, the integers $\nu_0 < \nu_1 < \ldots$ are increasing and $\mu \geq 1$ is a natural number (it is clear that the coefficients α_i are in their turn Puiseux series in ϵ_k and so on).

Finally, let us note that in the field $\mathbf{R}_k (k \geq 0)$ there is an analogy $\mathbf{Q}_k$ of the field $\mathbf{Q}$, defined by the same characteristic function as $\mathbf{Q}$ (because all formulas of $\mathcal{L}_0$ are also formulas of $\mathcal{L}_k$).

Consider the formulation of the corollary to lemma 1. For all E-polynomials of a fixed number of variables and of fixed degree the assertion of the corollary can be expressed by a formula of the language $\mathcal{L}_o$. In this, the existential quantifiers for γ are treated as a N_1-member disjunction. According to the transfer principle, the corollary is true for E-polynomials with coefficients from $\mathbf{Q}_k$ (for any $k \geq 0$).

We shall need an auxiliary algorithm for decomposing semi-algebraic curves, defined by polynomials with coefficients from the field $\mathbf{Q}(\epsilon_1, \ldots, \epsilon_k)$ into connected components. Before that, however, we ought to have given a definition of connected components because no topology in the space $(\mathbf{Q}(\epsilon_1, \ldots, \widetilde{\epsilon_k}))^n$ was introduced before (note, by the way, that is not easy to introduce a suitable topology, for example, in the topology with basis consisting of all open balls the segment $\{0 \leqq X \leqq 1\} \subset \mathbf{Q}(\epsilon_1, \ldots, \widetilde{\epsilon_k})$ is not connected).

Such a definition was given in [4], it was based on the fact that for semi-algebraic sets, defined with the help of polynomials with coefficients from $\mathbf{Q}$ of fixed degree and of fixed number of variables the connected components are defined by formulas of the language $\mathcal{L}_0$ (even by formulas of the first order theory of the field $\tilde{\mathbf{Q}}$). In the case of the field $\mathbf{Q}(\epsilon_1, \ldots, \widetilde{\epsilon_k})$ for the semi-algebraic sets in $(\mathbf{Q}(\epsilon_1, \ldots, \widetilde{\epsilon_k}))^n$ defined by the same formulas we shall use the same term - connected components (see details in [4]).

Lemma 3. *Consider the semi-algebraic set*

$$W = \{H_1 = \ldots = H_{n-1} = 0 \ \& \ H_n \geqq 0 \ \& \ \ldots \ \& \ H_{n+m} \geqq 0\}$$

where the polynomials H_i satisfy

$$H_i \in \mathbf{Q}[\epsilon_1, \ldots, \epsilon_k][X_1, \ldots, X_n]$$

with

$$\deg_{x_1, \ldots, x_n}(H_i) < d_1, \quad \deg_{\epsilon_1, \ldots, \epsilon_k}(H_i) < d_2, \quad \ell(h_i) \leqq M_1,$$
$$1 \leqq i \leqq n + m \ , \ n \geqq 4$$

Let the Jacobian matrix of the system of equations $H_1 = \ldots = H_{n-1} = 0$ have maximal rank at every root of the system and suppose that, for a

certain $0 < R \in \mathbf{Q}(\epsilon_1, \ldots, \epsilon_k)$, *the inclusion* $\{H_1 = \ldots = H_{n-1} = 0\} \subset$
$\mathcal{D}_0(R)$ *is valid. There is an algorithm which decomposes the set* W *into connected components, producing each of them in the form* $\{\bigvee_j (\ \& \ _i Q_i^{(j)} \geq 0)\}$, *where for the polynomials* $Q_i^{(j)} \in \mathbf{Q}[\epsilon_1, \ldots, \epsilon_k][X_1, \ldots, X_n]$ *the following bounds are valid:*

$$\deg_{X_1, \ldots, X_n}(Q_i^{(j)}) < d_1^{O(n^3)}$$
$$\deg_{\epsilon_1, \ldots, \epsilon_k}(Q_i^{(j)}) < d_2 d_1^{O(n^3)}$$
$$\ell(Q_i^{(j)}) \leq (M_1 + (n+k)d_2)d_1^{O(n^3)}$$
$$i, j \leq d_1^{O(n^3)}$$

The running time of the algorithm is less than $\mathcal{P}(M_1, (md_2 d_1^{n^3})^{n+k})$.

The algorithm from lemma 3 finds a 3-dimensional subspace of the space $(\overline{\mathbf{Q}(\epsilon_1, \ldots, \epsilon_k)})^n$ (where the bar denotes an algebraic closure) such that the projection of the curve

$$C = \{H_1 = \ldots = H_{n-1} = 0\} \subset (\overline{\mathbf{Q}(\epsilon_1, \ldots, \epsilon_k)})^n$$

on this subspace is injective. Then the algorithm constructs with the help of [10] a projection C' of the curve C on $(\overline{\mathbf{Q}(\epsilon_1, \ldots, \epsilon_k)})^3$, decomposes the semi-algebraic set $C' \cap (\widetilde{\mathbf{Q}(\epsilon_1, \ldots, \epsilon_k)})^3$ into connected components with the help of [2,3] and, finally, reconstructs the desired components according to their injective projections.

Consider the formulation of lemma 2. For all systems of the form (3) with E-polynomials of a fixed number of variables and of fixed degree the assertion of the lemma can be expressed by a formula of the language $\mathcal{L}_0$. According to the transfer principle, the lemma is true for the systems of E-polynomials with coefficients from $\mathbf{Q}_k$ (for any $k \geq 0$).

The consistency decision agorithm

Let us proceed to a description of an algorithm for deciding consistency of (1), satisfying bounds (2) and defining in $\mathbf{R}^n$ the set V (maybe empty).
Introduce a new variable X_{n+1}, let $f_{k+1} = X_{n+1} \cdot f_{k_1+1} \cdot \ldots \cdot f_k - 1$; $f_{k+2} = -f_{k+1}$ and write

$$V' = \{f_1 \geq 0 \ \& \ \ldots \ \& \ f_{k_1} \geq 0 \ \& \ f_{k_1+1} \geq 0 \ \& \ \ldots \ \& \ f_k \geq 0$$
$$\& \ f_{k+1} \geq 0 \ \& \ f_{k+2} \geq 0\} \subset \mathbf{R}^{n+1}.$$

It is clear that $V' \neq \emptyset$ iff $V \neq \emptyset$. Using [4], [5] the algorithm checks consistency of the system of polynomial equations $P_{f_1} \geq 0 \ \& \ \ldots \ \&$ $P_{f_{k+2}} \geq 0 \ \& \ U = 1 \ \& \ h = 0$ and in case of consistency stops.

Introduce elements $0 < \epsilon_i \in \mathbf{Q}_i \subset \mathbf{R}_i$ $(1 \leq i \leq 4)$ where the element ϵ_j is infinitesimal relative to elements of the field $\mathbf{R}_{j-1}(2 \leq j \leq 4)$. Introduce new variables X_{n+2}, X_{n+3} and let

$$g = \Pi_{1 \leq i \leq K+2}(f_i + \epsilon_1) - \epsilon_1^{k+2},$$

$$P = P_g^2 + (e^h - U)^2 + (X_{n+2}((1 - U)^2 + h^2 - 1)^2,$$

$$f = P + (\sum_{1 \leq i \leq n+3} X_i^2 + e^h - (\epsilon_2)^{-1})^2,$$

$$V^{(1)} = \{P_{f_1+\epsilon_1} > 0 \ \& \ \ldots \ \& \ P_{f_{k+2}+\epsilon_1} > 0 \ \& \ f - \epsilon_3 = 0\} \subset \mathbf{R}_3^{n+y}$$

One can prove that $V^{(1)} \neq \emptyset$ iff $V' \neq \emptyset$. Furthermore, zero is a regular value of the E-polynomial $f - \epsilon_3$, and

$$\{f - \epsilon_3 = P_{f_1+\epsilon_1} P_{f_2+\epsilon_1} \ldots P_{f_k+\epsilon_1} = 0\} = \emptyset$$

where

$$P_f \in \mathbf{Q}(\epsilon_1, \epsilon_2, \epsilon_3)[U_1, U, X_1, \ldots, X_{n+3}]$$

and if for some point $(u, x_1, \ldots, x_{n+3}) \in \mathbf{R}_3^{n+4}$ the equality $h(x_1, \ldots, x_n) = 0$ is valid then $P_f(1, u, x_1, \ldots, x_{n+3}) \neq 0$.

The algorithm looks through vectors $\gamma = (\gamma_2, \ldots, \gamma_{n+4}) \in \mathbf{Z}^{n+3}$ where $1 \leq \gamma_j \leq N_2^2$, $\gamma_i = (\gamma_i')^2$, $\gamma_i' \in \mathbf{Z}$, $N_2 = 2(8kd(n+4))^{(n+4)}$. For every vector γ the algorithm constructs the corresponding nonsingular matrix $\mathcal{B}$ (see the formulation of the corollary to lemma 1) and consideres the system of equations $F_1 = \ldots = F_{n+y} = F_{n+s} = 0$. Here $F_1 = f(\mathcal{B}^{-1} \cdot Y) - \epsilon_3$, $F_j = \frac{\partial f(\mathcal{B}^{-1} \cdot Y)}{\partial Y_j}$ with $2 \leq j \leq n + y$, $F_{n+s} = Y_{n+s} \cdot J_1(Y) - 1$ where $Y_1, \ldots, Y_{n+s}$ -variables, vector $Y = (Y_1, \ldots, Y_{n+y})$, $J_1(Y)$ is the Jacobian of the system $F_1 = \ldots = F_{n+y} = 0$.

According to the corollary to lemma 1 (in its "non-standard" version) if $\{f - \epsilon_4 = 0\} \neq \emptyset$ then for a certain γ the system of equations $F_1 = \ldots = F_{n+s} = 0$ is consistent and its Jacobian does not vanish at any root of the system.

Introduce a new variable Y_{n+6} and consider in $\mathbf{R}_3^{n+5}$ a set

$$V^{(2)} = \{P_{f_1+\epsilon_1}(\mathcal{B}^{-1} \cdot Y) > 0 \ \& \ \ldots \ \& \ P_{f_{k+2}+\epsilon_1}(\mathcal{B}^{-1} \cdot Y) > 0$$

$$\& \ F_1^{(1)} = \ldots = F_{n+5}^{(1)} = F_{n+6}^{(1)} = 0\},$$

where $F_i^{(1)}$ $(1 \leq i \leq n + 5)$ denotes the result of substituting the variable Y_{n+6} for the term $(\beta N_2(n + 4))^{-1/2}$ into the E-polynomial F_j, $F_{n+6}^{(1)} = Y_{n+6}^2(\beta N_2(n + 4))^{-1}$. One can prove that $V^{(2)} \neq \emptyset$ iff $V^{(1)} \neq \emptyset$ i.e. iff $V \neq \emptyset$.

Introduce two more new variables Y_{n+7}, Y_{n+8} and two polynomials $P_{n+7} = Y_{n+7}(M_1^2 + \ldots + M_{n+7}^2) - 1$, where $M_1, \ldots, M_{n+7}$ are all maximal minors of the Jacobian matrix of the system $PF_1(1) = \ldots = PF_{n+6}^{(1)} = 0$ and $P_{n+8} = \sum_{1 \leq i \leq n+8} Y_i^2 + U_2^2 - (\epsilon_4)^{-1}$. Then the Jacobian matrix of the system

$PF_1(1) = \ldots = PF^{(1)}_{n+6} = P_{n+7} = 0$ has maximal rank at every root of the system, and all roots of the system $PF_1(1) = \ldots = PF^{(1)}_{n+6} = P_{n+7} = G = 0$ lie in the ball $\mathcal{D}_0(\epsilon_4^{-1})$, where $G = U_2 - e^{h(\mathcal{B}^{-1} \cdot Y)}$. Besides, the rank of Jacobian matrix of the system $PF_1(1) = \ldots = PF^{(1)}_{n+6} = P_{n+7} = P_{n+8} = 0$ is maximal at every root of the system. Thus, the system of inequalities defining the set $V^{(2)}$ satisfies all the conditions of lemma 2 (in its "nonstandard" version).

The algorithm applies the procedure from lemma 3 to the set

$$A = \{P_{f_1+\epsilon_1}(\mathcal{B}^{-1} \cdot Y) \geqq 0 \ \& \ \ldots \ \& \ P_{f_{k+2}+\epsilon_1}(\mathcal{B}^{-1} \cdot Y) \geqq 0$$

$$\& \ PF_1(1) = \ldots = PF^{(1)}_{n+6} = P_{n+7} = P_{n+8} = 0\}.$$

As a result all connected components of this set will be found. After that, using [4], [5] the algorithm finds finite representative sets for the sets T_1, T_2 described in the formulation of lemma 2 applied to our case. The procedure from [4], [5] produces elements of the set $T_1 \cup T_2$ in the following way. For every point $y = (u_2, y_1, \ldots, y_{n+8})$ an irreducible — over the field $\mathbf{Q}(\epsilon_1, \epsilon_2, \epsilon_3, \epsilon_4)$ — polynomial $\varphi \in \mathbf{Q}(\epsilon_1, \epsilon_2, \epsilon_3, \epsilon_4)[Z]$ and polynomials $\eta, \xi_1, \ldots, \xi_{n+8} \in \mathbf{Q}(\epsilon_1, \epsilon_2, \epsilon_3, \epsilon_4)[Z]$ are constructed, so that for a certain root Θ_0 of φ the equality $y = (\eta(\Theta_0), \xi(\Theta_0), \ldots, \xi_{n+8}(\Theta_0))$ is true. Besides, the signs of values of derivatives of φ of all orders in the point Θ_0 are indicated. This determines exactly the root Θ_0. Then the algorithm determines the sign of element $G(y) \in \mathbf{R}_4$.

Denote by $h' \in \mathbf{Q}(\epsilon_1, \epsilon_2, \epsilon_3, \epsilon_4)[Z]$ the result of substituting in the polynomial h of $\eta, \xi_1, \ldots, \xi_{n+8}$ instead of the variables $U_2, Y_1, \ldots, Y_{n+8}$ respectively. Thefore it remains to decide the sign of element $\eta(\Theta_0) - e^{h'(\Theta_0)}$. With the help of Newton's diagram the algorithm finds the lowest terms of Puiseux series $\eta(\Theta_0), h'(\Theta_0)$.

At first the algorithm considers all cases when at least one of elements $\eta(\Theta_0)$, $h'(\Theta_0)$ is infinitely large to $\mathbf{R}$. Then the algorithm considers the possibility of both values not being infinitely large relative to $\mathbf{R}$.

For this goal the algorithm at first computes the standard parts $st_{\mathbf{R}}$ of the values $\eta(\Theta_0)$, $h'(\Theta_0)$ and analyzes separately two remaining possibilities: a) $st_{\mathbf{R}}\eta(\Theta_0) = 1$ and $st_{\mathbf{R}}(h'(\Theta_0)) = 0$; b) $st_{\mathbf{R}}(\eta(\Theta_0)) \neq 0$ or $st_{\mathbf{R}}(h'(\Theta_0)) \neq 0$. The consideration of both possibilities essentially involves the lower bound obtained in [7] for $|\alpha - e^\beta|$ for algebraic numbers α and β in terms of the degrees and heights of these numbers. The important peculiarity of this bound is that it is polynomial in heights of numbers. Besides, it can be extended to algebraic elements over $\mathbf{Q}(\epsilon_1, \epsilon_2, \epsilon_3, \epsilon_4)$ with the help of the transfer principle. It follows that in the case a) it is sufficient to consider certain initial segments of the Puiseux series for $\eta(\Theta_0)$ and $h'(\Theta_0)$ and in the case b) the corresponding rational approximations of numbers $st_{\mathbf{R}}(\eta(\Theta_0))$, $st_{\mathbf{R}}(h'(\Theta_0))$.

Therefore, for any polynomial φ and each of its roots Θ_0 such that

$$y = (\eta(\Theta_0), \xi_1(\Theta_0), \ldots, \xi_{n+8}(\Theta_0)) \in T_1 \cup T_2$$

the sign of the element $G(y)$ will be found and the connected component of the set containing y will be determined. If on at least one of the components, the function G changes sign, then the system (1) is consistent. Otherwise it is inconsistent. Here the algorithm stops.

The upper complexity bound of the algorithm can be obtained straightforwardly.

REFERENCES

[1] Tarski, A., "A Decision Method for Elementary Algebra and Geometry," University of California Press, 1951.

[2] Collins, G. E., *Quantifier elimination for real closed fields by cylindrical algebraic decomposition*, in "Second GI Conf. Automata Theory and Formal Languages," Lect. Notes Comp. Sci. 33, 1975, pp. 134-183.

[3] Wüthrich, H. R., *Ein Enttscheidungsverfahren für die Theorie der reall-abgeschlossenen Körper*, in "Komplexität von Entscheidungsproblemen: ein Seminar," Lect. Notes Comp. Sci. 43, 1976, pp. 138-162.

[4] Grigor'ev, D. Yu., Vorobjov, N. N., Jr., *Solving systems of polynomial inequalities in subexponential time*, J. Symbolic Comp. 5 (1988), 37-64.

[5] Renegar, J., *A faster PSPACE algorithm for deciding the existential theory of the reals*, Technical report N 792, College of Engineering, Cornell University.

[6] Hovansky, A. G., *On a class of systems of transcendental equations*, Soviet Math. Dokl. 22 (1980), 762-765.

[7] Waldschmidt, M., *Simultaneous approximation of numbers connected with the exponential function*, J. Austral. Math. Soc. (series A) 25 (1978), 466-478.

[8] Davis, M., "Applied Nonstandard Analysis," J. Wiley., New York, 1977.

[9] Milnor, J., *On the Betti numbers of real varieties*, Proc. Amer. Math. Soc. 15(2) (1964), 275-280.

[10] Chistov, A. L., Grigor'ev, D. Yu., *Complexity of quantifier elimination in the theory of algebraically closed fields*, in "Mathematical Foundations of Computer Science, Praha 198," Lect. Notes Comp. Sci. 176, 1984, pp. 17-31.

Nikolaj N. Vorobjov (Jr.)
Leningrad State University
Universitetskaya emb., 7/9
Leningrad, 199164, USSR

Progress in Mathematics

Edited by:

J. Oesterlé
Département de Mathématiques
Université de Paris VI
4, Place Jussieu
75230 Paris Cedex 05
France

A. Weinstein
Department of Mathematics
University of California
Berkeley, CA 94720
U.S.A.

Progress in Mathematics is a series of books intended for professional mathematicians and scientists, encompassing all areas of pure mathematics. This distinguished series, which began in 1979, includes authored monographs and edited collections of papers on important research developments as well as expositions of particular subject areas.

We encourage preparation of manuscripts in such forms as LaTeX or AMS TeX for delivery in camera-ready copy which leads to rapid publication, or in electronic form for interfacing with laser printers or typesetters.

Proposals should be sent directly to the editors or to: Birkhäuser Boston, 675 Massachusetts Avenue, Cambridge, MA 02139, U.S.A.

A complete list of titles in this series is available from the publisher.